D1602859

THE KAISER WILHELM INSTITUTE FOR ANTHROPOLOGY, HUMAN HEREDITY, AND EUGENICS, 1927–1945

BOSTON STUDIES IN THE PHILOSOPHY OF SCIENCE

VOLUME 259

THE KAISER WILHELM INSTITUTE FOR ANTHROPOLOGY, HUMAN HEREDITY, AND EUGENICS, 1927–1945

CROSSING BOUNDARIES

HANS-WALTER SCHMUHL
University of Bielefeld

Hans-Walter Schmuhl
University of Bielefeld
Germany

Originally published – Kalorien, Kautschuk, Karrieren. Pflanzenzüchtung und landwirtschaftliche Forschung in Kaiser-Wilhelm-Instituten 1933–1945, Wallstein (Göttingen), 2003, translator: Sorcha O'Hagan

ISBN 978-1-4020-6599-6 e-ISBN 978-1-4020-6600-9

Library of Congress Control Number: 2007935799

Printed on acid-free paper

9 8 7 6 5 4 3 2 1

springer.com

CONTENTS

ACKNOWLEDGMENTS

This work originated in the context of the Max Planck Society's research program "The History of the Kaiser Wilhelm Society in National Socialism." My special thanks go to the presidential commission responsible for this program, especially to Reinhard Rürup and Wolfgang Schieder, as well as the directors of the research program, first Doris Kaufmann, then Carola Sachse, and finally Susanne Heim. All accompanied the genesis of this book with critique, suggestions, and tips, as did my other associates from the research program, above all Bernd Gausemeier, Rüdiger Hachtmann, Helmut Maier, Florian Schmaltz, Michael Schüring, Alexander von Schwerin, and Achim Trunk. I owe much to the cooperative exchange among colleagues within the research program.

Other colleagues and associates have also helped me with selfless words and deeds. The first of these who deserves mention here is Thomas Beddies, whom I would like to thank sincerely for his magnanimous support in the search for the children and youths of Wittenau. I also owe thanks to Thomas Barow, Matthias Benad, Friedrich Brenzel, Karl Dienst, Karl Ditt, Wolfgang Freund, Jochen-Christoph Kaiser, Uwe Kaminsky, Jürgen Peiffer, Volker Roelcke, Holger Starke, Katrin Stoll, Matthias M. Weber, and Christine Wolters for their valuable tips. Special thanks are due to Kazuko Kibata, Yasushi Maruyama, and Takashi Nakagawa, who tirelessly traced the tracks of the Japanese guest scholars at the KWI-A. I sincerely thank Baron Helmut von Verschuer and Michael Wörle for their information about Baron Otmar von Verschuer.

Part of the archive material used in this book was screened by Benoît Massin during his activity for the research program. I was further assisted in the study of the archives by Dagmar Vorbeck and Thomas Sandkühler. They also deserve my sincere thanks, as does Birgit Monhof-Halbach, who took on the self-denying work of reconstructing the institute's library. I would also like to thank the staff of the

Archive on the History of the Max Planck Society. Christine Rüter edited the book with the extreme precision she calls her own. I would like to thank her sincerely as well. My last thanks, finally, go to my wife Regina Geitner, who not only accompanied me, once again, through all highs and lows of the writing process, but this time also assisted me with her accumulated medical expertise.

LIST OF ABBREVIATIONS

ADW	Archiv des Diakonischen Werkes (Archive of the Protestant Church Welfare Organization)
AfRGB	Archiv für Rassen- und Gesellschaftsbiologie (Archive for the Biology of Race and Society)
BArch.	Bundesarchiv (Federal Archives)
BDC	Berlin Document Center
BDM	Bund deutscher Mädel (League of German Girls)
DFA	Deutsche Forschungsanstalt für Psychiatrie (German Research Institute for Psychiatry, a Kaiser Wilhelm Institute)
DFG	Deutsche Forschungsgemeinschaft (German Research Association)
DNVP	Deutschnationale Volkspartei (German National Party of the Volk)
DVP	Deutsche Volkspartei (German Party of the Volk)
EGG	Erbgesundheitsgericht (Hereditary Health Court)
EGOG	Erbgesundheitsobergericht (Hereditary Health Appellate Court)
EK	Eisernes Kreuz (Iron Cross)
Fs.	Festschrift
GzVeN	Gesetz zur Verhütung erbkranken Nachwuchses (Law on the Prevention of Genetically Deficient Progeny)
HJ	Hitlerjugend (Hitler Youth)
IFEO	International Federation of Eugenic Organizations
IfZ	Institut für Zeitgeschichte (Institute for Contemporary History)
IUSIPP	International Union for the Scientific Investigation of Population Problems
KPD	Kommunistische Partei Deutschlands (Communist Party of Germany)
KWG	Kaiser-Wilhelm-Gesellschaft (Kaiser Wilhelm Society)

KWI	Kaiser-Wilhelm-Institut (Kaiser Wilhelm Institute)
KWI-A	Kaiser-Wilhelm-Institut für Anthropologie, menschliche Erblehre und Eugenik (Kaiser Wilhelm Institute for Anthropology, Human Heredity and Eugenics)
KZ	Konzentrationslager (concentration camp)
MPG-Archiv	Archiv zur Geschichte der Max-Planck-Gesellschaft (Archive on the History of the Max Planck Society)
MPIP-HA	Max-Planck-Institut für Psychiatrie, Historisches Archiv (Max Planck Institute for Psychiatry, Historical Archive)
Ms.	Manuskript (manuscript)
ND	Neudruck (reprint)
NL	Nachlaß (estate)
NSDÄB	Nationalsozialistischer Deutscher Ärztebund (National Socialist League of German Physicians)
NSDAP	Nationalsozialistische Deutsche Arbeiterpartei (National Socialist German Workers' Party)
NSV	Nationalsozialistische Volkswohlfahrt (the National Socialist welfare organization)
R	Rückseite (verso)
RM	Reichsmark (currency of the German Reich)
RuSHA	Rasse- und Siedlungshauptamt (Main Office for Race and Settlement)
SA	Sturmabteilung (Storm Troopers)
SD	Sicherheitsdienst (Security Service)
SPD	Sozialdemokratische Partei Deutschlands (Social Democratic Party of Germany)
SS	Schutzstaffel ("protection unit")
USSR	Union of Soviet Socialist Republics
uk	unabkömmlich (in a reserved occupation = excused from service on the front)
USA	United States of America
Uschla	Untersuchungs- und Schlichtungsausschusses (Investigative and Arbitration Committee)
VfZ	Vierteljahrshefte für Zeitgeschichte (Quarterly Journal for Contemporary History)
WVHA	Wirtschafts- und Verwaltungshauptamt (Main Office of Economics and Administration)
ZIAVL	Zeitschrift für induktive Abstammungs- und Vererbungslehre (Journal for Inductive Theory of Descent and Heredity)
ZMA	Zeitschrift für Morphologie und Anthropologie (Journal for Morphology and Anthropology)
ZMVKL	Zeitschrift für menschliche Vererbungs- und Konstitutionslehre (Journal for Human Genetics and Theory of Human Constitution)

Introduction

From its founding in 1927 until its dissolution in 1945, the Kaiser Wilhelm Institute for Anthropology, Human Genetics, and Eugenics (KWI-A) in Berlin-Dahlem transgressed many a boundary; indeed, the transgression of boundaries was in a sense its *raison d'être* from the outset. Initially this applied to the boundaries within the disciplinary canon of the human sciences. Even from its basic conception, the institute, centered around the person of its founding director Eugen Fischer (1874–1967), was to unify anthropology, genetics, *and* eugenics under one roof. In keeping with the understanding predominant in Germany between the wars, anthropology went beyond the scope of the framework of the ascendant "race theory" to cover not only physical anthropology, including paleoanthropology, but also elements of what we today would call cultural and social anthropology. Thus, this anthropology extended far into the fields of archeology, paleontology, prehistory and early history, history and sociology, and especially into ethnology and folklore. Human genetics, in turn, was more than the attempt to apply to humans the genetics developed by Thomas Hunt Morgan (1866–1945) and his school in the USA on the model of *drosophila.* In Germany, Morgan's genetics, which concentrated on investigating the dissemination of genetic traits on the chromosomes and their morphological structure, was received with skepticism for two reasons. The first criticism was that this special orientation of the genetics neglected the problem of how genetic traits are manifested, so that the labyrinthine path from the genotype to the phenotype and the forces involved other than genes remained in the dark. Second, German geneticists – probably a consequence of the holistic style of thinking in the sciences predominant in Germany – found it unsatisfying that Morgan's genetics studied, so to speak, development without evolution. German genetics, in contrast, attempted to link together genetics and evolutionary biology. Eugenics, finally, was located from its very establishment in the boundary area between many human sciences: biology, genetics, medicine, psychiatry, hygiene, anthropology, demography, political economy, and sociology. What was undertaken in Dahlem was thus nothing less than the attempt to develop a new, transdisciplinary key science of humans.

Yet, scientific praxis initially remained far behind these lofty goals. The phase of setting up the institute fell in the period of economic depression and political disintegration at the end of the Weimar Republic. The implementation of its ambitious plans was postponed due to scarce financial resources. This did not change until the consolidation phase of the "Third Reich" from 1934 to 1939, after a turbulent phase of *Gleichschaltung* and "self-alignment" when the Nazis took over. But not even during this period did the integration of the research fields worked on at the KWI-A proceed as planned, and, paradoxically, not *despite*, but rather *because*, the institute was so successfully established in the National Socialist state. Although funding now flowed abundantly, the prescientific and nonscientific services performed by scientists at the KWI-A for the National Socialist state tied up considerable resources. What weighed more heavily was that, as a consequence of its intimate interconnections with politics, the practice of research, especially in the conventional paths of race theory and eugenics, was in danger of paralysis. The two departments assigned to these fields visibly drifted apart, while the field of human genetics nearly lay fallow after the departure of its department director, Otmar Freiherr von Verschuer (1896–1969), in 1935. From 1938 onward, Fischer, the founding director, and his pupil Verschuer, who returned to Berlin to take over as director in 1942, strived to reorganize the research under the new paradigm of phenogenetics. This shifted the problem of the manifestation of genes into the center of scientific interest, that is, the development of the genome into a phenome, and thus the proteome, the level of proteins, enzymes, and hormones, as well as the complex process of interaction between hereditary predisposition and environmental factors. Besides being able to generate new research perspectives, the new paradigm also made it possible to merge disciplines like experimental genetics, evolutionary biology, embryology, medicine, and anthropology, and to combine methods like twin research, dermatoglyphics, blood group research, and animal models. What is interesting is that the complex of questions linked with the concept of phenogenetics, which was already a subject of discussion under the label "epigenetics" in the USA at the beginning of the 1940s, appears highly topical from today's perspective, as the completion of the human genome sequence increasingly is redirecting the interest of geneticists back to the intermediate steps between the genes and the completely developed organism.

At its founding in 1927, the KWI-A set out with more than the objective of transgressing the boundaries between scientific disciplines. From the outset it consciously endeavored to cross the boundaries between science and politics, although, in distinction from older forms of race hygiene and race anthropology, which were clearly located in the nationalist camp, the express goal of the KWI-A was not to perform politicized science, but rather to lay the foundations for scientific policy. This goal was based on a technocratic model of political consulting, in which "scientific expertise" decomposes politics into pure "practical constraints," political decision-making processes congeal into a "rational" solution of "factual issues," and, ultimately, science and technology replace politics. Because there was a broad political consensus for such social engineering in the Weimar Republic, it is no wonder that the founding of the institute was supported by a coalition extending from the Social Democrats to the Catholic Center Party, all the way to the far right of the party spectrum. In fact, research at the KWI-A was by no means as "free of political values" as the scientists claimed. Fischer and his staff attempted to preform political decision-making processes – with notable success, in the case of the Prussian law for eugenic sterilization drafted in 1932; yet, it was hardly the case that the Weimar state implemented each and every one of the recommendations from science on a one-to-one basis.

Nevertheless, the KWI-A and its director Eugen Fischer initially found themselves under political pressure in 1933. The National Socialists, suspicious that the institute had been too closely bound up with the Weimar welfare state, forced the KWI-A to "align itself." Yet, it is hardly the case that politics first invaded the sphere of "pure science" in the year 1933. The KWI-A had always attempted to extend its influence into the realm of political action. After a short phase of provocation and friction, it was quite willing to place its expert knowledge at the service of the regime, which, for its part – to a greater degree than the governments of the Weimar Republic – was ready to ground its political decision-making processes on scientific expertise, and quickly recognized the political value of the KWI-A. The result was *mutual* exploitation, albeit on the basis of a basic ideological consensus: science and the state were in agreement about the utopian goal of a comprehensive biological policy, whose objective was nothing less than the control of human evolution.

The consequence of the institute's collaboration with the National Socialist state was that the scientists of the KWI-A participated in various positions and functions in the crimes this state committed: mass sterilizations in the interest of "race hygiene" in accordance with the Law for the Prevention of Genetically Deficient Progeny (*Gesetz zur Verhütung erbkranken Nachwuchses*, GzVeN), the "euthanasia" campaign, the persecution and annihilation of Jews, Sinti, and Roma, and the plans for a new ethnic order in occupied Eastern Europe. In return, the KWI-A profited significantly from the National Socialist regime's policy of genetic health and race policy. Research projects received generous support, the institute progressively expanded. Both the political prestige and the social status of the KWI-A and the scientists employed there increased constantly. However, the start of World War II threatened to push the KWI-A into the background and rob it of its value as a political resource. The reorganization of research under the paradigm of phenogenetics renewed the institute's value as a political partner, which was expressed not least of all in the fact that the KWI-A – as the last Kaiser Wilhelm Institute in Greater Berlin – was ultimately recognized as important for the war. Of particular interest for the rulers in this context was the KWI-A's research on the hereditary disposition for tuberculosis and on the development of a biochemical "race test." In the middle of the war the KWI-A was able to expand its budget and solicit additional funds for research. But above all, it profited enormously from the unfettered access to human subjects in the National Socialist "prerogative state."

And this takes us to a third level of boundaries transgressed, the boundaries of scientific ethics. This boundary transgression, although neither conscious nor intentional, began in the institute's early years, as participation in the large-scale inventory of the hereditary biological stock of the German people at the end of the Weimar Republic violated what we understand today to be the fundamental rules of personal privacy. In a sense, the Third Reich constituted a land of unlimited opportunities for the science of humans. The research group around Fischer and Verschuer now received unhampered access to sensitive data; they could sound out humans who were the object of their interest even without their consent and against their will, without having to take any consideration of human or civil rights. All precautions based on law, morals, or professional ethics were rescinded. The scientists of the KWI-A performed investigations on human beings who were held prisoner in the National Socialist camp system and could no longer control what happened to their own bodies. They researched samples taken from the victims of genocide committed on the Jews, Sinti, and Roma. And they performed experiments for the benefit of others on patients who were not capable of giving their consent. As such, the guidelines on scientific ethics formulated in the early 1930s were violated even by those very researchers whose attitude toward National Socialism was marked by serious reservations. This obsession with scientific feasibility resulted not from political indoctrination, but rather from the hubris of research.

The transgression of these three kinds of boundaries constitutes the vanishing point of this study. It is conceived as a comprehensive history of an institute, which looks at structural conditions – the crisis of the welfare state in the late Weimar Republic and the reshaping of biopolicy in National Socialism, as well as institutional developments – financing, personnel development and policy, erection and expansion of the material infrastructure for scientific work, and finally scientific practice – paradigms and issues, fields of research, epistemic objects, and applied methods guiding research. Special attention is also due to the interfaces between science and politics: the institute's Board of Trustees, the manifold relationships to the ministerial bureaucracy and to the power centers of the National Socialist regime, the institute's contributions to the practice of genetic health policy and race policy. Against this background, the institute's participation in the crimes of the National Socialist state as well as the ethically reprehensible research projects embedded within it will be subjected to analysis. By placing these issues in a comprehensive institutional history it should be possible to make clear how it came to this fateful degeneration into the zone of crime.

Although by now there is a broad spectrum of literature on partial aspects of the institute's history, such a comprehensively conceptualized study has yet to be submitted. The closest anyone has come to a total history of the institute is the biography of Eugen Fischer published by Niels C. Lösch.[1] Lösch created solid ground for further research – passages of this work, too, are based on his pioneering study, although the portrayal and especially the interpretation of what happened there often arrives at different conclusions. The first brief overview portrayals of the history of the institute were presented by Paul Weindling as well as Anna Bergmann, Gabriele Czarnowski, and Annegret Ehmann.[2] Long passages by Peter Weingart, Jürgen Kroll, and Kurt Bayertz in their overview of the history of eugenics and race

[1] Lösch, Rasse.

[2] Weindling, Weimar Eugenics; Bergmann/Czarnowski/Ehmann, Menschen.

hygiene in Germany deal with the KWI-A, as does Paul Weindling in his large-scale study *Health, Race, and German Politics Between National Unification and Nazism, 1870–1945* – with an emphasis on the period of the Weimar Republic.[3] Kristie Macrakis was the first to deal in greater detail with the *Gleichschaltung* and "self-alignment" of the KWI-A in the Third Reich, in her book about the history of the Kaiser Wilhelm Gesellschaft (KWG) in National Socialism, *Surviving the Swastika*.[4] As to the relationship between the institute and the National Socialist state in the years from 1933 to 1939, Lösch's work set a new standard. The analysis and interpretation of this complex of problems was driven further by Sheila F. Weiss in her recent, excellent study, *Humangenetik und Politik als wechselseitige Ressourcen*. In an additional, thoroughly inspiring study, Weiss investigated the role played by leading scientists at the KWI-A in international conferences during the period of the Third Reich.[5] More valuable information on this period is supplied in the first chapters of the outstanding postdoctoral work by Hans-Peter Kröner, who portrayed in detail the dissolution of the KWI-A after 1945, the history of its successor institutes and the postwar careers of its leading scientists.[6] With this the history of the institute in the first years of the Third Reich, along with its "posthistory" in the period after 1945, could be regarded as quite well researched; yet, for the period during World War II great gaps in research are apparent. This work thus places its emphasis primarily on the years from 1938 until 1945, while its discussion of the phase of the Nazi's seizure of power and *Gleichschaltung* are based for the most part on the exhaustive portrayals by Lösch, Weiss, and Kröner; the period after 1945 is left out completely.

The research gaps regarding the period of World War II appear astonishing at first glance, as such a great number of authors have dealt with the "complicity" of the KWI-A in the crimes of the National Socialist state and the criminal research associated with them. Benno Müller-Hill, in his study *Tödliche Wissenschaft*, earned the distinction of being the first to point out the institute's important role in the genetic health and race policies of the Third Reich, especially in its policies toward Jews.[7] Müller-Hill and Ute Deichmann were the first to report about an experiment in which epilectic children from the "euthanasia" facility in Brandenburg-Görden were subjected to a partial vacuum by Hans Nachtsheim (1890–1979).[8] Hans Hesse exhaustively investigated the biography of Karin Magnussens (1908–1997) and her research on the eyes of Sinti and Roma from the "Auschwitz Gypsy Camp."[9] Bernd Gausemeier and Achim Trunk recently shed light on the "specific

---

[3] Weingart/Kroll/Bayertz, Rasse, pp. 239–246, 407–424; Weindling, Health, pp. 430–439.

[4] Macrakis, Surviving the Swastika, pp. 125–130.

[5] Weiss, Humangenetik; ibid., Sword.

[6] Kröner, Von der Rassenhygiene zur Humangenetik.

[7] Müller-Hill, Tödliche Wissenschaft; ibid., Blut.

[8] Müller-Hill, Genetics after Auschwitz; Deichmann, Biologen, pp. 308–314; ibid., Hans Nachtsheim, pp. 146–148. See also Koch, Humangenetik, pp. 120–148.

[9] Hesse, Augen.

proteins" research project so shrouded in secrecy, which worked with blood samples sent to Dahlem from Auschwitz by Josef Mengele.[10] In his book, *Auschwitz, die NS-Medizin und ihre Opfer*, Ernst Klee studied the "Dahlem-Auschwitz" connection more closely.[11] Benoît Massin succeeded in convincingly reinterpreting Mengele's research against the background of the twin research performed in Dahlem, although this study does not agree with Massin's critical thesis that Mengele's research empire amounted to the "Auschwitz branch" of the KWI-A.[12]

Even today, the research praxis of the KWI-A is still *terra incognita*. Benoît Massin, with his attempt at a quantitative assessment of the publications that emerged from the institute, made first inroads into this broad field of research.[13] Several works have been produced on individual departments of the institute. In two shorter papers, Mitchell G. Ash gave insight into the work of the Department for Genetic Psychology directed by Kurt Gottschaldt (1902–1991).[14] The Department for Experimental Hereditary Pathology, led by Hans Nachtsheim, is portrayed exhaustively in an outstanding dissertation by Alexander von Schwerin.[15]

This study is intended to spin the threads of collective knowledge on the current state of research into a common strand. However, painstaking review of the central source materials – although many of these have been reviewed frequently in the past – revealed a surprising multitude of, as yet, unknown details. The systematic screening of the institute's publication proved equally fruitful. In this manner it was possible to fill in a number of gaps in the total picture of the institute's history, to accentuate some nuances differently, to give sharper contours to some fuzzy lines, and to deepen the background, so that the picture, the author hopes, takes on a sharper focus.

---

[10] Gausemeier, Radikalisierung; Trunk, Zweihundert Blutproben.

[11] Klee, Auschwitz, pp. 449–491.

[12] Massin, Mengele.

[13] Massin, Rasse, und Vererbung. Unfortunately, this essay is of only limited significance due to a number of methodological deficiencies: the publication list upon which it is based (which is by and large the same as the list compiled by Niels C. Lösch on the basis of the institute's work reports) contains a number of errors; the attribution of the titles to the individual fields of research is often uncertain, the research fields are demarcated quite arbitrarily, and a temporal differentiation is largely missing, but would seem imperative, because the institute's research emphases shifted considerably on several occasions.

[14] Ash, Erbpsychologische Abteilung; ibid, "Positive Eugenics."

[15] Schwerin, Experimentalisierung.

Chapter 1
A "Purely Theoretical Institute for the Study of the Nature of Man": The Founding of the Kaiser Wilhelm Institute for Anthropology, Human Heredity, and Eugenics, 1920–1927

The linkage of race hygiene, science, and politics in the early Weimar Republic was the context in which the initiative to found an extra-university research institution for the fields of anthropology, human genetics, and eugenics took shape. This triangle was the result of converging interests: race hygiene conceived of itself as an applied science, aspiring to a scientific foundation for its practical agenda, and hoping to establish itself as a new discipline between biology, genetics, medicine, psychiatry, hygiene, anthropology, demographics, national economy, and sociology, and to influence state and society through scientific policy consulting. The modern interventionist state, in turn, demanded expertise from the human sciences to the extent that it laid claim to the control over birth and death, sexuality and reproduction, body and germline, variability and evolution, as part of a comprehensive *biopolicy*.[1] After all, the human sciences ultimately offered the modern interventionist state their expertise in exchange for funding. In this process race hygiene functioned as a kind of relay switched between scientific knowledge and biopolicy interest.

1.1 World War I, the Crisis of the Weimar Welfare State, and the Rise of Race Hygiene

1.1.1 The Inceptions of Race Hygiene

Race hygiene originated as a German variety of eugenics at the close of the nineteenth century.[2] Initially the field constituted a loose lattice of friendships and relations surrounding the young physician Alfred Ploetz (1860–1940).[3] At first, hardly

---

[1] Foucault, Wille. See also: Stingelin (ed.), Biopolitik; Schwartz, Wissen; Raphael, Verwissenschaftlichung; Byer, Rassenhygiene; Reyer, Eugenik; Usborne, Frauenkörper; Kaufmann, Eugenik; Hahn, Modernisierung.

[2] Survey portrayals: Kröner, Eugenik; Kroll, Entstehung; Weingart/Kroll/Bayertz, Rasse; Weindling, Health; Schmuhl, Rassenhygiene, Nationalsozialismus, Euthanasie; Proctor, Race Hygiene. On the state of research: Trus, Heiliger Krieg. On parallel developments abroad and the international context: Adams (ed.), Science; Kühl, Internationale; Weingart, Science; Schmuhl, Rassenhygiene in Deutschland.

[3] For a biography: Doeleke, Alfred Ploetz; Labisch/Tennstedt, Weg, vol. 2, pp. 467–470; Klee, Personenlexikon, p. 466.

more than a hermetic circle of young academics in the wake of the "Life-Reform Movement" (*Lebensreformbewegung*), the circle around Ploetz began taking on a more solid organizational structure in the first decade of the twentieth century. Yet the consolidation process proceeded laboriously. A first step was made with the founding of the "Archive for the Biology of Race and Society" in the year 1904.[4] In 1905 Ploetz founded the "Berlin Society for Race Hygiene;" even before World War I, additional societies emerged in Munich (1907), Freiburg (1908–1910), and Stuttgart. In 1910 these groups merged to form the "German Society for Race Hygiene." From 1907 on the local organizations were also members of an "International Society for Race Hygiene," which maintained contacts to Sweden,[5] Norway, the Netherlands, the USA, and Great Britain. Do not be misled, however, on the eve of World War I, the network of race hygienists, which understood itself to be not only the crystallization point of an emerging scientific community, but also a scientifically grounded movement for the reform of life and society, amounted to only around 400 men and women.[6]

1.1.2 The Disruption of the World War I

At the beginning of the war, the German Society for Race Hygiene increasingly directed its energies to the outside world. The prospects for legislation along the lines of race hygiene appeared good, as it was expected that the war would explode the encrusted structures of state and society: "An opportunity […] like the one presented by the reordering of all circumstances after the war may never return […]."[7] However, the mood of renewal soon dissipated. Toward the end of the war, resignation took hold among the race hygienists. Their proposal to prescribe by law that partners exchange health certificates before marrying, and their demand that marriage be prohibited for those with mental illnesses and disabilities were as fruitless

[4] On this, with an international comparative perspective: Heuer/Propping, Comparison.

[5] The first "country group" of the International Society for Race Hygiene was founded in 1909 in Sweden, the *Svenska sällskapet för rashygien*. The majority of its founding members already belonged to the International Society for Race Hygiene. In 1920, these 46 Swedes constituted the largest group of its foreign constituents: Broberg/Tydén, Eugenics in Sweden, p. 83. Slightly different numbers in: Weindling, Health, pp. 145, 140f., ibid., International Eugenics, p. 192. For additional German-Swedish relations in the field of eugenics, see Bär, Schwedische Eugenik.

[6] Weindling, Health, p. 145, gives the number of *German* members of the International Society for Race Hygiene in December 1913 as 425 (including nine members from Austria-Hungary, Switzerland, Romania, and Germans from foreign provinces). The numerical data vary in the literature, but give an impression of the approximate order of magnitude. Cf. Fischer, Geschichte, p. 2, Weingart/Kroll/Bayertz, Rasse, pp. 188–208; Weber, Ernst Rüdin, p. 71. Physicians dominated the Society for Race Hygiene in the period before 1914, representing around one third of the members. Women constituted one fifth of the membership. Cf. Weindling, Health, p. 148f. (table 39).

[7] Lenz, Bund zur Erhaltung, p. 555.

as their drive to legally allow voluntary sterilization.What is more, race hygienists did not even achieve their immediate objective of imposing a *qualitative* population policy upon the *quantitative* population policy aspired to by the Reichstag and leading positions in the ministerial bureaucracy, in order to propagate the race hygiene program in the state and society in the wake of a visibly increasing population policy movement. In this case, their efforts actually backfired: the "German Society for Population Policy," founded in 1915, whose membership soon surpassed that of the German Society for Race Hygiene, deftly appropriated the latter's "Guidelines on the Birth Question," which it passed shortly before the outbreak of World War I, thus taking the wind out of the race hygiene movement's sails.[8] At the end of the World War I, the youthful German Society for Race Hygiene appeared to be on its last legs, and its work ground to a temporary halt.

Yet the devastating defeat in the World War I, the collapse of the old state order, the November Revolution, the subsequent chaos of the civil war, and hyperinflation proved to be fruitful soil that fostered the growth of the race hygiene movement. Societal upheaval and disruptions, compounded by the immense loss of human life, the drop in the birth rate caused by the war, the poor health of broad sectors of the population, and the economic privation of the postwar period, created a climate of fear, bewilderment, and violence, which very quickly turned against those who had been pushed beyond the fringes of society during the war. For social oppositions had erupted during the course of World War I that cut across the class structures of Wilhelminian society. First and foremost, starting around 1916, when the "hunger blockade" imposed by the Entente powers against the German Reich kicked in to its full effect, a new social class system of "food hierarchy"[9] emerged, based on the population's varying access to food. On its lowest level were the residents of the large cities, white-collar employees and civil servants, families of soldiers on the front, seniors living alone, and – at the very bottom – those *institutionalized* in war prison camps, prisons and penitentiaries, welfare homes, nursing homes and rest homes, sanatoriums and clinics. Despite the fact that more than 70,000 of the disabled and infirm in closed institutions starved, froze, or succumbed to contagious diseases between 1914 and 1918,[10] this population became the target of a biologistic version of the "stab-in-the-back" legend after the war:

> The war protects the blind, the deaf and dumb, the idiots, the hunchbacks, the scrofulous, the imbeciles, the impotent, the paralytics, the epileptics, the dwarfs, the freaks. – All of this residue and dross of the human race can rest assured, for no bullets are whistling toward them [...]. For them the war thus constitutes nothing less than life insurance, for this physically and mentally 'cripple guard,' hardly able to hold its own against its able rivals in free competition during peacetime, now gets the most lucrative positions and is highly paid.[11]

---

[8] Weingart/Kroll/Bayertz, Rasse, pp. 216–230; Weindling, Health, pp. 295–301.

[9] Faulstich, Hungersterben, p. 29.

[10] Siemen, Menschen, p. 29. Summarized in: Kersting/Schmuhl, Einleitung, pp. 6–8.

[11] Nicolai, Biologie, vol. 1, p. 81.

This diatribe against the disabled, sick, and weak was penned by one of the most dogmatic pacifists in Germany: the physiologist Georg Friedrich Nicolai (1874–1964). During World War I Nicolai had been degraded from a civilian doctor with the rank of an officer to a medical orderly because of his critical position against the German conduct of war, and had eluded the attempt to force him into combat service by fleeing to Denmark. After the war he returned to Germany and was appointed to an associate professorship for physiology at the University of Berlin, but, because he had "fouled his own nest," was forced to succumb to the pressure of his students and colleagues and leave Germany for good in 1922. Thus, he was anything but a race nationalist,[12] whose book *The Biology of War*, written during the war and published in 1919, according to the preface was oriented toward the "humanitarian idea."[13] Nicolai was a cosmopolitan, who believed he had an obligation to humanity as a whole. And yet, on the authority of the Darwinist principle of natural selection, it was a matter of course for Nicolai to exclude from humanity all of the sick, handicapped, weak and decrepit. One root of his pacifism was the eugenic thesis of the contraselective effects of modern war. The two founding fathers of race hygiene, Alfred Ploetz and Wilhelm Schallmayer (1857–1919),[14] already had warned about the danger of adverse selection associated with the war, and even during World War I – once the initial euphoria about the "war for existence" (*Kampf ums Dasein*) between the nations of Europe had dissipated – prominent advocates of race hygiene like the Munich hygienist Max von Gruber (1853–1927)[15] and the Swiss psychiatrist August Forel (1848–1931)[16] had drawn attention to the devastating consequences of the life-devouring, static war for genetic health. This prophecy of doom was based on a distorted perception: the shock of the first industrial war was so deep that it deadened the senses to the death, hunger, and misery on the home front. The consequence was an abrupt break in mentality, as the psychiatrist Karl Bonhoeffer (1868–1948) was right to fear, indeed, a "transformation in the concept of humanity." At the annual convention of the German Association for Psychiatry in 1920 he hypothesized:

> [U]nder the difficult experiences of war, we were forced to value the individual human life differently than before, and that in the years of hunger during the war we had to come to terms with watching as our institutionalized sick died en masse of malnutrition, and nearly to sanction this in the thought that this sacrifice may have saved the lives of the healthy. This emphasis of the right of the healthy to self-preservation, as is entailed in a time of need, conceals a risk of exaggeration [...].[17]

---

[12] Zuelzer, Fall Nicolai; Schmiedebach, Sozialdarwinismus, pp. 110–116.

[13] Nicolai, Biologie, vol. 1, p. 5.

[14] Weiss, Race Hygiene, pp. 140–146. On his biography, see also: Labisch/Tennstedt, Weg, vol. 2, pp. 486 ff.

[15] Schmiedebach, Sozialdarwinismus, pp. 100–102; Eckart, "Versuch," pp. 314–316. With a different emphasis: Kudlien, Max von Gruber. On his biography, see also: Klee, Personenlexikon, pp. 205 f.

[16] Wettley, August Forel, pp. 140–144; Schmiedebach, Sozialdarwinismus, pp. 108–110. Cf. Kühl, Internationale, pp. 41–48.

[17] Bericht über die Verhandlungen psychiatrischer Vereine, p. 598. On its interpretation: Gerrens, Medizinisches Ethos, pp. 63 f.

Voices exhorting caution like this one found little resonance, however. Under the impression of the immense loss of human life in the "storms of steel" of the World War I, the argumentation of the race hygienists appeared compelling. It was picked up by warmongers and pacifists alike, by the *Alldeutsch* movement, liberals and socialists, Protestants and Catholics, all the more as the agenda of the race hygiene program must have appeared a moderate alternative to the demand voiced at the conclusion of the pamphlet by Karl Binding (1841–1920) and Alfred E. Hoche (1865–1943) for "clearance for the annihilation of life 'unworthy of life'."[18]

1.1.3 Racial Hygiene in the Weimar Republic

Race hygiene continued to receive impetus in the Weimar Republic. Paradoxically, this development was a consequence of the fact that the principle of the social state received constitutional status for the first time, and even achieved inclusion in the Weimar constitution's catalog of basic rights – an altogether cunning project, whose significance for the future development of the social state in Germany cannot be estimated highly enough. Yet in this the young republic also took on a heavy burden, for ultimately there was not enough economic power to advance the erection and expansion of the social state in a sufficient manner. The guarantee of the social state thus remained "an unredeemed promise." The Weimar Republic was "an overstrained welfare state."[19] From its inception, the reality of social policy lagged behind the high target specification, and the gulf between the state guarantee of social security and a reality marked by inflation, laborious stabilization, and depression could hardly be bridged. Especially in the beginning and final phases of the Weimar Republic, the social state seemed about to collapse under the constantly growing welfare burden. According to a contemporary estimate, in the first years after the war the circle of those deemed needy and eligible for welfare amounted to around 15 million at times, nearly one fourth of the Reich's population.[20] And in late 1932, at the nadir of the worldwide economic crisis, the official statistics office calculated that – in addition to the around 2.5 million unemployed, who received support transfers from unemployment insurance – a further nine million people hung on at mere subsistence, or often below, from public welfare for unemployed welfare recipients, small pensioners and retirees, war invalids, war widows, and war orphans. Beyond this there were an estimated 2.4 million individuals who had fallen through the mesh of all of the social security systems.[21]

---

[18] Binding/Hoche, Freigabe. Cf. Hoffmann, Inhalte, pp. 99–107; Rehse, Euthanasie, pp. 73–79; Hafner/Winau, Freigabe, pp. 233–245; Fichtner, Euthanasiediskussion, pp. 25–29; Nowak, "Euthanasie," pp. 48–52; Burkhardt, Euthanasie; Benzenhöfer, Der gute Tod.

[19] Abelshauser, Weimarer Republik, p. 30, 31.

[20] Schreiber, Deutsches Reich, pp. 12 f.

[21] Sachße/Tennstedt, Wohlfahrtsstaat, pp. 84–97; Schmuhl, Arbeitsmarktpolitik, pp. 162 f., 189. For a classic example: Lohalm, Wohlfahrtskrise.

Under these circumstances, race hygiene, characterized by its own tension between the apocalyptic vision of the final days and millenary promise of salvation, critique of civilization, and scientocratic obsession with feasibility, fell on fruitful soil.[22] On the one hand it provided a simple, seemingly scientifically grounded explanation for societal miseries: the social state, by abrogating the laws of natural selection, rears an army of the genetically inferior, cripples, idiots, mentally ill, neurasthenics, psychopaths, and hypochondriacs, who threaten to hopelessly overstrain the systems of social insurance and irretrievably damage the genetic health of the nation within just a few generations. On the other hand, race hygiene also indicated a way out of the scenario of decline it prophecied: with the instruments of positive and negative eugenics, scientifically grounded social engineering appeared possible, which could not only check the alleged degeneration process, but also raise the genetic health of the nation. At the same time, the social state was to be stabilized through restructuring toward a social welfare that differentiated among recipients on the basis of their genetic biology. Race hygiene therefore did not limit itself to criticism of the Weimar welfare state, but further designed the utopia of a technocracy that was to raise the social state to a new level – on the strength of comprehensive, scientifically guided biopolicy.

Against this background, the race hygiene movement in the Weimar Republic rose to become an influential pressure group, which succeeded in gradually anchoring race hygiene/eugenics as a research orientation and subject of instruction in the human sciences and in placing eugenic postulates and programs on the political agenda.

Since race hygiene/eugenics was located in the boundary region between the various human sciences, it was not clear from the outset which of these disciplines would develop into sciences of reference for race hygiene. Accordingly, at the first German Sociology Convention, held in Frankfurt in 1910, Ploetz was still attempting to establish race hygiene as a subdiscipline of sociology. While he lost this debate, driven from the field by Max Weber (1864–1920),[23] Ploetz found more favor at the meeting of the German Association for Public Health Care in the year 1911. Two years later, race hygiene was accepted into the main medical group of the Society of German Natural Scientists and Physicians. The emerging field of race hygiene thus oriented itself even on the eve of the World War I toward the field of medicine, a tendency that intensified in the Weimar Republic.

This was also demonstrated through the introduction of race hygiene as a subject of instruction at the institutions of higher education in the German Reich. Before World War I only a few university catalogs listed courses on race hygiene. After the war, the number of courses rose dramatically. Over the course of the 1920s, race hygiene gained access to the curricula of all German universities and nearly all polytechnic institutions. The tendency toward establishing race hygiene as a subdiscipline of medicine was unmistakable in this development. While in the early 1920s most courses on race hygiene were offered in the context of the *studium*

[22] Schmuhl, Rassenhygiene, Nationalsozialismus, Euthanasie, pp. 59–65.

[23] Verhandlungen des Ersten Deutschen Soziologentages, pp. 154f. Cf. Schmuhl, Max Weber.

generale for students of all faculties, over the course of the decade medical lectures, courses, laboratory courses, seminars, and colloquia took up an ever larger share of the courses offered on the subject of race hygiene.[24] Even where race hygiene was positioned within the field of medicine underwent a transformation. At the beginning of the 1920s it was still generally considered to be a specialized area of (social) hygiene, but it increasingly took on an orientation toward psychiatry, which soon became its most important science of reference. The psychiatric science of the postwar period, still proceeding from Wilhelm Griesinger's (1817–1868) axiom that "mental illnesses are brain illnesses," was directed toward the study of genetic influences in the emergence of mental illnesses and mental disabilities, while practical psychiatry was dedicated to eugenic prevention, because there were still very few therapeutic possibilities for psychic illnesses.[25] In this context the German Research Institute for Psychiatry (*Deutsche Forschungsanstalt für Psychiatrie* – DFA), which was admitted into the KWG in 1924, took on the important role of pacemaker, especially the Institute for Psychiatric Genetics (*Institut für psychiatrische Erblichkeitsforschung*), renamed to the Institute for Genealogy and Demography (*Institut für Genealogie und Demographie*) in 1924.[26]

In the 1920s race hygiene found an additional mainstay in the research and instruction of anthropology.[27] Even in the period between the wars, in Germany anthropology was primarily *physical* anthropology. There were some incipient steps toward *cultural* anthropology, but in Germany these cultural anthropological elements tended to flow into the predominant *race* anthropology – a very different situation than in the USA, where the influence of the school founded by Franz Boas (1858–1942) resulted in a demarcation between cultural anthropology and race anthropology.[28] Methodologically, the race anthropology of the 1920s was heading for a crisis – increasingly, classical anthropometry was regarded as insufficient. The trend was toward anthropobiology, which aspired to build a bridge to paleoanthropology and evolutionary biology, to genetics and research of the human constitution, to physiology and pathological anatomy, to psychiatric genetics, genealogy, and genetic family research as well as to hygiene, and also to social hygiene. In the mid-1920s, a king's road to combining anthropology and human genetic biology appeared emerge: blood-group research.[29] Thus, anthropology became increasingly interlinked

[24] Günther, Institutionalisierung. Cf. Weindling, Health, p. 339 (table 6). In addition to hygienists, anthropologists and psychiatrists, individual physiologists, internists, gynecologists, dermatologists, biologists, zoologists, historians, and theologians gave instruction in race hygiene. The center of race hygiene theory was the University of Munich, but the various emphases stood out at the universities of Berlin, Breslau, Dresden, Freiburg, Halle, Hamburg, Heidelberg, Jena, Leipzig, and Rostock as well.

[25] Kersting/Schmuhl, Einleitung, pp. 11–20.

[26] Weber, Forschungsinstitut; ibid., Psychiatrie; ibid., Ernst Rüdin; ibid., Harnack-Prinzip; ibid., Forschungen.

[27] Weingart/Kroll/Bayertz, Rasse, pp. 355–362.

[28] Cf. Kaufmann, "Rasse und Kultur."

[29] Geisenhainer, "Rasse," pp. 125–140.

with areas of knowledge in which race hygiene had already gained a firm foothold. As a consequence, the objects, concepts, and methods of anthropology – at least conceptually – became increasingly accessible to the race hygiene agenda.

By contrast, the connection between race hygiene and genetics in Germany was initially extremely casual.[30] There were several reasons for this: First, genetic research in Germany initially concentrated on the genetics of plants and animals in the context of breeding research – human genetics was positioned on the margins. Second, genetics was established predominately "as a theoretical science," which "was oriented to fields of practice only in individual cases."[31] Third, concepts of population genetics, which was, after all, the means of building a bridge to human genetics, initially found little resonance in Germany outside of psychiatric genetics. Nevertheless, the circumstance that nearly all members of the leading circle of the German Society for Race Hygiene joined the newly founded German Society for Genetics (*Deutsche Gesellschaft für Vererbungswissenschaft*) in 1922[32] indicates that race hygiene aspired to connect with the field of genetics.

As the field became more scientific, the race hygiene movement changed as well: its social basis broadened and shifted toward the universities, the administrative bodies of the Reich, the states and provinces, the districts and municipalities, the churches, the Protestant and Catholic relief organizations, the parliaments, the parties, associations, and clubs. Race hygiene was now no longer a matter for a handful of intellectuals more or less freelancing for the cause; now it found support in a broad group of sponsors from the educated middle bourgeoisie – university professors, preparatory school teachers, medical officers, physicians, pastors, jurists. Influential interest groups like the "Reich Alliance of Germans with Many Children" (*Reichsbund der Kinderreichen Deutschlands*), founded in 1923, picked up on the race hygiene program. Politically, race hygiene disengaged itself from the nationalist corner, became "presentable," even finding growing acceptance among the Catholic Center party, the Protestant relief organization, the women's movement, and the Social Democrats.[33] This development entailed many inner conflicts and tension.

At its general meeting in 1922, the German Society for Race Hygiene underwent a major reorganization. A change of the guard took place at the top: the leadership of the Munich contingent was relieved by the association's Berlin chapter of the group. Ploetz and Gruber were appointed honorary chairmen, with the medical councilor Otto Krohne (1868–1928)[34] elected as chairman of both the German and

[30] Weingart/Kroll/Bayertz, Rasse, pp. 351–355.

[31] Ibid., p. 352.

[32] For example: Erwin Baur, Eugen Fischer, Max von Gruber, Fritz Lenz, Hermann Muckermann, Alfred Ploetz and Ernst Rüdin. ibid., p. 352.

[33] On the churches: Nowak, "Euthanasie;" Kaiser, Innere Mission; Schwartz, Milieus; Schleiermacher, Sozialethik; Richter, Katholizismus. On the women's movement: Herlitzius, Frauenbefreiung; on the political left: Schwartz, Eugenik; ibid., "Proletarier."

[34] For a biography, see Labisch/Tennstedt, Weg, vol. 2, pp. 445 f.; Stürzbecher, Otto Krohne; Weindling, Health, p. 217, 349; Saretzki, Reichsgesundheitsrat, pp. 17 f.

the Berlin Societies for Race Hygiene. The personnel changes signified a change in course, which was also expressed by the society's passing a new set of guiding principles, in which the old "life reform" elements were discarded once and for all. After this shift, new local chapters were founded in rapid succession.[35] By 1931 the society had raised its total membership to 1,085.[36]

The upswing was marred by internal oppositions, however. As early as the first half of the 1920s, an old line of conflict was thrown into bold relief. This tension had become apparent at the very outset of the race hygiene movement and was rooted in the different standpoints on the race question. From the outset, the boundaries between race hygiene and race anthropology had been blurred. While Wilhelm Schallmayer advocated a strict demarcation of what he called *Rassehygiene* from all forms of race anthropology,[37] Alfred Ploetz soon developed an affinity for the Nordic idea. After moving to Munich in 1907, Ploetz started a "Nordic Ring," later the "Munich Bow Hunters," which left an enduring mark on the Munich branch of the German Society of Race Hygiene. During the Weimar period, tensions increased between this "Munich line" and the "Berlin line," which was closely associated politically with the Center Party and with elements of the Social Democratic Party and clearly distanced itself from all nationalist and Nordic aspirations. They were openly manifest in the arguments about compulsory nomenclature. For the Munich line, the very term "eugenics" constituted "a kind of deviation to the left"[38] away from race hygiene, while the Berlin line, in order to avoid any misunderstandings, favored the term eugenics over "race hygiene."

This opposition was also behind the temporary split in the race hygiene movement. Ostensibly it was about questions of the organizational structure. In 1924, the around 7,000 members of the "Reich Association of German Registrars"(*Reichsverband der deutschen Standesbeamten*) proposed to the German Society for Race Hygiene that they form a working group together; the latter refused, pointing to the limited possibilities of claiming nonprofit status. As a response, the "German Association for Volkish Improvement and Genetics" (*Deutscher Bund für Volksaufartung und Erbkunde*) was founded in 1925. The new organization enjoyed the solid support of the Reich Ministry of the Interior, the Prussian Welfare Ministry, the Berlin City Council and, of course, the Association of Registrars, whose director Edwin Krutina (1888–1953),[39] the driving force behind the new founding, was elected secretary of the association. Its chairman was the jurist and founder of the German and Prussian Association for Infant Welfare, Carl von Behr-Pinnow

---

[35] Dresden (1922), Kiel and Bremen (1923), Tübingen (1924), Württemberg (1925) Münster and Osnabrück (1926). Fischer, Geschichte, p. 4; Kroll, Entstehung, pp. 167f.; Kröner, Eugenik, pp. 80f., 84, 87. Cf. also Weindling, International Eugenics, p. 183 (table 1).

[36] Weingart/Kroll/Bayertz, Rasse, p. 215. Fischer, Geschichte, p. 4, listed the number of members in 1930 as 1,300.

[37] Schallmayer, Vererbung, pp. 375–387. Cf. Weiss, Race Hygiene, pp. 100–104.

[38] Saller, Rassenlehre, p. 72. Cf. Lenz, Auslese, pp. 249–254.

[39] For a biography: Labisch/Tennstedt, Weg, vol. 2, pp. 446f.

(1864–1941);[40] the ministerial councilor in the Prussian Welfare Ministry, Arthur Ostermann (1876–1943) became deputy chairman.[41] After 1 year, the new association had 1,500 members, and was thus already larger than the Society for Race Hygiene. Officially, the Association for Volkish Improvement and Genetics portrayed itself as an extension of the Society for Race Hygiene, which it characterized as an institution that performs "pioneering work of a more scientific and theoretical nature," while the association was supposed to "cultivate and propagate" eugenics "in an entirely popular form understandable to everyone."[42] The primary vehicle for this was the *Zeitschrift für Volksaufartung und Erbkunde* ("Journal for Volkish Improvement and Genetics"), called *Eugenik, Erblehre, Erbpflege* ("Eugenics, Genetics, Care of Heredity") from 1930 onward, which had a circulation of 5,000 and thus reached five times the audience of the *Archiv für Rassen- und Gesellschaftsbiologie* ("Archive for the Biology of Race and Society"). However, the German Association for Volkish Improvement and Genetics also contrasted with the German Society for Race Hygiene as regards its content, in that it emphasized the "generally biological." As its chairman Behr-Pinnow explained, the association intended to "restrain the manifestations of degeneration and be of service to the positive improvement of the people [...] without consideration of any particular race, be it Nordic, Dinaric or other."[43] Accordingly, the statute claims that the association is to serve "the totality of the German people without any differentiation among political parties or confessions."[44] The German Society for Race Hygiene viewed the new organization skeptically as a "competing movement,"[45] especially since the Association for Volkish Improvement and Genetics, contrary to its appeasing announcement that it would restrict itself to propaganda, also regarded it as its duty to make suggestions to the "organs of legislation and administration,"[46] with good chances of success thanks to the new organization's close ties to the Prussian ministerial bureaucracy.

Thus we can summarize that different lines of development came together in the first half of the 1920s, with the consequence that race hygiene, science, and politics became closely entangled. First, the catastrophe of the World War I catapulted race hygiene out of its sectarian corner to the middle of society – the race hygiene agenda appeared to offer a direction toward a new form of biopolicy to guide state and society out of the many overlapping social problems and crises the war had left behind. Second, the race hygiene paradigm increasingly trickled into the human sciences – entanglements under the auspices of heredity, above all psychiatry's turn toward genetics, anthropology's transition to anthropobiology and the first steps

[40] For a biography: ibid., pp. 380f. During World War I, Behr-Pinnow was a board member of the Berlin Society for Race Hygiene.

[41] For a biography: ibid., pp. 463f.

[42] Behr-Pinnow, Deutscher Bund für Volksaufartung, p. 411.

[43] Ibid.

[44] Quoted in Lenz, "Deutscher Bund für Volksaufartung," p. 349.

[45] Ibid.

[46] Quoted in ibid.

toward extending animal and plant genetics toward human genetics, contributed to race hygiene becoming more scientific. Third, the social basis of the race hygiene movement broadened: in the political center and to its left, a new form of eugenics emerged, which, in contrast to the old race hygiene linked with the nationalist camp, was an acceptable partner for cooperation with the Weimar coalition governments. Fourth, the "overburdened welfare state" ultimately relied more and more on the human sciences to find a way out of the crisis of the social systems, and in this framework of scientific political consulting, a key role fell to eugenics. From this constellation emerged the initiative for a state-dominated research institute dedicated to the areas of anthropology, human genetics, and eugenics.

1.2 The First Attempt: A Reich Institute for Human Genetics and Demographics

1.2.1 Genetics, Race Hygiene and Scientific Politicy Consulting

In 1919, responsibility for medical affairs in Prussia was transferred from the Ministry of the Interior to the newly founded Ministry for the Welfare of the People (*Ministerium für Volkswohlfahrt*), headed by the Center politician and Christian trade unionist Adam Stegerwald (1874–1945), who was open to the eugenic idea. The Department for People's Health (*Abteilung für Volksgesundheit*) was delegated to the Berlin physician and city councilman Adolf Gottstein (1857–1941),[47] who was in charge of Prussian medical affairs until his retirement in 1924. From the outset, Gottstein demonstrated a marked interest in genetics, genetic family research, and eugenics.[48] It is telling that

[47] For a biography: Labisch/Tennstedt, Weg, vol. 2, pp. 416f.; Weindling, Health, p. 171, 219; Saretzki, Reichsgesundheitsrat, p. 16.

[48] For example, cf. the explicit decree of 2 September 1920, in which Gottstein urged the Prussian President (*Oberpräsident*) and the president of the district government to set up information centers for the "nervous and mentally ill." He justifies his proposal with the argument that "The difficult experiences during the war period and the chronic malnourishment frequently disturbed the mental balance of sensitive individuals, especially those of high intellectual development. Under the unusually strict restrictions of personal freedom existing during the war period, and the relaxation of social sentiences and bonds that occurred after the revolution, it was thus possible for a schism to emerge in people disposed to illness, in which the initiatives suppressed for so many years liberated themselves from the powerful impulse of dissatisfaction, to the detriment of the general public. The intense excitability of affect and the psychopaths' conviction of the truthfulness of their view, which persists even when strong exaggerations and distortions of facts are apparent to the mentally healthy, secured these ill individuals a disproportionate influence on the mass of the nation, an influence that is all the more threatening because the extraordinarily miserable economic situation and the political inattentiveness of wide sectors of the population raised the general level of dissatisfaction, thus creating unusual receptiveness to influence by suggestion. Just as in earlier nationalist movements, it has been observed now, too, that these youthful psychopaths are the ones at the foreground of the political extreme orientations." Cited in Kersting/Schmuhl (eds.), Quellen, pp. 145f. Cf. also Weindling, Health, pp. 381–383.

almost immediately a unit was formed within the Department for People's Health, which – in addition to "care for the crippled," hospitals, nursing care, medical statistics, midwife affairs, infant and toddler care – was also responsible for population policy and race hygiene. Initially this unit was directed by the medical councilor Otto Krohne. Having worked in the Medical Department of the Prussian Ministry of the Interior since 1911, Krohne had concerned himself with the issues of quantitative population policy even before World War I; in the immediate postwar period he also adopted aspects of race hygiene. As mentioned above, in the Weimar Republic, he advanced to the top of the Berlin and German Societies for Race Hygiene.[49]

At Krohne's initiative a "Commission for Race Hygiene" was convened in May 1920, which was to advise the Welfare Ministry on the question of "whether and to what extent our nation appears endangered by the weighty consequences of the war in terms of race hygiene as well, and what kind of proposals should be made to combat such a danger."[50] In 1921, this commission was renamed the "Committee for Race Hygiene and Demography" and subordinated to the newly founded Prussian State Health Council.

The composition of this body[51] is evidence of the state's endeavor to apply scientific expertise to the solution of population policy and health policy issues, especially expertise from the area becoming established as genetics. In May 1920 the members of the commission – besides three officers of the ministry – were the five geneticists Erwin Baur (1875–1933),[52] Agnes Bluhm (1862–1943),[53] Carl Erich Correns (1864–1933), Richard Goldschmidt (1878–1958),[54] and Heinrich Poll (1877–1937), the medical statistician Eugen Roesle (1875–1962), the two gynecologists Ernst Bumm (1858–1925) and Max Hirsch (1877–1948), the anatomist Hans Virchow (1852–1940),[55] the Pathologist Max Westenhöfer (1871–1957), the anthropologist Felix von Luschan (1854–1924),[56] the psychiatrist Karl Bonhoeffer, and the physiologist Emil Abderhalden (1877–1950). At the time of their appointment, three of the five geneticists were working under the auspices of the KWG;

[49] Krohne made sure that the Prussian Medical Administration paid contributions to the society, subsidizing a small research project the society had initiated on the genealogy of crime, a propaganda film and an exhibition. The German Association for Volkish Improvement and Genetics also received financial support from Krohne's department. Weindling, Health, pp 406 f.; idem., Preußische Medizinalverwaltung, pp. 682 f.; Weingart/Kroll/Bayertz, Rasse, pp. 272 f.

[50] According to the draft of the invitation letter by the Prussian Minister for the Welfare of the People of May 1920. Cited in Gilsenbach, Erwin Baur, p. 186.

[51] Weindling, Health, pp. 340 f.; idem., Weimar Eugenics, p. 304; Weingart/Kroll/Bayertz, Rasse, p. 240; Lösch, Rasse, p. 168.

[52] For a biography: Gilsenbach, Erwin Baur; Kröner/Toellner/Weisemann, Erwin Baur; Hagemann, Erwin Baur; Labisch/Tennstedt, Weg, vol. 2, p. 380; Klee, Personenlexikon, pp. 32 f.

[53] For a biography: Vogt, Wissenschaftlerinnen, pp. 26 f.; Labisch/Tennstedt, Weg, vol. 2, p. 384; Klee, Personenlexikon, p. 54.

[54] Goldschmidt, Ivory Tower, esp. pp. 230–232.

[55] Labisch/Tennstedt, Weg, vol. 2, p. 507.

[56] For a biography: Grimm, Felix von Luschan; Geisenhainer, "Rasse," pp. 51–57.

a further joined them later: Correns was the director of the KWI for Biology founded in Berlin-Dahlem in 1914; from 1914 until he was forced to emigrate in 1935, Goldschmidt headed the Department for the Genetics and Biology of Animals at the KWI for Biology; Bluhm had been a visiting scientist in Corren's department since 1919. In 1928, Baur would take over as director of the newly established KWI for Breeding Research in Müncheberg. Baur, Correns and Goldschmidt founded the "German Society for Genetics" (*Deutsche Gesellschaft für Vererbungswissenschaft*) in 1922. It is obvious how closely the young science of genetics in Germany was bound up with the KWG, and through the KWG, already able to exert influence on policy.

With the Commission for Race Hygiene and Population, race hygiene also advanced into the sphere of scientific policy consulting. Agnes Bluhm had belonged to Ploetz' inner circle since completing her study of medicine in Switzerland. She was one of the first members of the Berlin Society for Race Hygiene and was on the board of directors of the German Society for Race Hygiene until 1932. Moreover, she was also represented on the editorial board of the *Archiv für Rassen- und Gesellschaftsbiologie*. Erwin Baur joined the Berlin Society for Race Hygiene in 1907; from 1917 until 1919 he was the chairman of the Berlin branch, and in 1933 a member of the *Archiv für Rassen- und Gesellschaftsbiologie* editorial board. Felix von Luschan was also an early, active member of the Berlin Society for Race Hygiene and its chairman for a time.[57] During World War I, Max Westenhöfer served as its deputy chairman; in 1920 he outlined the "Tasks of Race Hygiene in the New Germany."[58] Hans Virchow also joined the Berlin Society for Race Hygiene. Max Hirsch, one of the fathers of social gynecology, was among the leading figures in the Society of Physicians for Sexology and Eugenics (*Ärztliche Gesellschaft für Sexualwissenschaft und Eugenik*) founded in 1913, and played an influential role in the founding of the *Archiv für Frauenkunde und Eugenetik* ("Archive for Gynecology and Eugenics").[59] In 1921 the social hygienist Alfred Grotjahn (1869–1931),[60] one of the main exponents of socialist eugenics, was appointed to the commission as well.

1.2.2 Erwin Baur as the Central Figure

Erwin Baur advanced to become the central figure in the prehistory of a state-dominated research institute for anthropology, human genetics, and eugenics. He distinguished himself clearly from the mandarins predominant in the community of

[57] Doeleke, Alfred Ploetz, pp. 44 f.; Weindling, Health, pp. 134 f.

[58] Westenhöfer, Aufgaben. Weingart/Kroll/Bayertz, Rasse, pp. 289 f.

[59] Schneck, Entwicklung, pp. 37 f.

[60] Cf. Schwartz, Eugenik, pp. 70–80. For a biography, see also: Tutzke, Alfred Grotjahn; Dieckhöfer/Kaspari, Tätigkeit; Kaspari, Eugeniker; Weindling, Soziale Hygiene; Roth, (Schein-) Alternativen; Labisch, Alfred Grotjahn; idem./Tennstedt, Weg, vol. 2, pp. 419 f.

geneticists, as in every other scientific community.[61] The son of a pharmacist, from a small town in Baden, first studied medicine. After earning his Ph.D. in medicine from the University of Kiel in 1900, he worked as a scientific assistant in the field of bacteriology, traveled to Brazil as a ship's doctor, and worked as an intern at the University Psychiatric Clinic in Kiel. Then he turned to botany. He took a Ph.D. in this subject at the University of Freiburg in 1903. From there he went to the Botanical Institute in Berlin, where he was promoted to professor on the basis of his postdoctoral work on bacterial physiology. As a private lecturer, he was passed over for a professorial chair on several occasions before being appointed to the chair for botany at the Berlin Agricultural College (*Landwirtschaftliche Hochschule*) in 1911. Baur's career path was thus far from a straight-line procession; rather, he took winding paths to arrive "at a chair in a new, barely established subject at an institution that did not enjoy university status."[62] Although he was a trained botanist, Baur was not interested in general botany – increasingly, he directed his work toward experimental genetics. In 1908, he was the cofounder and editor of the *Zeitschrift für induktive Abstammungs- und Vererbungslehre* ("Journal for Inductive Theory of Descent and Genetics"), the world's first journal for genetics. In 1913, he became a professor for genetics and received his own institute for genetic research. Baur was not the universal scholar type – far from it, he championed specialization, while at the same time remaining very open to interdisciplinary collaboration. He, too, was among the advocates of applied research. He was doubly committed to establishing a Kaiser Wilhelm Institute in the area of human genetics and eugenics: As a geneticist and researcher of breeding, he had great interest in bringing the genetics of plants, animals, and humans closer together to found a comprehensive science of heredity. As a race hygienist, he was interested in linking the new research area of human genetics with the application-oriented approach of race hygiene in order to make it useful for state and society. In this he saw a close connection to the breeding research performed at his own institute. Baur's approach can be designated "agri-eugenics."[63] The point of departure for his considerations was *Ammon's law* – formulated in 1893 by Otto Ammon (1842–1915), an engineer who dabbled in physical anthropology, in the book *Natural Selection in Man*, according to which the phenomena of land flight and urbanization lead to a constant outflow of the best genetic material from the country to the city, where it gradually dies out as a consequence of the diminishing fertility associated with the social advancement of those that bear it, such that, *in balance*, the genetic material of the entire population deteriorates ever further.[64] This trend, Baur had insisted since 1917,[65] can be countered by a combination of measures from the area of *positive* eugenics,[66] like a child allowance and tax benefits, and agricultural policy

[61] On this: Ringer, Decline. What follows is according to Harwood, Analyse, pp. 342–347.

[62] Ibid., p. 344.

[63] Kröner/Toellner/Weisemann, Erwin Baur, p. 27.

[64] Ammon, Natürliche Auslese.

[65] Cf. e.g. Baur, Erhaltung; idem., Rassen; idem., Untergang; idem., Biologische Bedeutung.

[66] On elements of negative eugenics in Baur's "agri-eugenic" concept, cf. Kröner/Toellner/Weisemann, Erwin Baur, pp. 50–57.

measures to encourage medium-size farming, such as land reform, the encouragement of settlements, protective tariff barriers, and the institution of agricultural autarchy. "For Baur [...] race hygiene – just like the application of agricultural science research – was a reasonable method for organizing human life. His technocratic vision of the integration of science into a planned economy made necessary the close coordination of eugenics with the agricultural policy of the state."[67]

In 1917, at the very same time Baur was developing his concept of "agri-eugenics," he proposed that the KWG found an institute for the research of useful plants. Parallel to this, through the Berlin Society for Race Hygiene, whose chairmanship he had since taken over, he also advocated a research institute for human genetics and eugenics. On June 15, 1917, the society sent a circular signed by the three chairmen Baur, Westenhöfer, and Behr-Pinnow to "outstanding men of the German nation" requesting material support. "It is time for the decision in the struggle among nations," the letter introduces, "and the German nation is faced with the difficult question of how it should defend the achievements of the world war against its open and secret enemies. Everything connected with this, every hope for the future depends on whether the German people will have sufficiently numerous and capable progeny." With reference to the generous support for race hygiene in Great Britain and the USA – what was meant here were the "Eugenics Record Office" established by Francis Galton (1822–1911) in London in 1905, from which the "Galton Laboratory for the Study of National Eugenics" later proceeded, and the research institute for experimental evolution research set up by Charles B. Davenport (1866–1944) in Cold Spring Harbor in 1910, which also had a "Eugenics Record Office"[68] – the society solicited funds for the "founding of a popular science journal, the erection of a generous organization, and, if possible, a research institute of its own."[69] The Berlin society even brought its petition to a wider public in a pamphlet entitled "What Race Hygiene Wants."[70]

These appeals found no resonance during the war. The political climate in the immediate postwar period made a renewed advance seem more promising, however. The Commission for Race Hygiene and Population in the Prussian State Health Council offered a suitable platform. Baur was very active in this body until 1926[71] and was involved in consulting on the matters of sterilization, marriageability, emigration, and human settlement. He was also responsible, along with Goldschmidt, Luschan, and Roesle – for an expert opinion of June 7, 1922, by the Prussian State Health Council about "the necessity of a Reich institute for human genetics and population." This institute was to be staffed with four scientists or physicians and one statistician, and supplied with a budget of two million Reichmarks. Six departments were proposed: one microscopic-cytological; one

[67] Harwood, Analyse, p. 344, 345.

[68] Blacker, Eugenics; Kevles, In the Name; Mazumdar, Eugenics.

[69] Quoted in Kröner/Toellner/Weisemann, Erwin Baur, p. 31.

[70] Berliner Gesellschaft für Rassenhygiene (ed.), Rassenhygiene. Cf. Kröner/Toellner/Weisemann, Erwin Baur, p. 32.

[71] The circumstances leading to his resignation are depicted in Saretzki, Reichsgesundheitsrat, pp. 313–316.

department for genetic family research; one for race doctrine (with special consideration to the "tribe of the German people"); one for research of the human constitution and the genetics of abnormal conditions; one – literally – "genotherapeutic" department; and, finally a statistical department. The commission further demanded support for instruction in genetics at the universities.[72] In the context of this advance, Luschan turned to his friend Eugen Fischer, whom he had been trying to obtain to succeed him as ordinary professor for anthropology at the University of Berlin since 1921, and dangled before him the prospect of a chair for "social biology" and a research institute with the same objectives.[73]

Yet the further negotiations progressed in a way that was rather unexpected and disappointing for Luschan and Baur. The Prussian Ministry for Education forwarded the expert opinion to the Reich Health Office, the Prussian Ministry for Science, Art and Education, and the Reich Ministry of the Interior. Carl Hamel (1870–1949),[74] President of the Reich Health Office, agreed in principle with the expert opinion. The Prussian Science Minister approved the founding of an institute, but, for financial reasons, rejected the suggestion to authorize instruction for biology and genetics at every university. The Reich Ministry of the Interior ultimately declared itself willing in principle to continue pursuing the matter, but demanded that a representative of the "Central Office for the History of German Persons and Families" (*Zentralstelle für Deutsche Personen- und Familiengeschichte*) be involved in the planning, so that the needs of genetic family research would be taken into consideration.[75]

1.2.3 The Alliance Between Genealogy and Psychiatry

Where did this suprising player come from, and what role did he play? In the final third of the nineteenth century, the subject of genealogy, which had largely disappeared from the curricula of the German universities after the French Revolution and the Napoleonic wars, experienced a renaissance, which was closely connected with the awakening interest in family history among large sectors of the middle class. It was no coincidence that the *Gothaischen Genealogischen Taschenbücher* ("Gotha Geneological Paperbacks") were complemented by the *Genealogische Handbuch bürgerlicher Geschlechter* ("Genealogical Manual of Bourgeoisie Families") in 1889. At the same time, genealogy visibly disengaged itself from its traditional functional context as an auxiliary science of heraldry and sought a new science to guide it. Three of the human sciences presented themselves for this task, and thus, starting in the 1890s, a division arose within the field of genealogy as regarded the question of the discipline in which it should be positioned. This division

[72] Weingart/Kroll/Bayertz, Rasse, pp. 240f.; Weindling, Weimar Eugenics, p. 308; Lösch, Rasse, p. 168.

[73] Ibid., p. 164.

[74] For a biography: Labisch/Tennstedt, Weg, vol. 2, pp. 426f.

[75] Weingart/Kroll/Bayertz, Rasse, p. 241; Lösch, Rasse, p. 168.

deepened increasingly over the following decades: one faction had a conventional conception of genealogy as an auxiliary science of history, another attempted to bind genealogy closely to sociology, and a third, initiated by Ottokar Lorenz (1832–1904) and his *Lehrbuch der gesamten wissenschaftlichen Genealogie* ("Manual of the Entire Scientific Genealogy") in 1898, wanted to establish genealogy as an auxiliary science of a Darwinistically oriented human genetics. The "Central Office for the History of German Persons and Families" (*Zentralstelle für Deutsche Personen- und Familiengeschichte*) founded in Leipzig in 1904, and its chairman Hans Breymann (1873–1958), planned the erection of a "German Institute for Genetic Family Research, Genetics and Regeneration" (*Deutsches Institut für Familienforschung, Vererbungs- und Regenerationslehre*), which soon worked closely together with the German Society for Race Hygiene.[76]

"Genetically oriented race hygiene," commented the group of researchers around Peter Weingart, "had conquered its first auxiliary science."[77] Nevertheless, considering the balance of power, initially it was no means clear whether genealogy would become an auxiliary science of race hygiene, or race hygiene an auxiliary science of genealogy. In any case, the genealogists pursued their own objectives. In the negotiations about founding a "Reich Institute for Human Genetics and Population" in 1922, the Central Office for the History of German Persons and Families found an ally in the psychiatrist Ernst Rüdin (1874–1952), who had been the director of the Genealogical-Demographic Department of the DFA since 1917.

Rüdin, too, hailed back to the very beginnings of the race hygiene movement. A prospective psychiatrist, since the marriage of Alfred Ploetz to Pauline Rüdin (1866–1942) in 1890, he was the brother-in-law of the man who was to become the forefather of race hygiene with his main work, *Die Tüchtigkeit unserer Rasse und der Schutz der Schwachen* (1895) ("The Fitness of Our Race and the Protection of the Weak", 1905). Rüdin was also a member of the circle of pupils around the Swiss psychiatrist August Forel, whose influence on Ploetz and emerging race hygiene can hardly be overstated.[78] In Berlin from 1901 on, Ploetz and Rüdin worked together closely on the theoretical grounding and practical organization of the young race hygiene movement; Rüdin's commitment was "indispensable" to its continuity and further development in the first years after the turn of the century. Matthias M. Weber correctly assesses that "Rüdin's role in the early days of the race hygiene organization [...] as compared to Ploetz's contribution tends to be underestimated rather than overstated."[79] From 1905 until 1907, he was an editor of the *Archiv für Rassen- und Gesellschaftsbiologie* and later, as coeditor-in-chief, had a say in all editorial decisions. In 1905, Rüdin was one of the founding members of the Berlin Society for Race Hygiene. From 1907 on, Rüdin, by this time an assistant at the psychiatric clinic of the University of Munich under Emil Kraepelin (1856–1926), held the office of

[76] Sommer, Organisation; Breymann, Notwendigkeit; Rüdin, Wege und Ziele.

[77] Weingart/Kroll/Bayertz, Rasse, p. 204.

[78] On this also: Küchenhoff, Denken.

[79] Weber, Ernst Rüdin, p. 53.

business manager of the newly founded Munich branch of the Society for Race Hygiene, which was chaired by Gruber. In 1911, Rüdin and Gruber played a major role in conceptualizing the special exhibition on race hygiene at the International Hygiene Exhibition in Dresden, with which the Society for Race Hygiene approached a larger public for the first time.[80] The exhibition in the Dresden Hygiene Museum shows Rüdin's close collaboration with the genealogists. His approach, which combined genealogy, the statistics of heredity, and population genetics came much closer to the interests of the genealogists than Baur's human genetics approach oriented toward plants and animals. In fact, Rüdin's Genealogical-Demographic Department at the DFA meant the realization of the genealogists' demands for a Reich Institute for Genetic Family Research. This was the first appearance of the future line of conflict between Rüdin and the DFA in Munich on the one side, and Fischer and the KWI-A in Berlin on the other, which was to persist until 1945.

Initially, of course, it looked as if the "Munich line" might prevent the founding of a competing institute in Berlin. Invited to the further consultations about a Reich institute for human genetics, which took place on January 22, 1923, in the Reich Ministry of the Interior – in addition to the representatives of the Reich Finance Ministry, the Reich Health Office, the Prussian ministries for the Welfare of the People, for Science, and for Finances – were Breymann and Stephan Kekulé von Stradonitz (1863–1933)[81] of the Central Office for History of German Individuals and Families, Rüdin of the DFA, and Heinrich Poll, who had been appointed to the newly created chair for genetics at the University of Berlin. Baur, by contrast, was not invited. Rüdin's lofty institute plans were soon tabled after the representatives of the Reich and Prussian Finance Ministries expressed their veto. In view of the strained budget situation, it was first agreed to create a "working group" of the existing institutions, which would be linked loosely into a "Reich Institute," "which essentially [was to be] a money distribution office."[82] However, even this plan ultimately collapsed due to the meager finances of the Reich, and Baur, previously ousted, reentered the stage.

1.3 The Second Attempt: The Kaiser Wilhelm Institute for Anthropology, Human Heredity, and Eugenics

1.3.1 Involving the Kaiser Wilhelm Society

Erwin Baur had, as Niels C. Lösch formulates pointedly, "a sort of hinge function [...]. Over and again, he succeeded in liberating the project from deadlocked situations, using his disparate contacts to scientific organizations and state institutions to

[80] Gruber/Rüdin (ed.), Fortpflanzung.

[81] For a biography: Labisch/Tennstedt, Weg, vol. 2, p. 436.

[82] Quoted in Lösch, Rasse, p. 169. Cf. Weingart/Kroll/Bayertz, Rasse, pp. 241 f.; Weindling, Weimar Eugenics, p. 308.

find new allies and supporters."[83] After the approach via the State Health Office had run aground, Baur concentrated his activities on the KWG. In 1917 Baur, as mentioned above, had proposed to the KWG the founding of an institute for the research of useful plants, a project that he had initially not been able to realize because funding was lacking. Now he campaigned at the KWG for an institute for human genetics – presumably with the ulterior motive of thus indirectly supporting his own project, the founding of a KWI for breeding research, by these means. In his memoirs, Friedrich Glum (1891–1974), chief executive officer of the KWG at the time, portrays Baur as a whirlwind lobbyist, who skillfully exploited his extensive high-level political contacts to force the hand of the KWG administration.[84] However, it happened, Baur was successful in winning over the president of the KWG, the Protestant theologian and church historian Adolf von Harnack (1851–1930), to back his project.

In this, too, the disruptions caused by the World War I become apparent. For before the war, Harnack had persistently rejected all proposals to create a Kaiser Wilhelm Institute in the area of human genetics, (race) anthropology and race hygiene.[85] Géza von Hoffmann, Austrian-Hungarian Vice Consul in Berlin, an active member of the International Society for Race Hygiene and author of a book about the American sterilization laws,[86] had made the most recent attempt in April 1914, approaching Harnack with the idea of a Reich institute for race biology.[87] After the war, Harnack became no less than a driving force in the realization of Baur's plans for an institute within the KWG.

1.3.2 Eugen Fischer, the Designated Director

Baur's recommendation for the directorship of the new institute was the anthropologist Eugen Fischer, a close friend of his since their college days, when they skied together in the Schwarzwald Ski Club.[88] Fischer had completed his study of medicine at the University of Freiburg, graduating with a Ph.D. and a physician's license in 1898. He became an assistant at the Anatomical Institute at the University of Freiburg. After qualifying as a professor in 1900, he was appointed associate professor for anatomy and anthropology in 1904. In 1908, he took a sabbatical to

[83] Lösch, Rasse, p. 169.

[84] Glum, Wissenschaft, pp. 368–373. Glum's version, that Baur did not submit the proposal to found the KWI for Anthropology until his own KWI for Breeding Research had been wrapped up (p. 371), cannot be correct.

[85] Cf. e.g. Gruber, Organisation.

[86] Hoffmann, Rassenhygiene. Cf. Weingart/Kroll/Bayertz, Rasse, pp. 286–288.

[87] Wendel, Kaiser-Wilhelm-Gesellschaft, pp. 116–122, 175–178, Weindling, Health, pp. 239–241.

[88] For a biography: Lösch, Rasse; Gessler, Eugen Fischer; Horst/Maier, Eugen Fischer; Labisch/Tennstedt, Weg, vol. 2, pp. 403–406; Geisenhainer, "Rasse," p. 474; Klee, Personenlexikon, pp. 151f.

German Southwest Africa, the results of which he published under the title *Die Rehobother Bastards und das Bastardisierungsproblem beim Menschen* (1913) ("The Bastards of Rehoboth and the Problem of Miscegenation in Man").[89] In this book, Fischer uses the example of the "bastards of Rehoboth," a mixed population that resulted from the combination of the Boers and the female "Hottentots" in German Southwest Africa, to demonstrate that the heredity of human (race) characteristics can be proven without a doubt by the Mendelian rules. Although this assertion rested on fairly dubious argumentation, it was generally recognized after World War I and was the basis of Fischer's fame as an anthropologist.[90] In 1912 Fischer became a prosector, first in Würzburg and then in Freiburg. During World War I he worked as a military surgeon. In 1918, he was appointed to full professor and director of the Anatomical Institute in Freiburg. Fischer also played an important role in the race hygiene movement, however. He joined the Society for Race Hygiene in 1908; in 1909/10 he founded the Freiburg branch, from which he developed an important byline in the network of relationships that spanned the community of race hygienists.[91]

Among Fischer's students was Fritz Lenz (1887–1976),[92] who became editor of the *Archiv für Rassen- und Gesellschaftsbiologie* after moving to Munich in 1913, and in 1923 received a associate professorship for race hygiene at the University of Munich, the first in Germany. In 1921, Baur, Fischer, and Lenz wrote what was the fundamental German-language textbook about the theory of human heredity and race hygiene, *Menschliche Erblichkeitslehre und Rassenhygiene*, known for short by the three authors' names: "Baur-Fischer-Lenz."[93] Thus it was no surprise that Baur, in his thrust for a research institute for human heredity, would advocate his friend Fischer. Especially since Fischer, like no other, stood for the combination of anthropology, genetic pathology, human genetics, and eugenics.

However, Fischer tells a different story in his memoirs, of reluctantly giving in to his friend Baur's pressure:

> Well, Erwin Baur wrote me a long letter laying out this plan. But I refused. I had no occasion at all to leave my […] happy circumstances in Freiburg. […] But Baur did not relent. Insistently, he presented me the scenario that the founding would not happen at all if I refused. There was no other man who linked race science with genetics, and at the same time would research the large area of medical genetics, so to speak, and ultimately, who really appreciated the importance of race hygiene for the future of our people. Now we had the unique chance to help anthropology and genetics find complete acceptance and to secure them a place of their own among the old disciplines.[94]

---

[89] Fischer, Rehobother Bastards.

[90] On the basics of this, see: Mai, Humangenetik, pp. 53–60; Lösch, Rasse, pp. 53–81.

[91] Weindling, Health, pp. 143 f.

[92] For a biography: Rissom, Fritz Lenz; Weiss, Race and Class; Labisch/Tennstedt, Weg, vol. 2, p. 453; Geisenhainer, "Rasse," p. 480; Klee, Personenlexikon, pp. 366 f.

[93] Baur/Fischer/Lenz, Erblichkeitslehre. Cf. Fangerau, Etablierung.

[94] Quoted in Lösch, Rasse, p. 170.

According to Fischer, after some vacillation, for the love of science he decided to take to the "treadmill" that "Berlin meant for everyone."[95] Fischer's version is confirmed by the protocol of the Administrative Council meeting on May 20, 1926, which states that the plan for the founding of an institute for anthropology and eugenics submitted by Baur was "welcomed" by the Administrative Council of the KWG, and the General Administration was authorized to "undertake further steps to clarify the financing and location of the institution, under the prerequisite that it was successful in winning Professor Eugen Fischer as director."[96]

The General Administration set about its work with determination. Requesting funding, it turned to the Reich Ministry of the Interior and the Prussian Ministry for Science, Art, and Education. The signals it received from these bodies of the state appear to have been encouraging: in any case, in June 1926, a new building for the KWI-A was upgraded to urgent in the construction budget of the KWG. For June 19, 1926, the day on which the Senate of the KWG was to decide about the plan for a KWI for Anthropology, Human Genetics, and Eugenics, Harnack invited the designated director to Berlin for a personal interview. In the evening – he did not yet know that the Senate had already given its approval for the founding of the institute that afternoon – Fischer held a presentation of his agenda before the assembled senators, which not only denoted the scientific conception of the future institute, but also reflected the strategy with which this conception was to be presented to the political decision makers in the Weimar Republic.

Proceeding from the anthropobiological approach, Fischer stated that anthropology had overcome the stage of classical anthropometry: "Anthropology and the theory of heredity [have] become inseparable from each other." Fischer distinguished these decidedly from Hans Friedrich Karl Günther (1891–1968) [97] and his "race theory of the German people": while the book of this name does contain "very many excellent observations and exquisite remarks," these were placed "next to strongly tendential things" – Günther's race theory leads to the mistaken conclusion that one could, on the basis of skull shape, hair and eye color, "make race diagnoses without further ado." In contrast, Fischer, "of whom one truly [...] cannot claim that he [believes] the influence of heredity to be minor," emphasized that in the "development of man," that is, phenogenesis, an important role is played by the "peristatic influences," "environment, 'milieu,' if only social milieu" in addition to heredity. At this juncture Fischer referred explicitly to Franz Boas, who had demonstrated that the children of Eastern European Jews and Southern Italian immigrants in New York showed significant differences from their parents in terms of the shape of their skulls. "The very question of the Jewish population that lives among non-Jews," Fischer continued, "is a problem that must be seriously

[95] Ibid.

[96] Quoted in ibid., p. 172.

[97] For a biography: Hoßfeld, Jenaer Jahre; idem., Rassenkunde, pp. 524–526; Zimmermann, Berufung; Labisch/Tennstedt, Weg, vol. 2, pp. 420–423; Geisenhainer, "Rasse," p. 476; Klee, Personenlexikon, pp. 208 f.

addressed by anthropology, free of any tendential attitudes."[98] Fischer's comments regarding "mixed marriage" between Jews and non-Jews were very cautious.[99] As far as its basic idea is concerned, Fischer's argumentation already points to the paradigm of phenogenetics formulated in the late 1930s.[100] At this juncture the reference to the power of the environment was more likely motivated by strategy, as was the entire lecture: as would become apparent in the following years, Fischer was by no means as critical of Günther's "race theory," and by no means so open to Boas and the Boasians; in the Jewish question he did not take as unbiased a stance as he tried to make the senators of the KWG believe.[101]

In the passages on "pure hereditary theory," too, Fischer accommodated his audience by emphasizing strongly the potential practical utility of genetic science. The research objective is to study "lines of heredity" in which "the disposition to cretinism, to criminality, to idiocy, to constitutional anomalies like diabetes or lack of resistance to tuberculous infection," but also "special talents"[102] are contained. In order to make the findings obtained through this study useful in practice, anthropology should expand to become "social anthropology."[103] The concept of "race hygiene" is easily misunderstood, for it does not concern "race breeding," but rather the "genetic force" of the "entire nation;" the concept of "eugenics" is thus preferable.[104] In this speech Fischer tacitly placed himself in the "Berlin line" of the race hygiene movement. The major eugenic problems – land flight, urbanization, decline in the birth rate in the higher social classes – were located "on the boundary between sociology and national economics, anthropology and eugenics" and could "only be resolved through work from both sides."[105] Fischer's presentation closed with an emphatic appeal to the assembled elite from state, industry, and society, to conceive of biopolicy as the mandate and task of science:

> So often we claim of ourselves: gradually we have made ourselves the master of nature. With a great deal of technology, today we control infinitely much: for us, space [...] has become almost nothing, when we think of the most modern airplanes, wireless telegraphy and the like. We cope with a great number of illnesses with great competence. We have been able to eliminate for our European population epidemics that were the scourge of humanity for centuries, and, one may hope that for a number of other scourges still plaguing us, means of containment – the optimist would say: complete extermination is possible.

[98] Fischer, Aufgaben, pp. 749 f. The lecture is based for the most part on a memorandum Fischer had written for the administrative committee of the KWG in May 1926. Lösch, Rasse, pp. 171 f., 174 f.

[99] Fischer, Aufgaben, p. 754: "[...] I hold a moderate cross of two suitable races to be more advantageous than race purity [...]." Here Fischer refers to the heterosis theory he developed during his research on the bastards of Rehoboth. Cf. Chapter 3.

[100] Cf. Chapter 4.

[101] Cf. Chapter 2.

[102] Fischer, Aufgaben, p. 751.

[103] Ibid., p. 752. It would be even more fitting, Fischer continued, to speak of a "political anthropology" in the sense in which Aristotle conceived of political – man as *zoon politicon*.

[104] Ibid., p. 754.

[105] Ibid., p. 753.

But what we have *not yet* even *begun* to master and comprehend are these biological areas, is everything which damaged our culture biologically. Working on the preservation of lines of heredity, studying them and influencing them favorably, keeping them free or liberating them from the damages of our cultural measures has not yet begun at all! That is the actual and final task inherent in all of this research, and this task is of indispensible importance [...] for the salvation of our entire nation, it cannot be delayed, it requires all of our care and power."[106]

1.3.3 The Decision to Found the Institute

When Fischer held his presentation before the senators, the decisive session of the KWG Senate had already taken place. In it, Harnack emphatically supported the founding of the institute, listing as his reasons:

> After Sweden, the United States, France and England went ahead before us, it turns out that in Germany, too, it is imperative to create a scientific center for anthropology, human genetics and eugenics, all the more because insufficient and amateurish attempts in these areas must be countered.[107]

In this passage the unusual naming of the institute to be founded catches the eye. The discussion in the early planning phases always concerned an institute "for anthropology, human *genetics* and eugenics"– "human *heredity*" was a new term created by Fischer.[108] What is striking is that the concept of *race* did not appear in the name of the institute. No doubt it was avoided consciously out of consideration for the political constellation.[109] The triad "anthropology, human genetics and eugenics" was supposed to express that the new institute, according to Fischer's idea, was to be based on a conception that envisioned anthropology opening up in two directions: On the one hand, anthropology – as anthropo*biology* – was to meld with human genetics; on the other – as *social* anthropology – with eugenics. What was intended, therefore, was an anthropology based on theoretical research in genetics, directed toward eugenic application.

Harnack's final half-sentence was a clear allusion to Günther, whose "race theory of the German Volk" had set off an avalanche of pseudoscientific literature about the Nordic people. Apart from that, "the core of the arguments was restricted to the theme that Germany may not fall behind the developments abroad."[110] In fact, Fischer did have a model in mind: the Swedish State Institute for Race Theory in

[106] Ibid., pp. 754 f. (emphases in the original).

[107] Excerpt from the Senate protocol of 6/19/1926, MPG Archive, Dept. I, Rep. 1 A, No. 2411, pp. 5a–5b v, quote: p. 5a.

[108] Cf. Lösch, Rasse, p. 189.

[109] Looking back, Fischer stated in 1941: "The words race and race hygiene could not be used vis-á-vis the Socialist-Centrum dominated government." Anlage zur Niederschrift über die Sitzung des Kuratoriums des KWI-A am 9/1/1941, MPG Archive, Dept. I, Rep. 1 A, No. 2400, p. 188.

[110] Lösch, Rasse, p. 172.

Uppsala founded by Hermann Lundborg (1868–1943) in 1921.[111] Fischer had visited this institute on a trip to Sweden in 1924 and was very impressed. The international competition argument appeared regularly in correspondence with Reich authorities; it appears to have developed great appeal.

Harnack was already committed to his choice for the position of director: Fischer, he let the senators know, "has declared himself willing to take over direction." This indication of previously concluded pre-negotiations precluded any personnel discussion from the outset. The only location that came in question for the new institute was Berlin-Dahlem, for which three arguments were advanced: First, a connection was to be forged with the KWI for Biology and Baur's Institute for Genetics at the Agricultural College; second, savings were expected to be achieved through the close collaboration with the Reich Statistics Office; third, there was no alternative to Berlin as a location "because of the ideal and material support expected from the Prussian provincial administrations."[112]

The investment costs were estimated by Harnack and Glum at around 600,000 RM. The Reich was supposed to bear 300,000 RM of this; Prussia would supply 100,000 RM and provide the building site as well as an anthropology professorship at the University of Berlin for Fischer. The remaining financing of 200,000 RM was "to be expected [...] from the private sector" – more on this presently. In accordance with Harnack's proposal, the operating costs of the new institute, estimated at 80,000–100,000 RM, were to be covered by the contributions from the Prussian provinces and the city of Berlin to the KWG. Finally, Harnack exerted time pressure on all involved by indicating how important it would be for the institute's opening to take place during the 5th International Congress for Genetic Science, which was to meet in Berlin in September 1927. Once the representative of the Prussian Ministry for Science, Art, and Education had signalized the support of his ministry – which basically meant that Fischer was to receive Luschan's chair – the senators gave Harnack's plan their seal of approval. Just one reservation was put forward, by the representative of the business office of the unified Prussian provinces: The new institute must not be permitted "to harm [...] the interests of the German Research Institute."[113] The background of this reservation was the

---

[111] Broberg/Tydén, Eugenics in Sweden, pp. 88–91; Weindling, International Eugenics, p. 185 (table 5), pp. 192–194. In Switzerland the *Julius-Klaus-Stiftung für Vererbungsforschung, Sozialanthropologie und Rassenhygiene* (Julius Klaus Foundation for Genetic Research, Social Anthropology and Race Hygiene) was founded by the anthropologist *Otto Schlaginhaufen* in 1922. Keller, Schädelvermesser, p. 108; Schweizer, Eugenik, pp. 124f.

[112] Temporarily the Government of Baden expressed interest in bringing the institute to Freiburg, but this remained a short intermezzo, as Baden was not able to raise the necessary financing. Harnack used the hesitant proposal from Baden to convince a number of representatives of the Prussian Parliament of the necessity of generous subsidies for the planned institute. In fact, the KWG had committed itself to the location Berlin from an early date. Cf. Lösch, Rasse, p. 176; Weindling, Health, p. 435.

[113] Excerpt from the Senate protocol of 6/19/1926, MPG Archive, Dept. I, Rep. 1 A, No. 2411, pp. 5a–5b v, quotes: pp. 5a v, 5b.

circumstance that a significant portion of the membership dues of the Prussian provinces and the city of Berlin had previously been allotted to the DFA and the provinces were hardly disposed to approve massive financial support for the new institute in Berlin if it meant the research already sponsored in Munich would suffer. Through the question of financing, the rivalry between Berlin and Munich flowed into the Senate session as well. However, the misgivings of the Prussian provinces were quickly allayed. After Glum had completed negotiations in Munich and made clear that the DFA, after the completion of its new building financed by the Rockefeller Foundation, would require only 40,000 RM from the membership dues of the Prussian provinces and would raise no further demands from the provinces, it was quickly agreed that the future KWI-A should receive 60,000 RM toward its operating costs from the Prussian contributions.[114]

1.3.4 The Role of Hermann Muckermann

What is interesting about Harnack's cost calculations is the incidental remark that the financing gap in the investment costs was expected to be covered by contributions from the private sector. In fact the 200,000 RM were supposed to be raised by Hermann Muckermann (1877–1962).[115] Muckermann entered the stage as a "white raven."[116] In 1896 he joined Societas Jesu. He received his first philosophical and theological training at Ignatius College in Valkenburg (Netherlands). At the Jesuit College of the Sacred Heart in Wisconsin (USA) he graduated with a degree in philosophy, in which he was awarded a Ph.D. in 1902. He remained at the Jesuit University until 1906, teaching mathematics and natural sciences. In 1906 he returned to Valkenburg and continued his study of theology until 1910. In 1909 he was ordained as a priest. By this time his natural science interests came ever more clearly to the fore. From 1910 until 1913 he studied zoology in Leuven and Tronchiennes, from 1913 to 1925 he worked as a lecturer for biology in various colleges of the order. From 1913 until 1916 he was also the editor of *Stimmen aus Maria Laach* ("Voices from Maria Laach"), which he renamed to *Stimmen der Zeit* ("Voices of the Age") in 1915. From 1916 onward, Muckermann concerned himself exclusively with eugenics issues, which he had come in contact with through Alfred Ploetz, and became the most important representative of its "Catholic wing."

[114] At this time the Prussian provinces paid 100,000 RM and the city of Berlin 20,000 RM to the KWG in annual membership fees. Thus in the future the KWI-A was to receive 60,000, the DFA 40,000 and the KWI for Foreign Law Applying to Public Bodies and International Law (*KWI für ausländisches öffentliches Recht und Völkerrecht*) 20,000 RM. Glum to ret. Ministerial Councilor Dr. Moll, 14/7/1926, ibid., pp. 8–10; notice by Glum of 10/12/1926, ibid., p. 55.

[115] On Muckermann's biography: Grosch-Obenauer, Hermann Muckermann, pp. 1–22; Labisch/Tennstedt, Weg, vol. 2, pp. 458–460; Richter, Katholizismus, pp. 52f.; Klee, Personenlexikon, pp. 417f.

[116] Lösch, Rasse, p. 178.

From 1921 until 1933 he was the editor of the journal *Das kommende Geschlecht* ("The Coming Race"). In the Weimar Republic, Muckermann published and lectured on a massive scale, holding over 100 lectures annually, in Germany and abroad. In the 1920s, the elegant, worldly and eloquent Muckermann, who did not in the least correspond with the image of a Jesuit priest, advanced to become a kind of media star, who contributed like no other to the propagation of the eugenic idea, not only in Catholic circles, but also in the general public. Muckermann also maintained excellent contacts to the Prussian Ministry for the Welfare of the People, which had been expanded to an even stronger bastion of the Center under the direction of the Center politician and Christian unionist Heinrich Hirtsiefer (1876–1941), who took over from Stegerwald in 1921. Because Hirtsiefer, as his predecessor before him, was a convinced disciple of eugenics, the ministry continued to support the race hygiene movement as best it could. In 1926, when his move to the new institute was already imminent, Muckermann left the order of the Jesuits, but remained a Catholic priest.

Muckermann's role in the perimeter of the institute's founding is quite controversial. Paul Weindling emphasizes Muckermann's great importance, judging unequivocally that the KWI-A is "the brain-child of a Roman Catholic."[117] Niels C. Lösch contradicts this, by extolling – justifiably – the importance of Baur's initiative. The fact that Muckermann was incorporated into the founding of the institute, Lösch argues, based on Fischer's memoirs, can be traced back to the massive pressure of the Papal House Prelate and church historian Georg Schreiber (1882–1963), who – as a parliamentary representative of the Catholic Center – was a member of the Reichstag budget commission from 1920, where he served as the correspondent for the budget of the Reich Ministry of the Interior, making him "the authoritative and power-conscious consultant for the KWG and the needy community [of German science]" – "nothing happened without his approval."[118] According to Lösch, in the decisive negotiations of the budget commission in March 1927, Schreiber, in an attempt to infiltrate a confidant of the Catholic church and the Center[119] into the new institute so active in such a politically delicate field of research, drew up a tandem measure which could certainly be interpreted as a mild form of blackmail. Namely, he linked his approval for the Reich subsidy for the KWI-A with the "urgent wish"[120] that Muckermann be entrusted with the direction of the department for eugenics. Harnack, for his part, vigorously recommended to Fischer that this wish be granted. Muckermann's appointment was hard for Fischer to stomach, as the evangelist of eugenics was regarded "by experts to be a dilettante or at most a

[117] Weindling, Weimar Eugenics, p. 310; idem., Health, p. 433.

[118] Lösch, Rasse, p. 177. Kröner, Von der Rassenhygiene zur Humangenetik, p. 16, takes an intermediate position, characterizing Baur and Muckermann as the "founding fathers" of the institute.

[119] Muckermann may not have been a member of the Center party, but was very closely associated with it. Richter, Katholizismus, p. 296.

[120] Quoted in Lösch, Rasse, p. 179.

scholar of popular science, who had certainly done good through his illuminating little pamphlets about marriage and family, but could not be recognized as a scientific member of the KWG"[121] – and in fact, Muckermann's later appointment was not as a scientific member. One must also keep in mind that his installation as departmental head had an inestimable advantage – besides his great popularity, which certainly would benefit the external image of the institute: it immediately dispersed any potential misgivings the Center and Social Democrats had with regard to the institute's political neutrality. Further, Muckermann offered to raise the missing 200,000 RM through a donation drive in the course of his lecture tours through the German Reich – in a sense, Muckermann bought his way into the institute.

While this reconstruction of events is certainly quite plausible, Lösch's attempt to play down Muckermann's importance in the founding of the institute places the horse-trading between Schreiber, Muckermann, Harnack, and Fischer at too late a point in time. It seems highly improbable that they did not reach an agreement until the negotiations of the Reichstag budget committee were already underway in March 1927. It is much more likely that this had already occurred during the period between the session of the Administrative Council in May and the Senate session in June 1926. Harnack's incidental remark at the Senate meeting that the financing gap would be closed by the private sector was not the only indication of this. More relevant, the first donations collected by Muckermann were deposited in the KWG account as early as July 1926.[122] Muckermann had joined the German Society for Race Hygiene in 1922. There is no reason not to assume that, as a member of this quite manageable circle, he was constantly up to date on the progress of the plans to found a research institute for anthropology, human genetics and eugenics – be it through Otto Krohne, who was promoted to director of the Department for People's Health in the Prussian Ministry for National Welfare, or through Arthur Ostermann, who had taken over the group responsible for population policy and race hygiene from Krohne. As both were also leading representatives of the eugenics-race hygiene movement – Krohne as chairman of the Society for Race Hygiene, Ostermann as acting chairman of the German Association for Volkish Improvement and Genetics, all of the threads came together in the hands of these two ministry officials. They were closely associated with Muckermann, the central figure of Catholic eugenics. What could be more natural than establishing contact to the most successful propagandist of the eugenic idea – at the very latest, during the decisive phase when a financing plan was being assembled? Which side took the initiative in this contact is impossible to say. In his memoirs, Glum recalls that Muckermann's offer of financing was relayed by Baur.[123] The zoologist and geneticist Hans Nachtsheim (1890–1979), at that time a member of Baur's Institute for

[121] Glum, Wissenschaft, p. 372.

[122] The assertion in Weindling, Health, p. 434, that the first donations were deposited in April 1926, is incorrect. Cf. Liste der auf dem Konto KWI-A eingegangenen Beträge, MPG Archive, Dept. I, Rep. 1 A, No. 2411, pp. 4–4 v.

[123] Glum, Wissenschaft, p. 371.

Genetic Research at the Berlin Agricultural College, testified after World War II that Muckermann approached Baur with the idea of a national eugenics institute in 1925[124] – this could certainly be true, yet Bauer had been pursuing this plan since 1922. What can be established is that Muckermann very probably played an important role in the planning of the institute founding from around mid-1926 – although primarily with regard to the political realization of the institute plans and in securing the financing. The conception of the institute's work, however, clearly bears the signature of Eugen Fischer.

1.3.5 Financing Problems

After the senate session of June 19 the task was to resolve two complexes of problems inextricably linked together: First the public subsidies to the investment costs foreseen in the financing plan had to be adopted; second, sufficient political support was needed to get these subsidies through the Reichstag. First of all this required a declaration of support from the governments of the Reich and of Prussia. The Reich Ministry of the Interior demanded from the Prussian Ministry for Science, Art and Education a binding promise that the Prussian government would fulfill the obligations it had agreed to take on: payment of a subsidy of 100,000 RM, provision of the construction land and a chair for Fischer.[125] Only after fulfillment of this prerequisite was the Reich Ministry of the Interior willing to enter into negotiations with the Reich Finance Ministry about the Reich subsidy of 300,000 RM. Once the two governments had reached an agreement, the compromise had to be approved by the budget committees of the Reich and of Prussia.

Political resistance was expected there. Harnack endeavored to disperse misgivings before the vote by sending a memorandum to interested groups, especially to representatives of the Reichstag and the Prussian parliament. The KWG, its president assured them, assumes "full responsibility" to ensure that the new institute works "on a purely scientific foundation." He continued, "The institute has as little to do with the race conflicts of the present as with the cleavages of political and confessional nature." The German family stands at the center of its work, "which currently exhibits such great degeneration [...] that the future of the entire nation appears endangered." The task at hand was nothing less than "the containment of the tremendous, growing need for social welfare in the German nation and the preservation and procreation of healthy bearers of the German future."[126] As the discussions in the Reichstag Budget Commission showed, Harnack had hit upon

[124] Nachtsheim, Notwendigkeit, p. 323.

[125] In fact, the Prussian Ministry for Science, Art and Education succeeded in appointing Fischer despite attempts by the faculty in Berlin to delay and prevent his professorship. Cf. Lösch, Rasse, pp. 184–186.

[126] Quoted according to ibid., p. 175.

the right tone. The founding of the institute was supported not only by the Center – that was no surprise after Schreiber's intervention – but also by the SPD. Its representative Julius Moses (1868–1942)[127] did express the concern of the working classes about increasing nationalism in science, especially in the areas in which the new institute was to work, but nevertheless welcomed the establishment of the institute and approved the subsidy, explaining that he had become convinced that the new KWI would perform research "from a strictly scientific foundation, regardless of the nationalist or nonnationalist, anationalist or international results of its research."[128]

With this the public subsidies were secured for the moment. Meanwhile, the soliciting of private donations was creating problems, as contributions were coming in much more sluggishly than Muckermann had expected due to the gradually deteriorating economic situation. By the end of 1926 he had taken in much less than what he promised – just under 82,000 RM.[129] Because construction was already in full swing by this time, the KWG had liquidity problems, for many of the funds approved by the Reich and Prussia were not due until the fiscal year 1927/28. In February 1927 – the total donations solicited had since risen to just over 92,000 RM – Glum thus sent Muckermann a first reminder about the donations missing contributions he'd promised.[130] The financing gap weighed all the heavier as it became apparent around the middle of the year that the costs of construction would be around 200,000 RM higher than planned. The situation was aggravated by the fact that the Prussian government refused to pay out the approved subsidy of 100,000 RM unless the Reich valued the construction land at 200,000 RM – and in fact, the subsidy from the Prussian state was not paid out until the institute was completed. In this situation the General Administration had no choice but to apply to the Reich Ministry of the Interior for an extraordinary grant of 200,000 RM in August 1927, which the ministry approved without delay in order to avoid a scandal – the opening of the institute at the International Congress for Genetics was about to take place. For its part, of course, the KWG committed itself to procuring the still missing total of around 93,000 RM in donations.[131] At this stage the General Administration vigorously demanded the remaining amount from Muckermann,[132]

---

[127] For a biography: Nadav, Julius Moses; idem., Julius Moses und Alfred Grotjahn.

[128] Debates of the German Reichstag, Reich Budget Committee, IIIrd Term, 229the session of 4/3/1927, p. 2.

[129] Cf. Liste der eingegangenen Beträge, MPG Archive, Dept. I, Rep. 1 A, No. 2411, pp. 4–4 v. The sum Lösch gives in Rasse, p. 180, of 21,468 RM deposited by December 1926, is based on a note by Franz Arndt, the office director of the KWG General Administration, on 12/2/1926, which, however only lists the amounts deposited in November 1926. Ibid., p. 54.

[130] Glum to Muckermann, 2/21/1927, ibid., p. 72. In his response, Muckermann bemoans broken promises and refers hopefully to an imminent lecture tour to the Rhine-Westphalian industrial region. Muckermann to Glum, 2/26/1926, ibid., pp. 73–73 v.

[131] Glum to Reich Ministry of the Interior, 9/17/1927, ibid., pp. 89–89 v.

[132] Cf., e.g. Morsbach to Muckermann, 11/26/1927, MPG Archive, Dept. I, Rep. 1 A, No. 2412, pp. 97–97 v.

who initially responded evasively,[133] but then took the standpoint that his promise of raising 200,000 RM for the construction of the institute must be regarded as fulfilled through the approval of the extraordinary aid from the Reich: Reich Finance Minister Heinrich Köhler (1878–1949) – at that time still president of the state of Baden – had promised him financing, but had been unable to keep his promise at the time. Therefore the approval of the additional Reich subsidy was a success in which "my campaign was not without decisive influence."[134] The KWG did not go into this argumentation, however, and insisted, without "exerting pressure," as Adolf Morsbach, Glum's representative emphasized, that Muckermann fulfill his obligations. Hereby Morsbach expressly acknowledged "that circumstances external to your person prevented your adhering to your promise, that this is rather the fault of the general situation."[135] In April 1928, a compromise was proposed by the General Administration. The complete construction costs had been concluded with a deficit of around 35,500 RM. Muckermann, who by this time had solicited over 107,000 RM in donations, declared himself willing to cover this deficit; in return, the KWG waived any further demands.[136] However, the drastically deteriorating economic situation thwarted all of Muckermann's endeavors to solicit additional donations.[137] Ultimately he was compelled to pay the remaining amount from his personal assets, by bequeathing to the KWG a small villa he owned in Berlin-Schlachtensee in January 1930, in the form of a "Hermann Muckermann Endowment for Eugenics" effective upon his death, under the condition that the property or the proceeds from its sale be used for the purposes of eugenics research. In the eyes of the KWG, this endowment settled Muckermann's debt. "With this Muckermann himself, if not entirely voluntarily, had made the largest single donation for the institute from the private sector."[138]

Looking at the list of contributors Muckermann mobilized,[139] clear geographic emphases can be recognized in Upper and Lower Silesia as well as in Westphalia and the Rhine province. The generous donations from Silesia were traced back to the intercession of the priest and Papal House Prelate Carl Ulitzka (1873–1953), the

---

[133] Muckermann to Morsbach, 12/5/1927, ibid., pp. 98–98 v.

[134] Muckermann to Morsbach, 2/9/1928, ibid., pp. 106–106 v, quote: p. 106.

[135] Morsbach to Muckermann, 23/2/1928, ibid., pp. 113 f., quotes: pp. 113 v, 114.

[136] Glum to Muckermann, 4/2/1928, ibid., p. 117; Muckermann to Glum, 4/12/1928, ibid., pp. 118–118 v.

[137] In June 1928 Muckermann once again paid the net proceeds from a lecture tour to Freiburg to the KWG, a total of 1,500 RM. One month later, the district of Cosel, which had contributed regular grants for the project since 1926, discontinued its payments. With this the flow of funds was definitively exhausted. Liste der auf dem Konto des KWI-A eingegangenen Beträge, MPG Archive, Dept. I, Rep. 1 A, No. 2411, pp. 4–4 v.

[138] Lösch, Rasse, p. 181.

[139] Liste der auf dem Konto des KWI-A eingegangenen Beträge, MPG Archive, Dept. I, Rep. 1 A, No. 2411, pp. 4–4 v, as well as numerous individual receipts in the same file. Compliations of the donated funds – although incomplete and imprecise – are presented in Weindling, Health, p. 435 (table7), and Lösch, Rasse, p. 183 (Table 1).

"uncrowned king of Upper Silesia" who sat in the Reichstag for the Center Party.[140] In the Rhine-Westphalian industrial region, Muckermann himself had passed the hat during a several week-long lecture tour in 1926. The predominant share of the donations came from public funds. The provinces of Upper and Lower Silesia, Rhineland and Westphalia donated a total of 42,500 RM, cities and districts were represented with 26,225 RM,[141] state insurance institutions contributed 15,868 RM.[142] A sum of 14,300 RM came from private industry.[143] Private donors, finally, brought in 8,600 RM.[144] Even if one takes the difficult economic circumstances into account, it becomes apparent that above all the provincial, district and city administrations, in their function as state and district welfare associations, were interested in eugenics in view of the constantly increasing burden of public welfare. "This fact underlined the importance of the institute for political consulting, above all in the area of a eugenic social policy, which promised to minimize costs in the area of social policy."[145]

1.3.6 The Inauguration of the Institute

Despite all of the financial turbulence, the institute was able to be inaugurated as planned in September 1927, in time for the 5th International Congress for Genetics led by Erwin Baur, the first scientific congress of any kind to be held in Germany since the war. The institute, which had been erected in record time – barely 15 months had passed between the Senate resolution and its opening, and construction

[140] Muckermann to Landrat Neustadt/Oberschlesien, 8/8/1926, MPG Archive, Dept. I, Rep. 1 A, No. 2411, p. 23. On Ulitzka's biography: Richter, Katholizismus, p. 152 (note 78).

[141] The most important donors here were the districts of Beuthen (5,300 RM), Oppeln (4,000 RM), Cosel (2,400 RM) and Groß-Strehlitz (2,000 RM), and the cities of Essen (5,000 RM), Ratibor (4,000 RM) and Duisburg (2,500 RM).

[142] Generally with minor sums, the state insurance institutions of Baden, Berlin (main office), Brandenburg-Berlin, Brunswick, Grenzmark Posen-West Prussia, Hanover, the Hanseatic Cities, Hessia-Darmstadt, Hessia-Nassau, Mecklenburg-Schwerin, Lower Bavaria, Oldenburg, East Prussia, Palatinate, Rhine Province, Saxony, Schleswig-Holstein, Thuringia, Lower Franconia, Westphalia, Württemberg.

[143] Gräflich-Schaffgottsche Werke, Gleiwitz (5,000 RM); Oberschlesische Holz-Industrie AG, Beuthen (3,000 RM); Henckel von Donnersmarck-Beuthen Estates Limited, Carlshof-Tornowitz (3,000 RM); Schlesische Bergwerks- und Hütten AG, Beuthen (1,000 RM); Oberschlesische Portland-Zementfabrik (1,000 RM); Gräflich von Gottsteinsche Güterdirektion, Gleiwitz (1,000 RM); Preußische Bergwerks- und Hütten AG (300 RM).

[144] Noteworthy amounts in this area came from the Ruhr industrial magnates Julius and Hans Thyssen (2,500 RM each), the manufacturer and Center politician Rudolf ten Hompel (1878–1948) of Münster (1,500 RM) and the merchant Theodor Althoff (1858–1931), also of Münster (1,000 RM). Ten Hompel and Althoff were members of the KWG. Cf. Bergemann, Mitgliederverzeichnis, vol. 1, p. 22, 116.

[145] Kröner, Von der Rassenhygiene zur Humangenetik, p. 17.

concluded within 11 months – was a prestige object of international importance for the KWG, for German science, for the Prussian state and for the entire Weimar Republic. Harnack held the inaugural address before an assembled audience of over 900 congress participants, representatives of Berlin's institutes of higher education, and of the Prussian government. He deplored the easily misunderstood use of the term "race theory" in "everyday politics" and emphasized that "true race theory" will "bring the groups of a nation closer to each other, not divide them into enemy camps." Of course race theory "also must teach how our species [can] reach a higher level of biological development."[146] Then Harnack ceremoniously handed Fischer the keys to the new institute building at Ihnestrasse 22/24. Presumably with a special nod to the international community of geneticists, who were skeptical toward race anthropology and at this time also starting to distance themselves critically from eugenics as well,[147] Fischer declared that the new institute will "have to research in the most general sense all problems that can be summarized with the problematic term of 'race theory' at this time, [...] of course, on a purely natural scientific foundation and free of any other kinds of reasoning." It was to be "a purely theoretical institute for the study of the nature of man."[148] The *Deutsche Allgemeine Zeitung* cited Fischer:

> Today it is no longer possible for the scientist to completely avoid the questions so notorious in politics, of the meaning of race for nation and humanity. But he must study them objectively, not attack them as in an agitory fashion. In reality we are still infinitely far from any certain findings on these issues. However, it is certainly justfied to ask what race and belonging to a race means or does not mean for nations and for groups of cultures. It is imperative that these questions be illuminated scientifically by anthropologists. The point is not to create race prejudice, but race knowledge.[149]

In the name of the Prussian government, the Culture Minister Carl Heinrich Becker (1876–1933) and Ministry Director Otto Krohne conveyed salutory addresses; congratulations from the congress participants were conveyed by the American eugenist Charles Davenport, director of the Eugenics Record Office in Cold Spring Harbor, Long Island and president of the International Federation of Eugenic Organizations, and the Swiss anthropologist Otto Schlaginhaufen (1879–1973), director of the Julius Klaus Foundation for Genetic Research, Social Anthropology and Race Hygiene.[150]

Under critical scrutiny from abroad, the *Berliner Tageblatt* commented on Fischer's inaugural address with a skeptical undertone that the new institute will "have no easy job." For "race research" had become "race fantasy;" it could no

[146] Kaiser-Wilhelm-Institut für Anthropologie. Der Einweihungsakt, in: Deutsche Allgemeine Zeitung, 9/16/1927. Copy in: MPG Archive, Dept. IX, Rep. 2, KWI-A.

[147] Kühl, Internationale, pp. 103 f.

[148] Von wissenschaftlichen Instituten, p. 315. Cf. also Fischer, KWI-A. For more on this: Weingart/Kroll/Bayertz, Rasse, p. 244; Lösch, Rasse, pp. 186 f.

[149] Kaiser-Wilhelm-Institut für Anthropologie. Der Einweihungsakt, in: *Deutsche Allgemeine Zeitung*, 9/16/1927. Copy in: MPG Archive, Dept. IX, Rep. 2, KWI-A.

[150] Jahresbericht 1927/28, 4/21/1928, MPG Archive, Dept. I, Rep. 3, No. 4.

longer be regarded as "an unconditional science." The article continued, "Germany is, to date at least under the tacit toleration by the experts called upon to defend it, flooded by a tendential race literature from certain Volkish positions. Now they are trying in vain to banish the spirits they summoned." The point here is "to wipe the slate clean."This, in turn, provoked a vicious commentary by the *Deutschland-Tageblatt*, which remarked derisively that Fischer was being "quoted as a pro-Jewish helper by the Jewish Mosse paper."[151] In this short exchange of blows it already became apparent that Fischer, even if he had seriously wanted to, would have had a hard time sticking to the concept of a "purely theoretical institute" in the heated atmosphere of the late Weimar Republic.

[151] Ein neues Rassenforschungsinstitut. Einweihungsfeier in Dahlem, in: *Berliner Tageblatt*, 9/16/1927; Die "richtige" Rassenlehre, in: *Deutschland-Tageblatt*, 1/17/1927. Copies in the MPG Archive, Dept. IX, Rep. 2, KWI-A.

Chapter 2
"The Human of the Future Under the Scrutiny of Research": The Kaiser Wilhelm Institute for Anthropology, Human Heredity and Eugenics in the Weimar Republic, 1927–1933

The development of the Kaiser Wilhelm Institute for Anthropology, Human Genetics, and Eugenics in the years from 1927 to 1933 was linked closely with the glory and misery of the Weimar welfare state. While the founding of the institute took place in the last, narrow window of relative stabilization, in the period when the expansion of the social state reached its tentative zenith and end point with the introduction of unemployment insurance, the phase during which the institute was actually set up was characterized by the existential crises of the social state in the context of the economic depression in the late Weimar Republic. This was indeed fertile soil for the institute, whose *raison d'être* from the outset was, after all, to raise the social state to a new level on the basis of a comprehensive, scientifically guided biopolicy, and this task was all the more urgent in the period of crisis. Social, political, and economic events shifted the institute in Dahlem to the focal point of the biopolicy debates. At the close of the Weimar Republic it had become firmly established at the interface between science, the public, and politics.

2.1 Internal Structures

2.1.1 Building and Equipment

On October 1, 1927, the Kaiser Wilhelm Institute for Anthropology, Human Genetics, and Eugenics started its work in the institute building designed by the Munich architect Carl Sattler (1877–1966) at Ihnestrasse 22/24 in Berlin Dahlem. The centrally located main entrance was crowned by the bronze head of Pallas Athena, the goddess of the arts and sciences, which had served as the emblem of the KWG since 1911. The east wing of the ground floor housed the director's offices: a study and a reception room, a laboratory with microscope, microtome, and anthropological instruments; further, a measurement and examination room with two changing rooms, an instrument cabinet and desk for anthropometric examinations, and studies of the human constitution. To these were connected a laboratory for the technical assistant, a fully equipped photographic studio with an electrically powered turnplate for taking anthropological photographs, and a dark room. For photographs using natural light there was a terrace located on the garden side, partially covered with glass and screened off by curtains, where aquariums and terrariums were to be erected in the future for experiments on fish, amphibians, and reptiles. Finally, the east wing of the ground floor also contained two offices for assistants, equipped as private laboratories. The west wing was dominated by the large lecture hall, which seated an audience of up to 50.

The first floor housed the departments for human genetics and for eugenics. The directors of these two departments had one office and one laboratory each. This floor also had two offices for assistants, a large common laboratory with workplaces for the scientific staff and doctoral students, a library and reading room, an archive, a small operating room with anteroom, and even a modest casino for the assistants.

The attic story served as a collection room: here is where the skull collection of the University of Berlin was to be housed. The cellar story contained "rooms for a collection of moist materials (in part stoneware troughs, for larger items like monkey corpses, embryos, etc., in part, enamel vessels for brains, skin, small embryos, organs, etc.)." These facilities were supplemented by a plastering room, a room for chemicals, a workshop with carpenter's bench and lathe, a laundry and ironing room, a bath for assistants and a shower room, as well as quarters for the caretaker. Finally, the yard contained a small stall with exercise space for monkeys, dogs, rabbits, and chickens; facilities to prepare and store feed; a dog bath; and an additional small operating room for inoculations and injections.[1] At the other end of the yard the KWG had erected a representative villa as a residence for the director of the institute.[2]

Despite its increasingly constricted financial latitude, the first years of the institute were marked by the construction and expansion of its material infrastructure. Because it was not possible to finance all of the equipment and furnishings from the construction budget, some items had to be acquired using funds from the operating budget. Among these items were furniture and apparatus, but above all books and journals for the institute library. In April 1929, Fischer complained to the General Administration:

> Not even today do I [...] have all of those new acquisitions which we could no longer afford when the institute was opened. To name a few examples, I refer to the fact that I took over the animal stall without any cages or similar equipment installed, the collection room with not quite a third of the necessary cabinets, and the library empty, the laboratories without any chemicals. The Emergency Association held out the prospect of 40,000 RM for the library, but I have received fewer than a dozen books and not a penny in cash.[3]

---

[1] Fischer, Kaiser-Wilhelm-Institut für Anthropologie, quote: p. 150.

[2] Both the institute building and the director's villa remained intact. Today they house parts of the Otto Suhr Institute of the Free University of Berlin and a department of the university administration.

[3] Fischer to KWG, 15/4/1929, MPG Archive, Dept. I, Rep. 1 A, No. 2405, pp. 22–22 v, quote: p. 22 v. Cf. Aufstellung über Rechnungen, die bis zum 31/3/1928 vom Jahresetat 192728 nicht bezahlt werden konnten, ibid., p. 4. The only apparatus listed here are an icebox and a dioptograph (719.60 RM). For the anthropological collection 1,753 RM were spent, for books, 2,441.75 RM. The remainder went for utilities like light, water, gas, coke, telephone, and drainage.

In June 1929, Fischer again spoke of the "altogether deplorable condition" of the library. "More than half of the books listed," according to Fischer, are private property, but "despite this the most necessary items of a reference library" are lacking.[4] In spite of the difficult financial situation, Fischer continued to invest in the library. At the end of the 1930/31 fiscal year it was valued at 29,200 RM, increasing in value through new acquisitions worth a bit over 6,000 RM in the 1931/32 fiscal year and 5,000 RM in 1932/33 – allowing for depreciation – to 36,500 RM.[5]

Setting up the skull collection proved to be another financial burden. Once the decision to found the institute had been made, the University of Berlin pressed Fischer to take over the skull collection begun by Rudolf Virchow (1821–1902) and expanded by Felix von Luschan, which had led a shadowy existence in the basement of the pathological department of the university up to that time, as a loan to his institute.[6] While Fischer was less than thrilled by this demand, having clearly stated his desire to distance himself from anthropometry, he did not dare to refuse the offer due to his already strained relationship with the department of philosophy, which was offended that the ministry had neglected to include it in its decision to appoint Fischer to the chair for anthropology. Moreover, if he refused to take over the collection, there was serious danger that the University of Berlin would lose it forever, as Emma von Luschan had announced that she would bequeath the skull collection of her late husband to the Anthropological Institute in Vienna under the direction of Otto Reche (1879–1966) if the university could not manage to establish the collection in an institute dedicated to anthropology. As an ordinary professor for anthropology, Fischer could not risk such a loss of prestige. Under pressure he thus submitted an application to cede the collection to the Prussian Ministry for Science, Art and Education in 1927. "Almost as a kind of service in return for accepting the collection,"[7] in November 1928, Fischer succeeded in bullying the ministry into creating a "fund to buy up skulls" with 2,000 RM allotted to completing the collection. The ministry provided additional funds for the acquisition of skulls and skeletons in the succeeding period as well.[8] Nevertheless, the collection remained a financial burden for the institute: The acquisition of furnishings for the

---

[4] Fischer to KWG, 14/6/1929, MPG Archive, Dept. I, Rep. 1 A, No. 2405, pp. 29–29 v, quote: p. 29 v.

[5] Jahresrechnungen des KWI-A für die Rechnungsjahre 1930, 1931, 1932, MPG Archive, Dept. I, Rep. 1 A, No. 2406, pp. 111–117, here: Vermögensübersichten zum 31/3/1931, 31/3/1932 und 31/3/1933, pp. 112, 114, 116.

[6] The collection also included objects that had been brought back from the colonial navy expeditions to Melanesia in the years 1907 and 1909. Fischer himself had brought back a collection of Hottentot skeletons from Southwest Africa. They were subjected to anthropometric examination by Rudolf Uhlebach in 1914. Cf. Fischer, Kaiser-Wilhelm-Institut für Anthropologie, p. 116 f.; Uhlebach, Messungen, p. 449; Bergmann/Czarnowski/Ehmann, Menschen, p. 129 f.

[7] Lösch, Rasse, p. 195.

[8] Cf., for instance, Preußischer Minister für Wissenschaft, Kunst und Volksbildung to Präsident der KWG, 30/3/1931, MPG Archive, Dept. I, Rep. 1 A, Nr. 2412, p 159 (authorization to allocate 1,054 RM to the KWI-A "for the cost of acquiring the Ochos jaw, two skeletons from Rio de Oro [Morocco] and a series of 17 related monkey skulls from Cameroon" for the skull collection).

collection room entailed additional costs, as did the transport of the purchased objects – and beyond this, cataloguing the collection tied up manpower.[9]

All in all it can be maintained that the construction and expansion of the institute proceeded rapidly until the fiscal year 1930/31, despite the budget cap in the face of the brewing global economic crisis. New workplaces were set up for a constantly growing number of scientific staff and doctoral students, apparatus and material inventories of the laboratories were completed, animal breeding grounds were set up, scientific collections expanded. At the same time an effective bureaucracy emerged to take care of the extensive correspondence in connection with research on genetic pathology and eugenics. The library continued to grow. All of this demonstrated the talent of Fischer, who was able to keep budget cuts to a minimum in tough negotiations with the General Administration, and who also succeeded in obtaining funds from other sources.

2.1.2 Finances

The institute's financial situation, as is clear from above, was strained from the very beginning. Because the funds provided for additional acquisitions in the first 5 months after the founding of the institute were not sufficient, such that Fischer defrayed these expenses from the operating budget, the accounts for the fiscal year 1927/28, which showed a balance of 106,500 RM,[10] closed with a deficit of around 7,650 RM. Fischer urgently requested that the General Administration make up for this deficit and further grant a larger amount for additional one-off acquisitions. "If I have to cover this deficit from the budget of the new business year," Fischer warned, "my operating funds would never ever get back in order; every quarter I would have to drag debts into the next quarter [...]."[11] In April 1929 Fischer sounded the alarm again. The General Administration had just ordered that 4,500 RM be cut from the total to be paid to the KWI-A in the first quarter of 1929. Fischer not only protested against this cut, he even demanded that the KWI cover his deficit from the 1928/29 business year, which amounted to 4,700 RM. Although his demands were not met, Fischer pushed on ahead with the expansion of the institute unswervingly, in defiance of all financial straits.

Over the course of time, however, it became increasingly difficult to defray the costs of books, furnishings, and apparatus from the funds budgeted for operating

[9] While the custodian of the skull collection, the secondary school teacher Hans Weinert (on leave from school service), received no salary from the KWG, but a scholarship from the university, in whose possession the skull collection remained, he paid his deputy out of his own pocket and was then reimbursed by the KWG. However, these funds were not part of the KWI-A's budget. Ibid., p. 196. Cf. Fischer to Generalverwaltung, 1/11/1927, MPG Archive, Dept. I, Rep. 3, No. 31.

[10] Niederschrift über die Sitzung des Kuratoriums des KWI-A am 6/6/1928, MPG Archive, Dept. I, Rep. 1 A, No. 2405, pp. 9–10.

[11] Fischer to KWG, 14/4/1928, ibid., pp. 2–2 v, quote: p. 2 v.

costs, because these increased as well as research work proceeded. This applied first of all to personnel costs, which inevitably rose in step with the rapid expansion of the institute staff. Regular personnel costs climbed to 55,460 RM in the 1930/31 fiscal year, dropping to 46,950 RM as a consequence of rigorous salary cuts in 1932/33. But material costs also took up an increasing share of the budget, although hefty cuts were made in the face of the global economic crisis. The regular material costs, too, peaked in the 1930/31 fiscal year at 30,400 RM, then falling to 22,250 RM in 1932/33 due to a drastic, linear reduction. Fischer fought vigorously for every individual expenditure in the budget negotiations. The increasing postage costs, for instance, he justified with the fact that "tens of thousands of survey forms sent out and returned after completion [...] constitute not office correspondence, but incoming scientific material." He defended the rising travel costs with the argument that the "entire work of the institute is, so to speak, organized around examining people who must be sought out."[12]

Because of the major cuts in the budget in the face of continuously high costs, until 1933 the KWI-A continued to carry the debts it had accumulated into the future.[13] What is more: only through a significant reduction in capital could the accounts be balanced.[14] For in contrast to the costs, which could be damped only with great effort, the KWG's contributions to the institute budget were slashed as a consequence of the global economic crisis: In the fiscal year 1929/30, the General Administration had transferred to the KWI-A a contribution of around 91,000 RM, along with 3,800 RM from the general fund.[15] In the 1930/31 fiscal year the contribution was still 90,600 RM, but not even 500 RM flowed from the general fund. In 1931/32 the contribution was then reduced to 78,700 RM; in 1932/33 to 75,500 RM.[16]

Of course, it must be taken into consideration that these were not the only funds available to the institute. The Emergency Association of German Science (*Notgemeinschaft der Deutschen Wissenschaft*) supported the large-scale project to take stock of the genetic biology of the German nation with 17,000 RM annually up to the 1929/30 fiscal year; the Reich Ministry of the Interior and the Prussian

[12] Fischer to KWG, 14/6/1929, ibid., pp. 29–29 v, quote: p. 29 v. Similar: Fischer to KWG, 25/6/1930, ibid., p. 48 f.

[13] Jahresrechnungen des KWI-A für die Rechnungsjahre 1930, 1931, 1932, MPG Archive, Dept. I, Rep. 1 A, No. 2406, pp. 111–117, here: Vermögensübersichten zum 31/3/1931, 31/3/1932 und 31/3/1933, p. 112, 114, 116. The fiscal year 1930/31 ended with debts of 3,950 RM; in 1931/32 they were just under 5,800 RM; finally, in 1932/33 around 7,100 RM.

[14] Ibid. In the fiscal year 1930/31 the reduction amounted approximately 12,000 RM; in 1931/32 it was 18,150 RM; finally, in 1932/33 30,700 RM. The balance in the fiscal year 1930/31 was around 115,150 RM; in 1931/32 only 98,600 RM; in 1932/33 it had increased back up to 107,300 RM.

[15] Anlage zur Jahresabschlußrechnung 1929/30, June 1930, MPG Archive, Dept. I, Rep. 1 A, No. 2405, p. 51.

[16] Jahresrechnungen des KWI-A für die Rechnungsjahre 1930, 1931, 1932, MPG Archive, Dept. I, Rep. 1 A, No. 2406, pp. 111–117, here: Einnahmen- und Ausgabenrechnungen, p. 113, 115, 117.

Welfare Ministry subsidized tuberculosis and eugenics research with 10,000 RM a year. In the final years of the Weimar Republic, the institute received grants to the material budget from the ministries of the interior of both the Reich and Prussia amounting to 7,800 RM for tuberculosis research, 12,000 RM for eugenics and twin research, and a further 20,000 RM or so from the Emergency Association of German Science. Moreover, in addition to its regular personnel, the institute employed two secretaries paid with funds from the Emergency Assocation, one statistician[17] and one temporary assistant funded by the Rockefeller Foundation, as well as a temporary assistant funded by the Prussian Welfare Ministry.[18]

Although important additional funds flowed to the institute in this manner, the financial situation became dramatic at times. In mid-May 1929, the institute's account had a balance of just 137 RM. The Darmstädter und Nationalbank refused to allow the institute's account to be overdrawn, commenting that "the situation at the Kaiser Wilhelm Society is 'precarious'," [19] and neither would it accept a personal guarantee from Fischer. Thus the payment of salaries and wages was at risk, and even the telephone was in danger of being cut off. Such liquidity crises appear to have been the exception, however.[20] Nevertheless the institute's financial situation, because Fischer expanded research activities without any consideration for the financial framework, was quite strained over the entire period from 1927 to 1933. Not until the fiscal years of 1931/32 and 1932/33 did the process of expansion reach its financial limits. This had consequences for research, not only in the sense that the research agenda had to be restricted toward the end of the Weimar Republic, but to the effect that the contexts transformed in which the research projects were embedded – more on this below.

[17] The person in question was Erna Weber (1897–1988). From 1931 to 1935 she attended to the statistical evaluation of Otmar von Verschuer's studies of twins in the Department for Human Genetics. For her biography, cf. Vogt, Wissenschaftlerinnen, pp. 157 f.

[18] Tätigkeitsbericht Fischers vom 29/7/1933, MPG Archive, Dept. I, Rep. 1 A, No. 2406, pp. 143–145, here: pp. 143–143 v. The Prussian Welfare Ministry contributed nearly 2,000 RM to the institute budget in the 1930/31 fiscal year for a "Eugenic Museum." Moreover, it paid the institute compensation for the training courses held in Dahlem for medical officials – in the 1929/30 fiscal year, for example, 5,000 RM. Further, the institute had at its disposal funds from the province of Westphalia for the construction of a photographic archive, which may have been bound to this task – as were subsidies from the "International Committee for the Study of Population Problems," but these did increase the liquidity of the institute's coffers. Finally, the Prussian Ministry for Science, Art and Education, as mentioned above, contributed smaller sums for the expansion of the skull collection. Fischer to KWG, 25/6/1930, MPG Archive, Dept. I, Rep. 1 A, No. 2405, p. 48; Anlage zur Jahresabschlußrechnung 1929/30, June 1930, ibid., p. 51; Aufstellung über künftig wegfallende Fördermittel, January 1931, ibid., p. 83; Einnahmen- und Ausgabenrechnung 1930/31, MPG Archive, Dept. I, Rep. 1 A, No. 2406, p. 113. Starting in the 1931/32 fiscal year, the grants from the ministries came via special clearing accounts and no longer appeared in the regular KWI-A budget.

[19] Fischer to KWG, 14/5/1929, MPG Archive, Dept. I, Rep. 1 A, No. 2412, pp. 141–141 v, quote: p. 141 v.

[20] A similar situtation arose in November 1931. Fischer to Generalverwaltung, 11/11/1931, MPG Archive, Dept. I, Rep. 1 A, No. 2405, p. 79.

From 1931 on, Fischer had to fight hard to consolidate his budget. "My scientific reputation is at stake," he complained in May 1931, "and it is impossible to achieve anything if my funding is cut."[21] After being forced to agree to a rigorous reduction in material costs nonetheless, in October 1931 Fischer warned:

> Of course, it can be done only at the expense of scientific achievements. I have already aborted several animal experiments and since July, the entire twin study, in which our institute [...] is without a doubt the international leader, has been throttled completely for the time being. I refuse to accept responsibility for the reduction in our prestige this entails.[22]

In his report about the fiscal year 1931/32, Fischer ultimately painted a gloomy picture:

> The work of the past year was possible solely because I succeeded in obtaining significant grants from individual ministries for certain projects. This possibility, of course, no longer exists. The regular funds of K.W.G., taken by themselves, will in no way make it possible to continue operating the institute beyond the summer. Unless new sources are opened for funding I will have to close it.[23]

Nonetheless, the institute director's calculated pessimism is deceptive. While Fischer did have to fight with massive financial problems starting in the 1931/32 fiscal year, and was obviously dependent on grants from third parties to maintain research operations in their entirety, and, although some research projects were shelved temporarily because they could not be financed, the situation of the institute was not as dramatic as he portrayed it. Not a single important project was interrupted due to missing funds, and although dramatic salary cuts proved unavoidable, no positions were eliminated.

2.1.3 Scientific Personnel

In October 1927 the institute began its work with five scientists: Director Eugen Fischer, the two department directors Hermann Muckermann and Otmar Freiherr von Verschuer, and two assistants, Lothar Loeffler (1901–1983)[24] and Fritz Kiffner (1899–1969).[25] Additional personnel were the technical assistant Suse Lüdicke and the secretary Dorothea Michaelsen.[26]

---

[21] Fischer to KWG, 1/5/1931, ibid., p. 66 f., quote: p. 67.

[22] Fischer to KWG, 1/10/1931, ibid., pp. 75–75 v, quote: p. 75 v.

[23] Jahresbericht Fischers vom 1.4.1932, MPG Archive, Dept. I, Rep. 3, No. 8. Similar: Fischer to KWG, 29/6/1932, MPG Archive, Dept. I, Rep. 1 A, No. 2405, p. 91.

[24] For a biography: Lebenslauf vom 13/7/1927, MPG Archive, Dept. I, Rep. 3, No. 30; Klee, Personenlexikon, p. 376.

[25] Cf. the employment contracts in the MPG Archive, Dept. I, Rep. 3, No. 25.

[26] Dorothea Michaelsen had taken the college entrance examination and studied medicine for several semesters. She was responsible for the entire accounting of the institute and also took care of the institute director's correspondence. In addition, as Fischer emphasized in his repeated applications to promote her to a higher salary scale, she was "able to make scientific

As discussed above, Hermann Muckermann had been "placed" from the start as the director of the Department for Eugenics. His political influence, which was to reach its zenith in the era of the Brüning administration, contributed significantly to the fact that the KWI-A played a leading role in the debates about the legalization of eugenic sterilization in the late 1920s and early 1930s – more on this, too, below. Muckermann remained controversial as a scientist, however. By far the most important personnel decision Fischer made during his preparations to found the institute was thus the appointment of the key position of a director of the Department for Human Genetics, which was supposed to comprise the core of the new institute. As the field of anthropology expanded to become anthropo*biology*, human genetics shifted to the center of attention as a "leading science" of both physical anthropology and genetic pathology and eugenics. It was clear to Fischer that his "bastard biology" alone could not create a foundation that would support the human genetics he wanted to establish at his institute – and even if it could have played this role, the loss of the German colonies had made field research much more difficult. Thus, no significant impulses to drive the development of human genetics were to be expected from Fischer's Department of Anthropology; even less so from Muckermann's Department for Eugenics. Consequently, it was of decisive importance in terms of research strategy to appoint a scientist to head the Department for Human Genetics who would be able to set a course as a human geneticist and shape a paradigm, to provide direction and indicate objectives for the institute's research agenda.

Considering the circumstances, his appointment of Otmar von Verschuer to this post, at the tender age of just 31, is at first glance surprising. As the fate of the institute was so intimately connected with Verschuer's biography for the entire duration of its existence, at this juncture we will illuminate the circumstances of his appointment in greater detail. It is particularly important to examine Verschuer's previous life path because his past already revealed fundamental strategic orientations for the future development of the institute as regarded its ideological outlook, scientific character and personnel networks. Ultimately, Verschuer's importance for the history of the institute is no less than that of the founding director, Eugen Fischer. In fact, Fischer and Verschuer set the institute's course together. The relationship that developed between these two men was downright symbiotic – Fischer was the teacher, patron, and fatherly friend of Verschuer, developing ideas into a comprehensive research agenda and pulling the strings in the background as a scientific manager, providing for political support, soliciting funds, influencing job placements, controlling associations, congresses, and publications. Verschuer occupied a preferential position in the circle of Fischer's pupils, playing the role of

contributions as well." Fischer to KWG, 9/3/1929; Fischer to KWG, 28/7/1932, MPG Archive, Dept. I, Rep. 1 A, No. 2405, p. 19 and/or 94 (quote). At its founding, the personnel team was rounded out by a caretaker and a cleaning lady. In the 1929/30 fiscal year the institute's nonscientific employees included three technical assistants, two secretaries, a caretaker, a stoker, a telephone operator, and assorted temporary typists. It shared a gardener with the KWI for Biology. Gehaltsliste, 12/6/1930, ibid., p. 55.

the young star, and was systematically groomed as Fischer's successor. He executed Fischer's research agenda, developed a fine-tuned methodology and kept up with the latest trends in research. In terms of research policy, he rode slipstream behind his mentor until 1935.

Verschuer was born the son of an officer in 1896.[27] In July 1914 he completed his final examinations at the nonclassical secondary school in Karlsruhe. He fought as an officer in World War I; for him, as for the majority of the men born in the years between 1880 and 1900, the "adventure on the front" was a formative experience.[28] He began to study medicine in Marburg in 1919, where he took so many extra courses that he was able to complete the four semesters of preclinical study required in just 13 months, and quickly passed a special version of the preliminary examination needed to begin clinical training. During this time he met Karl Diehl (1896–1969), a fellow medical student with whom he shared not only what was to become a lifelong friendship, but also similar professional interests: from 1927 until 1945, Diehl and Verschuer worked together closely on tuberculosis research. During his period in Marburg Verschuer was active in the Association of German Students (*Verein Deutscher Studenten*). Called upon to participate in the Kapp putsch, the student association in Marburg remained aloof – upon Verschuer's advice, as he later asserted. However, when Communist revolts erupted in the Ruhr region, Saxony and Thuringia in the wake of the putsch, and Reich Military Minister Gustav Noske (1868–1946) appealed to the students at German universities to enlist in the Reichswehr as temporary volunteers on March 19, 1920, 1,800 students in Marburg formed the Marburg Student Corps. They elected as their commander the retired naval captain Bogislav von Selchow (1877–1943), and Verschuer became his adjutant. "It never came to military action," Verschuer wrote in his memoirs. "Yet in the rebellious areas my feeling of being in enemy territory was stronger than I had ever sensed before during my years in France [...]."[29] Indeed, during its advance to Gotha, the Marburger Student Corps behaved as if "in enemy territory": near Mechterstädt, 15 rebels who had been taken prisoner were shot "escaping."[30]

After this military intermezzo, "things got too hot" in Marburg for Verschuer.[31] He and his friend Karl Diehl, who wanted to continue his clinical studies at the university hospital in Eppendorf, switched to the University of Hamburg and from there to the University of Munich, where Verschuer completed his studies. He spent

---

[27] The following is based primarily on Verschuer's 1945 autobiography, *Erbe – Umwelt – Führung*, MPG Archive, Dept. III, Rep. 86 A, No. 3–1. Also on this: Wess, Humangenetik; Thomann, Otmar Freiherr von Verschuer. Titles of nobility have been left out below – except in quotes and mentions of the first and last name.

[28] Cf. Peukert, Weimar Republic, pp. 25–31.

[29] Verschuer, Erbe – Umwelt – Führung, "Vorklinische Semester in Marburg (1919–1920)" section, p. 5, MPG Archive, Dept. III, Rep. 86 A, No. 3–1.

[30] Selchow, Hundert Tage, pp. 326–338. Also: Weingartner, Massacre; Brunck, Deutsche Burschenschaft.

[31] Verschuer, Erbe – Umwelt – Führung, "Klinisches Studium in Hamburg, München, Freiburg und wieder München (1920–1923)" section, p. 8, MPG Archive, Dept. III, Rep. 86 A, No. 3–1.

winter semester 1921/22 at the University of Freiburg, where he met his future mentor, Eugen Fischer. Verschuer attended Fischer's anthropology lecture and also took part in his anthropology laboratory course, through which "the first personal bridge […] was built."[32] Returning to the University of Munich in summer 1922, Verschuer sought out the acquaintance of the race hygienist Fritz Lenz – reading *Baur-Fischer-Lenz* had roused the medical student's interest in anthropology, human genetics, and eugenics[33] – and through Lenz, also came in contact with Alfred Ploetz and his *Widarbund*.[34] Yet young Verschuer was not attracted to the sage of race hygiene and his circle. "The intellectually outstanding figure of the race hygiene movement was Fritz Lenz. But nothing about him thrilled or electrified. He was the cool, expert scientist and critic."[35] Although Lenz was the most important sponsor for the young scientist Verschuer until 1927, the relationship between the two remained detached. During his student days in Munich, Verschuer joined the Thule Society, sympathized with the nationalist movement and – according to his own testimony – once took part in a mass event of the German nationalist *Freiheitspartei*, where he heard Adolf Hitler (1889–1945) speak. After 1945 Verschuer reported about this experience:

> Savonarola or Calvin? Hitler impressed me as a fanatic possessed by an idea, and the question I submitted after this impression at the time was: will he be successful and create something permanent, or end up burned at the stake to perish with what he wanted? Because my impression tended toward the latter assumption, I did not join Hitler's movement despite my generally nationalist leanings […].[36]

A blue-blooded academic with a national Protestant background, Verschuer felt repulsed by the plebian populism, by the open fanaticism and the open brutality in Hitler's demeanor – despite a high degree of ideological congruence.

In July/August 1922 Verschuer passed the state medical examinatons in Munich and started as an intern at the 1st Medical Clinic of the university there, where he also started work on his dissertation on the effect of caffeine on the tumefaction of serum proteins.

> With all of the fire of my scientistic enthusiasm I plunged upon the task I had set for myself, worked intensively in the clinic laboratory for months, performed experiments on patients and ultimately also on myself, taking blood from my veins and capillaries every two hours from the early morning until the afternoon on two separate days, on an empty stomach on both occasions and after administering caffeine on one. Afterward I worked late into the night, using all kinds of methods to investigate the samples myself.[37]

---

[32] Ibid., p. 12.

[33] Lösch, Rasse, p. 202.

[34] The *Widarbund* was a later stage of the Munich Bow Hunters Club, which was characterized by nationalist ideas and the body of thought surrounding hereditary biology and the life reform movement. Doeleke, Alfred Ploetz, p. 46.

[35] Verschuer, Erbe – Umwelt – Führung, "Klinisches Studium in Hamburg, München, Freiburg und wieder München (1920–1923)" section, p. 13, MPG Archive, Dept. III, Rep. 86 A, No. 3–1.

[36] Ibid., p. 14.

[37] Ibid., p. 15.

While at the university Verschuer had developed a particular interest in psychiatry, which had been aroused by the lectures of Emil Kraepelin. After receiving his medical license and completing the laboratory tests for his dissertation, he thus started as a volunteer at the University Psychiatric Hospital in 1923. Kraepelin had retired by then, and the lecture was being held by Ernst Rüdin, for whom Verschuer felt "litte attraction."[38] Since there was no prospect of a paid position at the University Psychiatric Hospital and the young doctor felt ever more drawn to the fields of heredity and race theory – in 1923 he participated as a "rookie and minor beginner"[39] in the 3rd Annual Convention of the German Society for Genetics (*Deutsche Gesellschaft für Vererbungswissenschaft*) – he endeavored to find a position as an assistant which would offer him the opportunity to perform genetic studies on human beings. He ultimately found such a position through Lenz, whose brother-in-law Wilhelm Weitz (1881–1969)[40] was looking for an assistant for the Medical University Polyclinic in Tübingen. On October 1, 1923, Verschuer began his job in Tübingen. In addition to his work at the clinic, which also included consultation hours as a physician for the poor, he worked with his boss on scientific projects. The internal specialist Weitz, associate professor in Tübingen since 1918, had been investigating the hereditary transmission of internal and neurological diseases since the beginning of the 1920s. His first publication on hereditary science issues appeared in 1921, a comprehensive study on the hereditary transmission of progressive muscular distrophy on the basis of exhaustive genealogical research on 15 families.[41] In 1923, the same year he brought Verschuer to his polyclinic, Weitz began his genetic pathology research on pairs of twins. This work was divided as follows: Weitz made the clinical reports, and Verschuer took care of the anthropometric investigations. While Weitz presented the results obtained from their collaborative work on a total of 45 pairs of twins at the next convention of specialists in internal medicine and then temporarily receded from this field of research,[42] Verschuer continued with twin research, whereby he turned to the genetic foundations of research on twins. He lectured on his considerations for the first time at the 4th Annual Convention of the German Society for Genetics in Innsbruck in September 1924 – this was Verschuer's first scientific address.[43] No fewer than seven scientific and popular science papers on twin research followed in 1925/26, not to mention a number of articles on race hygiene and race science.[44]

---

[38] Ibid., p. 16. Verschuer, Studien über den Blutserumeiweißgehalt.

[39] Verschuer, Erbe – Umwelt – Führung, "Klinisches Studium in Hamburg, München, Freiburg und wieder München (1920–1923)" section, p. 17, MPG Archive, Dept. III, Rep. 86 A, No. 3–1.

[40] For a biography, and on the following: Mai, Humangenetik, pp. 100–115.

[41] Weitz, Vererbung.

[42] Weitz, Studien.

[43] Verschuer, Umweltwirkung. Cf. idem., Erbe – Umwelt – Führung, "Assistenzjahre in Tübingen (1923–1927)" section, pp. 2 f., MPG Archive, Dept. III, Rep. 86 A, No. 3–1.

[44] On twin research: Verschuer, Wirkung der Umwelt; idem., Fall von Monochorie; idem., Stand der Zwillingsforschung; idem., Anthropologische Studien; ders., Fragen der vererbungsbiologischen Zwillingsforschung; idem., Anteil von Erbanlage und Umwelt; idem., Review: Gunnar Dahlberg, Twin Births. On race hygiene, e.g. Verschuer, Rassenhygiene; idem., Vererbung, Auslese und Rassenhygiene.

In February 1924 a local chapter of the German Society for Race Hygiene was founded in Tübingen. Weitz was one of the founding members and was elected deputy chairman; Verschuer was elected secretary.[45] Verschuer became a zealous propagandist of race hygiene, held adult education courses on the subject in Tübingen and gave talks to students – for instance, at a student rally in the former Cistercian monastery in Gaming, Austria in spring 1924, where he not only met up with the philosopher Othmar Spann (1878–1950), whose ideas of a corporative state were to make a lasting impression on him, but was also reunited with Eugen Fischer and consolidated his acquaintance with him.[46]

The polyclinic was attached to the Medical Clinic of the Tübingen University Hospital, where the next free position as an assistant was reserved for Verschuer. But since none of the assistants or head physicians left the Medical Clinic, Verschuer remained at the polyclinic for the next 4 years. Again it was Lenz who helped break the deadlock in the young doctor's career. Verschuer maintained a close scientific exchange with Lenz, to whom he sent all of his manuscripts "for critical persual" and from whose "objective critique" he profited.[47] In September 1926 Verschuer held an address at the convention of the German Anthropological Society in Salzburg.[48] On the way back he stopped at Lenz's villa in Herrsching. Lenz advised him to send a carbon copy of the manuscript of his lecture in Salzburg and his previous papers to Fischer, as he had heard that Fischer was about to found a large research institute in Dahlem, and there was the prospect of Verschuer finding a scientific position there.[49]

Verschuer had hoped for an assistantship and was surprised when Fischer visited him personally to offer him the position of department director. Fischer's great interest was not unfounded, as Verschuer embodied the ideal candidate for this post: He had training as a physician with emphases on internal medicine and psychiatry, had also attained prominence in the field of human genetics, and, above all, brought with him experience in twin research, which was regarded as the most promising methodological approach in human genetics around the mid-1920s. Furthermore, Verschuer was an unshakeable advocate of eugenics, whereby he – like Fischer – resisted clear classification to either the Munich or the Berlin line of the race hygiene movement, having publicly distanced himself from Günther's "race theory," but remaining sympathetic to the *Nordic idea*.[50] His political outlook was German national, skeptical of democracy, and anti-Socialist. Verschuer certainly harbored sympathies for the nationalist camp, but was careful to expose himself too much. Besides, he was firmly anchored in the Protestant community.

---

[45] Tübinger Gesellschaft für Rassenhygiene.

[46] Verschuer, Erbe – Umwelt – Führung, "Assistenzjahre in Tübingen (1923–1927)" section, pp. 4f., MPG Archive, Dept. III, Rep. 86 A, No. 3–1.

[47] Ibid., pp. 9–9a.

[48] Verschuer, Anteil von Erbanlage und Umwelt.

[49] Verschuer, Erbe – Umwelt – Führung, "Assistenzjahre in Tübingen (1923–1927)" section, p. 9a, MPG Archive, Dept. III, Rep. 86 A, No. 3–1.

[50] Verschuer, Review: Hans F. K. Günther, Nordischer Gedanke.

In other words, Verschuer could be regarded as a moderate eugenicist, which was of inestimable importance in view of the institute's interconnections to the Weimar welfare state.

First of all, however, Fischer had to convince the General Administration of the KWG to accept his candidate. General Director Glum expressed reservations because Verschuer was not yet a full professor, and proposed installing him as an assistant initially. As department chief, Glum believed, sooner or later Verschuer would have to be appointed as a scientific member. However, an internal agreement was reached to make such appointments dependent on whether the scientist in question, had he set out on a university career, at least had the prospect of an associate professorship. The General Administration's intention here, it was stated, was to raise the standards for selecting scientific members in the future, in order to enhance the status of the scientific members and to attract top-class scientists to even the positions below the directorial level. Scholars like Richard Goldschmidt, Max Hartmann (1876–1962) and Otto Warburg (1883–1970) could feel discredited if "entirely too young and not yet acadmically established gentlemen were [to be] made scientific members."[51] Fischer must have foreseen such difficulties, as he urged Verschuer to complete his postdoctoral *Habilitation* in Tübingen. He advised against qualifying in anthropology, believing that the *venia docendi* for genetics would be ideal, although internal medicine would be fine, too; what was important above all, he stated, was that he come to Berlin with a professorial qualification at all.[52] Ultimately, the problem was solved when a longer study on genetic twin research, which Verschuer had written for the renowned *Ergebnisse der Inneren Medizin und Kinderheilkunde* ("Results of Internal Medicine and Pediatrics") and was at press at the time, was recognized as his professorial dissertation.[53] On February 1, 1927, he defended his dissertation, and on May 10, 1927, he held his inaugural lecture on the "Tasks and Objectives of Human Genetics." As Verschuer proudly recorded in his memoirs, it was the first time "that a lecturer had qualified for a professorship at a German university in the new subject of human genetics."[54]

The three department chiefs were in charge of a continually increasing staff of assistants, scientific staff, and doctoral students. The number of "working scholars" increased from the initial five to 15 (1928/29), 17 (1929/30), 28 (1930/31), 30 (1931/32), and finally 36 (1932/33).[55] Even back in his first annual report for 1930/31, Fischer established "that the workplaces are occupied to the very last corner and space is getting scarce."[56]

---

[51] Glum to Fischer, 29/11/1926, MPG Archive, Dept. I, Rep. 86 A (Münster), No. 9.

[52] Fischer to Verschuer, 28/9/1926, ibid. At this juncture Fischer also called attention to Verschuer's status relative to Muckermann and his assistant, which would improved considerably by the *Habilitation*.

[53] Verschuer, Vererbungsbiologische Zwillingsforschung.

[54] Verschuer, Erbe – Umwelt – Führung, "Assistenzjahre in Tübingen (1923–1927)" section, p. 10f., quote: p. 10, MPG Archive, Dept. III, Rep. 86 A, No. 3–1. Cf. idem., Aufgaben und Ziele.

[55] Cf. also the activity reports of the KWI-A in the journal *Die Naturwissenschaften*.

[56] Jahresbericht 1930/31, 14/4/1931, MPG Archive, Dept. I, Rep. 3, No. 7.

The two assistant positions experienced a great deal of personnel turnover in the beginning phase. The physician and anthropologist Fritz Kiffner, a pupil of Luschan's, left the institute after just 1 year, on September 30, 1928. His position was taken over by the physician Friedrich Curtius (1896–1975),[57] until then an assistant at the Clinic for Internal Medicine of the University of Bonn. Curtius, too, left the institute after just 1 year, on September 30, 1929, at the same time as Lothar Loeffler, who had occupied the second assistant position since the institute's founding. Curtius and Loeffler remained connected with the institute, constituting the first nodes in a network of former staff and doctoral students that became denser and denser: Curtius as a doctor at the Polyclinic, from 1934 as director of the Department for Genetic Pathology at the 1st Medical University Clinic of the Charité in Berlin, Loeffler initially (from 1931) as a lecturer at the University of Kiel, from 1934 on as associate professor and director of the Institute for Race Biology at the University of Königsberg, and finally, from 1942, as director of the Institute for Race Biology at the University of Vienna.

In October/November 1929, the assistantships were occupied by the physician Hans Glatzel (1902–1990)[58] and the French physician Marie-Thérèse Lassen – the first woman in this position at the KWI-A, albeit only as a "voluntary assistant."[59] These two did not stay long, either. Glatzel returned to the Medical University Clinic in Heidelberg in February 1931; in 1942 he became a professor at the University of Kiel. Lassen left after 1 year. In her stead came the physician Heinrich Kranz (1901–1979),[60] who was forced to leave the institute, after the National Socialists took power. In April 1931, the second assistantship went to the zoologist and anthropologist Wolfgang Abel (1905–1997),[61] who was to succeed Fischer in taking over the Department for Race Science and the chair of anthropology at the University of Berlin in 1942.

The custodian of the skull collection, the secondary school teacher Hans Weinert (1887–1967) had a largely independent position.[62] He not only took care of the collection, but also held the "minor anthropological laboratory course" at the university

[57] For a biography: Lebenslauf vom 31/7/1928, MPG Archive, Dept. I, Rep. 3, No. 27; Klee, Personenlexikon, p. 98. Curtius, a brother-in-law of the psychiatrist Viktor von Weizsäcker (1886–1957), had already worked with the twin method. Also interesting for the KWI-A were his studies on congenital partial hypophysia.

[58] For a biography: ibid., p. 185 f. On this, see also the cover letter of his application of 28/6/1929, MPG Archive, Dept. I, Rep. 3, No. 29.

[59] Jahresbericht 1929/30, Personalzusammensetzung, 18/3/1930, MPG Archive, Dept. I, Rep. 3, No. 6. The status of "voluntary assistant" meant that Lassen did not receive a salary from the KWG. From 1929 a number of female doctoral students worked at the KWI-A: Ida Frischeisen-Köhler, Irawati Karvé, Gisela Meyer-Heydenhagen, Brigitte Richter, Luise Brauns, Hertha Busse, Maria Frede, Lore Schröder. On their biographies, cf. Vogt, Wissenschaftlerinnen, pp. 28, 30, 41–43, 101, 116, 129.

[60] For a biography: Klee, Personenlexikon, p. 335.

[61] For a biography: ibid., p. 9.

[62] For a biography: ibid., p. 662; Schmuhl (ed.), Rassenforschung, p. 344 f. Also, the cover letter of Weinert's application of 30/10/1927, MPG Archive, Dept. I, Rep. 3, No. 31.

for Fischer, thus relieving the institute director of his instruction duties. In addition, he was very active in research and publication on the area of paleoanthropology and blood group research. He worked at the KWI-A until 1935 and then accepted an appointment as professor and director of the Anthropological Institute at the University of Kiel – a further important link in the chain of relations among former members of the institute.

The professional quality of the three scientific staff members under contract in 1927 was extremely varied. The hiring of the retired director of the Wiesloch Sanatorium and Nursing Home in Baden, Max Fischer (1862–1940), who was presented the sinecure of a scientific staff position in the Department for Eugenics by his cousin Eugen Fischer, can be regarded as nepotism.[63] The physician Konrad Kühne,[64] on the other hand, whom Fischer also brought from Freiburg, with his projects on the anomalies of the spinal column, contributed an important building block on the institute's research agenda. Fischer also obtained the blood group researcher Max Berliner (1888–1960 or 1965) as an external staff member. From 1929 on the fluctuation among the scientific staff of the institute was substantial. In the long view, the appointment in 1930 of the retired high-school principal Georg Geipel (1871–1973),[65] a specialist on the technique of dactyloscopy, who made a decisive contribution to the fact that the genetics of the hand and finger ridges – in combination with twin research, genetic pathology, and race anthropology – played a central role at the KWI-A in the 1930s and 1940s.

Worth emphasizing here is the surprisingly high number of foreign guest scholars and doctoral students from about 1929: With Marie-Thérèse Lassen from France, Sei Hara from Japan,[66] Valentina Bosca from Romania, Thordar Quelprud from Norway, Bozo Skerlj (1904–1961) from Yugoslavia, Baeckpyeng Kim from Korea, Yun-kuei Tao from China, Eduardo Fleury Cuello from Venezuela, Fritz Schrijver from the Netherlands, and both Miguel Carmena and Luis de Lazerna from Spain, 11 scientists can be mentioned by name. International contacts were made here, some of which remained intact well into the period after 1945.

2.1.4 The Board of Trustees

Each Kaiser Wilhelm Institute had its own Board of Trustees. In some cases these boards of trustees were very active and had a major voice in determining the institute's course, in other cases they existed more or less only on paper. In every case, however, the composition of the Board of Trustees is illuminating as to the objectives of the General Administration of the KWG and the institute director in terms of the

[63] Massin, Rasse und Vererbung, p. 200 (note 53).

[64] For a biography: Lösch, Rasse, p. 568.

[65] For a biography: Biographische Daten, MPG Archive, Dept. III, Rep. 48, No. 1.

[66] Actually *Tadashi* Hara. In Germany he worked and published under the name of Sei Hara.

institute's dovetailing with state, society, industry, and science. In the case of the KWI-A, Friedrich Glum contacted Eugen Fischer in June 1927 to discuss appointments to the Board of Trustees.[67] Fischer left the composition of the board completely up to the General Administration, however,[68] and thus the final list of board members, introduced in February 1928, corresponded to Glum's suggestions with few exceptions.

According to the charter passed on June 25, 1927, the KWI-A Board of Trustees had to consist of at least 19 members, of whom 14 were to be elected by the senate of the KWG – among them the chairman, while the unified Prussian provinces were to send three representatives, and the Reich Ministry of the Interior and the Prussian Ministry for Science, Art and Education one representative each.[69] The Chairmanship of the Board was assumed by the KWG president, Adolf von Harnack. The General Administration of the KWG was further represented by General Director Friedrich Glum, Director Adolf Morsbach and Deputy Vice President and former Prussian Minister of Culture Friedrich Schmidt-Ott (1860–1956), who played an outstanding role in the science sponsorship of the late Weimar Republic as founder and chairman of the Emergency Association of German Science (from 1934: German Research Association/DFG). As external board members, the institute succeeded in winning the Social Democratic Reichstag deputy and former Reich Minister of Finance Rudolf Hilferding (1877–1941), Lord Paul Schottländer (1870–1938), and Albert Vögler (1877–1945), the General Director of Vereinigte Stahlwerke AG, the largest steel concern in Germany at the time. The composition of the board suggests that Schmidt-Ott and Hilferding were to guarantee the connection to politics, while Schottländer and Vögler opened up access to industry. This last category also included Gustav Krupp von Bohlen und Halbach (1870–1950), who was slated as deputy chairman of the Board of Trustees. He rejected the post, however – his particular interest was directed to the KWI for Brain Research under the direction of Oskar Vogt (1870–1959).[70]

In the place of Krupp von Bohlen und Halbach, the senate elected the Berlin Medical Councilor Wilhelm von Drigalski (1871–1950).[71] During the meetings about the draft of the charter, Glum originally had suggested giving the public health officer of Berlin a seat and vote in the Board of Trustees as a function of his office,[72] but this provision had been dropped. Even though Drigalski was not appointed to the board expressly due to his official function, it subsequently

[67] Glum to Fischer, 16/6/1927, MPG Archive, Dept. I, Rep. 1 A, No. 2403, pp. 1a-1a v.

[68] Fischer to Glum, 21/6/1927, ibid., p. 3.

[69] Satzung des Kaiser-Wilhelm-Instituts für Anthropologie, menschliche Erblehre und Eugenik, 25/6/1927, § 3, MPG Archive, Dept. I, Rep. 1 A, No. 2402, pp. 14–16, here: p. 15.

[70] Cf. Schmuhl, Hirnforschung.

[71] For a biography: Labisch/Tennstedt, Weg, vol. 2, p. 399 f.; Klee, Personenlexikon, p. 120. Cf. Harnack to Senatoren der KWG, 24/10/1927, MPG Archive, Dept. I, Rep. 1 A, No. 2403, p. 29.

[72] Glum to Fischer, 16/6/1927, ibid., pp. 1a-1a v; draft charter (undated), MPG Archive, Dept. I, Rep. 1 A, No. 2402, pp. 5–8, here: p. 5.

became customary to accord a seat and vote in the KWI-A Board of Trustees to the Public Health Officer of the capital of the Reich. Upon the intervention of the Chairman of the German Trade Union Council (*Allgemeiner Deutsche Gewerkschaftsbund*), Theodor Leipart (1867–1947),[73] the senate of the KWG nominated a further board member on February 10, 1928, the Social Democratic Social Hygienist Alfred Grotjahn, at the time Dean of the Department of Medicine at the University of Berlin and consultant to the Hygiene Section of the League of Nations in Geneva, thus enlarging the Board of Trustees by one seat.[74] Apparently the trade union chief was concerned that social hygiene and eugenics be represented in the body not only by Drigalski, who was firmly in the bourgeois camp as a member of the German Democratic Party, but also by Grotjahn, who is associated with the right wing of the Social Democrats. The result of these party-political machinations was that the Berlin line of Weimar-era eugenics was represented on the board by two prominent physicians.

At one point during the consultations about the draft charter it was proposed that four seats on the Board of Trustees be reserved for the two directors of the KWI for Biology, a representative of the Institute for Genetics at the Berlin Agricultural College, and a representative of the DFA in Munich. Although this provision had been dropped from the final version of the charter, the senate nevertheless appointed to the board the Director of the KWI for Biology, Carl Erich Correns, as well as the Deputy Director and Department Head Richard Goldschmidt, along with Erwin Baur as director of the Institute for Genetics at the Berlin Agricultural College. A representative to the board was also granted to the DFA in Munich – none other than Ernst Rüdin. Baur, Correns, and Goldschmidt, who had vigorously promoted the founding of his institute, must have been more than welcome to Eugen Fischer; however, he must have acknowledged Rüdin's election with mixed feelings, as a rivalry and competition for resources between the KWI-A and the DFA had begun to take shape even during the preparations for the institute's founding.

In consideration of the fact that their contributions would cover the lion's share of the operating costs of the institute, the Prussian provinces had been allotted three seats on the board. These were occupied by the former undersecretary August von Schenck, head of the Association of the Prussian Provinces; the director of the University Mental Hospital and Province Sanatorium and Nursing Home in Göttingen, Dr. Ernst Schultze (who had achieved some degree of fame as psychiatric expert in the trial of the serial murderer Fritz Haarmann), as well as the governor of the Rhine Province, Johannes Horion (1876–1933). Horion, closely connected with the Catholic Center party, was very interested in questions of public relief – it

[73] Leipart to Harnack, 31/10/1927, MPG Archive, Dept. I, Rep. 1 A, No. 2403, pp. 35–35 v; Harnack to Leipart, 14/11/1927, ibid., p. 41. Leipart's intervention can be traced back to a suggestion by the medical director of the Department for Vocational Hygiene and Health Matters in the board of directors of the German Trade Union Council (*Allgemeiner Deutscher Gewerkschaftsbund*), Dr. Meyer-Brodnitz. Meyer-Brodnitz to Leipart, 31/10/1927, ibid., p. 36f.

[74] Auszug aus dem Senatsprotokoll vom 10/2/1928, MPG Archive, Dept. I, Rep. 1 A, No. 2402, p. 18.

was presumably his service that the Provincial Association of the Rhineland had contributed 10,000 RM to the institute's founding.

Particular interest was also declared by the Reich Ministry of the Interior, which asked to appoint to the board not only the undersecretary responsible for the KWG, Max Donnevert (1872–1936), but also Undersecretary Max Taute (1878–1934),[75] an experienced colonial physician who had worked as a consultant for physician's affairs in the Reich Ministry since 1924 and also for the Reich Health Council from 1927. Finally, the Prussian Ministry for Science, Art and Education sent to the Board of Trustees Undersecretary Werner Richter (1887–1960).[76]

Up to the beginning of the Third Reich the board of trustees demonstrated practically no activity. It is doubtful that the board met at all between its constitutive session on June 6, 1928 and the session held on July 5, 1933, with a thoroughly altered composition.[77]

2.2 Research Agenda and Research Praxis

From 1927 until 1945, a total of 675 publications proceeded from the KWI-A. Of these, however, only 341 can be conceived of as scientific works in the proper sense, that is, original papers that presented new results on the basis of research performed at the institute, provided important contributions to the discourse on methodology and theory formation, or, in the form of overview surveys – manuals and articles for manuals, survey papers at scientific conventions, textbooks – bundled the state of knowledge in the field of research, thus opening up new, structured perspectives and developing new conceptions. Excluded from these, on the other hand, are nonedited reprints of earlier publications, as well as works that presented material that had already been published elsewhere, obituaries, popular science articles, and political pamphlets.[78] If the scientific publications in the strict sense

[75] For a biography: Labisch/Tennstedt, Weg, vol. 2, p. 504.

[76] Reichsinnenministerium to KWG, 27/8/1927, MPG Archive, Dept. I, Rep. 1A, No. 2403, p. 21; Preußisches Ministerium für Wissenschaft, Kunst und Volksbildung to Harnack, 31/8/1927, ibid., p. 23.

[77] Niederschrift über die Sitzung des Kuratoriums des KWI-A am 6/6/1928, MPG Archive, Dept. I, Rep. 1 A, No. 2404, pp. 1–1a. No further protocols are found before 1933. At the board meeting on July 5, 1933 the annual accounts for the fiscal years 1930/31, 1931/32, and 1932/33 were submitted, a certain indication that the board had not met since 1930. Cf. Niederschrift über die Sitzung des Kuratoriums des KWI-A am 5/7/1933, ibid., pp.. 11–13.

[78] My quantitative analysis is oriented on Verschuer's retrospective classification on the basis of the omnibus volumes of *Arbeiten aus dem Institut*, whereby, as Lösch, Rasse, p. 489, correctly emphasizes, several titles must be added, which Verschuer left out of his sample because they apparently all too clearly indicated the complicity of the KWI-A in the state crimes of the National Socialist regime. The standard I use is clearly stricter than that of Kröner, *Von der Rassenhygiene zur Humangenetik*, p. 49, or of Massin, Rasse und Vererbung, p. 199 f. The lesser number of scientific publications I determined essentially can be attributed to the fact that, to avoid distortions,

are put in the order of their date of publication, one can establish that the starting phase in the years from 1927 to 1933, despite the difficult economic conditions, represented quite a productive period, with no less than 103 of the 341 publications, that is, almost a third, emerging during this time.

2.2.1 *The Concept of Anthropobiology*

The new institute, as Eugen Fischer had announced proudly, would no longer occupy itself with "mere skull measuring." In the sense of opening up anthropology toward human genetics, which Fischer outlined with the catchword of anthropo-*biology*, the conventional, static, taxonomically organized concept of race that proceeded from morphological features was to be abandoned in favor of a dynamic concept of race conceived in terms of evolutionary biology and grounded in genetics. With this, Fischer gradually disavowed the idea of pure systems of race that were given a priori, like those upon which his study about the "bastards of Rehoboth" had been based. Under the influence of Walter Scheidt (1895–1976), Fischer embraced the idea of "local races" or "breeds of men," groups with a relatively frequent occurrence of specific genetic dispositions, which, in turn, were understood to be the product of selection and adaptation in geographic isolation.[79] Yet, as Fischer emphasized, a human being's external appearance, the phenotype, resulted from the sum of these parts disposition *and* environment – and this was also true for the occurrence of features on which common race typologies were oriented.[80] Therefore, in order to categorize mankind into races that made any sense, anthropobiology had to pervade the external appearance and advance to become the genetic image of man, the genotype. As a logical consequence, the central group of issues that emerged from Fischer's early research program was thus the question as to the relative importance of genetic disposition as opposed to the environment in the occurrence of specific elements of the phenotype: To what degree do certain clearly demarcated features and characteristics of humans depend on genetic disposition, to what degree are they influenced by the environment? These questions were decisive not only with a view to physical anthropology and the classification of human races, but also, since they could be expanded to pathological features as well, with regard to medicine and eugenics. Only if the influence

---

I did not count works based on research performed at the institute if they had been covered for the most part by previously published articles. Moreover, I count such works as Eugen Fischer's contributions to the *Handwörterbuch der Naturwissenschaften* just once, while other authors counted each individual article as a separate scientific publication. For all deviations in detail, I agree with the Kröner and Massin's tendency: A large number of the institute's publications cannot be assessed as scientific publications in the strict sense.

[79] On the basics of this: Gausemeier, Walter Scheidt.

[80] Hair shape, pigment ratios in skin, hair and iris, skull shape, body size and proportions, and shape of nose and lips.

of genetic disposition and that of the environment on the emergence of diseases and handicaps could be demarcated from each other, so the consideration behind these questions, was it possible to make informed decisions as to whether and to what degree the instruments of medicine – prophylaxis, therapy and rehabilitation – should be supplemented by eugenic prevention.

The question guiding the search for knowledge about the relative shares of genetic disposition and the environment in the implementation of the phenotype opens up a broad field of research. It could be applied to a multitude of topics, practically *all* anatomical, morphological, physiological, pathological, and psychological features and characteristics – from the dimensions of the skull to peculiarities in the structure of the spine, red hair, the shape of the ear, the pattern of fingerprints, hemogram, or disposition to tuberculosis, all the way to conceptions of morals, criminality, performance in school or talent for playing chess. It could be researched on a wide variety of objects: on living human beings, prepared human samples, fossil samples, x-ray images, and family trees – but also, particularly important from the start, on animal models. And it could also make use of a wide variety of methods: in addition to the methods introduced from the areas of anthropometry, "bastard research," clinical diagnostics and pathology, embryology, genealogy, and genetic family research, and psychological suitability diagnostics, new methods were utilized at the KWI-A even in the period from 1927 to 1933, many of which were to become significant in the future: experiments on animals, blood group research, dactyloscopy, and above all twin research.

2.2.2 *Twin Research*

The idea of investigating the influence of heredity and environment – nature and nurture – on the basis of twins can be traced back to Francis Galton.[81] However, for a long time the systematic investigation of twins failed because there was no reliable method of differentiating with any certainty between identical and fraternal pairs of twins (zygosity testing). Heinrich Poll, for instance, performed comprehensive investigations on the fingerprints of twins, endeavoring in vain to establish the papillary pattern as a certain characteristic for differentiating between identical and fraternal twins.[82] The decisive breakthrough was not achieved until Hermann Werner Siemens (1891–1969)[83] introduced his "polysymptomatic similarity diagnosis" in 1923, which was based on the combination of a multitude of anthropometrically determined characteristics that were very often consistent among identical twins, or very often different among fraternal twins: hair color and shape, skin color, color of lanugo hairs,[84] freckles, telangiectasia, cornification in hair follicles, tongue creases, and the further –

[81] Galton, History of Twins.

[82] Poll, Zwillingsforschung.

[83] For a biography: Klee, Personenlexikon, p. 583.

[84] Fine hairs on the fetus in the second half of pregnancy.

less reliable – characteristics of the face, shape of the ear, form of the hand and body type. At the same time, Siemens provided a better understanding of another critical methodological problem of twin research – putting together a representative sample of twins – by systematically recording pairs of twins with the cooperation of school authorities. This meant significant progress with regard to older studies, which were based on single cases that had become known by chance, and had a correspondingly high selection error in favor of the twins who shared many characteristics. Additional innovations introduced by Siemens included the statistical observation of large groups as well as the systematic comparison between identical and fraternal twins.[85] On the basis of the standards set by Siemens, twin research began to be performed in Germany on a large scale from the mid-1920s.

The twin method was based on a simple approach: identical twins, which proceeded from the cells of a single fertilized egg cell, were held to be genetically identical – accordingly, any differences between identical twins were conceived of as exclusively environmental. Fraternal twins, on the other hand, which proceeded from different egg cells fertilized at the same time or in short succession, shared just over half of their genomes – like other siblings,[86] so that differences between fraternal twins were understood to be the result of the coaction of a genetic *and* an environmental factor. If the serial investigation of twins with regard to a certain characteristic proved that fraternal twins show a lesser degree of consistence (concordance) – or, expressed inversely: a greater degree of deviation (discordance) – than identical twins, the difference was then traced back to the genetic factor. The environmental influence was assumed to be constant for all pairs of twins.[87] The twin method represented a "nearly universally utilizable instrument for the practical investigation of the problem of heredity vs. environment," as every characteristic that could be measured or counted could be tested "for rates of concordance/discordance."[88]

It must be kept in mind that twin research was more than just a method; on the contrary, it constituted a paradigm shift in the Kuhnian sense, which, proceeding from a specific method, stated premises, introduced concepts, demarcated epistemic objects, provided a theoretical framework, created an experimental culture, and prefigured classic examples, forms of representation and symbolic generalizations.[89]

The paradigm constructed by twin research was distinguished by a marked conceptual reductionism in four respects: First, it presupposed genetic disposition and environment as analytical categories without demarcating them precisely from each other.[90] Second, the paradigm of the twin method did not itemize the two components

---

[85] Siemens, Zwillingspathologie; idem., Eineiigkeitsdiagnose; idem., Bemerkungen über die Ähnlichkeitsdiagnose; idem., Geschichte der Zwillingsmethode.

[86] This surplus resulted from potential genetic consistencies among ancestors.

[87] Lenz/Verschuer, Bestimmung, p. 425.

[88] Mai, Humangenetik, p. 1.

[89] Kuhn, The Structure of Scientific Revolutions.

[90] Thus, for instance, the influence of the physical environment on the manifestation of genes in twin research was sometimes attributed to heredity – in the form of "modification genes" (Diehl/Verschuer, Zwillingstuberkulose, p. 103), and sometimes to external influences – as an "internal environment" (Luxenburger, Zwillingsforschung, p. 234).

of heredity and environment into any subordinate components. The urgent interest was not in the individual genes, their placement on the chromosomes, or the mechanisms of their propagation, not the reciprocal actions they exerted upon each other, and not the complex connections between individual genes and phenotypical characteristics (expressivity, penetrance, specificity) – at least not initially. Rather, the subject of interest was *the* genome, and also *the* environment, conceived of as black boxes. Third, the paradigm of twin research proceeded from the assumption that the two components of heredity and environment interacted *additively* in the development of characteristics, and that consequently it is possible to break down the process of phenogenesis according to magnitudes of influence and determine the respective importance of heredity and environment *quantitatively*. The complex processes of interaction between hereditary factors and environmental conditions, and the effects of synergy and emergence that result from this interaction, are ignored completely in this approach – the question was not even posed as to whether it makes sense at all to conceive of heredity and environment as bundles of factors that can be clearly differentiated, and effective in and of themselves. Fourth and finally, the idea that the elements of the phenotype are dependent variables, which ultimately can be traced back over a complex causal chain to two independent variables, the genome and the environment, resulted in an arbitrary definition of dependent and independent variables used in twin research to address the highly complex characteristics of human beings. In so doing it ran the risk of superficially assigning a *gene for* – be it for musical talent, sensation of taste, moral instability, criminality, or schizophrenia.

These critical theoretical weaknesses in twin research, which were recognizable in principle even considering the state of knowledge at the time, raise the question as to why Fischer and Verschuer oriented the new institute so asymmetrically toward this methodology. Christoph Mai proposed the thesis that there was a close connection between the boom in twin research and the strengthening of the race hygiene movement in the 1920s. "Leading German human geneticists," according to Mai, "explicitly determined the goals and practical application of their research under the aspect of their eugenic-race hygiene – i.e. sociopolitical – usability [...]."[91] Human genetics, he continues, stood under significant "pressure to succeed," to prove the heritability of eugenically relevant characteristics of humans, and thus resorted to "questionable methods."[92] Because the Mendelian genetics corresponding to the state of the science was only suitable for this purpose when sloppy methodology was applied, as Fischer's study about the "bastards of Rehoboth" had shown, twin research was the clear means of choice for eugenically oriented human geneticists, stated in slightly exaggerated terms, precisely because of its theoretical weaknesses and sources of practical error – the heritability of those pathological phenomenon at the focus of biopolitical interest could be proven more or less at will. It follows that the emphasis on twin research was a politically motivated recourse to a concept that was already scientifically obsolete – in short, Mai characterizes twin research as a pseudoscience.

[91] Mai, Humangenetik, p. 44 f.

[92] Ibid., p. 59.

However, this interpretation oversimplifies the context: twin research at the KWI-A before 1933 was by no means directly oriented toward eugenics. Only the research on the heritability of tuberculosis on the basis of the twin method had eugenic implications. On the contrary, before 1933 the KWI-A did not produce a single twin-based study concerning any of the mental or degenerative nerve diseases, or any of the forms of mental or physical disability that played a role in the eugenic debates of the late Weimar Republic. At the KWI-A twin research was understood as *theoretical* research.

In fact, Verschuer believed that he had found in twin research the king's road for theoretical research on human genetics – and Fischer shared this view, as Verschuer's appointment to the key position as head of the Department for Human Genetics demonstrated. In a text published in 1934 for advanced training for physicians, Verschuer developed a three-step model to illustrate the superiority of the twin method over both classical anthropometry and Mendelian genetics:

> The history of human genetics shows three phases: the pre-Mendelian, the Mendelian and that marked by developmental physiology. In the first period, biometrical methods were used: The similarity between blood relatives was measured and compared (Galton). The second period (experimental genetic research) brought a new direction: analogies for humans to the regularities observed in plants and animals. Limits: Mendelian analyses presuppose simple relationships between genes and external characteristics. This is what designates the new problems for genetic research: How great are the kind and degree of the influence of peristasis (environment) on the developmental tendencies given by genetic disposition? How can and does the gene react under certain circumstances (mission of the third phase)? Some epistemic potentials are presented by the results of general genetics, others demand methods specific to twin research (picking up on Galton again).[93]

Against the background of Jonathan Harwood's comparative analysis of genetic research in the German Reich between the world wars, it becomes clear that Verschuer's approach fit in well with the mainstream of German genetics.[94] While genetic research in the USA after World War I, under the predominant influence of the research group around Thomas Hunt Morgan, chiefly devoted itself to the transmission of genetic dispositions and the morphological structure of the chromosomes, it largely ignored questions about the manifestation of genetic factors, about the complex processes that led from the genotype to the phenotype. Morgan, who had worked in the field of developmental physiology himself in his youth, was well aware of this gap; in his Nobel Prize acceptance speech in 1934, he even related how genetics and embryology must combine to take on the problems of developmental physiology.[95] In his own research, however, he consciously restricted himself to a question he believed could be solved: that of genetic transmission. This research strategy, which determined the course of US genetics into the 1930s,

[93] Verschuer, Zwillingsforschung im Dienste der Erblehre, p. 189.

[94] Harwood, Styles.

[95] Morgan, Relation. Cf. Allen, Thomas Hunt Morgan and the Split. Examples for developmental genetics in the USA in Harwood, Styles, p. 96.

ultimately proved to be a very successful one that made possible rapid gains in knowledge in a narrowly limited scientific field, albeit at the cost of contracting the field's perspective.[96] The majority of German geneticists may have accepted the chromosome theory further developed by Morgan and his school,[97] yet many critical voices were raised, which, in the best tradition of German science, were skeptical in view of the danger of the excessive specialization and fragmentation of the science of humans, and rejected the conceptual reductionism of the *transmission genetics* performed by the group around Morgan on the model organism of the *drosophila* fruit fly. Thus, in sharp contrast to the USA, comparatively little research was performed on *drosophila* and very little on transmission genetics in Germany in the 1920s and 1903s. Instead, German genetics in the 1920s turned its attention all the more strongly to evolutionary biology, developmental physiology, and embryology. The resulting forms of developmental genetics constituted the main strand of genetic research in Germany during the period between the wars and – together with the research on plasmatic heredity[98] and evolutionary biology[99] directed against the "core monopoly" of the Americans – comprised the particular signature of the German genetics of this period.

As such, Verschuer's skepticism with regard to transmission genetics and his call for building a bridge between human genetics and developmental physiology was consistent with the common sense of those working on genetics in Germany. For all that, what is striking is that the method of twin research was *not* capable of achieving this, since – as discussed above – it was a method that looked not at the *dynamics* of development, which resulted from the complex interaction of heredity and environment, but rather proceeded from a cumulative effect of both development factors, each of which had effects independent of the other. In practice, the original question: *how do genes and environmental factors interact?* was shortened to: *how important is the influence of the genome?* "An epistemological interest narrowed to this degree hardly offers starting points from which new and fruitful questions can be posed," Mai establishes correctly, "its potential for explanation is restricted *eo ipso* to answering this single question."[100]

The fact that the question guiding research had been so narrowed certainly had something to do with the philosophies of race hygiene and race anthropology at the time – albeit not so much in the sense of a conscious, biopolitical instrumentalization of human genetic research as in the sense of an unconscious incorporation of pre- and extra-scientific interests and mentalities into the conceptualization of research. In the longer perspective, this conceptual reductionism was to prove a dead end. By the mid-1930s many questions had accumulated that could not be resolved by twin research, and

---

[96] Allen, Thomas Hunt Morgan.

[97] Harwood, Styles, pp. 38–45.

[98] For comprehensive information on this: Harwood, Styles, pp. 61–84. The concept of "core monopoly" is from Hans Winkler, Rolle.

[99] Harwood, Styles, pp. 99–137.

[100] Mai, Humangenetik, p. 63.

the critique of the twin method grew stronger. As a reaction to these aporias, from 1938 on Eugen Fischer formulated the paradigm of *phenogenetics*, which steered research back to the original questions of the institute – by this time on a broader methodological foundation. All the same, the twin method was still suitable as a kind of paradigm to guide the scientific practice of the institute well into the 1930s.

According to the will of Fischer and Verschuer, the KWI-A was supposed to conduct "twin research on a grand scale." For this purpose work was begun on setting up a register to record all twins in the greater Berlin area, with name and address as well as detailed anthropometric data, further with information about illnesses they had survived and on genetic diseases in the family or special talents. With the support of the Provincial Teaching College (*Provinzial-Schulkollegium*) of Brandenburg and Berlin, Verschuer sent out over 1,200 questionnaires, in order to find twins among the pupils at schools in Berlin. In this manner Verschuer quickly traced around 1,000 pairs of twins, who then were invited to the institute for an examination, on a voluntary basis. In 1931, the *Rheinische Zeitung* depicted the procedure of the investigation vividly and in great detail:

> There is no reason to fear the examinations to which the twins are subjected here. First they come on the "electric chair" – only to be photographed, of course. This chair is a high stool, which is rotated by an electric motor in precisely fixed angles to the camera, guaranteeing that the head is in exactly the same position in the photographs of both twins. Then the hands are examined. Prints are made of all ten fingers [...]
>
> Next a droplet of blood is taken from the twins and subjected to a test that is supposed to illuminate the ratio of red to white blood cells. This is followed by thorough anthropological measurements. A metal box is opened; twelve glass eyes of different colors stare spookily into space. The eye color of the twins is checked against these. The hair color and color of the skin are determined in a similar manner. Also important, by the way, is the examination of the ears, which are usually of the same shape among "one-celled" twins, further that of the hairline and of freckles, which the twins often have in exactly the same places, down to the millimeter. All vital statistics are recorded painstakingly, from tongue creases to the half-moon of the fingernails; with special apparatus the electric currents evoked by the heartbeat and the depth of breathing are measured.
>
> The intelligence test comes last. Subjects fill out a questionnaire with 50 questions. Simple words must be defined, like island, rent, greed; abstract concepts explained, incomplete sentences completed, proverbs interpreted. Even the sense of humor is tested scientifically, for instance by the question "What is funny about the sentence: 'When push comes to shove, I shove off?' Logic is tested by other questions, for instance: 'Why are houses built higher in the city than in the country?' Then the subject must also reveal his ethical and social attitudes: 'What would you do if you won the lottery?' – 'When would you feel happy, when unhappy?' Finally, the twins are shown a series of ink blots, certain figures like the ones children make on paper with ink that has run; here they are used to detect the twins' differences and commonalities of emotion, sense of shape, fantasy and perceptive faculty from their different interpretations of the shapes seen. The curious researcher makes a note of the subject's life philosophy, temperament, character, moral attitudes and then asks whether the parents can tell the twins apart, what characteristics they use to recognize them and when they had their first shave.
>
> The older twins are sent home with a warm 'Thank you very much,' the younger ones with a bar of chocolate.[101]

---

[101] Das Geheimnis der Zwillinge. Neue Wege der Erblichkeitsforschung – Klecksographie, großes Los und elektrischer Stuhl, in: Rheinische Zeitung (Cologne), 11/8/1931, MPG Archive, Dept. IX,

By 1928, 40 pairs of twins had been examined in this manner.[102] By the close of 1930, detailed information had been collected on 163 identical, 145 same-sex, and 56 different-sex fraternal pairs of twins[103] – thus, as the press proudly proclaimed, the institute had at its disposal "the largest card index in this area and the most copius material in the entire world."[104] And despite all cuts to the material budget, the construction of the twin register proceeded rapidly until 1933.[105] Besides the school authorities and the Statistical Office of the Reich (*Statistische Reichsamt*),[106] the Berlin City Health Office and the directors of various hospitals in Berlin assisted in collecting the data. Verschuer had stamps made with the message: "Is the patient a twin?", and dispatched them to the hospitals. The stamp was to be printed on the medical records so that the search for pairs of twins would not be forgotten in the anamnesis. On reply postcards, also printed and stamped by the KWI-A, the names and addresses of twin patients were sent to the institute, which then endeavored to find the siblings.[107] Maternity hospitals regularly reported twin births to the institute, and usually even sent the placenta to the institute for zygosity testing.[108] However, the KWI-A also addressed the public directly, which was highly unusual at the time, courting understanding for genetic research and asking pairs of twins to make themselves available to the institute for testing.[109] In this the scientists took advantage of the circumstance that the perfect examples of human genetics made available by twin research restored a great deal of the clarity that was being lost as Mendelian transmission genetics advanced. The amazing similarity of identical twins visualized the power of the gene, making it directly apparent even to the broad public.[110]

With the collected data material, the twin method quickly became established as the dominant methodological approach at the KWI-A. Of the institute's 103 publications that were scientific in the stricter sense in the period from its founding

---

Rep. 2, KWI-A. The intelligence questionnaire corresponded to the one used later in the genetic health trials according to the "Law on the Prevention of Genetically Diseased Offspring" (*Gesetz zur Verhütung erbkranken Nachwuchses*) down to the last detail. The "ink blots" referred to here were the Rohrschach tests developed in 1921.

[102] Tätigkeitsbericht 1928, MPG Archive, Dept. I, Rep. 3, No. 4.

[103] Glatzel, Anteil. Cf. also Verschuer, Kaiser-Wilhelm-Institut für Anthropologie, p. 152.

[104] 500 Zwillinge werden geprüft. Ein interessanter Forschungszweig im Kaiser-Wilhelm-Institut in Berlin-Dahlem, in: Berliner Illustrierte Zeitung No. 30, [1930 ?], p. 1333f., MPG Archive, Dept. IX, Rep. 2, KWI-A.

[105] Contemporary publications present conflicting information. Cf., e.g.: 600 Zwillingspaare werden beobachtet, in: Der Bergfried No. 13, 1931, p. 4f., MPG Archive, Dept. IX, Rep. 2, KWI-A. Bergmann/Czarnowski/Ehmann, Menschen, p. 134, state that 700 pairs had already been entered in the card incex by summer 1933. Cf. also Verschuer, Erbe – Umwelt – Führung, "Freier Forscher in Dahlem (1927–1935)" section, pp. 2f., MPG Archive, Dept. III, Rep. 86 A, No. 3–1.

[106] Lassen, Frage, p. 268 (note 1).

[107] Lösch, Rasse, p. 204.

[108] 500 Zwillinge werden geprüft.

[109] In addition to the newspaper articles listed above: Was lernt die Wissenschaft von Zwillingen?, in: JZ [= Jenaer Zeitung ?], Nr. 12, [1930?], p. 270.

[110] Cf. Mai, Humangenetik, p. 65.

until the seizure of power by the National Socialists, not fewer than 34, or almost exactly one third, fell to the share of the twin research area, be it because they applied the twin method to problems in genetic biology, genetic pathology, or genetic psychology; or because they dealt with the methodological problems of twin research.[111] In fact, in the first years of its existence, the KWI-A made a significant contribution to the further development of the twin method.

First, the polysymptomatic similarity diagnosis introduced by Siemens was developed further in Dahlem.[112] Important progress was achieved with the approach of combining similarity diagnoses and blood-group research performed independently, applied for the first time by Verschuer in collaboration with the blood-group researcher Fritz Schiff (1889–1940). Of the 202 pairs of twins determined to be identical on the basis of the morphological similarity diagnosis, not a single pair showed blood-group discordance, while the blood group results for pairs of fraternal twins were the same as for other pairs of siblings – an important confirmation of the assumption that identical twins are actually genetically identical clones.[113]

Second, the relationship between similarity diagnosis and placenta diagnosis was investigated in a large-scale comparative test. Kiffner developed a new method for the test using stereo x-ray images of the twins' placentae. Curtius and Lassen reported about the ongoing comparative tests. Their result gave Verschuer and his staff proof that, in contrast to the conventional doctrine, identical twins who separate at a very early embryonal stage can develop their own amnia.[114]

Third, researchers at the KWI-A occupied themselves exhaustively with the question as to whether the mirror asymmetry is a typical characteristic of identical twins, as American twin researchers like Horatio H. Newman († 1957) assumed.[115] This question was important because it touched on one of the assumptions, upon which the twin method was based, namely the assumption that identical twins are genetically identical clones. Through the systematic investigation of the twins in Berlin, Verschuer concluded that mirror asymmetry was not a regular phenomenon, and that it was thus possible to proceed from the assumption that identical twins are genetically equivalent.[116]

Fourth, at the KWI-A research was performed to pursue the question as to whether a genetic factor plays a role in the genesis of twin pregnancies. Curtius and

[111] These papers made up the majority of the total of 21 scientific publications from the KWI-A between 1927 and 1933 concerning methodological and theoretical questions.

[112] In summary: Verschuer, Ähnlichkeitsdiagnose; Diehl/Verschuer, Zwillingstuberkulose.

[113] Schiff/Verschuer, Serologische Untersuchungen; idem., Serologische Untersuchungen II.

[114] The similarity diagnosis yielded the finding that, of 100 same-sex twins whose placentae had two amnia and two chorions each, only 76 were fraternal and 24 identical. This disproved the previously valid doctrine that identical twins always possessed only one placenta with a single chorion and one aminon, usually double, while fraternal twins had either separate or fused placentae, and always two chorions and amnions. Cf. Kiffner, Stereoröntgenbefunde; Curtius, Nachgeburtsbefunde; Lassen, Nachgeburtsbefunde.

[115] Cf. Newman et al., Twins.

[116] Diehl/Verschuer, Zwillingstuberkulose, p. 84 f.

Verschuer affirmed this question on the basis of genealogical studies on the twins examined in Dahlem, yet right away, Lenz vehemently disputed this view. Verschuer further studied the question as to whether twin births occur more frequently at certain times of year or times of day.[117]

Fifth, studies on the genesis of twins in animals were performed at the KWI-A, including by Wolfgang Abel on the sacred baboon, and by Eugen Fischer on rodents and the badger.[118]

Sixth and finally, Verschuer developed his twin method further by dividing the twins he examined into three groups: (1) same genes–same environment, (2) same genes–different environment, (3) different genes–same environment.[119] If differences were greater between the first and second groups, he believed the environmental influence could be determined quantitatively; if differences were greater between the third and the first group, the same was true for the genetic influence. With this differentiation he came closer to the procedure favored by American twin research, which preferred investigating identical twins who had grown up in different surroundings.

2.2.3 Blood Group Research

Verschuer combined the twin method with other promising methodological approaches. Besides dactyloscopy, the most important of these was blood group research.[120] Around 1900 the Jewish Austrian physician Karl Landsteiner (1868–1943), working at the Institute for Pathological Anatomy at the University of Vienna at the time, discovered the ABO system of human blood groups. In 1910 the Polish physician Ludwik Hirszfeld (1884–1954),[121] an assistant in the serological department of the Institute for Cancer Research in Heidelberg, together with that institute's director, Emil von Dungern (1867–1961), successfully proved the Mendelian transmission of the ABO blood groups. Their mathematically complex mode of transmission was illuminated conclusively in 1924 by Felix Bernstein (1878–1956), a statistician in Göttingen.

At the very dawn of blood group research, a connection to race science was established. In Macedonia during World War I, Ludwik and Hanna Hirszfeld had tested the blood groups of soldiers from 16 nations, concluding that the blood group A was very common in Europe, whereas the blood group B was primarily found in Asia.

---

[117] Curtius/Verschuer, Anlage zur Entstehung von Zwillingen; Lenz, Frage der Ursachen von Zwillingsgeburten; Verschuer, Geburtsstunde.

[118] Abel, Zwillinge bei Mantelpavianen; Fischer, Frage der Zwillingsbildung; idem., Entwicklungsgeschichte des Dachses.

[119] Verschuer, Biologische Grundlagen.

[120] For general information, cf. Mazumdar, Species; Geisenhainer, "Rasse"; Massin, Rasse und Vererbung, pp. 210–214.

[121] After the occupation of Poland, Hirszfeld was deported to the Warsaw ghetto, from which he later managed to escape.

They subsequently formulated an equation, designated the "biochemical race index," which was supposed to be able to calculate the distribution of the blood groups within a group of the population. While this aspect of the correlation between race and blood groups did not achieve terribly much for the Hirszfelds or for blood group research in general,[122] a small group of scientists in Germany and Austria picked up on it eagerly. These researchers congregated around the anthropologist and ethnologist Otto Reche and the Navy physician Paul Steffan (1885–1957), who founded the "German Society for Blood Group Research" (*Deutsche Gesellschaft für Blutgruppenforschung*) in Vienna in 1926. When Reche was appointed to the chair for ethnology at the University of Leipzig in 1927, the headquarters of the society were moved to Leipzig as well. By publishing the *Zeitschrift für Rassenphysiologie* ("Journal for Race Physiology"), Reche attempted – with moderate success overall – to anchor more firmly the connection between race and blood group research in the fields of anthropology, ethnology, and human genetics. What is interesting in this context is that Eugen Fischer, although he joined the German Society for Blood Group Research, remained skeptical. As early as 1926, Reche complained that Fischer were "not really to be convinced [...] because he has the 'instinctive feeling' that blood groups do not have anything to do with 'race'."[123] Thus it comes as no surprise that the topics of race and blood groups were dealt with in only one paper produced at the KWI-A before 1933: In 1929, the external staff member Max Berliner, proceeding from the study by Ludwik and Hanna Hirszfeld, studied the blood of the cattle at the Institute for Animal Breeding at the Veterinary College in Berlin, whereby he reached the conclusion that belonging to a blood group is not a reliable criterion for race determination.[124] Hans Weinert's blood group studies on the evolutionary biology of apes, which were largely unconnected with the remainder of the institute's research agenda, were welcomed vigorously by Reche. Reche, who had heard that Landsteiner, the sage of blood group research, was studying apes, urged Weinert to hasten his studies so that he could publish before Landsteiner – with the clear intention of rolling back "the Jewish influence in this area in which primarily Jews had worked so far."[125] Verschuer, by contrast, although he had been interested in serological race diagnostics back in the Weimar Republic, had meanwhile distanced himself from this field of research, restricting himself to the above-mentioned combination of blood group

[122] Cf. Mazumdar, Blood and Soil, p. 188f.; Weindling, Health, p. 464f.

[123] Otto Reche in a letter to Oswald Streng, 23/11/1926, quoted in Geisenhainer, "Rasse", p. 132. In 1933 Staffan complained to Reche: "Lenz and Fischer, too, are simply disinterested because *others* had worked on the subject so far. As regards Fischer, I cannnot even say 'without asking;' for in 1923 I visited him in Freiburg for this reason (or was it back in '22) and there he calmly pronounced the great judgement: I don't think much of blood groups. I don't think that he ever concerned himself with it at all. And now his pride probably won't allow the subject to find respect after all. Lenz probably feels the same way." Steffan to Reche, 4/11/1933, quoted in: ibid., p. 172f. (original emphasis).

[124] Berliner, Blutgruppenzugehörigkeit.

[125] Hesch, Otto Reche, p. 12.

research and polysymptomatic similarity diagnosis in the context of twin research. Worth noting here was his collaboration with Fritz Schiff, director of the bacteriological department at the City Hospital of Friedrichshain in Berlin and a lecturer on hygiene and bacteriology at the University of Berlin. As the Jewish physician, one of the most important German blood group researchers, was in open conflict with the German Society for Blood Group Research,[126] and by working with Schiff (and also Berliner), Verschuer took a clear stand against the group around Reche and Steffan. Blood group research achieved practical importance in connection with paternity tests, which were requisitioned from the KWI-A by the Berlin courts starting in 1928.[127]

2.2.4 *Fields of Research*

When the scientific publications that originated during the period from 1927 until 1933 are ordered according to fields of research, the first thing that becomes apparent is that the work of each individual department was not carefully separated from that of the others. The division of the institute into three departments, as Fischer made clear from the outset, was not supposed to signify "any kind of mutual closure or [set] any strict limits."[128] Considerable overlap was found in the fields of research between the departments of anthropology and of human genetics. Of the anthropological publications, only the 11 works dealing with the subjects of paleoanthropology or geographical anthropology had no connection or only a very loose internal one with the research on the field of human genetics. In contrast, 11 others, among them the "bastard studies," were clearly related to the genetics of normal attributes. And in the inverse, the works from the area of human genetics – obviously, the projects that dealt with "race characteristics," but also other projects located entirely in the area of genetic pathology – could be related at least indirectly to the anthropological research at the institute. The research in Verschuer's department oriented toward genetic pathology, for its part, extended into the area of eugenics, although there was no appreciable overlap with the research in Muckermann's department, whose publications stood out in clear relief from the other research at the institute in terms of conception and methodology, and which produced only very few studies that could satisfy the demands of science to any degree. In the following, the first step will be to outline the large complex of network research projects so central to work at the institute, from the areas of anthropology, human

[126] Geisenhainer, "Rasse", pp. 133–136.

[127] Verschuer, Kaiser-Wilhelm-Institut für Anthropologie, p. 144.

[128] Fischer, Kaiser-Wilhelm-Institut für Anthropologie, p. 147 f. As Verschuer, too, established in retrospect, there was no need to "set limits," the "intellectual unity of the institute" was expressed by the fact "that the wide variety of research problems of the entire area could be worked on in every department." Verschuer, Kaiser-Wilhelm-Institut für Anthropologie, p. 128 f.

genetics, and genetic pathology. The second step will be to take a look at the research more at the margins of the institute, in the areas of anthropology, paleoanthropology, and eugenics.

2.2.5 The Genetics of Normal Attributes

The institute's main field of work, with a total of 37 publications,[129] was constituted by the genetics of normal anatomical, morphological, and physiological attributes and characteristics of humans. Fischer's most pressing interest in this work was in *varieties*, the small, not abnormal deviations from the normal type. He even founded a "Commission for the Statistics of Varieties," which was to gather information about the anomalies established in the autopsies of the deceased, but so little material was submitted that the project soon petered out.[130]

In classical anthropology, skull and skeleton were considered to be the most important constant attributes of humans and human races; however, Franz Boas' studies of Eastern European and Southern European immigrants and their children in New York had brought down the elaborate structure of classifying human races on the basis of skull measurements. As mentioned above, Fischer had acknowledged Boas' study approvingly, and he thus accorded a high priority to the question as to the influence of the environment on bony structures in the research practice of his own department of the institute. Especially since he rejected the conventional practice of measuring skulls, Fischer's department concerned itself intensively with craniometry, whereby initially the phenomenon of southern German brachycephaly (short-headedness) was at the focus of interest. On the basis of the craniometric study of skulls from Medieval and Early Modern graves, anthropological research had formulated the conception that the population of southern Germany from the early medieval period until the end of the first millennium AD had been primarily mesocephalic or dolichocephalic, while the brachycephalic type had dominated in this region since. Studies by Eduardo Fleury Cuello,[131] Wilhelm Emil Mühlmann (1904–1988),[132] and Eugen Fischer[133] on skulls from Baden appeared to confirm

[129] Including the eleven anthropological publications mentioned above related to the genetics of normal attributes.

[130] Lösch, Rasse, p. 193. Cf. Fischer, Varietätenforschung.

[131] Eduardo Fleury Cuello, presumably a Venezuelean, was a guest scholar at the KWI-A in 1931. He examined twenty skulls from row graves near Oberrotweil/Baden, which were dated to the period between 600 and 800 A.D., and observed a predominance of dolichocephaly. Fleury Cuello, Untersuchungen.

[132] The anthropologist and ethnologist Wilhelm Emil Mühlmann was a doctoral student on the scientific staff of the KWI-A in 1932 (for a biography: Klee, Personenlexikon, p. 419). He described 71 skulls from different regions of Baden from the period between the sixteenth and eighteenth centuries, establishing a predominance of brachycephaly. Mühlmann, Untersuchungen.

[133] Fischer, Untersuchungen. Zur Vorgeschichte: Gessler, Eugen Fischer, pp. 91–93.

this view. Fischer interpreted this "rounding" as an environmentally conditioned change in the shape of the skull. Fischer – in connection with "I.G. Farbindustrie in Elberfeld"[134] – performed "experiments with vitamins and poisons on rats to study […] how the shape of the skull could be influenced":[135] "A third of the laboratory animals receives feed low in vitamins, another third receives feed rich in vitamins, and a third a normal, mixed diet." Then the animals were x-rayed, killed, and dissected. "With this it was possible to approach the mystery of short and long skulls, for the animals with a vitamin-poor diet always had shorter skulls, while long skulls can be traced back to an abundance in vitamins."[136]

Fischer was also interested in varieties of skull shapes. The doctoral student Irawati Karvé (* 1905)[137] used the skull collection available at the institute to develop a methodology for measuring the asymmetries of the human skull and applied it to three series of 50 skulls each – "Negro, South Seas, mixed group." She obtained consistently negative findings: there was not a single skull without asymmetries, the asymmetries of the individual points showed no correlations with each other, they did not concentrate on one of the two sides of the skull, and no "race differences" could be detected. In contrast, the Korean doctoral student Baeckpyeng Kim, based on the examination of the embryonal skulls of pigs, came to the conclusion that clear race differences could be detected between wild boars, long-snouted and short-snouted domestic pigs from the very first development of the cartilaginous skull at an early embryonal stage.[138]

A further important subject of investigation with regard to the ability of the environment to influence bony structures were the varieties of spines – from enlarged or diminished extensions and arches of the vertebrae to missing or additional vertebral bodies or ribs – and the heritability of such anomalies. Fischer himself began with extensive breeding experiments on toads and tooth-carps, in order to influence the number of vertebrae as a function of environmental factors like oxygen supply, light, heat, and food supply. He had to abandon these experiments without success, however, along with his attempts to cross large and dwarf races of chickens to influence the number of vertebrae.[139] While these research projects of his own ran

[134] What was probably meant here was the Bayer Institute for Chemotherapy in Elberfeld founded in 1910, which belonged to the I.G. Farben concern. Cf. Plumpe, I.G. Farbenindustrie, p. 474. It also supported the projects by Lore Schröder and Heinz Boeters by supplying chemical substances – more on this below.

[135] Stichwortartiger Tätigkeitsbericht Fischers, 8/6/1928, MPG Archive, Dept. I, Rep. 1 A, No. 2404, p. 2f., here: p. 2 v.

[136] Der Zukunftsmensch unter der Lupe der Forschung. Wie die deutsche Wissenschaft arbeitet (around 1929, source unknown), MPG Archive, Dept. IX, Rep. 2, KWI-A (with photographs of the measurements of rat skulls).

[137] Irawati Karvé, born in Burma, was a citizen of Lübeck. From 1926 until 1928 she had studied in Bombay and graduated with a Master of Arts; from 1928 until 1930 she studied in Berlin. She was listed as a doctoral student at the KWI-A from 1929/30 on. She went on to become a professor of anthropology in Pune, India (for a biography: Lösch, Rasse, p. 568). Karvé, Asymmetrie.

[138] Baeckyeng Kim was listed as a doctoral student at the KWI-A from 1930/31 on. Kim, Rassenunterschiede.

[139] Lösch, Rasse, p. 192.

aground – since 1927 Fischer had been so overextended by his functions as university professor, institute director, science manager, and political consultant that he hardly worked on scientific projects of his own any more – as a mentor and organizer he showed considerable skill, for instance by bringing Konrad Kühne,[140] whom he knew from back in his days in Freiburg, to his institute as a scientific staff member. Kühne had taken his doctorate at the University of Dorpat in Estonia with a dissertation about vertebral varieties in lizards and had then researched on rabbits at a lung sanatorium near Freiburg, where he came to Fischer's attention. In Dahlem, Kühne initially occupied himself with experiments on the heritability of the number of vertebrae and ribs, with over 1,200 rats as subjects.[141] He achieved his breakthrough with another research project about the varieties of the human spinal column, however. The large-scale study took over 4 years – from 1928 to 1932 – was financed by the Emergency Committee of German Science, and still would not have been possible without the generous support of a private company that provided 10,000 x-ray films free of charge. Kühne evaluated x-rays of the spinal columns of a total of 23 families, allowing him to follow the occurrence of spinal column varieties across several generations. He was not able to prove the heritability of specific anomalies, yet his studies appeared to furnish evidence for the fact that, as far as the localization of varieties was concerned – that is, whether spinal column anomalies occurred in the neck, breast, or lumbar region – a "tendency" was hereditary. Kühne interpreted his findings by presuming the existence of a dominant gene for the *cranial* (toward the head) and a recessive gene for the *caudal* (toward the feet) tendency.[142] In the Weimar period Kühne continued his studies using the twin method, through which, as Fischer emphasized in his annual report of 1932/33, "for the very first time what are known as the penetrance differences of normal hereditary attributes were established beyond a doubt in humans," which, in turn, was "of importance for the interpretation of as yet unexplained phenomena of certain nervous and mental illnesses."[143] In 1933, Fischer placed Kühne's research in a comprehensive, phylogenic context by comparing the variability of the human spinal column with that of the orangutan, the gorilla, the chimpanzee, the gibbon, and other monkeys.[144]

The variability of morphological attributes caused by heredity and environment was also researched at the KWI-A. In his lecture at the 5th International Congress for Genetics in the year 1927, Verschuer had already used the development

---

[140] For a biography: ibid., p. 568.

[141] Stichwortartiger Tätigkeitsbericht Fischers, 8/6/1928, MPG Archive, Dept. I, Rep. 1 A, No. 2404, p. 2f., here: p. 2 v; Cf. also the newspaper article "Der Zukunftsmensch unter der Lupe der Forschung. Wie die deutsche Wissenschaft arbeitet" (around 1929, source unknown), MPG Archive, Dept. IX, Rep. 2, KWI-A (with photographs showing Kühne at work on his project).

[142] Kühne, Vererbung.

[143] Jahresbericht 1932/33, MPG Archive, Dept. I, Rep. 3, No. 9.

[144] Fischer, Genetik. Upon Fischer's suggestion Kühne was later awarded the silver Leibniz Medal of the Prussian Academy of Sciences. Cf. Fischer, Tätigkeitsbericht 1939/40, MPG Archive, Dept. I, Rep. 3, No. 17; Lösch, Rasse, p. 194.

curves showing the body weight and height of identical and fraternal twins to study the question as to the "variability of the human body."[145] In 1928, Berliner investigated the correlations between body height and width.[146] The doctoral student Harry Conitzer (* 1905) occupied himself with the heritability of red hair;[147] Kranz reconstructed the family tree of a family in which spirally curled hair had been passed down through five generations.[148] Max Fischer, a private scholar dabbling in genetics, published on the "Shaping of the human nose during puberty."[149] From 1932, the Norwegian guest scholar Thordar Quelprud studied the shape of the ear using the methods of twin and family research – his publications constituted the foundation for the fact that the shape of the ear became one of the criteria for the genetic biological paternity test.[150] In 1931 Irawati Karvé published a paper about the eye color of 584 Brahmins from Bombay.[151] In only one case was the shape of an internal organ the subject of a study: in 1929 Verschuer – in collaboration with V. Zipperlein of the Medical University Clinic in Tübingen – published an essay about the "genetically and environmentally conditioned variability in the shape of the heart," which was based on the x-rays of the hearts of 28 identical and 27 same-sex fraternal pairs of twins.[152]

The heritability of morphological attributes – under the aspect of human races and their "crossing" – was the concern of the "bastard studies" as well, which Fischer continued in Dahlem. In 1931, the sculptor Hans Lichtenecker traveled to Southwest Africa on behalf of the institute, to perform a follow-up examination of the "bastards of Rehoboth" in accordance with Fischer's precise instructions. Lichtenecker supplemented the family trees compiled by Fischer, photographed numerous of the Rehoboths examined by Fischer and their children, took handprints of many, determined hair and eye color and, in a few cases, even made plaster casts of their faces. This material was processed by Fischer in Dahlem – however, the first information about his studies did not

[145] Verschuer, Variabilität.

[146] Berliner, Hochwuchs.

[147] Conitzer, Rothaarigkeit. Conitzer, himself a Jew, also considered the propagation of red hair among Jews in his work – a typical set piece of the classical anti-Semitic stereotype. In this question Conitzer's conclusions were extremely cautious, due to uncertain statistical foundations. The only assumption he held to be justified was that the attribute of red hair was rarer among the Sephardic Jews than among the Ashkenazi (cf. ibid., pp. 105–107, 125 f., 129).

[148] Kranz, Vererbung.

[149] Max Fischer, Formung.

[150] Quelprud, Untersuchungen; idem., Über Zwillingsohren; idem., Zwillingsohren. Photographs from Quelprud's paper are included in the article "Das Rassenbild im Stammbaum. Aus der Arbeit des Kaiser-Wilhelm-Instituts für Anthropologie," in: Illustrierter Beobachter, Folge 31, 1933, p. 988 f. (MPG Archive, Dept. IX, Rep. 2, KWI-A).

[151] Karvé, Beobachtungen.

[152] Verschuer/Zipperlein, Variabilität. The studies upon which this publication was based had been performed at the instigation of Weitz in winter 1926/27 at the Medical University Polyclinic in Tübingen.

appear until 1938 and 1942.[153] In 1930 Fischer published a family study on the "European-Polynesian crossbreed."[154] In the early 1930s, one of his pupils, the Chinese doctoral student Yun-kuei Tao, performed a large-scale anthropometric study about children from marriages between Chinese men and European women, using twenty cases from Berlin and 50 cases from Paris. Tao painstakingly analyzed skin, hair and eye color, shape of hair, and the occurrence of the "Mongolian spot"[155] and "Mongolian fold" among the crossbred children.[156] Wolfgang Abel published a paper about the teeth and jaws of "bushmen," "Hottentots," "Negros," and their "bastards" in 1933.[157]

A first evaluation of the patterns of the fingerprints and handprints taken regularly in the twin examinations was performed in 1932 by the Japanese guest scholar Sei Hara, associate professor of anatomy at the Medical College in Nagasaki,[158] on the basis of 45 pairs of identical twins and 48 pairs of fraternal twins of the same sex.[159] Based on the methodology further developed by Georg Geipel, Fischer took stock of the use of dactyloscopy for twin research for the first time in 1933, with regard to both the heritability diagnosis and issues of asymmetry and environmentally conditioned variability.[160] In two publications of 1933, Abel dealt with the race anthropology aspects of fingerprint and handprint patterns. This work was based primarily on fingerprints and handprints that had been taken from the Inuits of eastern Greenland on the arctic expedition led by Alfred Wegener (1880–1930) in 1929/30. Abel established a striking similarity between the fingerprint and handprint patterns of the Inuits and the Japanese Ainu, contrasted by clear differences in comparison with the prints of Chinese and North American natives.[161] The prenatal

[153] Fischer, Neue Rehobother Bastardstudien; idem., Neue Rehobother Bastardstudien II. For the acquisition of the "casts of living Hottentot and Bushman heads for the continuation of the anthropological skull collection," the Prussian Ministry of Science, Art and Education approved 540 RM. Cf. Preußischer Minister für Wissenschaft, Kunst und Volksbildung to Präsident der KWG, 8/3/1932, MPG Archive, Dept. I, Rep. 1 A, No. 2413, p. 4.

[154] Fischer, Europäer-Polynesier-Kreuzung.

[155] Also called the sacral spot: skin discoloration in the area of the sacrum.

[156] Tao, Chinesen-Europäer-Mischlinge.

[157] Abel, Zähne.

[158] Sei Hara (actually: Tadashi Hara) graduated from the Medical Academy in Osaka in 1905 and then worked as a scientific assistant at various medical academies, finally – from 1915 to 1917 – at the Japanese Medical Academy in Tokyo. Afterward he switched to the Medical College of Nagasaki as an associate professor. In the years from 1920 to 1925 he published a number of papers about the skulls and brains of criminals who had been sentenced to death. After returning to Japan, he was again employed by the Medical College of Nagasaki. He died in 1937. Friendly information from Takashi Nakagawa.

[159] Hara, Untersuchung.

[160] Verschuer, Erbbiologie der Fingerleisten.

[161] Abel, Frage; idem., Finger- und Handlinienmuster. A photograph of Abel studying handprints is included in the article "Das Rassenbild im Stammbaum. Aus der Arbeit des Kaiser-Wilhelm-Instituts für Anthropologie", in: Illustrierter Beobachter, Folge 31, 1933, p. 988 f. (MPG Archive, Dept. IX, Rep. 2, KWI-A).

origination and development of the lines of the hand were pursued by the doctoral student Johannes Schaeuble (1904–1968) in the year 1933, in a macabre examination which involved dissecting the hands of 33 embryos at different stages of development in series of sections.[162]

Physiological phenomena also met with obvious interest at the KWI-A – above all hemograms, as mentioned above. Glatzel published on the "Share of Genetic Factors and Environment on the Variability of the Normal Hemogram." Here, he compared findings from 44 pairs of identical twins and 48 pairs of fraternal twins. Further, in 12 pairs each of identical and fraternal twins, Glatzel investigated the importance of genetic and environmental factors for the normal functioning of the stomach by means of irrigating the stomach after the ingestion of alcohol.[163] On the basis of family studies Fritz Schrijver, a guest scholar from the Netherlands, attempted to prove that the sensations of taste were inherited.[164] The dissertation by Heinz Boeters (* 1907), a doctoral student at the KWI-A from 1929 to 1930, dealt with the influence of the anterior pituitary hormone prolan on the development of gonads in rats.[165] The Yugoslavian guest scholar B. Skerlj reported in a lecture to the Berlin Anthropological Society in 1931 about the relationship between menarche and climate in Europe.[166] In 1929, Max Berliner used a method he had developed himself to examine the hardness of the soft tissues on various parts of the body (shoulder, back, trigonum, abdomen, hollow of the knee). One year later Berliner published the results of his studies of the puncture fluids of 12 patients suffering from cancer and eight control subjects.[167]

2.2.6 Genetic Pathology and Tuberculosis Research

The boundaries to genetic pathology become fluid at this point. In the institute's early years there was only sporadic, rather unsystematic research in this field – a total of just 14 papers from this area were published.[168] Curtius, for instance, published a family study about "diffuse sclerosis and familial spastic spinal paralysis" in 1932; Max Fischer used the example of the English Queen Victoria's family to deal with the heredity of hemophilia.[169] While these studies still proceeded in a genealogical fashion, the combination of genetic pathology and twin research

---

[162] Schaeuble, Entstehung.

[163] Glatzel, Anteil; idem., Erbanlage.

[164] Schrijver, Erforschung.

[165] Boeters, Hypophysen-Vorderlappenhormon. The chemical substances required were provided by the "I.G. Farbindustrie Elberfeld."

[166] Skerlj, Menarche.

[167] Berliner, Untersuchungen über die Weichteilhärte; idem., Untersuchungen über optisch wahrnehmbare Phänomene.

[168] Including the two projects mentioned below in the border area between genetic pahology and eugenics.

[169] Curtius, Sklerose; Max Fischer, Hämophilie. Cf. also Berliner, Besonderheiten des Kleinkindesalters.

opened up new horizons. At the Genetics Congress in Munich in 1931 Kranz introduced the results of data he had collected on tumors in twins.[170] In 1931 Glatzel presented an initial assessment of the twin pathological findings.[171]

The most important research project in this area by far was performed by Karl Diehl and Otmar von Verschuer, on twins who had contracted tuberculosis.[172] Diehl had enlisted at the beginning of World War I as a high-school student and was sent to the front; in October 1914 he was seriously wounded near Ypers, Flanders, and ended up an English prisoner of war. Released in a prisoner exchange, he started studying medicine in Marburg, where, as mentioned above, he became friends with Verschuer. As a young assistant he contracted the pulmonary tuberculosis that was to bring him to tuberculosis research. After a course of treatment in Davos he became an assistant at a lung clinic there. After consulting with his friend Verschuer, in 1927 he moved to Berlin as the directing physician of the surgical department of the tuberculosis hospital of the city of Berlin, "Waldhaus Charlottenburg," located in Sommerfeld near Beetz in the eastern Havel region.[173]

Based on genealogical and statistical material, other researchers had already pointed out various indications suggesting the influence of a hereditary component on the emergence, type and the course of tuberculosis. Diehl and Verschuer had undertaken to produce irrefutable proof for the genetic disposition for tuberculosis using the twin method. By pointing to the practical importance of tuberculosis research, Fischer raised considerable additional funding from the Prussian Ministry for Welfare and the Reich Ministry of the Interior. At the Annual Convention of the German Tuberculosis Society on the island of Norderney in 1930, Diehl and Verschuer reported about the results of their investigation for the first time publicly. On the basis of 75 pairs of twins (19 identical, 56 fraternal), they established higher concordance among the identical pairs of twins than the fraternal ones: The risk of *both* siblings contracting tuberculosis, was, statistically seen, greater among pairs of identical twins than for fraternal twins, and, when both siblings were affected, the probability that the illness would take a similar course was, again – statistically speaking – higher for identical twins than for fraternal twins. Thus, Diehl and Verschuer asserted, "*unequivocal* proof has been produced that the hereditary disposition is *of substantial importance* for the course of tuberculous events."[174]

This thesis caused a furor. In the debate erupting in the late Weimar Republic about whether and to what extent tuberculosis should be regarded as an infectious or as a hereditary disease, Diehl and Verschuer took the extreme position for a far-reaching genetic disposition. For this, as Fischer wrote in an interim report to

[170] Kranz, Tumoren.

[171] Glatzel, Beiträge zur Zwillingspathologie.

[172] Diehl/Verschuer, Erbuntersuchungen; idem., Zwillingstuberkulose; Verschuer, Erbuntersuchungen an tuberkulösen Zwillingen.

[173] Verschuer, Erbe – Umwelt – Führung, "Assistentenjahre in Tübingen (1923–1927)" section, p. 4, "Freier Forscher in Dahlem (1927–1935)" section, pp. 8–10, MPG Archive, Dept. III, Rep. 86 A, No. 3–1.

[174] Diehl/Verschuer, Erbuntersuchungen, p. 214 (original emphases).

the Prussian Ministry of Welfare at the close of 1931, they reaped "enthusiastic recognition, but also sharp opposition."[175] The controversy was fought out in such harsh tones not least of all because it was closely linked with the question as to which path tuberculosis control should take – consistent prevention, serial examinations, and expansion of clinical treatment, or rather a eugenic program to eliminate the hereditary disposition for tuberculosis in coming generations. Spokesman of the opposition was the tuberculosis physician Franz Redeker (1891–1962).[176] He found fault with the period of investigation, which was too short, and found that too few pairs of twins were examined. He also called into question the interpretation of the data and granted the hereditary factor at best a subordinate role in the course of the disease. Above all, however, he insisted that tuberculosis was an infectious disease, railing against the "pipe dream of eugenic tuberculosis control."[177] In their voluminous monography *Zwillingstuberkulose. Zwillingsforschung und erbliche Tuberkulosedisposition* ("Twin Tuberculosis. Twin Research and Hereditary Disposition for Tuberculosis") published in 1933, Diehl and Verschuer had paid tribute to the methodological critique by performing large-scale follow-up examinations and enlarging their sample to 127 pairs of twins. In the interpretation of the data, however, they held fast to their view that tuberculosis was somewhere between the infectious and the hereditary diseases, but closer to the hereditary ones. They dismissed the reproach that they intended to reduce tuberculosis control to the eugenic aspect, but reaffirmed their standpoint that tuberculosis should be addressed in eugenic marriage counseling, eugenic aspects taken into consideration in tuberculosis relief, and voluntary sterilizations of tuberculosis sufferers permitted in accordance with the bill submitted by the Prussian State Health Council in July 1932.[178] The sterilization of tuberculosis sufferers was a common demand at this time, extremists called for obligatory sterilization, even of cured "bacteria excreters."[179] Diehl and Verschuer held back with any eugenic demands. Hermann Muckermann, on the other hand, in his public lectures around 1930, is said to have denounced tuberculosis relief as misguided and wasteful, calling for "the elimination of the embryonic mass predisposed for tuberculosis."[180] In any case, Fischer rushed to send the work by Diehl and Verschuer to the Prussian Minister of the Interior on January 18, 1933, with the request to continue his financial support for the tuberculosis project, which was "of vast importance […] for the health of the nation and the control of tuberculosis."[181]

[175] MPG Archive, Dept. I, Rep. 1 A, No. 2412, pp. 165–165 v, quote: p. 165 v. On the following: Kelting, Tuberkuloseproblem, pp. 15–33.

[176] For a biography: Labisch/Tennstedt, Weg, vol. 2, pp. 474–476.

[177] Redeker, Tuberkulosevererbung.

[178] Diehl/Verschuer, Zwillingstuberkulose, pp. 462, 478; idem., Tuberkulose und Eugenik.

[179] Münter, Lungentuberkulose, p. 353.

[180] Braeuning/Redeker, Lungentuberkulose, p. 120.

[181] Fischer to preußischer Innenminister, 18/1/1933, MPG Archive, Dept. I, Rep. 1 A, No. 2413, pp. 8–8 v.

2.2.7 Research on Genetic Damage

Located between the genetics of normal attributes, genetic pathology and eugenics were two research projects about genetic damage through environmental influences. That alcohol had the effect of damaging the genes was an important topic of eugenic literature from the outset – the reason for this was that race hygiene in the German-speaking world was originally closely linked with the abstinence movement – and Muckermann and Max Fischer, too, repeatedly expressed their views on the problem of alcoholism.[182] Yet the projects investigating genetic damage emerged not only in the Department for Eugenics. Well beyond their eugenic implications, they were also closely connected with the problem of hereditary disposition at the focus of research in the Departments of Anthropology and Human Genetics. In her dissertation, Lore Schröder, a doctoral student of Fischer's in the years 1931/32, discussed the origination of genetic damage in humans through different poisons, performing her own series of tests on rats with a combination of lead and prolan.[183] Lothar Loeffler began with a project on the experimental generation of hereditary damage in mice through arsenic substances. Parallel to this subject, he began – like Konrad Kühne – a project on humans, in this case a "poll about the genetic damage caused by X-rays in radiologists and radiology laboratory assistants."[184] This research plan emerged in the context of the controversial discussion about the temporary sterilization of women by irradiating their ovaries, a frequent practice at the time, which Fischer repeatedly acted against as a physician, geneticist, and eugenicist.[185] Aside from the genetic pathology–eugenics aspect, the question of radiation damage to the genotype was of principle interest in view of the fundamental questions guiding research: the extent to which the phenotype of an organism is subject to imprinting by heredity and environment. The American geneticist Hermann Joseph Muller (1890–1967), at the 5th International Congress for Genetics in the year 1927, so to speak at the hour of the KWI-A's birth, had introduced his pioneering experiments on the artificial generation of mutations through x-rays in the *Drosophila* fruit fly. Muller's method was picked up immediately in Germany, quite close to Dahlem, by Nikolaj V. and Elena A. Timoféeff-Ressovsky (1900–1981, 1898–1973), a married couple researching at the KWI for Brain Research in Berlin-Buch. At the KWI-A there was no interest in the genetic analysis of mutation processes, but the importance of radiation as an environmental factor affecting phenogesis was a subject worth researching. Of course, when the institute presented its work to the outside world, it emphasized the practical relevance for radiation medicine of Loeffler's study, upon which his reputation as a radiation

---

[182] Max Fischer, Alkoholmißbrauch; Muckermann, Wirkungen des Alkoholgenusses.

[183] Schröder, Frage. The chemical substances required were provided by I.G. Farbindustrie Elberfeld.

[184] Stichwortartiger Tätigkeitsbericht Fischers, 8/6/1928, MPG-Archives, Dept. I, Rep. 1 A, No. 2404, p. 2 f., here: p. 2 v. Loeffler, Röntgenschädigungen.

[185] Fischer, Strahlenbehandlung; idem., Erbschädigung. On the discussion about temporary x-ray sterilization, cf. von Schwerin, Experimentalisierung, pp. 119–122; Proctor, Blitzkrieg, pp. 103 f.

biologist was based, and which was to lead him to the Working Group for Radiation Biology of the German Atomic Commission (*Arbeitskreis Strahlenbiologie der Deutschen Atomkommission*) in 1957.[186]

2.2.8 Genetic Psychology

From the very beginning, twin research at the KWI-A was also directed toward genetic psychology. Seven papers from this area were published before 1933. As early as 1930 Verschuer presented the results of intelligence and Rohrschach tests on 100 pairs of twins. In the same year he wrote the chapter about "Intellectual Development and Heredity" for the manual *Vererbung und Erziehung* ("Heredity and Education") published by Günther Just (1892–1950). A case study of the following year describes a pair of identical twins with an outstanding talent for chess.[187] Ida Frischeisen-Köhler (1887–1958) presented a paper about the grade cards of twins in 1930, before starting her large-scale research project to study what she called "personal tempo." For this she performed experiments on reaction speed on over 1,000 subjects, among them 85 families with 318 children, and on 118 pairs of twins. These experiments were supplemented by observations of everyday activities like walking, writing, and speaking. In the evaluation of her findings, Frischeisen-Köhler believed she had found indications of the "dependence on hereditary" of the personal tempo – supported especially by the high concordance between identical twins. To study a comparable question, she then undertook investigations on the "sensitivity to differences in speed."[188] Lassen distributed a questionnaire to twins to survey "social and moral dispositions of character."[189] In 1933 Kranz presented the first results of his studies on a nonselected series of criminal twins, which he had performed on prisoners of penal institutions in Berlin since 1932 with the permission of the Prussian Ministry of Justice and support from the Emergency Committee of German Science.[190] With this project, which had been begun originally by Friedrich von Rohden of the Nietleben State Sanatorium in Halle/Saale and then transferred to Fischer,[191] the KWI-A was already clearly

[186] Klee, Personenlexikon, p. 376.

[187] Verschuer, Erbpsychologische Untersuchungen; idem., Intellektuelle Entwicklung; idem., Erbgleiches Zwillingspaar.

[188] Frischeisen-Köhler, Untersuchungen; idem., Tempo; idem., Empfindlichkeit.

[189] Lassen, Frage.

[190] Kranz, Frage der Konkordanz; idem., Kriminalität bei Zwillingen.

[191] Rohden had been granted a credit of 4,000 RM for this project by the Emergency Association of German Science on December 6, 1930. Due to a transfer he had to abort the project in 1931. Upon Verschuer's intercession he made his preliminary notes available to the KWI-A, to which the funds approved by the Emergency Association were then transferred. Cf. Fischer to Notgemeinschaft, 30/4/1931; Notgemeinschaft to Fachausschuß für Gemeinschaftsarbeiten: Rassenforschung, 12/5/1931; Notgemeinschaft to Fischer, 12/5/1931, BAK, R 73/11.004.

"poaching" in the territory of Rüdin and his institute in Munich. When Rüdin applied in May 1932 to the Emergency Association for 7,000 RM for research by his staff member Friedrich Stumpfl (1902–1994)[192] on criminal twins, Fischer consented in an expert opinion, but only on the condition that the funds for Kranz be increased by over 2,000 RM so that his work could appear before Stumpfl's.[193] Stumpfl's twin study about "The Origins of Crime" (*Die Ursprünge des Verbrechens*) did not appear until 1936, but then it did raise a furor.[194]

While it was clear that the papers on the genetics of normal attributes, on genetic pathology and on genetic psychology, as well as a large share of the race anthropology works were related to each other, held together by the conception of genetic analysis as a whole,[195] other papers in the areas of eugenics, paleo-anthrolopolgy and geographic anthropology remained largely unconnected to this complex of issues.

2.2.9 Eugenics Research

Brisk activity emerged at the KWI-A in the field of eugenics in the final phase of the Weimar Republic – above all, by Muckermann, Fischer, and Verschuer, but this activity was restricted for the most part to public lectures, popular science writings, and science policy consulting. Genuine eugenics *research* – aside from the few studies already mentioned about neurodegenerative diseases and genetic damage through environmental influences, which also had eugenic implications, occurred at the KWI-A only to a minor extent. A generous estimate makes out a total of 13 publications from this area that can be imputed to the institute's scientific production. Evaluating the morbidity and mortality statistics in terms of population genetics was left up to Rüdin's institute in Munich.[196] Apart from the confidential, and thus nonpublished "Studies of the Family Statistics of the Officers of the Army and Navy of the Reich under Special Consideration of the Problem of Differences in Reproduction" by Günther Brandt (1898–1973),[197] only Hermann Muckermann's research project on the different rates of reproduction among different social strata can be understood as an original contribution to eugenics research. It was quite meager methodologically: Muckermann calculated the average number of children born to 3,947 families of German university professors, established that since the

[192] For a biography: Weber, Ernst Rüdin, pp. 173, 248 f.; Schmuhl (ed.), Rassenforschung, p. 340; Klee, Personenlexikon, p. 612.

[193] Expert opinion by Fischer of 24/5/1932; Notegemeinschaft to Fischer, 24/5/1932, BAK, R 73/11.004.

[194] For a detailed discussion: Wetzell, Kriminalbiologische Forschung.

[195] Fischer, Genanalyse.

[196] In summary: Verschuer, Umfang der erblichen Belastung.

[197] "Familienstatistischen Untersuchungen an den Offizieren des Reichsheeres und der Reichsmarine unter besonderer Berücksichtigung des Problemes der unterschiedlichen Fortpflanzung"; cf. Lösch, Rasse, p. 538.

turn of the century this number had been too low to secure self-recruiting, and contrasted this finding with the values for a village in Westphalia.[198] However, the publications do not reflect the entire scope of the statistics collected. In 1933 Fischer announced that studies had been concluded about the "families of around 5,000 university professors, 40,000 [sic!] civil servants and policemen and 4,800 officers of the Reich Army." Beyond this, in Westphalia "a town and its corresponding rural population [had been] included," further, "in the Rhineland a population of 1,200 families of workers living in a closed residential block." Finally, "460 families with feeble-minded children and pupils of special schools for the retarded" in Berlin were examined.[199] In January 1932, Fischer had reported to the Emergency Association of German Science that he was the dissertation advisor for a thesis about the workforce of the Gerrit van Delden & Co. factory in Gronau, Westphalia. The doctoral student Schulz, he reported, used questionnaires to collect statistics on 1,395 workers and their families, a total of around 8,000 subjects, collecting not only anthropological information, but also data "about immigration to and emigration from the area, about the particulars of their homes, economic circumstances, female labor, extra-professional activities, and further children's nutrition, education and activities."[200] Although Fischer reported that the dissertation was about to be completed, it apparently never appeared. Against the background of the economic, societal, and political crisis at the end of the Weimar Republic, many a statistical study is likely to have succumbed to a similar fate.

Although the projects kicked off by Muckermann on differential rates of reproduction could hardly fulfill scientific standards, they were extolled excessively for their practical relevance in the accounts of the institute's research. In fact, the numbers calculated by Muckermann, as will be demonstrated later, did play a certain role during consultations in 1932 about a bill to permit voluntary sterilization on the basis of eugenic indications. Regarded from the focus of the research at the institute, they were marginal. Fischer and his staff were more interested in laying the scientific foundations of human genetics than in the practical implementation of the eugenic program on the basis of social statistics.

2.2.10 Paleoanthropology, Geographic Anthropology, and the "German Race Science" Project

The situation was similar in the area of paleoanthropology, for instance, as regarded Fischer's research on the "Guanche problem." On his return trip from studying the "bastards of Rehoboth" in German Southwest Africa in 1908, Fischer stopped in Las Palmas on Tenerife to visit the skull collection of the local museum. He had

---

[198] Muckermann, Differenzierte Fortpflanzung; idem., Neue Forschungen.

[199] Fischer, Tätigkeitsbericht [July] 1933, MPG Archive, Dept. I, Rep. 1 A, No. 2404, pp. 14–17, quotes: p. 17.

[200] Fischer to Notgemeinschaft, 29/1/1932, BAK, R 73/170.

categorized the skulls of the original inhabitants of the Canary Islands, the Guanches, as of the Cro-Magnon type, which he also believed to recognize on occasion in the living population. In 1925, Fischer thus undertook a research expedition to the Canary Islands to look into this phenomenon. He saw his intuitive idea confirmed in his examinations of skeletal remains and anthropometric studies of living subjects: The Guanches belong to the "Cro-Magnon race," which is, surprisingly, also clearly preserved in the recent population.[201]

Most of the numerous publications by the custodian of the skull collection, Hans Weinert, were also far removed from the main fields of research at the KWI-A. This was true of his postdoctoral qualification thesis published in 1928 on *Pithecanthropus erectus*, an anthropological follow-up study of the vault of the cranium and femur discovered on Java by the Dutch physician Eugène Dubois (1858–1940) in 1890/91, and of his description of *Sinanthropus pekinensis* in 1931.[202] However, Weinert, as mentioned above, attempted to break through the boundaries of paleoanthropology to approach the issue of becoming human from another perspective. Through blood-group analyses of apes, Weinert was able to establish the same blood groups as for humans, O, A, B, and AB, while the findings for other kinds of monkeys clearly deviated from this classification.[203] In connection with this, Weinert opened to debate the question of whether the problem of the *missing link* could be clarified by crossing humans and apes, since "the evidence of living bodies [has] more to say than the best fossils of extinct transitional forms." Weinert had even considered how one would have to proceed in practice: a female chimpanzee should be inseminated with the sperm of an "African negro, probably best by a jungle pygmy."[204] The proposal was presumably meant quite seriously – the idea was in the spirit of the time. What Weinert did not know was that the Soviet biologist Ilja I. Iwanow (1870–1932), a specialist for the artificial insemination in animal breeding and for experiments on hybridization between species, had tested this very idea in practice in 1926/27 on an expedition to West Africa financed by the Soviet government.[205]

Finally, in contrast to the chief scientific interest of the KWI-A, which was oriented toward the phenogenesis of normal and diseased attributes of humans, there were works in the area of geographic anthropology. They were launched above all for strategic reasons, in order to profit from a research program sponsored by

[201] Fischer, Alte Kanarier. Cf. also idem., Frage einer äthiopischen Rasse; idem., Bemerkungen.

[202] Weinert, Pithecanthropus Erectus; idem., "Sinanthropus pekinensis." Cf. idem., Fossile Menschenreste; idem., Ursprung der Menschheit. Some papers – such as the study of two *Sakai* skeletons brought back from a research expedition to Sumatra by Dr. Max Moszkowski in 1907 and donated to the institute, were products of Kiffner's care for the unloved skull and skeleton collection. Kiffner, Beitrag.

[203] Weinert, Blutgruppenuntersuchungen; idem., Weitere Blutgruppenuntersuchungen; idem., Neue Blutgruppenuntersuchungen.

[204] Idem., Kreuzungsmöglichkeit, pp. 219 f., 222.

[205] Rossijanow, Beziehungen, pp. 345–351. Iwanow had even conducted negotiations with the governor of Guinea and the physician of the hospital in Konakry in order to procure permission to artificially inseminate African women with the sperm of chimpanzees.

the Emergency Association of German Science.[206] For the KWI-A had barely begun working when Friedrich Schmidt-Ott, president of the Emergency Association, invited a number of well-known anthropologists and geneticists to a meeting, among them Fischer, as well as officials of the Reich Ministry of the Interior. "For some time now the applications for support in the area of blood group research, of race research, and of anthropological studies have been increasing," Schmidt-Ott explained his invitiation. "With this abundance of individual applications," it was often difficult to select the projects to be funded. The purpose of the meeting with the invited specialists was thus to outline "the existing research tasks in this area" and sound out "the possibilities of working on them with the most extensive possible exploitation of the available funds."[207] For the scientists who participated in the meeting on December 17, 1927 – among them the leading genetic researchers Erwin Baur, Carl Erich Correns, Richard Goldschmidt, and Hans Nachtsheim, the anthropologists Eugen Fischer, Theodor Mollison (1874–1952),[208] and Otto Reche, along with the aspiring university lecturers Walter Scheidt and Karl Saller (1902–1969)[209] – this was a unique opportunity to stake their claims. In this situation Fischer provided proof of his talent as a science manager by immediately taking charge.[210] Upon his recommendation, the conference declared itself *against* mass statistical studies of cross-sectional groups like army recruits or schoolchildren in the style of Franz Boas' work and *for* compiling the selective, tendentially complete genealogic-genetic biological records of isolated populations. As a model of such a project recording all data on a particular group, Fischer submitted Walter Scheidt's article about the Elbe island Finkenwärder (1926),[211] which had created the basis for a new form of study: the "culture-biological village study," which combined the methods of genetic anthropology, social psychology, social and economic history, and folklore. The foundation of the study was the complete genealogy of the study area, so that the "lines of heredity" of the individual families could be retraced.[212] In this sense, Fischer claimed, all of Germany was to be "overlaid with an observation network that must become tighter and tighter." At

[206] On the following, above all: Gausemeier, Walter Scheidt, pp. 50–62; Geisenhainer, "Rasse," pp. 278–282. On this, also: Roth, "Erbbiologische Bestandsaufnahme," pp. 60–62; Lösch, Rasse, pp. 199–202; Weber, Ernst Rüdin, pp. 161 f.

[207] Letter of invitation from Schmidt-Ott to Erwin Baur, Carl Erich Correns, Eugen Fischer, Martin Hahn (* 1865, Hygiene Institute of the University of Berlin), Fritz Lenz, Theodor Mollison, Otto Reche, Karl Saller, Georg Thilenius and Walter Scheidt as well as the senior ministry officials Taute and Donnevert, BAK, R 73/169.

[208] For a biography: Klee, Personenlexikon, pp. 414 f.

[209] For a biography: ibid., p. 518. – The other participants were Schmidt-Ott, Donnevert, Hahn, Max Hartmann, Schmidt-Ott's employee Karl Stuchtey and Taute. Lenz and Thilenius were unable to attend.

[210] Bericht über eine Besprechung am 17. Dezember 1927 in den Räumen der Notgemeinschaft über Rassenforschung, Blutgruppenforschung und Anthropologische Untersuchungen, BAK, R 73/169.

[211] Scheidt/Wriede, Elbinsel Finkenwärder.

[212] For details, see: Gausemeier, Walter Scheidt.

the same time, Fischer was skilled at reining in the ambitions of Felix Bernstein and Otto Reche, who hoped to shift blood group research into the center of the research agenda under discussion: Since blood group research records just one single hereditary attribute, one might "just as well support 'nose research'."[213] Fischer ultimately succeeded in pursuading the group to entrust him with working out a research plan. With this plan, which he presented on February 2, 1928,[214] Fischer carved out a key role for himself in the large-scale research agenda from the outset. At least in the first years, he molded its conception, was responsible for coordinating the participating research institutions and took charge of setting up the "German Race Science" series at the Gustav Fischer publishing house in Jena, where the results of the research program were to be published. A total of 17 volumes appeared from 1929 to 1937.

The university and extra-university research institutes participating in the program – the research teams slated were from Göttingen (Karl Saller), Hamburg (Georg Thilenius, Walter Scheidt), Kiel (Otto Aichel), Leipzig (Otto Reche) and Munich (Theodor Mollison)[215] – were to serve as regional headquarters and train scientific assistants – the primary targets were elementary school teachers on location – to whom the collection of genealogical data and the anthropological examinations were to be entrusted. In order to standardize the methods, on April 19, 1928, the Society for Physical Anthropology adopted a list of the attributes that should be taken into consideration in the series of anthropological examinations:

1. The following measurements and information are *absolutely* necessary:

   Longest part of head – widest part of head – face height – face width – nose height – nose width
   Body height – height seated
   Hair color, head, and beard – eye color (both must be compared with chart) [...]
   Bridge of nose: convex (strongly, weakly), straight, concave, undulating
   Shape of back of the head: straight, convex, strongly convex. – constitution type (?)
   As many photographs as possible, as good as possible

2. The following measurements and information are *highly desirable*:
   Shortest width of forehead – width of lower jaw angle – ear height – arm length – armspan
   Skin color (?)
   Blood group

---

[213] Bericht über eine Besprechung am 17. Dezember 1927 in den Räumen der Notgemeinschaft über Rassenforschung, Blutgruppenforschung und Anthropologische Untersuchungen, BAK, R 73/169 (signed by Fischer, Mollison, Reche, Saller, Scheidt (the last two also as substitutes for Aichel and Thilenius).

[214] Fischer's exposé, 2/2/1928, BAK, R 73/169.

[215] The signatories of the exposé (cf. note 213).

Physiological attributes (age at first menstruation)
Psychological attributes. – pathological attributes

3. Genealogical materials, that is, the consultation of church records and other sources is imperative, but how far this should be pursued, to what extent the thorough perusal of church records is necessary and promising, depends on the conditions at the individual location [...]
4. The expansion and augmentation of these measurements and descriptions are unrestricted, of course[216]

The program was initially fast under the anthropologists' control, but it was Fischer's will that questions of genetic pathology and eugenics be investigated as well:

> New anthropological inquiry is not content with simply making assertions from which the distribution of such attributes as skull shape, blondness, eye color, etc. throughout Germany follows; on the contrary, it will provide rich material about the influence of social conditions on the affected populations, it will be able to cast bright light on any phenomena of degeneration, on the distribution of genetic pathological characteristics, on the issues of inbreeding, the fall in the birth rate, etc. and thus become not only work that is interesting in terms of pure science, but also an infinitely important collection of sources practical for administrative and regulatory issues of sanitation and population policy.[217]

From 1928, a plethora of projects started up at the research institutions involved – initiated above all by Walter Scheidt, Karl Saller, and Otto Reche. Additional projects were in the planning stage, among them one by Franz Weidenreich (1873–1948), at that time the only Jew to hold a chair for anthropology in Germany, about the Jewish population of Frankfurt.[218] Coordination and cooperation among the research locations existed only on paper. In fact the program lacked internal coherence – in principle it was nothing more than a broad framework to bundle a multitude of research projects and combine applications for funding. The first volumes of "German Race Science" – Willy Klenck and Walter Scheidt on farmers in the sandy uplands of the Elbe-Wesermünde area, Karl Saller on a small group in Franconia known as the "Keuperfranken," Hans August Ried on the rural population of Miesbach, and Saller again on the population of Fehmarn[219] – had no uniform methodology and were, by and large, quite conventional. In a collective review, Verschuer appeared disappointed, not least of all because none of the authors had attended to the genetic pathological statistics demanded.[220]

---

[216] Fischer, Anthropologische Erhebungen, pp. 21 f. (original emphases).

[217] Fischer's exposé of 2/2/1928, BAK, R 73/169.

[218] Fischer to Notgemeinschaft, 19/9/1929, BAK, R 73/169.

[219] Klenck/Scheidt, Niedersächsische Bauern; Saller, Keuperfranken; Ried, Miesbacher Landbevölkerung; Saller, Fehmarner.

[220] Verschuer, review of Deutsche Rassenkunde.

Therefore, in early 1932, a discussion ensued about the concept upon which the program was based. The occasion was an unexpected windfall. In September 1929, Schmidt-Ott had turned to the Rockefeller Foundation to solicit funds for a "large-scale examination of anthropological conditions in Germany," which was to concern issues of the distribution of genetic pathological characteristics, criminal biology, and practical eugenics. The Rockefeller Foundation immediately approved US$ 25,000 annually for the period from 1930 to 1934 – an enormous amount for the participating research institutions, considering the acute global economic crisis. It is thus not surprising that the struggle for resources intensified, a battle joined by not only Eugen Fischer, but also Georg Thilenius (1868–1937),[221] the doyen of German ethnology, Oskar Vogt, Hermann Werner Siemens and, above all, Ernst Rüdin.[222] Since the mid-1920s, Rüdin's staff had been collecting material in the Allgäu and other regions of southern Bavaria as a foundation for "empirical genetic prognosis." In this they combined methods of genealogy and mass statistics, the former by creating family trees using church records, and the latter by collecting the vital statistics of the living population by means of census forms. The goal was a "complete population policy register." Rüdin now wanted to integrate this research into the "German Race Science" program. This meant breaking away from the original concept, to be sure, because a statistically significant genetic psychiatric inventory, as envisioned by Rüdin, would require an area of investigation with a population of around 10,000 and a requisitely large area. In fact, Rüdin was successful in bringing to Munich a considerable portion of the funds provided by the Rockefeller Foundation, although the results of the projects financed with this money did not appear in the *German Race Science* series.[223]

Even before the decisive meeting of the participating institute directors on February 22, 1930,[224] Fischer cleared the field in the face of Rüdin's advance, to the extent that he cautiously distanced himself from his original idea of carrying out the anthropological, hereditary pathological, and hereditary psychiatric examinations of a given population group in one fell swoop. Instead he now proposed systematic "triple inquiries": The anthropologist was to analyze the population of a test region for its race attributes, the physician for general genetic pathological phenomena and, finally, the psychiatrist for psychopathological hereditary attributes, whereby Fischer wanted to leave up to the psychiatrist – with this he actually meant Rüdin – the decision as to which population group he believed he could "attack [sic!] with the prospect of success." For his own

[221] For a biography: Klee, Personenlexikon, p. 623.

[222] Cf. Schmidt-Ott to Vogt, 3/2/1930; Vogt to Schmidt-Ott, 12/2/1930; Thilenius to Schmidt-Ott, 12/2/1930; Siemens to Schmidt-Ott, 18/2/1930; Kallius to Schmidt-Ott, 18/2/1930; Fischer to Schmidt-Ott, 18/2/1930; Rüdin to Schmidt-Ott, 20/2/1930, BAK, R 73/169.

[223] Weber, Ernst Rüdin.

[224] Invited were the pathologist Ludwig Aschoff (1866–1942), the surgeon August Bier (1861–1949), Donnevert, the physiologist Arnold Durig (1872–1961), Fischer, Glum, Hahn, Erich Kallius (* 1867), the internist Friedrich von Müller (1858–1941), Roesle, Rüdin, Moritz Benno Schmidt (1863–1949, Pathological Institute of the University of Würzburg), Siemens, Taute, Thilenius and Vogt.

institute Fischer claimed the field of genetic biological research on Germans abroad, as Fischer believed that these isolated populations would allow rare hereditary traits, even those governed by a recessive gene, to be tracked over generations: "Normal characteristics of this kind would serve us as an example and proof, pathological ones would interest us." Fischer also attempted to siphon off funds from the "German Race Science" program for the project by Diehl and Verschuer on the heritability of tuberculosis, for Muckermann's project on differentiated reproduction, and even for the projects pursued in his department on experimentally influencing the shape of the skull in rats and tests on embryonal pig skulls.[225]

Here it becomes apparent that "German Race Science" program was ultimately little more than a money machine for Fischer and his institute. While Fischer's commitment initially may have been sustained at least in part by his interest in the data material such a large-scale project would make available, this interest extinguished visibly as it became apparent that this material remained highly fragmentary, that no uniform standards and methods were used, and that the findings were not terribly meaningful for thc questions pursued at the KWI-A. In fact the individual projects begun in the context of the program were pushed ahead only half-heartedly. Fischer's grandiose declarations cannot obscure the facts:[226] In the framework of the program, the KWI-A had taken on the anthropological investigation of the Markgräfler region in Oberbaden (Fischer), the Nagold valley in Württemberg (Loeffler), Westphalia/Lippe-Detmold (Fischer),[227]

---

[225] Fischer to Notgemeinschaft, 19/1/1930, BAK, R 73/169, reprinted in part in Roth, Erbbiologische Bestandsaufnahme, pp. 60 f.

[226] Cf. Niederschrift der Sitzung des Kuratoriums des KWI-A am 6.6.1928, MPG Archive, Dept. I, Rep. 1 A, No. 2404, pp. 1–1a, here: p. 1a; Jahresbericht 1927/28, 21/4/1928 (which also discusses investigations in parts of Mecklenburg, which were also not realized), MPG Archive, Dept. I, Rep. 3, No. 4; Jahresbericht 1928/29, 8/3/1929, MPG Archive Dept. I, Rep. 3, No. 5; Jahresbericht 1929/30, MPG Archive, Dept. I, Rep. 3, No. 6; Jahresbericht 1930/31, MPG Archive, Dept. I, Rep. 3, No. 7. Further, the enclosure "Übersicht über die anthropologischen Erhebungen" (Overview of the anthropological studies) to Fischer's letter to the Emergency Association, 19/8/1929, BAK, R 73/169: Here Fischer stated that the project on the Nagold parish in Württemberg was "concluded," projects on the district of Höxter (gentry), Kandern in Wiesental/Baden, Vogelsdorf in Silesia, Transsylvania and on the Markgräfler region were "in progress," additional projects on Wolfach in the Black Forest, the region of Münster (Adel), Bork in Westphalia, Krefeld, on rural communities near Paderborn, Detmold and Bielefeld, on the Fuggerei in Augsburg, on rural communities near Krumbach in Bavaria, and on a German village near Temesvar in Hungary were "in preparation."

[227] Fischer had come into contact with the Westphalian Province Association through Hermann Muckermann, who had held several lectures in Westphalia in 1927/28. Fischer began training elementary school teachers in the western Münster region and took the head measurements of Westphalian noblemen himself. The plan to put an assistant to Fischer on the budget of the Province Association and to found a branch office of the KWI-A in Westphalia was frustrated by the strained financial situation of the provincial administration. In the 1950s Ilse Schwidetzky (1907–1997), as part of the large-scale project "The Westphalian Region" initiated by Hermann Aubin (1885–1969) and Franz Petri (1903–1993), carried out studies on the "Westphalian race type" – a project in whose organization Otmar von Verschuer, who by then held the chair for human genetics at the University of Münster, played a leading role. Cf. Ditt, Raum, pp. 77 f.; idem., Kulturraumforschung, pp. 143–145; idem., Was ist 'westfälisch'?, pp. 83–85. I thank Karl Ditt for alerting me to this research.

Upper Silesia (Göllner), and Transsylvania (Fischer/Verschuer). By 1933 only the extended version of Herbert Göllner's dissertation, *Volks- und Rassenkunde der Bevölkerung von Friedersdorf* ("The Science of Volk and Race of the Population of Friedersdorf"), had appeared, a study of a village in the Lauban district of Silesia, the ninth volume of *German Race Science*.[228] The work was largely restricted to genealogical studies based on the village's church records and on an ethnological collection of materials – anthropometric studies would hardly have been possible in Friedersdorf, as Göllner allowed, "measurements of the naked bodies of members of the population [were] permitted only in the rarest of cases."[229] The study was thus practically worthless for the institute's study program. Fischer's project on Baden and Westphalia ran aground, as did the project by Loeffler on Württemberg. The project on Transsylvania, which Fischer and Verschuer had launched on a trip to the Carpathian Mountains in 1928,[230] was finally brought to a conclusion by Albert Hermann (1893–1975), the headmaster of a preparatory school in Kronstadt. The results appeared in 1937 under the title *Die deutschen Bauern des Burzenlandes* ("The German Farmers of Burzenland") as volume 15/16 of "German Race Science".[231] In the previous year, the 14th volume of the series had been published, with the study by Brigitte Richter (* 1907), *Burkhards und Kaulstoß, zwei oberhessische Dörfer* ("Burkhards and Kaulstoss, Two Upper Hessian Villages").[232] Also published in 1936 was a study by Heinz Boeters about *Familienaufbau und Fruchtbarkeitsziffern* ("Family Structure and Fertility Numbers") for 1,213 Russian German farming families,[233] who made a stop at the Hammerstein refugee camp during their flight from the Soviet Union to their final destination of the USA in 1929/30, where they had been "thoroughly examined anthropologically."[234] This was the total yield of the research performed in the framework of the "German Race Science" research program at the KWI-A – a meager balance, especially in comparison to other research locations like Leipzig (Otto Reche)[235] and Hamburg (Walter Scheidt),[236] not to mention Ernst Rüdin's large-scale projects in southern Germany.[237]

---

[228] Göllner, Volks- und Rassenkunde.

[229] Ibid., p. VIII.

[230] Verschuer, Kaiser-Wilhelm-Institut für Anthropologie, p. 133.

[231] Hermann, Deutsche Bauern.

[232] Richter, Burkhards und Kaulstoß.

[233] Boeters, Untersuchungen.

[234] Jahresbericht 1929/1930, MPG Archive, Dept. I, Rep. 3, No. 6.

[235] Cf. Zusammenstellung von Forschungsprojekten am Institut für Rassen- und Völkerkunde der Universität Leipzig in Geisenhainer, "Rasse," p. 282. According to this list, in the years 1929–1932, the physician Wilhelm Hilsinger performed examinations of the Sorb population in Upper Lusatian villages. Hilsinger is mentioned in the 1929/30 annual report of the KWI-A as an (external) member of the scientific staff. MPG Archive, Dept. I, Rep. 3, No. 6.

[236] Cf. Gausemeier, Walter Scheidt, p. 58.

[237] Weber, Ernst Rüdin, pp. 161 f.

In the research practice of the institute in Dahlem, the "German Race Science" program consequently left almost no traces. Nevertheless it was extremely important for the institute, not only because it primed the funding pumps of the Emergency Association and the Rockefeller Foundation in an economically precarious situation, but above all because it linked the KWI-A with a network of university and extra-university research institutions. This network extended from the University Skin Clinic in Leiden, the Netherlands, under Hermann Werner Siemens; to the chair for anatomy in Göttingen held by the university lecturer Karl Saller; the chair for ethnology and the Ethnology Museum in Hamburg under Georg Thilenius and Walter Scheidt; the Anatomical Institute in Heidelberg under Erich Kallius (1867–1935),[238] the chair for anthropology in Kiel under Otto Aichel (1871–1935); the chair for ethnology in Leipzig under Otto Reche; the German Research Institute for Psychiatry (*Deutsche Forschungsanstalt für Psychiatrie*) in Munich under Ernst Rüdin; the KWI for Brain Research in Berlin under Oskar Vogt; all the way to the University Neuropathic Clinic in Breslau under Johannes Lange (1891–1938).[239]

As to the degree to which research practice was embedded in the contexts upon which it was founded, it is striking that Fischer moved ever further away from his original intention to set up a "purely theoretical institute" as finances became scarce. In budget negotiations with the General Administration he always placed special emphasis on those projects whose practical relevance was directly apparent, even if they played a rather subordinated role in the institute's research plan. Thus, in June 1930 he justified the 10,000 RM increase in the budget plan for the 1931/32 fiscal year as follows:

> I have the conviction that the circle of tasks of the institute has grown to such a degree, and moreover, that all of the work performed, beyond any theoretical assessment, concerns all of the social problems so directly, that, in complete appreciation of the necessity to save as much as possible, I regard this request for an increase as my duty. Without the increase, the work already underway on the heritabiliy of tuberculosis, on the reduction in birth rate, on differentiated reproduction, on genetic damage in humans, on the propagation of pathological hereditary dispositions in the German nation would have to be simply abandoned.[240]

One year later, in May 1931, Fischer fought against further cuts to his material budget using similar argumentation:

> I may remark without arrogance that the achievements of the institute are not only generally recognized by the scientific community, but have also become and will continue to be of direct practical importance for our nation. I point to the proof of the heritability of tuberculosis which we supplied, to our examinations of X-ray damage among humans, which have received so much attention, to the conditions of propagation of certain important social groups, portrayed numerically for the first time.[241]

---

[238] Cf. Weindling, Health, pp. 468 f. Like Diehl and Verschuer, Kallius researched the genetic disposition for tuberculosis.

[239] For a biography: Klee, Personenlexikon, p. 356.

[240] Fischer to KWG, 25/6/1930, MPG Archive, Dept. I, Rep. 1 A, No. 2405, p. 48 f., quote: pp. 48 v–49.

[241] Fischer to KWG, 1/5/1931, ibid., p. 66 f., quote: p. 66.

The tuberculosis research by Diehl and Verschuer, Loeffler's survey on genetic damage through x-ray radiation, the genetic statistical surveys by Muckermann on differentiated reproduction and participation in the "German Race Science" research program – these were the resources Fischer exploited. With the exception of tuberculosis research, all of these projects were rather on the margins of the internal research plan. To its sponsors, Fischer presented a distorted portrayal of the institute's research focus in order to emphasize the practical application of the research performed in Dahlem. For in fact, Fischer regarded the theoretical research he had initiated as anything but removed from the biopolitical context – it was just *not directly* relevant to practice. When Fischer presented his examinations of rat skulls and Kühne's studies on the vertebral anomalies to the public as contributions to "designing the human of the future," this was no false labeling. In the internal logic of the research design, theoretical research, on the ability to shape the phenotype, and the influence of genetic disposition and environment, preceded the formulation of concepts for the biopolitical control of society. In this sense the KWI-A conceived of itself, to use the succinct formulation by Niels C. Lösch, as the "think tank"[242] of eugenics in the late Weimar Republic.

2.3 The Kaiser Wilhelm Institute for Anthropology, Human Heredity and Eugenics, the Eugenics Movement and the Crisis of the Weimar Welfare State

From the circumstance that eugenics played only a marginal role as a field of *research* at the KWI-A it nevertheless cannot be concluded that Fischer and his staff did not have any interest in eugenics as a field of *politics*. On the contrary, toward the end of the Weimar Republic Fischer and his institute were at the vanguard of the eugenics movement in Germany.

2.3.1 The German Society for Race Hygiene toward the end of the Weimar Republic

Shortly after moving to Berlin, Fischer took over the chairmanship of the Berlin branch of the German Society for Race Hygiene. In the elections to the Board of Directors of the German Society for Race Hygiene in 1929, Fischer became first chairman, Undersecretary in the Prussian Welfare Ministry Arthur Ostermann the deputy chairman, Muckermann and Verschuer the secretaries. This Dahlem faction energetically instigated the Society for Race Hygiene's approach to the German Association for Volkish Improvement and Genetics – the key figure in this was

[242] Lösch, Rasse, p. 218.

Ostermann, who, as deputy chairman of the Society for Race Hygiene *and* the Association for Volkish Improvement, was henceforth able to prepare the ground for a unification of the two organizations. Upon the request of Fischer and Muckermann at its general meeting in February 1930, the Berlin branch of the German Society for Race Hygiene voted unanimously to change its name to "Berlin Society for Eugenics" (*Berliner Gesellschaft für Eugenik*). This step was motivated by the fact that "the word 'race hygiene,' as manifold experiences have shown, encounters misunderstandings that retard the progress of the cause itself."[243] A whole slew of local branches followed the Berlin parent society; other branch groups founded in 1929/30 operated as eugenic societies from the outset, such as those in Barmen-Elberfeld, Solingen, Cologne, Leverkusen, Cloppenburg, Vechta, and Greifswald, as well as the "German Eugenic Society of Prague (*Deutsche Eugenische Gesellschaft Prag*) and the "German Eugenic Alliance" (*Deutscher Eugenischer Bund*) of the same city.[244] This was no coincidence either, as most of these organizations were founded in response to lectures held by Muckermann, Fischer, and Verschuer.[245] This nomenclature was bound up with a programatic avowal to the Berlin wing with its strict eugenic course without the ballast of race ideology, to "impartiality and cosmopolitanism."[246] The Munich wing around Ploetz, Rüdin, and Lenz, sympathetic to the "Nordic idea," found itself forced into a defensive posture; powerless, it could do no more than bemoan what the nationalist publisher Julius Friedrich Lehmann (1864–1935) called the "infiltration" of the Berlin Society for Race Hygiene by the "Berlin Jews."[247]

At the general meeting of the German Society for Race Hygiene in September 1931, the Berlin wing went on the offensive. The assembly voted to rename the organization "The German Society for Race Hygiene (Eugenics)." A new charter was passed as well, which emphasized the eugenic character of the organization more strongly: The Society, the new charter read, not least with a view to the KWI-A, sponsors "scientific work on the area of genetics and eugenics." The objective of "enhancement of the race" became instead the "eugenic design of family and nation."[248] The new guidelines were restricted almost exclusively to postulates of a positive eugenics and quantitative population policy. On this basis the Society for Race Hygiene was able to merge with the Association for Volkish Improvement. In October 1930, Ostermann – together with Fischer, Lenz, Muckermann, Rüdin, and

---

[243] Dokumente aus der eugenischen Bewegung, p. 20.

[244] Weingart/Kroll/Bayertz, Rasse, p. 252, note 136.

[245] "The extensive lecture activity of Professor Muckermann and in part Professor Fischer is to thank for the founding of a great number of eugenic societies in all parts of Germany, whose activity for the elimination of eugenic damage in our nation (rise in the number of mentally ill and feeble minded, drop in the birth rate of the genetically healthy) will hopefully be of decisive importance." Jahresbericht 1930/31, MPG Archive, Dept. I, Rep. 3, No. 7.

[246] Lilienthal, Rassenhygiene, p. 117.

[247] Lehmann, J. F. Lehmann, pp. 128, 403 f. For a biography, see also: Labisch/Tennstedt, Weg, vol. 2, pp. 451 f.

[248] Aus der Gesellschaft für Rassenhygiene und Eugenik, p. 95, 101.

Verschuer – founded the journal *Eugenik, Erblehre, Erbpflege* ("Eugenics, Genetics, Heredity Care"), which took the place of the *Zeitschrift für Volksaufartung und Erbkunde*. It became the de facto official organ of the "Fischer wing" of the German Society for Race Hygiene and thus competed with the *Archiv für Rassen- und Gesellschaftsbiologie*. "Eugenic propaganda"[249] was also the purpose of the organ published by Muckermann, *Das kommende Geschlecht. Zeitschrift für Familienpflege und geschlechtliche Volkserziehung auf biologischer und ethischer Grundlage* ("The Race of the Future: Journal for Family Care and Racial Education of the Volk on a Biological and Ethical Foundation").[250]

2.3.2 "Eugenic propaganda" at the Kaiser Wilhelm Institute for Anthropology, Human Heredity, and Eugenics

These publications from the KWI-A, which were directed to a broader audience, were supplemented by brisk lecturing activity, particularly by Muckermann, but also by Fischer and Verschuer. They pursued, as stated back in the 1929/30 annual report, the purpose "of procuring for the idea of eugenics in the entire nation, above all in the politically and legislatively influential circles, the prestige to which it is entitled. The institute is deliberately leading the fight for the qualitative improvement of the propagation of the nation."[251] In the 1932/33 annual report Fischer became even more concrete on this topic: "The director of the institute as well as both department heads held lectures [...] (aside from generally public lectures) before numerous domestic and international medical associations, then before the Association of Teachers of the Deaf-Mute (*Vereinigung von Taubstummenlehrern*), the German League of Physicians (*Deutscher Ärztebund*), before associations of clergymen of both confessions."[252]

Of particular importance for the penetration of the Weimar welfare state with eugenic ideas were training sessions at the KWI-A. The initiative for this came from Undersecretary Ostermann, who had come to Fischer back in 1929 with the request to hold further education courses about genetics and eugenics for Prussian medical officials at the KWI-A, which Fischer "booked [...] as a particular success."[253] The first of these courses, with 50 participants comprised of public health officers and officials from the medical administration, took place for 10 days in mid-January 1930. It was repeated the following year. In February 1932, at the

[249] Jahresbericht 1931/32, MPG Archive, Dept. I, Rep. 3, No. 8.

[250] The annual report of 1930/31 states that the journal *Das kommende Geschlecht* was published by Muckermann "in connection with Fischer and Verschuer." Jahresbericht 1930/31, MPG Archive, Dept. I, Rep. 3, No. 7.

[251] Jahresbericht 1929/30, MPG Archive, Dept. I, Rep. 3, No. 6.

[252] Jahresbericht 1932/33, MPG Archive, Dept. I, Rep. 3, No. 9.

[253] Jahresbericht 1929/30, MPG Archive, Dept. I, Rep. 3, No. 6.

request of the Reich Labor Ministry, a corresponding course was held for physicians in the civil service.[254] One year later, finally, training courses took place in the institute for 60 civil servants of the Reich Labor Ministry, the Reichswehr Ministry, and the Prussian Ministry of the Interior.[255]

These courses were the external expression of the KWI-A's increasing integration into the Prussian ministerial bureaucracy. The key figure in this development – more so than Fischer and Verschuer – was Hermann Muckermann. His expansive network of contacts was to thank for the fact that the contacts of both the KWI-A and the German Society for Race Hygiene extended all the way to Carl Heinrich Becker, the Prussian Minister for Science, Art and Education; and to Heinrich Hirtsiefer, the Prussian Minister for National Welfare. The Department for National Health in the Prussian Welfare Ministry was firmly established in the camp of the Berlin wing of the eugenics movement and was closely linked with the institute in Dahlem. This was true not only of Undersecretary Heinrich Schopohl (1877–1963),[256] who took over the direction of the Department for National Health in 1928 as the successor of the deceased Otto Krohne, and the position of President of the Prussian State Health Council, but also of Arthur Ostermann, the director of the department responsible for population policy and race hygiene, and of Carl Hamel, president of the Reich Health Office since 1926. At the general meeting of the German Society for Race Hygiene in 1931, Muckermann maintained with great satisfaction that his interviews with ministers and ministerial officials had shown that "the inclination to assimilate eugenic ideas experienced a significant increase": "In a number of meetings, about which no newspaper or journal reports, there were many opportunities to remove prejudices and direct attention to the genetically healthy family."[257]

2.3.3 *The Crisis of the Welfare State, Open Care, and Eugenics*

The background of the ministerial bureaucracy's growing interest in eugenics constituted the crisis of the Weimar welfare state under the portent of economic crisis. The social security systems were hopelessly overtaxed by the mass poverty caused by the sharp rise in unemployment. Payments for the unemployed had to be slashed ever further, whereby the welfare unemployed, who had to be supported by the cities and districts, were hit especially hard. Some of the communities had to cease payments altogether due to empty coffers, as millions of people who were disqualified from unemployment insurance and their families fell between the mesh of the nets of all social security systems and struggled to survive without any assistance

[254] Jahresbericht 1931/32, MPG Archive, Dept. I, Rep. 3, No. 8.

[255] Jahresbericht 1932/33, MPG Archive, Dept. I, Rep. 3, No. 9.

[256] For a biography: Labisch/Tennstedt, Weg, vol. 2, pp. 493 f.; Richter, Katholizismus, p. 294 (note 33).

[257] Muckermann, Aus der Hauptversammlung, p. 100.

at all.[258] In view of this catastrophic situation, a broad discourse began in both the public and the nonprofit welfare organizations about a "differentiated welfare," which basically concerned the restriction of institutionalization that was dubious under eugenic and population policy aspects, and furthermore quite costly, namely assistance for the mentally ill, feeble minded, epileptics, cripples, drunkards, nomads, neurotics, and psychopaths.[259]

Indeed, back in the mid-1920s, the system of institutionalization had begun to open toward society. The model of "external care" proceeding from the sanatoriums and clinics developed by Gustav Kolb (1870–1938) at the State Sanatorium in Erlangen, and the "Wittenau scale system" developed by Emil Bratz (1868–1934),[260] which allowed a step-by-step transition from stationary treatment to partially stationary facilities all the way to release, were designed to integrate the institutions into a comprehensive system of prevention, therapy, and rehabilitation and thus embed them in society.[261] The standard work, *Die offene Fürsorge in der Psychiatrie und ihren Grenzgebieten* ("Open Care in Psychiatry and its Boundary Areas") (1927)[262] by Hans Roemer (1878–1947), the head of the medical division of the Ministry of the Interior of Baden, Gustav Kolb and his senior physician Valentin Faltlhauser (1876–1961) marked the breakthrough of open care on a broad front in the second half of the 1920s. Combined in this development were the pressure of empty coffers, changed sociopolitical conditions, and professional policy aspects with the push for reform in the field of psychiatry. In the administration's view, "the policy of savings" became "the vehicle of the reformers":[263] The compulsion for efficiency was one important reason, but not the only one for setting up and expanding open care for the mentally ill. Beyond this it paved the way for the integration of institutionalized psychiatry in the welfare state – the introduction of an open system of care for the mentally ill proceeding from the institutions appeared to be a prerequisite for the realization of a new right to care and the "aspects anchored therein, of individualizing procedures, evaluating social circumstances, foresighted prevention, recovery of the ability to work independently, and social reintegration."[264] In this concept the institution lost "its monopoly as an instance of psychiatric care," however, and became a "transit station" for the mentally ill.[265] In return, institutionalized psychiatry secured primacy in the open care of the mentally

---

[258] Cf. Schmuhl, Arbeitsmarktpolitik, pp. 161–193.

[259] Cf. for example: Harmsen, Bevölkerungspolitische Neuorientierung; idem., Eugenetische Neuorientierung.

[260] For a biography: Damm/Emmerich, Irrenanstalt Dalldorf-Wittenau, pp. 41 f.; Beddies, Geschichte, p. 63.

[261] On the Erlangen model: Kolb, Reform; Siemen, Menschen, pp. 34–37, 52–57, 80–85. On the Wittenau scale system: Damm/Emmerich, Irrenanstalt Dalldorf-Wittenau, pp. 40 f.

[262] Roemer/Kolb/Faltlhauser (ed.), Offene Fürsorge.

[263] Siemen, Menschen, p. 48.

[264] Walter, Psychiatrie, p. 283.

[265] Ibid., pp. 298 f.

ill, immediately opening up a number of boundary areas: "Through open care, psychiatry should become established as a leading science with regard to the care of neuropaths, psychopaths, the vulnerable, alcoholics and prisoners, which could be run appropriately 'not *as* psychiatry, but indeed *using* psychiatry'."[266] Open care ushered psychiatry into society, which, in turn, entailed a progressive "psychiatrization" of deviant behavior. Directly linked with open care was the issue of eugenic sterilization, as the dismantling – or at least the opening up – of closed institutions meant that the institutions increasingly lost their function as asylums. From the eugenic perspective there was thus a risk that increasing amounts of inferior genetic material would succeed in propagating. Therefore, voices from the eugenic camp, though restrained at first, demanded eugenic sterilization as a necessary supplement to differentiated care – including those from the avant-garde of the psychiatric reform movement, who had assembled under the umbrella of the "German League for Mental Hygiene" (*Deutscher Verband für Psychische Hygiene*) under the chairmanship of Roemer, Kolb and the director of the sanatorium in Gütersloh, Hermann Simon (1867–1947).[267]

This combination of differentiated care and eugenics was also advocated by Fischer,[268] Verschuer, and Muckermann. Verschuer, for instance, with express reference to Alfred Grotjahn, placed the critique of the Weimar welfare state at the center of his "eugenic propaganda" from 1927/28 onwards. In his tract *Sozialpolitik und Rassenhygiene* (Social Policy and Race Hygiene) (1928), Verschuer raked the social insurance system over the coals. While from the race hygiene perspective he welcomed care for the unemployed, especially the productive care of the unemployed, he sharply rejected unemployment insurance with its legal claim to support payments: Its "beneficiaries" were "to no small extent lazy, mentally and ethically inferior people." This, he continued, was not only a social/ethical issue, but also dubious in terms of race hygiene, as "the unemployed supported by the state [were] particularly at the mercy of sexual excesses"; "the idle life strengthens the drives and lames the inhibitions." The consequence was an "increase in births among the dregs of the nation." Facing this development was "the decline among the remaining part of the nation, so that this process of degeneration [was] further accelerated."

Verschuer further deplored "an improper application" and "too great an extent and too little differentiation" in health and accident insurance. Therefore, Verschuer demanded a curtailment of the social insurance system, the return from the principle of insurance to the principle of care, the socialization of health care – in particular the appointment of doctors as public officials, and a segmentation of the health insurance providers according to profession. Finally, he demanded an increase in

[266] Ibid., p. 285 (Quote in quote by Hans Roemer, emphases in the original).

[267] Cf. Kersting/Schmuhl (ed.), Quellen, pp. 377–385.

[268] On Fischer's cautious remarks about social anthropology and race hygiene, see Gessler, Eugen Fischer, pp. 107–114. A similar position was taken by his cousin Max Fischer in 1925, cf. Müller, Sterilisation, p. 65.

the salary of high-level officials, along with income-dependent subsidies for children and families.[269] Race hygiene, so Verschuer stated candidly, must benefit the educated and propertied middle classes above all:

> The dispositions that qualify an individual for intellectual work – be it as teacher, pastor or doctor, as judge, administrative official or officer, as artist, philosopher or scientist, as manufacturer, mine owner or landowner, [...] are required to secure and support the state, for the life of our national characteristics. They are the noble treasure of the race. Guarding this precious treasure is the actual task of race hygiene.[270]

2.3.4 Verschuer and Muckermann on Eugenic Sterilization

As to demands for negative eugenics, above all sterization due to a eugenic indication, Verschuer was markedly reserved before 1933. While back in his inaugural lecture about the "Tasks and Goals of Human Genetics" he had advocated the "sterilization of the mentally and morally diseased and inferior," he had done so only "on the basis of hereditary scientific facts."[271] At the first Protestant Symposium for Eugenics, which took place from May 18–20, 1931 in Treysa, Hesse, at the initiative of the physical and national economist Hans Harmsen (1899–1989),[272] director of the department for health care in the Central Committee for the Inner Mission, Verschuer argued in principle *for* the eugenic sterilization of the feeble minded, schizophrenics, the manic depressive, epileptics, psychopaths, chorea sufferers, the congenitally blind and deaf-mute, whereby he qualified his statement by referring to the uncertain prognosis of heredity in such cases as manic-depressive insanity, epilepsy, and deaf-mutism. In appraising these conditions, he emphasized, every kind of "schematism" must be rejected. In each case this appraisal must be performed "by physicians trained in genetic biology." Verschuer advocated "only voluntary sterilization, not compulsory," however, he also gave consideration to the possibility of whether "an indirect compulsion [could] be exerted." In closing, Verschuer went into the moral, theological aspect of the issue. Eugenic sterilization, which Verschuer equated with medical curative treatment, is no "unauthorized intervention into the natural process of creation;" the willingness to make oneself sterile is rather a command of Christian charity. The fulcrum and hub of the argumentation comprised the concept of "sacrifice":

> It is demanded of us Christians, who follow the example of our master, that we be willing to sacrifice our lives in the service of Christian charity. Christian charity extends to children who will be born. We are therefore obligated to extend the circle of humans to those who

---

[269] Verschuer, Sozialpolitik und Rassenhygiene, quotes: pp. 17 f., 20 f., 23 f.

[270] Ibid., p. 26.

[271] Verschuer, Aufgaben und Ziele, p. 1001.

[272] Schleiermacher, Sozialethik; Schmuhl, Krankenhäuser, pp. 31 f., 63–70.

> are not yet born, and I believe it is justified to demand from people a lesser sacrifice than the sacrifice of life, namely to forego having children, for the love of children that are expected to be diseased, so that from the perspective of Christian charity sterilization must be regarded as justified.[273]

According to his own account, in 1931 Verschuer held several lectures at "pastor conferences" held by the Apologetic Headquarters of the Inner Mission in the *Evangelisches Johannesstift Spandau*.[274] "After all," he wrote in retrospect, he found "the cultivation of relations to the Protestant Church particularly suitable" in the late the Weimar Republic, "as a counterweight to Muckermann's activity in the circles of the Catholic Church."[275]

Hermann Muckermann underwent a remarkable change of heart in the sterilization question. In 1924 he clearly rejected sterilization as "violent intrusion."[276] In 1929, however, he changed his opinion "under the impression of the nation's growing need for care and after studying the increased probability of the heredity prognosis."[277] In 1931, he still advocated permitting voluntary sterilization under three conditions: secured prognosis of heredity, existence of a state of emergency, best possible protection from abuse.[278] Shortly thereafter Muckermann and other Catholic eugenicists found themselves in an extremely difficult situation, as *Casti conubii*, Pope Pius XI's

---

[273] From the discussion protocol of the first Symposium for Eugenics, 18–20/5/1931 in Hephata near Treysa, Archives of the Diakonisches Werk Berlin, CA, G 381, pp. 43–94. My warmest thanks to Uwe Kaminsky and Jochen Christoph Kaiser for allowing me to examine their transcript. – A voice in Treysa more skeptical than Verschuer's was that of Carl Schneider (1891–1946), at the time head surgeon at Bethel. From 1940 Schneider – by then professor for neurology and psychiatry at the University of Heidelberg – took a major role in the mass murder of the mentally ill and mentally retarded (cf. Schmuhl, Ärzte in Bethel, pp. 76–80). Verschuer held Schneider, as he stated in his memoirs, to be his "most vehement opponent […] in the very animated discussion" in Treysa; Verschuer, Erbe – Umwelt – Führung, "Freier Forscher in Dahlem (1927–1935)" section, p. 13, MPG Archive, Dept. III, Rep. 86 A, No. 3–1. Schneider's and Verschuer's reservations were incorporated directly into the Treysa resolution, which recommended sterilization on a voluntary basis "for a eugenic indication": "The results of family genealogical research, especially the observation of identical twins, allow no doubt about the great importance of hereditary factors for the development of the entire course of life. What remains unclear for many cases even today is the importance of environmental influences, including the question as to damage of egg and sperm cells before their fusion and the issue of their ability to regenerate." From the Treysaer Resolution des Central-Ausschusses für Innere Mission, printed in: Kaiser et al. (ed.), Eugenik, pp. 106–110, quote: p. 107. Cf. Kaiser, Protestantismus, pp. 324 f.; idem., Innere Mission, p. 205.

[274] Head of the Johannesstift from 1926 to 1932 was the theologian Helmuth Schreiner (1893–1962), who turned out to be a decisive opponent of the "destruction of unworthy life" in the "euthanasia" debate of this period. Cf. Bräutigam, Angehörige der Ostvölker, p. 16; Schmuhl, Rassenhygiene, Nationalsozialismus, Euthanasie, pp. 123, 312.

[275] Verschuer, Erbe – Umwelt – Führung, "Freier Forscher in Dahlem (1927–1935)" section, p. 12, MPG Archive, Dept. III, Rep. 86 A, No. 3–1.

[276] Muckermann, Kind und Volk, p. 110.

[277] Muckermann, Wesen der Eugenik, p. 32.

[278] Muckermann, Ehe, p. 719.

encyclical "On Christian Marriage"[279] was published – backdated to December 31, 1930, which endorsed measures of positive eugenics, and did not clearly condemn the "soft" forms of negative eugenics, but left no doubt about its rejection of eugenic sterilization.[280] Now Muckermann, if he wanted to remain true to his Catholic basis, was left with nothing but cautious "wriggling."[281] On the one hand, he now condemned both eugenic sterilization on behalf of the state and voluntary sterilization,[282] on the other hand he was on the lookout for starting points that might yet allow measures of negative eugenics to be reconciled with Catholic doctrine "through the back door."[283] Nevertheless, in 1932, together with the Catholic Center Party in Prussia, he turned about-face once again, playing a major role in creating a draft law for voluntary sterilization.

2.3.5 The Kaiser Wilhelm Institute for Anthropology, Human Heredity and Eugenics and the Debate About the Legalization of Eugenic Sterilization in the Context of Criminal Law Reform

The year 1929 marked a turning point in the sterilization debate that had been under way since the early 1920s.[284] In view of the intensifying crisis of the Weimar welfare state, the organized race hygienists saw their chance to succeed in instituting a policy for the legalization of eugenic sterilization in the context of criminal law reform. According to the legal practice of the Reich courts in Germany until 1933, sterilization and castration, independent of the indication, counted as aggravated assault according to sections 224 and 225 of the criminal law statute. However, there had been efforts to soften this legal position since the beginning of the Weimar Republic. As early as 1919, in contradiction to the legal practice of the Reich courts, a draft by the criminal law commission for a new German criminal law statute anchored the principle that an action performed in accordance with the rules of the medical art for the purpose of healing can never be assault. The official draft of a General German Penal Code (*Allgemeines Deutsches Strafgesetzbuch*) of 1925 went a step further, by eliminating from the stipulation about the legality of medical interventions the term "purpose of healing." Accordingly, section 238 then read: "Interventions and modes of treatment which are consistent with a conscientious

[279] The English subtitle of this encyclica.

[280] Richter, Katholizismus, pp. 263–267.

[281] Ibid., p. 276.

[282] Muckermann, Enzyklika Casti conubii. Cf. idem., Eugenik und Katholizismus, p. 48 f.

[283] Richter, Katholizismus, p. 277.

[284] On this – and the following – see first: Müller, Sterilisation; Weingart/Kroll/Bayertz, Rasse, pp. 283–306; Weindling, Health, pp. 388–393, 441–457; Schwartz, Eugenik, pp. 264–327; Richter, Katholizismus, pp. 197–256, 288–311.

doctor are not assaults in the legal sense."[285] As regards sterilization, the following section 239 was also of importance: "He who undertakes an assault will only be punished when the deed, despite consent, is morally offensive (*contra bona mores*)."[286] These two sections were integrated as sections 263 and 264 into the Reich Council draft for a German Penal Code, which was submitted to the Reichstag in 1927. According to these statutes, sterilizations with a medical indication would remain exempt from punishment, as they had been previously, with impunity granted on slightly different grounds. Whether sterilizations on the basis of social or eugenic indications were to be regarded as allowed in accordance with section 263 (as the practice of a conscientious doctor), or in accordance with section 264 (as not *contra bona mores*), was a hotly debated question of interpretation. What also remained open was whether the consent of a legal representative was to be considered sufficient for minors or those declared not legally responsible. From the eugenic perspective it was an obvious opportunity to present the legislators proposals in order to clarify these points, especially since the consultations of the criminal law commission dragged on for years.[287]

Alfred Grotjahn had made a first advance in the Prussian State Health Council in 1925, by advocating the compulsory sterilization of the "feeble-minded" and epileptics – but his petition was rejected by a large majority.[288] The Criminal Law Committee of the Reichstag concerned itself with eugenic sterilization for the first time in October 1928, after a petition from members of the Deutsche Volkspartei, the Bayerische Volkspartei and the Deutsche Demokratische Partei submitted a petition according to which court decisions about the early release of offenders declared mentally incompetent and held in preventive detention in asylums and santoriums was to be made contingent on their consent to sterilization before release. As grave reservations were raised against this proposal, however, it was "buried silently."[289]

In this situation, the organized race hygienists became active, and the KWI-A played a key role in this activity. In February 1929, the "Eugenic Working Group Elberfeld" (*Eugenische Arbeitsgemeinschaft Elberfeld*) took the initiative.[290] On July 4, 1929, a committee meeting of the Berlin Society for Race Hygiene took place in the rooms of the KWI-A, whose participants included – in addition to Fischer, Muckermann, and Verschuer – Alfred Grotjahn, Ernst Rüdin, and Carl von Behr-Pinnow. In his opening speech, Rüdin emphasized that there were now "reliable foundations for speedy, expedient laws [… on the] psychiatric indication for sterilization." For tactical reasons, however, he advised against demanding compulsory measures. All in all, the assembly came to an extraordinarily positive assessment of eugenic sterilization. At the end of consultations, Muckermann submitted the "double petition": (1) "certified doctors should have impunity […]

[285] Quoted in Söhner, Recht, p. 52.

[286] Quoted in ibid., p. 53 (note 1).

[287] Müller, Sterilisation, pp. 72 f.

[288] Saretzki, Reichsgesundheitsrat, pp. 325–327.

[289] Schwartz, Eugenik, pp. 293–299, quote: p. 299; Richter, Katholizismus, pp. 230–238.

[290] Dokumente aus der eugenischen Bewegung, pp. 20 f. Cf. Richter, Katholizismus, pp. 238–241.

under corresponding requirements, even if the sterilization were to be performed because of the health or life of progeny; (2) compulsory intervention is to be advocated at least in those cases that concern genetically encumbered criminals."[291] Remarkably, Muckermann, one of the most decisive representatives of the Berlin wing of the race hygiene movement, championed the application of force, at least in those cases which were considered in the Criminal Law Commission, while Rüdin, an exponent of the more radical Munich wing, urged caution, so that the demand for the decontrol of voluntary sterilization would not be put at risk.[292] Compulsory sterilization could not be achieved politically at this time. The assembly thus resolved to defer the issue of the compulsory sterilization of "genetically encumbered criminals" for the time being.

The Berlin branch of the German Society for Race Hygiene (Eugenics), chaired by Fischer and Muckermann, finally formulated a petition to the Criminal Law Committee of the Reichstag at its session on July 4, 1929 to add the following article to section 238 of the penal code draft:

> An assault in the sense of this law does not exist when an accredited doctor performs the artificial sterilization of a human with his or her consent, when the intervention is required according to the rules of the medical art to avert a serious risk for the life or health of the individual or his or her progeny.[293]

The crisis of the Weimar welfare state was listed as a reason:

> The need for public care in the German nation has taken on a scale many times greater than in the period before the war and takes from the healthy, hard-working bearers of the German future ever more funds that would be indispensable for the maintenance of their health and labor. One of the many causes of this calamity is found in the genetically encumbered, who are held in sanatoriums and clinics or in prisons.

With allusion to the Erlangen system of open care, the complaint was voiced that "many who appear permanently or temporarily normal, or who had served out their prison sentence [were] released from institutions," which led to an "increase in the demands on the national wealth." For this reason one had to either "avert the marriage of genetically encumbered families" or "prevent their reproduction."[294] This could, as was clearly conceded, also be achieved with a rehabilitation law, but sterilization was seen to be the more economical route.

The Munich branch of the German Society for Race Hygiene, under the central control of Rüdin and Lenz, concurred in this position with a petition to that effect submitted on August 31, 1929, and the general assembly of the German Society for

[291] Quoted in ibid., p. 241.

[292] See also Rüdin, Psychiatrische Indikation, p. 16: "Some consider a legal regulation and compulsory application as imperative in specific cases in view of the malevolence, lack of insight or indifference of certain inferior humans. For my part, for various reasons I believe that the time for compulsory sterilization is still in our distant future [...]."

[293] Dokumente aus der eugenisccchen Bewegung, p. 21.

[294] Ibid., pp. 21 f.

Race Hygiene later took over the proposal, remitting it with one slight change as a petition to the Criminal Law Committee.[295]

Ultimately this drive ran aground. Over the course of the last reading of the penal code draft, the view prevailed that section 263 of the 1927 draft was applicable for sterilization on the grounds of a medical indication, whereas for the eugenic – and social – indication, section 264 was valid, whereby it must be considered in each individual case whether sterilization was "morally offensive." Once the majority of the committee members had reached the conclusion that the question of eugenic sterilization was regulated in a satisfactory form by the – slightly altered – section 264, it was passed on January 28, 1932 with the votes of the SPD, the Staatspartei and the DVP, while the Center, the Bayerische Volkspartei, the Bayerischer Bauernbund, and the Protestant Christlich-Soziale Volksdienst voted against it. This voting result was, of course, only possible because the DNVP and NSDAP were boycotting the committee sessions due to the Harzburger Front. "The Reichstag's passing the altered section 264 as part of the entire penal code reform package was to that extent anything but certain"[296] – in fact, the impending crisis of the political system prevented the vote from taking place.

2.3.6 *Muckermann's Exposé on Positive Eugenics*

Around this time Muckermann, assisted by Fischer, made a final attempt to influence Reich policy in the the spirit of the eugenic movement.[297] In late 1931, he was empowered by the general assembly of the German Society for Race Hygiene (Eugenics) to write "an exposé about the question of how eugenicists could contribute to the reduction of the need for care and to the preservation of the genetically healthy family."[298] This memorandum, which developed a program of *positive* eugenics, was presented to Reich Chancellor Heinrich Brüning (1885–1970) via Fischer on December 19, 1931. Muckermann's influence on the Center Party and the Center bureaucracy had reached its pinnacle during the period of Brüning's administration. He enjoyed good contacts to the chancellor, and to his intimate confidant Hermann Pünder (1888–1976), state secretary and head of the Reich chancellory, who also belonged to the Center Party.[299] Yet Muckermann's initiative came too late; the days of Brüning's cabinet were already numbered. The Reich chancellor did indeed intend to work through Muckermann's exposé, but he did not manage to do so before he was ousted in May 1932. Neither was the exposé forwarded to the Reich Ministry of the Interior, as had been planned. Pünder did ask the new chancellor Franz von

[295] Ibid., pp. 23–25.

[296] Schwartz, Eugenik, pp. 305–311, quote: p. 310; Richter, Katholizismus, pp. 250–253.

[297] On the following ibid., pp. 67–73.

[298] Aus der Gesellschaft für Rassenhygiene und Eugenik, p. 231.

[299] Cf. Pünder, Politik, pp. 79 f.

Papen (1879–1969) to consider Muckermann's exposé, but neither Papen nor his successor Kurt von Schleicher (1882–1934) pursued the matter further. For the majority of the eugenic movement, Muckermann's memorandum did not go far enough anyway, because, as Fritz Lenz criticized, it "says nothing of sterilization."[300]

Once the attempt to carry through a clear regulation of eugenic sterilization on the Reich level had been exhausted in the "formulaic compromise of the Reichstag," the political activity over the course of 1932 – as at the beginning of the 1920s – shifted back to the state level. Various states saw endeavors to achieve a sterilization law of their own. Decisive in these attempts was "the *sterilization policy shift of Prussian state politics* – and it was so extraordinarily important especially because it became possible only through a fundamental course change in the alliance between Center Party, welfare bureaucracy and scientific political consulting dominant on this level. [...] At this stage, not the Prussian SPD, but rather the Center Party and bureaucracy were the new protagonists – in complete renunciation of their previous positions."[301] The KWI-A made a considerable contribution to this development – above all the breathtaking reversal of the Center was closely connected with the entanglements between the Prussian Ministry for Welfare, the organized race hygiene movement, and the KWI-A.

2.3.7 The Prussian State Council's Resolution on Eugenics

In October 1931, an initiative proposal was submitted to the "Committee of Communities" (*Gemeindeausschuss*) of the Prussian State Council, the state representation of the provinces, districts and communities, with which the Prussian state government was called upon to become active in the field of eugenics.[302] The proposal was presented to the committee by the Kiel physician Wilhelm Struve,[303] member of the State Council for the Deutsche Staatspartei and former chairman of the Prussian State Parliament Committee for Population Policy. It was formulated as follows:

> Considering the fact that the drop in birth rates among the circles of the genetically healthy, familially responsible population has especially strong effects, and that through this an increasingly strong increase in the genetic-biologically inferior part of the population results, and that the expenditures for people with genetically conditioned, physical and mental deficiencies have already reached a level that cannot be borne by our depressed economic situation, the Local Authority Committee proposes: The State Council will ask the state government, in the first instance, 1. to introduce training in the field of human genetics and eugenics as an obligatory subject for doctors, teachers and theologians, and as

[300] Lenz, Denkschrift, p. 232.

[301] Schwartz, Eugenik, pp. 311 f. (original emphasis).

[302] On the following: ibid., pp. 312–316; Richter, Katholizismus, pp. 288–294.

[303] For a biography: Weindling, Health, p. 356.

> an examination subject for doctors, 2. to impart instruction about the essence and importance of human genetics and eugenics to the rising generation in the upper classes of secondary school, and of those institutions for professional, further and higher education, 3. to require an exchange of eugenetic health certificates before couples become engaged, 4. to encourage the propagation of knowledge in the field of genetic theory and genetic care, 5. to entrust the Reich Council with the responsibility for making sure that demands 1 through 3 are also fulfilled in the other states [...].[304]

In his substantiation of the initiative proposal, Struve advocated a “strong differentiation” of the entire system of welfare care, whereby he referred explicitly to the results of Muckermann’s investigation. In view of “the contemporary economic emergency” it was necessary to “learn to think eugenically.”[305] In the Committee of Communities session on December 3, 1931, the Prussian Working Group (*Preussische Arbeitsgemeinschaft*) under the control of Baron Wilhelm von Gayl (1879–1945)[306] of the right wing of the DNVP, subsequently submitted a proposal that shifted the economic aspect into the foreground: “The State Ministry at once will see to it that, with the greatest possible acceleration, the costs to be expended by the communities, districts and provinces and the state for the care and support of the mentally and physically inferior be sunk to a level which can be borne even by a nation that is impoverished and bled completely dry.”[307] Both proposals could be traced back to impulses from Muckermann and Verschuer.[308]

Over the course of the discussions in the Committee of Communities the proposals were condensed into one joint committee petition, whereby, probably with consideration for the Center, the concrete recommendations by Struve as regards academic training, instruction at institutions of higher education and the exchange of marriage health certificates were watered down to a vague prescription of orientation. The State Ministry, as it was now worded, was to “take [...] measures in close contact with the appointed authorities (physicians, teachers, theologians), in order to obtain a greater propagation and respect for the recognized teachings of eugenics.”[309] This edited version of the committee proposal was passed on January 20, 1932 with the votes of the Prussian Working Group, the SPD, the Center Party and the Staats- und Wirtschaftspartei – only the Communist KPD abstained from voting. On the very same day the proposal was submitted to the plenum of the Prussian State Council meeting under the chairmanship of Konrad Adenauer (1876–1967) and was accepted without any major discussion in multipartisan consensus – only the KPD rejected it. The critique of the Communists was directed not toward its eugenic contents (they would have supported the original proposal by

[304] Quoted in Harmsen, Verminderung, pp. 41 f.

[305] Ibid., pp. 42 f.

[306] For a biography: Richter, Katholizismus, p. 290 (note 12).

[307] Harmsen, Verminderung, p. 46.

[308] Dürre, Eugenische Bewegung, p. 80. Cf. Schwartz, Eugenik, p. 312.

[309] Harmsen, Verminderung, p. 46, also in: Kaiser et al. (ed.), Eugenik, pp. 64 f.

Struve), but was rather against its being linked with the petition by Gayl, which the KPD rejected with reference to the already catastrophic conditions in the sanatoriums and institutions of care.

The Committee of Communities had also touched on the question of eugenic sterilization. Opinions diverged on this subject. The decision was perceived as "especially difficult, because especially responsible." While there were some representatives who insisted that it was impossible to "go further" without eugenic sterilization, the majority of the committee advocated deferring the question for the time being.[310] Since the scientific debate had not been concluded, they argued, it would be better to await the upcoming discussion in the Prussian State Health Council. In the Prussian Welfare Ministry it was clear that, "even though the resolution [of the State Council … did not deal with] sterilization for political reasons, […] in the consultations before [there had certainly] been the disposition and discussion and discourse for it."[311] The fact that the Welfare Ministry signalized its agreement with the resolution showed how open it was to the sterilization issue. All in all, the resolution by the Prussian State Council marked a "turnaround in the political climate with regard to race hygiene in general and sterilization in particular."[312]

In his 1931/32 annual report, Fischer booked the resolution of the State Council as a success of his institute,[313] justifiably so, as Muckermann – and to a lesser degree Verschuer – had paved the way for eugenics in the denominationally bound milieus and at the same time gave concrete impulses that were reflected in the consultations of the State Council. Now attention had to be directed to the forthcoming session of the Prussian State Health Council. Back in November 1931, independent of the discussions of the State Council, Heinrich Schopohl, the president of the State Health Council, had announced a meeting of this body on eugenic issues. During Schopohl's tenure as director of the Department for National Health, the influence of the Center Party on the Prussian Welfare Ministry reached its pinnacle – and with this the Catholic eugenicist Muckermann took center stage in Prussian sterilization policy. As far as the discussions in the State Health Council were concerned, Schopohl had discretely set the course beforehand, by appointing Muckermann and Fischer to full membership.

---

[310] Quoted in Schwartz, Eugenik, p. 314.

[311] Ostermann, Eugenik, p. 76.

[312] Weingart/Kroll/Bayertz, Rasse, pp. 296f.

[313] Jahresbericht 1931/32, MPG Archive, Dept. I, Rep. 3, No. 8.

2.3.8 *The Prussian State Health Council and the Draft Law on Eugenic Sterilization of July 1932*

The network around the KWI-A constituted what actually must have been the most compact interest bloc when the Committee for Population and Eugenics (*Ausschuß für Bevölkerungswesen und Eugenik*) of the Prussian State Health Council convened on July 2, 1932, to consider the topic "Eugenics in the Service of National Welfare."[314] Among the 23 members of the State Health Council who belonged to the committee were – in addition to Fischer and Muckermann – the Berlin Medical Councilor Wilhelm von Drigalski and Deputy Director and Departmental Director at the KWI for Biology, Richard Goldschmidt, both represented in the board of trustees of the KWI-A, and further Undersecretary Ostermann and the gynecologist Max Hirsch, who was also one of the eugenic movement's earliest proponents. Among the 36 scientific experts invited were Otmar von Verschuer and Lothar Loeffler, who had worked as an assistant at the KWI-A until 1929, and in addition Erwin Baur, director of the KWI for Breeding Research – a close friend of Fischer's who had paved the way for the KWI-A – and Agnes Bluhm, Fischer's companion from the inceptions of the eugenic movement. Fischer had worked with the anthropologists Otto Aichel (Kiel) and Karl Saller (Göttingen) in the framework of the "German Race Science" program, and also with the psychiatrist Johannes Lange (Breslau). The KWI-A also must have enjoyed close contact with the psychiatrist Emil Bratz, director of the Wittenau Sanatorium, and Kurt Pohlisch (1893–1955),[315] Karl Bonhoeffer's chief assistant at the University Psychiatric Clinic of the Charité. Another member of the expert circle was Wilhelm Struve. Among the representatives of associations were former Undersecretary August von Schenck, who represented the Association of Prussian Provinces (*Verband der Preußischen Provinzen*) on the KWI-A board, as well as the representative of the Central Committee for the Inner Mission Hans Harmsen, who worked with Verschuer in many different ways through the Inner Mission's Committee for Eugenics.[316] Fischer's extremely strong position at the session of July 2, 1932, becomes even more prominent when one keeps in mind that neither Fritz Lenz nor Ernst Rüdin participated, the main representatives of the Munich wing of the race

[314] Die Eugenik im Dienste der Volkswohlfahrt. See also: Müller, Sterilisation, pp. 95–99; Weingart/Kroll/Bayertz, Rasse, pp. 294–298; Weindling, Health, pp. 454–456; Schwartz, Eugenik, pp. 318–322; Richter, Katholizismus, pp. 294–305; Saretzki, Reichsgesundheitsrat, pp. 336–344.

[315] For a biography: Klee, Personenlexikon, pp. 467 f.

[316] Cf. the list of participants in: Die Eugenik im Dienste der Volkswohlfahrt, pp. 5–8. Weindling, Health, p. 455; Schwartz, Eugenik, p. 321; Richter, Katholizismus, p. 302 asserts that Karl Diehl also participated in the meeting, but this is a mixup. The session on July 2, 1932 was attended by the Krefeld physician Dr. Emil Heinrich Diehl, who was a National Socialist deputy to the Prussian State Parliament.

hygiene movement. The German Research Institute for Psychiatry was represented only by the former director of its Clinical Department, Johannes Lange. While the sustained influence of Rüdin and his school was unmistakable, especially as regards the indications for eugenic sterilization,[317] neither Rüdin nor Hans Luxenburger (1894–1976), who had made a name for himself with his twin research on the hereditary conditionality of schizophrenia,[318] took part directly in the decision-making process. While the Munich and Berlin schools were in agreement on the sterilization issue, the composition of the committee showed the distribution of power within the race hygiene movement. At this point in time Fischer was at the zenith of his political influence.

The opening speech was held by Hermann Muckermann. First he presented an overview of the state of human genetics, whereby he took every occasion to emphasize the research of the institute in Dahlem: Fischer's "bastard studies," Verschuer's twin method, Diehl's and Verschuer's examinations on the heritability of tuberculosis, but above all his own genetic statistical studies on differentiated reproduction, and works on this by the former members of the institute's staff, Curtius and Loeffler. Under reference to calculations employed by "a geneticist so canny as Otmar von Verschuer," Muckermann estimated the number of "critically mentally ill" in Germany to be over 200,000.[319] Picking up on the initiative of Struve's proposal, over the further course of his talk Muckermann sketched out a program for eugenic education. Both in the study of medicine and in the training of all those "in the service of education and welfare," genetics and eugenics were to be codified as compulsory subjects in the curriculum. "Despite economic destitution," the "sites of research should be strengthened and increased," and eugenic literature sponsored. This education must strive to make the voluntary "exchange of health certificates before engagement a family custom."[320] Marriage counseling centers were to keep their core eugenic task in mind. In the further course of his remarks, however, Muckermann made clear that this education program, combined with measures of negative eugenics like the implementation of preventative means for eugenic purposes and eugenic asylum were not sufficient. Since he rejected out of hand the "destruction of unworthly life," eugenic sterilization remained the only effective measure. Using roundabout formulations, the Catholic eugenicist argued that eugenically indicated sterilization should be declared not "ethically permissible,"

[317] Cf. Weber, Ernst Rüdin, pp. 178 f.

[318] Cf. also Kaufmann, Eugenische Utopie.

[319] Die Eugenik im Dienste der Volkswohlfahrt, p. 13. In 1930 Verschuer had estimated the number of the physically and mentally ill in Germany whose condition was hereditary at 300,000 (Verschuer, Umfang der erblichen Belastung). Lenz (Auslese, 3rd edn. 1931, p. 272) estimated the number of mentally ill, feeble minded and idiots at over 2 million; the number of people who are "mentally inferior" due to their genetic material at 6 million, or 10% of the population. For their estimates the race hygienists could refer to the fact that the Social Democratic social hygienist Alfred Grotjahn (Soziale Pathologie, p. 475) had estimated this share even higher, at one third of the population.

[320] Die Eugenik im Dienste der Volkswohlfahrt, p. 19.

but rather "exempt from punishment."[321] This was remarkable to the extent that Muckermann in this statement, as mentioned above, flouted the papal encyclical letter *Casti conubii*. However, Muckermann remained vague on this point. The "pope of positive eugenics"[322] quickly turned to his actual field of interest and demanded work-creation measures, family compensation, the graduation of taxes by income and number of children, as well as differentiated care: "The expenditures for the genetically encumbered must be shifted into a just relationship to the means available to the average genetically healthy family,"[323] Muckermann warned, with reference to a study by the Prussian Center deputy Helene Wessel (1898–1969),[324] which caused a furor in the debate at the time. Finally, Muckermann demanded that the eugenic idea be considered in rural settlement, as "a genetically sound rural population [is] the source of all life and all competence."[325] In total, Muckermann's speech, as Michael Schwartz judges pointedly, served the "positive-eugenic smoke-screen"[326] of the political course change of the Center Party on the sterilization issue. The other speeches – Johannes Lange, a psychiatrist in Breslau, gave a lecture on the state of knowledge about hereditary mental illnesses; Eduard Kohlrausch (1874–1948), a criminal law expert in Berlin, spoke about the question of the statutory legalization of voluntary sterilization – also indicated where the journey was heading.

After the three opening lectures, the guidelines of the Prussian State Health Council were opened up for discussion. These guidelines had derived from Muckermann's lecture, further had been written by him, and were closely connected to the guidelines of the German Society for Race Hygiene of the year 1932. Fundamental critique came from two directions. The young physician and National Socialist representative to the State Parliament, Leonardo Conti (1900–1945),[327] who was later to advance to the office of *Reichsgesundheitsführer* ("Reich Health Leader"), expressed sharp critique that the procedure was to be voluntary – "as a remnant of a past liberalist epoch" – and prophesied "that […] the practical successes of such a measure will be reduced from the outset to practically nothing."[328] Through his demand to incorporate the "race question" into the debate, Conti maneuvered himself completely into the margins so that he had to tolerate a rebuke from Schopohl.

Benno Chajes (1880–1938),[329] health expert of the SPD faction in the Prussian Parliament, also expressed his "disappointment" with the speeches and guidelines.

---

[321] Ibid., pp. 21 f.

[322] Schwartz, Eugenik, p. 165.

[323] Die Eugenik im Dienste der Volkswohlfahrt, p. 24.

[324] For a biography: Kersting/Schmuhl (ed.), Quellen, p. 385 (note 274). Cf. Wessel, Lebenshaltung.

[325] Die Eugenik im Dienste der Volkswohlfahrt, p. 24.

[326] Schwartz, Eugenik, p. 166.

[327] For a biography: Kater, Conti; Labisch/Tennstedt, Weg, vol. 2, pp. 393–395; Kersting/Schmuhl (ed.), Quellen, p. 617 (note 578).

[328] Die Eugenik im Dienste der Volkswohlfahrt, p. 59.

[329] For a biography: Labisch/Tennstedt, Weg, vol. 2, pp. 390 f.

On the one hand he criticized that the discussion was focusing on the aspect of eugenic sterilization and bemoaned that "the questions of birth prevention and artificial abortion [were] completely neglected." Chajes found the guideline on sterilization formulated by Muckermann to be "well in need of amendment." The impunity of sterilization for a eugenic indication should, according to Chajes, be codified clearly and unambiguously. With reference to the already existing sterilization laws in the Swiss canton of Waadt and in a number of states in the USA,[330] he proposed including a passage in the guidelines stating that "for certain cases [...] under compliance with clearly outlined legislative provisions, compulsory sterilization [would come] into question." Moreover, Chajes wanted to induce the State Health Council to endorse the impunity of voluntary sterilization not only for medical and eugenic indications, but also for *social* ones – this suggestion, however, met with vigorous protest from all sides. As to the work-creation measurements and family compensation Muckermann had proposed as instruments of positive eugenics, Chajes was skeptical; the State Health Council was not supposed to deal with economic issues.[331]

Fischer did not speak until the end of the session. His contribution was a tactical masterpiece. He emphatically retreated to the position of the genetic biologist, who – quasi *sine ira et studio* – took a position on practical medical, political, and economic issues according to strict scientific aspects. Fischer's speech can be read as an exclusive plea for a seemingly value-neutral scientific political consulting. Skillfully he managed to accommodate the critique from left and right, to scold mildly in the style of an "elder statesman," and above all to fill in the trenches between the opposing points of view. He did this by painstakingly factoring out all of the controversial questions in which agreement appeared impossible from the outset – the impunity of eugenic sterilization, the exercise of compulsion in sterilizations, the permissibility of the social indication, and the abortion issue.

First of all Fischer addressed the psychiatrists on the panel of experts. He acknowledged expressly that, as a consequence of savings measures in institutionalized psychiatry, the costs of closed care had already been sunk significantly, even implying that the policy of savings had reached its limits.

> If we simply cannot reduce costs any further, if we arrive at conditions similar to those during the war, if these persons can, in fact, no longer be held in compliance with human dignity – whereby not only food, drink, housing, and hygiene, but also medical treatment must be considered –, then arises the serious, absolutely burning question for us, what can happen to curtail the burden of heredity.[332]

---

[330] On the reception of foreign sterilization legislation: Hoffmann, Rassenhygiene; Popenoe, Rassenhygiene; Blasbalg, Gesetze; Steinwallner, Gesetzgebung; Binswanger/Zurukzoglu (ed.), Verhütung. On the US: Haller, Eugenics; Reilly, Solution; Trombley, Right; Kline, Race. On Switzerland: Keller, Schädelvermesser, pp. 152ff.; Müller, Sterilisation, pp. 37–41.

[331] Die Eugenik im Dienste der Volkswohlfahrt, pp. 87–90, quotes: pp. 87, 88.

[332] Ibid., p. 96.

Consequently, in a roundabout way, Fischer opened up the alternative between *either* minimizing the expenses for closed care to below the existential minimum – and thus triggering mass mortality behind the institutional walls as during World War I – *or* continuing along the path to opening up the institutions, *but then* supplementing the open care for the mentally ill, mentally handicapped, and epileptic with eugenic prophylaxis – concretely: by means of a sterilization program. This second option, as mentioned above, is the one which had been discussed so intensively in the perimeter of the German Society for Mental Hygiene. For Emil Bratz, the creator of the Wittenau scale system, which provided for a step-by-step transition from institutionalized treatment via partially stationary institutions all the way to release, Fischer and his plea for eugenic sterilization must have been preaching to the converted,[333] and even a more critically disposed psychiatrist like Karl Bonhoeffer found it difficult to remain aloof to his argumentation.[334]

Then Fischer turned to the deprecating position speech by Benno Chajes, whereby he picked out just one single point of his critique. Certainly, eugenic sterilization must not remain the only measure that could be taken, rather, it must be supplemented by an entire range of instruments of positive eugenics. However, in opposition to Chajes, Fischer argued against "simply abandoning" economic, financial, and tax policy to the politicians and economists. Genetic biology is a young science and therefore largely unfamiliar, Fischer objected, which brought him to the courses for medical officials taking place at the KWI-A: "[...] my 'pupils' in these courses were nearly all older than I, but I [nevertheless] have never heard from my pupils so many appreciative remarks about how a new world has opened up for them [...]."[335] Therefore, Fischer continued, he believes it to be "absolutely necessary that we genetic biologists and eugenicists support economic leaders, legislators, parliamentary groups, etc. and make our findings available to them.[336] In this context Fischer did not neglect to indicate that he had been in contact with the Prussian Welfare Ministry even before 1928, during Krohne's term, and had attempted to incorporate eugenic aspects in legislation. Fischer actually claimed primogeniture for himself and the direction of eugenics he represented. Referring to Conti's contribution to the discussion, he stated:

> I am glad to recognize that, of all parties, the National Socialists have included a great number of good eugenic aspects into their agenda – in addition to others we rejected. This party alone has included them as long as it exists. But it has not existed as long as our eugenic movement.[337]

---

[333] Cf. the speech by Bratz (ibid., pp. 64–66) and idem., Versorgung.

[334] On Bonhoeffer's position – in connection with the draft law, he did not want to exclude the possibility of exercising compulsion – Cf. Gerrens, Medizinisches Ethos, p. 85; Neumärker/Seidel, Karl Bonhoeffer, pp. 273 f.

[335] Die Eugenik im Dienste der Volkswohlfahrt, p. 96.

[336] Ibid., p. 97.

[337] Ibid., p. 98.

With an eye to the radical race hygiene the National Socialists had argued for, Fischer took pains to liberate eugenics from the odium of *human breeding*. "It has been said," Fischer addressed this point, "it is a bold experiment to 'intervene' in something given by nature, the reproduction of human beings – the word 'breeding' was used in this context earlier." Fischer skillfully advanced the argument that there was no longer such thing as *natural* reproduction in modern industrial society. Rather, this phenomenon was subject to the constraints of economic relations, so that the hard economic necessity of women's labor compels "birth control," but that this takes place "entirely without plan and without any sense or consideration for the health of the nation and the future." As such, a "rationalization of births" was indeed necessary.[338]

Fischer closed his speech with the appeal to the assembly to approve the submitted guidelines "by and large and in principle [...] as the foundation of further consultation and the formulation of future petitions." He closed with thanks to Schopohl and his ministry, for having "given us the opportunity [...] to undertake a truly great scientific thrust in this direction for the first time."[339]

The spontaneous applause Fischer received showed that he had hit upon the right tone. Indeed, after Fischer's speech the discussion ebbed. Schopohl breezily noted that the discussions had resulted in "this committee of the State Health Office, with acknowledged unanimity, [sides with] the reports presented here and the guidelines experienced along with them." Thus Schopohl passed over the at times quite fundamental critique, not only on the part of the Social Democrats and National Socialists; further, he saw no reason not to interpret the "silence of a number of prominent specialists" as "affirmation."[340] The assembly established a commission, which was entrusted with compiling the guidelines and working out a draft law on voluntary sterilization. Its members, besides the president of the State Health Council Heinrich Schopohl, included Undersecretary Arthur Ostermann, Eugen Fischer, and Hermann Muckermann. "The dominance of the 'Center eugenics' in the commission was tangible."[341] The commission convened on July 30, 1932 and, after a short discussion, passed the draft of a law on voluntary sterilization. This draft bore the signature of Muckermann and the Prussian Center bureaucracy, whose days were already numbered at this time, 10 days after the "Prussian Putsch" of July 20, 1932. Thus the draft never became law. However, it constituted an important milestone on the road to the "Law on the Prevention of Genetically Diseased Offspring" (*Gesetz zur Verhütung erbkranken Nachwuchses*) of July 14, 1933.

[338] Ibid., p. 97.

[339] Ibid., p. 98.

[340] Ibid., p. 100.

[341] Richter, Katholizismus, p. 304.

2.4 Eugen Fischer and Nationalist Race Science

Fischer and his faction had positioned themselves at the head of the eugenic movement at the beginning of the 1930s and dominated the key positions at the interfaces between eugenics and science, as well as those between eugenics and politics. As the group of researchers around Peter Weingart justifiably emphasized, in this development it would have been possible to "see a victory of 'more moderate' eugenicists over the 'more radical' race hygienists," if the "political spectrum as a whole [had not] already shifted to the right."[342] Ostensibly, Fischer took pains to remain politically neutral. For this reason, from 1926 he withdrew from his political activities for the DNVP.[343] In fact, Fischer was widely considered to be an exponent of a moderate position – his admission to the time-honored "Wednesday Society" in November 1927 perhaps expressed this best.[344] In reality, however, Fischer was anything but clearly demarcated from radical race hygiene, nationalist race science and anti-Semitism.

2.4.1 The "Schemann Affair"

This became abruptly clear through the "Schemann affair."[345] On December 12, 1929, an article appeared in *Vorwärts* with the title "Pulp Fiction from Reich Funds?" (*Schundliteratur aus Reichsmitteln?*), which exposed that the nationalist scholar Ludwig Schemann (1852–1938),[346] who had become known for his German translation of *Essai sur l'inégalité des races humaines* (1853–1857) by the French race theorist Joseph Arthur Comte de Gobineau (1816–1882), had received grants from the Emergency Association of German Science from 1926 to 1929 to complete his three-volume work *Die Rasse in den Geisteswissenschaften* ("Race in the Humanities"), published by the nationalist publishing house J. F. Lehmann in Munich. The first volume, *Die Hauptepochen und -völker der Geschichte in ihrer Stellung zur Rasse* ("The Main Epochs and Nations of History in their Attitude toward Race"), published in 1927, abounded with anti-Semitic excursions: The German nation is "soured through with Jewish blood"; if a "political reversal [did not prevent] the worst from happening to the race," it would "perish in a pestilent air in which only the sub-human can still flourish."[347] From the Weimar Republic Schemann did not expect such a reversal, as the Reichstag and its members were,

[342] Weingart/Kroll/Bayertz, Rasse, p. 253.

[343] Proctor, Anthropologie, p. 157; Gessler, Eugen Fischer, pp. 24 f.

[344] Scholder, Mittwochs-Gesellschaft, pp. 21 f., 368.

[345] On the following: Nemitz, Antisemitismus; Lösch, Rasse, pp. 221–225; Massin, Rasse and Vererbung, pp. 192–194.

[346] For a biography: Klee, Personenlexikon, p. 530.

[347] Quoted in Nemitz, Antisemitismus, pp. 385 f.

in his eyes, the "rule of the inferior."[348] In the first instance it was the Social Democratic Reichstag representative Julius Moses who exposed the science policy scandal. Fischer was implicated as well, because, in his function as a consultant to the government, he had repeatedly approved Schemann's applications for funding. Fischer, who had been in contact with Schemann since before World War I (the latter had moved to Freiburg in 1897), had expressed his conviction in 1926 that the book to be sponsored would become "a scientifically highly important work."[349] Fischer knew very well that Schemann's work was to be understood "more politically than scientifically."[350] For all of his critique in detail, Fischer was nevertheless quite open to Gobinist race theory. He was also aware of the enormously broad effect of an author like Schemann. Ultimately he attached science *policy* relevance to Schemann's book project. In his expert opinion of August 13, 1927, on Schemann's first application to extend his grant, Fischer recommended sponsoring Schemann because of the expectation that the second volume of the trilogy would be "of particular importance for anthropological science to the extent that a name like that of Schemann's would force scholars of the humanities to finally grant anthropology the attention it has long deserved."[351] This reveals a sort of calculation for the imperialism of the discipline: Schemann's work was to function as a vehicle to carry the ideas of race anthropology to the domain of the humanities. Fischer then appeared less than impressed by the public stir that led to the withdrawal of Schemann's stipend. Fischer advised Schemann not to inflate the importance of the matter, consoled the venerable pioneer of the Nordic idea that the scandal would help the book obtain more publicity "among Semites and anti-Semites,"[352] and made his contribution by publishing a positive review in the *Zeitschrift für Morphologie und Anthropologie* ("Journal for Morphology and Anthropology").[353] In 1930, Fischer requested from Schemann a portrait of Gobineau, which then apparently hung in the rooms of the institute in Dahlem.[354] Fischer also remained in friendly contact with Schemann.

The "Schemann affair" was not an isolated incident. Fischer also reviewed the work of Houston Stewart Chamberlain (1855–1927)[355] positively[356] and collaborated with Hans F. K. Günther, from whom he had distanced himself demonstratively during the phase of the institute's founding. Notwithstanding, at this time Fischer and Günther had jointly published a short book about *Deutsche Köpfe nordischer Rasse*

[348] Lösch, Rasse, p. 225.

[349] Nemitz, Antisemitismus, p. 399.

[350] Lösch, Rasse, p. 224.

[351] Quoted in Nemitz, Antisemitismus, p. 401.

[352] Quoted in Lösch, Rasse, p. 225.

[353] Fischer, Review: Ludwig Schemann, Die Rasse in den Geisteswissenschaften.

[354] Massin, Rasse und Vererbung, p. 193.

[355] For a biography: Field, Evangelist.

[356] Fischer, Review: Houston Stewart Chamberlain, Rasse und Persönlichkeit.

("German Heads of the Nordic Race").[357] Günther's numerous pamphlets received positive reviews in Fischer's journal. On the occasion of the 9th edition of Günther's *Rassenkunde des deutschen Volkes* ("Race Science of the German Volk") (1926), Fischer wrote that the "brilliant success of the book" was to be "welcomed with great pleasure. Finally even broad circles understand what race is [...]." Upon its 12th edition (1928), Fischer confirmed, "That anthropology is gradually acknowledged today is due in part to this book; that must be honestly acknowledged [...]."[358] Fischer also found praise for Günther's *Rassenkunde des jüdischen Volkes* ("Race Science of the Jewish Volk) (1930).[359] Fischer's reviews make especially clear that he, in spite of a few reservations, concurred in essential characteristics with the nationalist views of Schemann, Chamberlain and Günther – even in the strict hierarchization of the human races, the connection between race and culture, the assumption of *mental characteristics of race*, and the esteem for the *Nordic race* – and that he made no secret of this concurrence. What Fischer criticized about the nationalist authors was the static, so to speak pre-Darwinian concept of race, from which he disengaged more and more – and Fischer always elaborated this critique in a tactically clever manner whenever this appeared opportune in the given political context, through which the convergence and congruencies faded into the background. However, arguing in terms of educating the people, Fischer always emphasized the immensely broad effect of the popular science works by Schemann, Chamberlain, and Günther, which were supposed to prepare the soil for the scientifically grounded anthropology of the German nation yet to come.

2.4.2 Eugen Fischer and the International Eugenics Movement

A look at the international level makes clear that the wing of the eugenic spectrum to which Fischer was assigned definitely depended on the standpoint of the observer. At any rate, Stefan Kühl places Fischer, on the basis of his activities in the international eugenic movement, in the "orthodox-racist"[360] faction, which united eugenics and racism. Fischer was a member of the International Federation of Eugenic Organizations (IFEO) founded in 1925, a "reservoir for eugenicists who linked their scientific interest in an internationally organized race research with a political propaganda for the enhancement of the 'white race'."[361] Within the IFEO, Fischer came to the fore as chairman of the commission founded in 1927 for the research of "bastardization" and "miscegenation." Assembled in this commission were the *crème de la crème* of the "bastard researchers": Charles Davenport from the USA, who had performed "bastard studies" on Jamaica with his colleague

[357] Fischer/Günther, Deutsche Köpfe.

[358] Fischer, Review: H. F. K. Günther, Rassenkunde des deutschen Volkes (1926, 1928 respectively). Cf. Massin, Rasse und Vererbung, pp. 193f.

[359] Fischer, Review H. F. K. Günther, Rassenkunde des jüdischen Volkes (1930).

[360] Kühl, Internationale, p. 75.

[361] Ibid., p. 71. On the following, cf. ibid., pp. 71–83.

Morris Steggerda (1900–1950) from 1926 to 1929; from Germany, the tropical physician Ernst Rodenwaldt (1878–1965),[362] who had published his classic work about the *Mestizen auf Kisar* ("Mestizos on Kisar") in 1922; the Swedes Hermann Nilsson-Ehle (1873–1949) and Herman Lundborg, who had worked on "race hybrids" between "Lapps" and the "Nordic Swedish population"; from Norway, Alfred Mjöen (1860–1939); from Great Britain, Ruggles Gates (1882–1962), who had researched on "Indian cross-breeds" in Canada. The initiative had come from Davenport, who aspired to a "World Institute for Miscegenations" and was working on a "world map" of the "mixed-race areas,"[363] which he introduced for the first time at a meeting of the IFEO in Munich in 1928. Fischer and Davenport published an article together in German, English, and American journals, in which they appealed for international cooperation. The contrast between the "minor scientific activities on the field of bastard research" and the "degree of miscegenation" in almost all parts of the world, they claimed, was "immense."[364] Davenport and Fischer prepared a questionnaire in English, German, French, Spanish, and Dutch, which was sent by the IFEO to doctors, missionaries, officials, merchants, farmers, and travelers abroad – without much resonance, however. Ultimately the project ran aground.

Fischer, together with Erwin Baur, Hermann Muckermann, Ernst Rüdin, Fritz Lenz, and Alfred Grotjahn, the demographers and statisticians Friedrich Zahn (1869–1946), Hans Harmsen, Friedrich Burgdörfer (1890–1967),[365] and Robert Kuczynski (1876–1947), as well as the two national economists Paul Mombert (1876–1938) and Julius Wolf (1862–1937), belonged to the German section of the International Union for the Scientific Investigation of Population Problems (IUSIPP), which had been founded at the conclusion of the first World Population Conference in Geneva in 1927.[366] The "Fischer faction" made sure that the German section – as opposed to other sections – remained fast in the grip of the "orthodox" eugenicists. In 1931 the IUSIPP threatened to split. The board of directors canceled a second international conference on population science, which was originally to take place in Rome, because it feared that its political neutrality could not be maintained in Fascist Italy. As a consequence, the Italian eugenicist Corrado Gini (1884–1965), with the support of the Fascist government and the IFEO (but without the blessing of the IUSIPP), held an international congress on population research in Rome. In this situation Fischer – as Davenport – sided with Gini and took his seat at the president's table in Rome.[367]

---

[362] For a biography: Klee, Personenlexikon, pp. 501 f.

[363] Kühl, Internationale, p. 81.

[364] Manuscript in: MPG Archive, Dept. I, Rep. 3, No. 23, pp. 159–162.

[365] For a biography: Klee, Personenlexikon, pp. 85 f.

[366] Cf. Kühl, Internationale, pp. 110–117; Ferdinand, Bevölkerungswissenschaft, pp. 65–71.

[367] Kühl, Internationale, pp. 118–120; Ferdinand, Bevölkerungswissenschaft, pp. 73 f.

The IFEO had already met in Rome once, in September 1929, with Benito Mussolini (1883–1945) also attending. Upon this opportunity, *Il Duce* was presented an address written by Fritz Lenz and edited by Eugen Fischer,[368] in which the IFEO, picking up on the imperial style of the Fascist government, urgently advised Mussolini to integrate eugenic elements in his population policy agenda:

> Here and now, in the oldest capital of the world, we ceremoniously articulate the hope that it might be granted to precisely those responsible men of the gifted Italian nation, to demonstrate first and as a model that an energetic will is capable of compensating for the damages which our culture inflicts upon the reproduction of the nation and the preservation of the talented. May it be granted what was denied to previous cultures, to grasp the wheel of fate, to face up to it and reverse it! Quality in addition to quantity! And it is high time, the danger is formidable. Videat consul!"[369]

Fischer set great hopes in Mussolini, whom he regarded as the only politician who could and would "really carry out eugenic measures,"[370] but saw his expectations disappointed, for in his opinion the population policy of *Il Duce* continued to pay too little attention to the qualitative aspect.[371] A few years later, Adolf Hitler's rise to power in Germany meant a leader who could and would fulfill Fischer's hopes, and Fischer readily tendered his specialized knowledge to the new rulers. Yet initially he fell in disfavor in the eyes of the National Socialists, he had been far too involved with the Weimar welfare state, such that the "dirty smell of democracy" adhered to him and his institute.

---

[368] This composition process is implied in: Lenz to Verschuer, 20/9/1929, MPG Archive, Dept. III, Rep. 86 B, No. 1, p. 7 f., here: p. 8.

[369] Draft with edits by Fischer in: MPG-Archives, Dept. I, Rep. 3, No. 23, pp. 262–265, quote: p. 265. Reprint: Archiv für Rassen- und Gesellschaftsbiologie 22, 1930, p. 435.

[370] So Fischer in a letter to Davenport of 19/7/1929, quoted in Kühl, Internationale, p. 266 (note 30).

[371] For the population policy of Fascist Italy: Cf. Ipsen, Dictating Demography; Schneider, Mussolini.

Chapter 3
The "Faustian Bargain": The Kaiser Wilhelm Institute for Anthropology, Human Heredity and Eugenics in the National Socialist Era, 1933–1938/1942

After a short phase of upheaval characterized by provocation and friction in the years 1933/34, a modus vivendi settled in between the Kaiser Wilhelm Institute for Anthropology, Human Genetics and Eugenics and the National Socialist "health leadership," which could best be described as the reciprocal instrumentalization of science and politics. To pick up on an interpretive figure drafted by Mitchell Ash and recently further developed by Sheila F. Weiss, politics and science represented "resources for each other," and this was the case for the KWI-A as well.[1] This was true not only for the period after 1933. As described above at length, from the very beginning the KWI-A had attempted to extend its sphere of influence into the realm of political activity, demanding public funding in return – the very prototype of a *state-dominated* Kaiser Wilhelm Institute. In the "Third Reich," too, it was willing to put its specialized knowledge into the service of the new regime, which for its part – to a higher degree than the governments of the Weimar Republic – was willing to ground its political decision-making processes on scientific expertise and legitimation. Important political actors soon recognized the potential value of the KWI-A as a resource for "knowledge, apparatus, personnel, institutional structure or rhetoric,"[2] which they exploited to realize their political objectives. Conversely, the interest of influential men in power in a scientific foundation for population, health, and social policy constituted a resource for the KWI-A, primarily in terms of financing. However, it is important to distinguish carefully between the two levels of means and ends: On the level of means – jockeying for power and influence, positions, and money – politicians and scientists contracted a strategic alliance, which was renegotiated over and again to strike a balance between both parties' interests. On the level of ends – in the biopolitical utopia of a scientifically guided dictatorship of development, which aspired to control birth and death, sexuality and reproduction, body and germ line, variability and evolution – there was vast convergence and congruence: politicians and scientists worked together out of conviction. This is where the actual basis of the alliance between science and politics lies.

[1] Ash, Wissenschaft und Politik; Weiss, Humangenetik.

[2] Ash, Wissenschaft und Politik, p. 32.

3.1 Pressure for Consolidation and Willingness to Conform in 1933/34

The first task, of course, was to clarify the balance of power within this alliance. Eugen Fischer, a learned mandarin still completely caught up in the attendant intellectual arrogance, continued to proceed from the assumption of a *technocratic* model of political consulting, which conceived of political decision-making processes as the solution to factual issues and ultimately aspired to replace politics with science and technology. The fascist rulers, however, soon showed him that this model was not their guide. Although they wanted to place their genetic health and race policy on a scientific foundation, they proceeded from a *decisionistic* model of political consulting, reserving the decision-making authority for themselves. The primacy of the political remained untouched. This was the paradigm according to which the roles of the alliance partners were distributed in a conflictual process of consolidation and conformity.

3.1.1 Eugen Fischer in Distress

In the final phase of the Weimar Republic, Eugen Fischer ostensibly had retreated to the position of political neutrality as a scientist. Although he had worked together closely with the government dominated by SPD and Center under Otto Braun (1872–1955) and Carl Severing (1875–1952), he had avoided any attachment to a political party and taken pains to remain equidistant to the parties of the Weimar coalition and the National Socialists. Even in the final months of the Weimar Republic he refused to work more closely with the National Socialists. Characteristic of this posture was his reaction to a letter of November 8, 1932, to the leading eugenicists and race hygienists of the German Reich, written by Hermann Alois Boehm (1884–1962),[3] the director of the Department of National Health, part of the Third Main Department of the NSDAP organization in the Reich. In the "Race Hygiene Subsection," Boehm, picking up on a proposal by Fritz Lenz, wanted to create "loose working groups" to create a sustainable basis for quantitative and qualitative population policy, and to win "specialists" for the program who were "in themselves distant" from the NSDAP: "Any existing differences of opinion regarding party politics must retreat in the face of the lofty goal of preservation and enhancement of the nation […]."[4] The moderate tone of the letter, its appeal to national responsibility, the offer of nonparty cooperation it included, the interest in it emphasized in scientific expertise, not to mention the appointment of the nonpartisan Fritz Lenz – Boehm made every effort to build golden bridges to the Berlin wing

[3] For a biography: Jakobi et al., Aeskulap and Hakenkreuz, pp. 161–169; Klee, Personenlexikon, pp. 58 f.

[4] Quoted in Weingart/Kroll/Bayertz, Rasse, p. 384 (original emphases).

of the race hygiene movement. While Fritz Lenz, Ernst Rüdin, and even Hermann Muckermann agreed to work with the Nazis, Fischer sent a negative response to Boehm's inquiry. Still, he followed "the endeavors of the NSDAP on the field of eugenics with the most vivid sympathy and great gratification [...]. Just recently, in a session of the Prussian Health Council, I mentioned publicly that this party is the only one to establish a eugenic program, most of which I could subscribe to (the other part concerns the question of foreigners)." Yet Fischer rejected cooperation extending beyond personal suggestions, in order to preserve the political neutrality of his institute. However, Fischer wrote:

> By no means does this entail an opinion against or even a mere value judgement about any party at all or the party system in itself; it is purely a result of the necessity to live with my research free of all commitments and through this, I am convinced, to serve our nation.[5]

Just as it had been a matter of course for Fischer to place his expertise in the service of the "black-red" coalition government in Prussia as recently as July 1932, when the government was preparing to formulate legislation to regulate eugenic sterilization, from February 1933, he was just as willing to do the preliminary work of formulating a race and genetic health policy for the "government of national concentration" in his function as a scientific policy consultant.

Against this background, Fischer held a public lecture on February 1, 1933, about "Miscegenation and Intellectual Achievement" at Harnack Haus, the representative venue and meeting place of the KWG. The fact that the director of the leading research institute in the field of anthropology, human genetics, and eugenics spoke on such an explosive subject a mere 2 days after Hitler's cabinet was appointed had to arouse the interest of the public – the audience expected Fischer to take a position on the policy of discrimination against the Jewish minority proclaimed by the National Socialists. Therefore, not only specialized colleagues from the KWG and the university appeared, but also representatives of government bodies and correspondents of both German and international newspapers and press agencies. "That a lecture on this topic held at this time would also be understood politically," Niels C. Lösch writes, "must have been clear to Fischer [...]."[6] And in fact Fischer must have foreseen quite well that political meaning would be attributed to his public appearance. What he had not understood, however, was that the role he had intended for himself – namely, that of the scientist located above the parties, who, in the national interest, publicly pointed out to the government when the genetic health and race policy it pursued was not backed by the findings of his science – would not be able to bear up in the "new Germany." Fischer's lecture on February 1, 1933, had been prepared and written at a point in time when there was no way to foresee that Hitler would be appointed Chancellor of the Reich – the press release was dated July 18, 1932. Nevertheless, the lecture must have been conceived from the outset as a message in the direction of National Socialism, in

[5] Fischer to Boehm, 25/11/1932, quoted in: Lösch, Rasse, p. 243.

[6] Ibid., p. 231.

order to induce Hitler and his party to a course correction in its race policy. Mind you: not to a *fundamental* course correction. The sole issue involved was *miscegenation*, in which Fischer took a different position than the party ideologues on the basis of his "bastard studies." After January 30, 1933, Fischer saw no reason at all to depart even an iota from the prepared lecture. On the contrary, now that the National Socialists were integrated into responsibility for the government, instruction about how they should handle race politics "correctly" seemed even more urgent than before. Fischer unequivocally expressed that he welcomed the "government of national concentration" in principle – before the campaign against him began – by signing an appeal by Berlin professors to vote for Adolf Hitler printed in the right-wing newspaper *Völkischer Beobachter* immediately before the parliamentary elections of March 5, 1933.[7]

Fischer had misread the signs of the time. While the new government pinned its hopes on scientific policy consulting, especially in the areas of population, genetic health, and race policy, the National Socialist rulers rewrote the rules of the game. A public lecture expressing scientific critique of the core of their race ideology, even if it was meant as constructive criticism, was something they were not willing to stand for. It was certainly clear to the representatives of the domestic and foreign press who attended the lecture that the National Socialists would have to understand Fischer's remarks as a provocation. In any case, the lurid headlines of the press articles – for example, "Bastardization of World History!" was the lead in the *Essener Allgemeine Zeitung*[8] – certainly indicate that the journalists had recognized the imminent showdown. In deviation from the usual conventions, Fischer's lecture did not appear in printed form, but its main theses can be reconstructed from the advance notices and press reports. The argumentation's point of departure was the conclusions that Fischer had drawn from his "Bastards of Rehoboth" studies:

> The crossing of intellectually very capable races with intellectually less capable ones, e.g., Europeans with Negroes, yields a product that is between these races intellectually […] Occasionally, an individual half-breed of this kind can also be strikingly intellectually capable (Pushkin, Washington), as is to be expected from the laws of heredity. The assumption that half-breeds are always worse than both parents intellectually – or even morally – is incorrect.[9]

This is even less true when two more or less equally capable races interbreed. Rather, the resulting hybrid population is superior to the initial races:

> Interbreeding may be the most important factor in the emergence of high cultures all over the world. It is incorrect to attribute the highest cultural achievements to relatively pure races. The Nordic race, where it is purest, has no cultural achievements to show for itself

---

[7] Ibid., p. 262.

[8] Ibid., p. 231.

[9] Magdeburger Zeitung No. 123, 8/3/1933, p. 6, quoted in ibid., p. 232. Aleksandr Sergeyevich Pushkin (1799–1837) was a great-grandson of Ibrahim Hannibal (*c.*1698–1781), Peter the Great's famous "Moor." The Washington meant here was not George Washington, but rather the African American civil rights activist Booker T. Washington (1856–1915).

> that are higher than those of many other cultures. In contrast, it shows its highest achievements in hybrid zones with related races apparently its equal or at least not significantly inferior.

As such, Fischer declared Central Europe to be a miscegenation "hot spot":

> The entire development of great art and culture from the Gothic era over the Renaissance and Baroque periods is located in a hybrid zone of the Nordic race mixed with the Alpine and Dinarian, extending to the north and south of the Alps over Germany and Italy, into France and down the Rhine.[10]

Even up to this point, Fischer's lecture must have been hard for some race ideologues bound to the Nordic idea to digest. But the ranks of the National Socialists also included voices who warned against an absolutization of the Nordic race,[11] and Fischer might not have become the target of a political campaign had he not taken his argumentation even further. However, over the further course of his lecture, Fischer expressly addressed the Jewish question. By applying the newer, dynamic race concept from population genetics to the Jewish races as well, he arrived at a momentous differentiation:

> As an example for the influence of lifestyle on the characteristics of children emerging from interbreeding, Professor Fischer used the example of crosses between Nordic races and Jews. He is of the opinion that it makes an enormous difference whether Nordic people cross with the offspring of old, cultivated Jewish families or those from recently immigrated Eastern Jewish families.[12]

In this, conclusions from anthropological human genetics research were in alignment with widely propagated anti-Semitic stereotypes against the "Eastern Jews." In contradistinction to the National Socialists, some academics from the national conservative camp wanted to accept the completely assimilated Jewish bourgeoisie from the special legislation for Jews, which they also endorsed. Fischer gave this position a scientific foundation by distancing the cultivated Western Jews quasi as a separate biological race from the philistine Eastern Jews. Thus, he had transgressed one of the points up to which the regime tolerated a discussion of its race policy. These points were not easy for a scientist to recognize, since the National Socialist race ideology was anything but internally coherent and consistent, and the regime tolerated a remarkable bandwidth of race concepts, albeit on the basis of a given intersection of basic convictions. One point about which the brown rulers did not allow any discussion, however, was exceptions for the acculturated Jewish bourgeoisie, like those Fischer proposed with anthropological arguments. The anti-Semitism of the National Socialists fed too strongly on antibourgeois feeling.

---

[10] Magdeburger Zeitung. Like note 5.

[11] Cf., e.g. Lilienthal, Rassenhygiene, pp. 121–123.

[12] From a report in the Bayerische Ärztezeitung No. 12, 1933, p. 141, about Fischer's lecture; quoted in Lösch, Rasse, p. 233.

3.1.2 The Changing of the Guard in the German Society for Race Hygiene

Sure enough, the first attack against Fischer came from his own ranks, from the Munich Society for Race Hygiene. In this case the initiative came from the younger generation around Karl Astel (1898–1945)[13] and Bruno K. Schultz (* 1901),[14] who had already drifted into the National Socialist camp before 1933. Astel, a former combatant in the Freikorps and member of the NSDAP since 1930, was working at the time as a race hygienist at the Reich Leadership Academy of the SA and in the Race and Settlement Office (*Rasse- und Siedlungsamt*) of the SS (from 1935: Main Race and Settlement Office [*Rasse- und Siedlungshauptamt*, RuSHA]). Schultz, a member of the SS since 1932, was also employed in the Race and Settlement office, as a "Technician for Race Science" (*Fachbearbeiter für Rassenkunde*). Since 1929, he had also served as secretary of the nationalists' popular science magazine *Volk und Rasse* ("Nation and Race") published by the Julius F. Lehmann publishing house, which advanced under his aegis to become the official organ of the German Society for Race Hygiene and the Reich Committee for National Health. Both Astel and Schultz, after earning their first spurs in the consolidation of the German Society for Race Hygiene, were to enjoy a rapid ascent through the offices of the Third Reich.[15]

At the beginning of the 1930s, the group around Astel and Schultz had resisted the German Society for Race Hygiene's course change towards eugenics, to no avail. When the new regime took power, the National Socialist faction found itself on top and prepared to disempower the board of the German Society for Race Hygiene (Eugenics), consisting of Fischer, Ostermann, and Muckermann. In late May 1933, the Munich branch, upon a petition submitted by Astel and Schultz, resolved to demand that the board be voted out. Astel and Schulz justified their petition as follows:

> The Munich Society for Race Hygiene [...], in consideration of the immense importance accorded to race hygiene in the new state unified under Adolf Hitler, believes that it is essential for the board of the Society to be comprised of men whose positive attitude to race

---

[13] For a biography: Klee, Personenlexikon, p. 20.

[14] Ibid., p. 565.

[15] In the same year, 1933, Astel became president of the "Thuringian State Office for Race Issues" (*Thüringisches Landesamt für Rassewesen*) in Weimar. One year later he was appointed professor and director of the University Institute for Human Breeding Theory and Genetic Research (*Universitätsanstalt für Menschliche Züchtungslehre und Vererbungsforschung*) in Jena. In 1936, he advanced to become director of Health and Welfare Issues in Thuringia. In 1939 he became rector of the University of Jena and took over the direction of the District Station of the Race Policy Office (*Gauamtsstelle des Rassenpolitischen Amtes*) of the NSDAP. Schultz was appointed assistant professor for race biology at the Reich Academy for Gymnastics in Berlin in 1938. In 1942 he received a professorial chair and the directorship of the Institute for Genetic and Race Hygiene (*Institut für Erb- und Rassenhygiene*) of the German Karl University in Prague and became chief of the Race Office in the RuSHA Prague.

> hygiene and affirmation of the state of Adolf Hitler is beyond a doubt. This is not the case for the current board. The Munich Society for Race Hygiene therefore demands a corresponding change in the board [...].[16]

The Munich branch further resolved to send to the next general meeting, which was to take place in Berlin on June 10, 1933, not Fritz Lenz, as originally planned, but a delegation headed by Alfred Ploetz, who would be accompanied by Astel and Schultz. This delegation was to bundle all votes of the Munich branch and deploy this voting bloc to oust the current board. According to yet another petition by Astel and Schultz, the delegation was to lobby intensively to have Ernst Rüdin appointed as the First Chairman of the German Society for Race Hygiene. The session of the Munich branch was apparently quite turbulent, and it seems that Astel and Schultz exerted massive pressure to achieve the unanimous acceptance of their petitions. In any case, they emphasized in their report to Reich Minister of the Interior Wilhelm Frick (1877–1946), "that, had we two not made corresponding preparations, all of the resolutions could just as easily have been decided unanimously in the opposite direction."[17]

Lösch divides the members of the Munich branch into two camps: one from the circle around the Julius F. Lehmann publishing house, the other from the German Research Institute for Psychiatry – in addition to Ernst Rüdin, this group comprised Bruno Schulz (1890–1958), Adele Juda (1888–1949),[18] Hans Luxenburger, Hugo Spatz (1888–1969), Friedrich Stumpfl, and Heinz Riedel (*1904).[19] While the KWI-A and the DFA were bitter rivals, and the relationship between Fischer and Rüdin continued to be charged (although it had relaxed somewhat since Rüdin's return to Munich), reservations against the frontal attack against Fischer must have been roused in the ranks of the DFA as well, and even Theodor Mollison and Fritz Lenz must have hesitated to take a position against Fischer.

However, the crucial vote intended for the general meeting of the German Society for Race Hygiene (Eugenics) on June 10, 1933 never came to pass. The Society was consolidated from the top down as part of the National Socialist *Gleichschaltung*. All members of the old board were compelled to resign, and Reich Minister of the Interior Frick installed Rüdin at the head of the Society as "Reich Commissioner." Frick's energetic actions presumably can also be traced back to a denunciation sent to the Minister of the Interior by Astel and Schultz on May 29, 1933. In this letter, the two renegades from Munich took advantage of the

---

[16] Lösch, Rasse, p. 234.

[17] Quoted in ibid., p. 235.

[18] For a biography: Weber, Ernst Rüdin, p. 144; Klee, Personenlexikon, p. 290.

[19] For a biography: Weber, Ernst Rüdin, pp. 263 f.; Klee, Personenlexikon, p. 496. Riedel worked at the Psychiatric Clinic of the University of Freiburg. As member of the NSDAP, the SS, and the NSDÄB, after 1933 he advanced to become director of the Race Office of the Student body in Freiburg as well as "Führer of the Assistant Professors" and was appointed to a chair for race biology. From 1935 on, he received a stipend from the DFG to investigate the children of "psychopathic personalities" at the DFA. From 1941, supported by the Reich Criminal Investigation Department (Reichskriminalpolizeiamt), he sniffed out homosexual pairs of twins.

circumstance that Fischer had been elected the rector of the Friedrich Wilhelm University in Berlin on May 2, 1933 – surprisingly, and to the great reluctance of the National Socialist faculty – after the previous officeholder, the jurist Eduard Kohlrausch, had abstained from running again in the election scheduled by the Reich Commissioner for the Prussian Culture Ministry, Bernhard Rust (1883–1945).[20] Astel and Schultz reported on a supposed rumor in race hygiene circles, purporting that Fischer intended to call on Hitler in his capacity as university rector and recommend himself as the director of a new "Ministry for Race Hygiene." Although this rumor probably lacked any foundation (no such indication is found in any other source), the denunciation did not miss its target. Frick, fearing that his area of competence would be restricted, followed Astel's and Schultz's recommendation, forced the board of the German Society for Race Hygiene to resign, and installed Rüdin at its head. Fischer was ordered to report to the ministry on June 13, 1933.[21]

Ernst Rüdin fulfilled all expectations of those who appointed him. He deleted the word "eugenics" from the name of the German Society for Race Hygiene and in 1934 gave it a new charter, through which the Society was incorporated into the Reich Committee for National Health Service (*Reichsausschuß für Volksgesundheitsdienst*) at the Reich Ministry of the Interior. Rüdin moved the Society's main headquarters to the DFA in Munich. Paul Weindling estimates that around 500 members of the Society were expelled due to their Jewish origins or their political attitudes.[22] Only around half the chairmen of the 20 branches existing in 1933, with a total of 1,300 members, were "reconfirmed in their office in pursuance of the consolidation now beginning [...]. In all remaining cases, after painstaking inquiries at the party offices in question, reliable National Socialists were entrusted with this responsible task [...]."[23] In May 1934, the Race Policy Office of the NSDAP decreed that its staff and appointees on location should take over the leadership of the branches of the German Society for Race Hygiene. In fact, in many locations the directors of the district and regional Race Policy Offices were the same individuals who chaired the local branches, while conversely, due to a labor agreement, the branch chairmen of the German Society for Race Hygiene were appointed as experts to the Main Science Office (*Hauptstelle Wissenschaft*) of the Race Policy Office.[24] Until 1938, the thus consolidated Society had grown to 58 local branches with 3,800 members. This number rose to 63 local branches with 4,500 members by 1939, whereby the escalation in membership was due primarily to the incorporation of the Viennese Society for Race Hygiene. Within 6 years, the German Society for Race Hygiene had tripled its membership[25] – yet politically

[20] For details on this: Lösch, Rasse, pp. 253–266.

[21] Cf. ibid., pp. 236f.

[22] Weindling, Health, p. 499.

[23] Bohn, Deutsche Gesellschaft für Rassenhygiene, p. 464.

[24] Ibid., p. 465.

[25] Ibid., p. 464. Cf. also the table in: Weindling, Health, p. 499.

it had sunk into meaninglessness; it no longer had any part in political decision-making processes. From the long-term perspective, it was thus hardly adverse for Fischer that he had been largely thrust aside by the Society in the course of its *Gleichschaltung*. In the current situation of the year 1933, however, the palace revolt by the race hygienists weakened Fischer's position in the negotiations with the Reich Ministry of the Interior about the future of his institute.

3.1.3 The "Gleichschaltung" of the Institute

On June 13, 1933, Fischer was received by Arthur Gütt (1891–1949), who had been appointed to head the National Health Division in the Reich Ministry of the Interior quite recently, in May 1933. Gütt was an "old warrior." Back in the early 1920s, he had joined the nationalist movement in East Prussia; in 1924, he came to the fore as the founder and district leader of the illegal German National Liberation Movement (*Deutschvölkische Freiheitsbewegung*) in the district of Labiau. Also in 1924, he published *Rassepolitische Richtlinien für die Nationalsozialistische Freiheitsbewegung* ("Race Policy Guidelines for the National Socialist Liberation Movement"). Since 1918, he had been working as a general physician in his own medical practice. After having passed the district medical examination in 1923, in 1925 he began his career in public health as a medical assessor in Waldenburg, Silesia. In 1926, he became district medical officer in Marienwerder; in 1931, in Wandsbek. He officially joined the NSDAP in 1932. As the head of a ministerial division, and since 1934 as Ministerial Director of the Committee for National Health Service and the State Academy of Public Health Service (*Staatsakademie des öffentlichen Gesundheitsdienstes*), and a member of the RuSHA, up to 1938 Gütt was the man to know in the state health leadership.[26] Fischer must have recognized quickly that he needed Gütt's protection to consolidate his position in Dahlem.

Several meetings between Gütt and Fischer took place in the second half of June 1933; once Frick even joined in personally. Fischer's position was all the more precarious because an informer was keeping Gütt up-to-date with internal information from the KWI-A. This informer was Günther Brandt,[27] who had served as an officer in the Imperial Navy in World War I, participated as a member of the "Ehrhardt Brigade" in the defeat of Communist revolts in Berlin and Upper Silesia in 1919, joined the NSDAP in 1921, and was an accessory to the murder of Foreign Minister Walther Rathenau (1867–1922) in 1922. Sentenced by the State Constitutional Court to 4 years in prison in 1925, Brandt was released before serving out his sentence and spent 1926–1932 studying medicine in Kiel, Berlin, and Munich.

[26] For a biography: Labisch/Tennstedt, Weg; vol. 2, pp. 423f.; Geisenhainer, "Rasse", p. 477; Klee, Personenlexikon, p. 210. In the SS Gütt advanced to the rank of Brigadeführer (brigade captain) (1938).

[27] For a biography: Lösch, Rasse, pp. 237f.; Klee, Personenlexikon, p. 70.

Immediately thereafter, he started at the KWI-A,[28] although nothing is known of the particulars. On June 13, right before the first meeting between Gütt and Fischer, a letter of denunciation by Brandt arrived at the Reich Ministry of the Interior – at the same time Brandt gave his notice at the KWI-A. Brandt, in his function as a National Socialist "works councilor," pointed out the institute's extraordinary importance for National Socialist genetic health and race policy, warning that there were "scientists at the forefront of such an institute who do not stand completely – even internally – on National Socialist ground." Therefore, Brandt demanded the dismissal of the department heads Muckermann and Verschuer. Fischer's importance, however, was "so paramount, that he alone continues [to come] into question for the direction of this institute." The director of the institute is "spiritually a nationalistic man through and through," who would "soon [...]be entirely pervaded by the National Socialist spirit [...]if he were only influenced properly."[29] In a personal conversation with Gütt, Brandt also denounced Heinrich Kranz and Ida Frischeisen-Köhler[30] as "politically unreliable," because they were so closely associated with the Center Party.

Considering the alternatives of being neutralized of any science policy influence or subjecting his institute to a personnel purge, Fischer did not hesitate long. As early as June 21, 1933, Gütt was able to make a note of the fact that Fischer had consented to Muckermann's resignation from the institute, "only he does not yet see the way to proceed in this."[31] Fischer apparently also agreed not to renew the contracts with Kranz and Frischeisen-Köhler. In contrast, he had taken Verschuer out of the line of fire.

3.1.4 The "Muckermann Affair"

As mentioned above, Hermann Alois Boehm, director of the Department for National Health in the Direction of the Reich Organization of the NSDAP, had turned to Hermann Muckermann in November 1932 to convince him to join one of the nonpartisan working groups planned on population policy and race hygiene.[32] Within the party, however, this met with immediate protest: The letter to Muckermann had hardly been dispatched when Boehm received a reprimand from

---

[28] It appears improbable that Brandt did in fact, as he himself claimed, work as an assistant at the KWI-A. In the personnel list he was not listed as such, but appears among the doctoral students.

[29] Quoted in Lösch, Rasse, p. 239.

[30] It is doubtful that Frischeisen-Köhler in fact, as asserted, had the status of an assistant at the KWI-A. After completing her dissertation she was more likely to have stayed on as a scientific staff member.

[31] Quoted in ibid., p. 241.

[32] The "Muckermann Affair" has been portrayed frequently in the literature. The following account is based on: Grosch-Obenauer, Hermann Muckermann, pp. 13–17; Weingart/Kroll/Bayertz, Rasse, pp. 385–388; Lösch, Rasse, pp. 300–302; Kröner, Von der Rassenhygiene zur Humangenetik, pp. 18–25; Richter, Katholizismus, pp. 314–320.

Walter Buch (1883–1949), Director of the Investigation and Arbitration Committee (*Untersuchungs- und Schlichtungsausschuss*, Uschla) of the NSDAP, who declared the cooperation with a former Jesuit to be "undesirable."[33] Buch advised Boehm to proceed with this issue in consultation with his highest superior, the director of the Reich Organization, Gregor Strasser (1892–1934). Because he was not able to reach Strasser, Boehm consulted with his direct superior and decided to await Muckermann's response before taking any further action. In early December 1932, Muckerman – in contrast to Eugen Fischer – indicated that he was willing to work with the party and justified his decision by pointing out that he, despite some different views, saw "a common goal in overcoming the degenerated biological genes of our ancestors and in the preservation and propagation of genetically healthy families of all occupational groups of the nation."[34] Boehm sent Muckermann's letter – and Fischer's – to Buch, requesting further instructions. Yet no one attended to the matter at first, because the direction of the Reich Organization of the NSDAP was distracted by the "Strasser Affair" in December 1932:[35] The Reich Chancellor appointed on December 2, Kurt von Schleicher, attempted to split the NSDAP by means of a "lateral front strategy," by courting its left wing around Gregor Strasser. On December 7, he offered Strasser the office of Vice Chancellor and Prussian Premier. Strasser refused on December 8, after Hitler had categorically forbid him to entertain the offer, but at the same time he resigned from all of his party offices. This entailed a restructuring of the Reich Organization, so that the "Muckermann Affair" rested. In the meantime, Theobald Lang (1898–1957)[36] had intervened, a member of Rüdin's staff who played the role of "militant National Socialist" at the Munich institute.[37] Lang was one of the first members of the NSDAP and was deputy chairman of the National Socialist League of Physicians (*Nationalsozialistischer Deutscher Ärztebund*, NSDÄB), an organization of which he was also a founding member. In an exacting protest letter of December 1932, he warned Boehm against contacting Muckermann, who "as an itinerant preacher [had worked] systematically to water down the race idea."[38] Once he found out that contact had already been established, Lang, this time together with Bruno K. Schultz, sent yet another protest letter to the Third Main Department of the Direction of the Reich Organization, and a few days later withdrew his agreement to join one of the planned working groups, remarking sardonically that Boehm, through the connection

---

[33] Quoted in Kröner, Von der Rassenhygiene zur Humangenetik, p. 18.

[34] Quoted in ibid.

[35] In assent with Hans-Peter Kröner, it must be stated that the "Muckermann Affair" must be seen in connection with the "Strasser Affair" (ibid., p. 21). In my view, Kröner's interpretation that the "Muckermann Affair" was "functionalized for the 'Strasser' affair" (ibid., p. 23), goes too far.

[36] For a biography: Weber, Ernst Rüdin, pp. 87, 146, 171, 245; Klee, Personenlexikon, p. 356. From 1926, Lang concerned himself with "goiter and cretin research" and became prominent in the Third Reich with his dubious theses on intersexuality and homosexuality. In 1941 he did not return from a trip to Switzerland.

[37] Weber, Ernst Rüdin, p. 171.

[38] Quoted in Kröner, Von der Rassenhygiene zur Humangenetik, p. 19.

with Muckermann, "now certainly [had] so much support from Center and Jewish quarters" that he probably would "have no difficulty doing without the cooperation of a National Socialist."[39] At the same time, Lang left the NSDÄB in protest, an organization he held to be too indulgent of political differences. Because the Third Main Department did not react either, on January 2, 1933 Lang and Schultz petitioned for "Uschla proceedings" to clarify whether the behavior of Boehm and his department could be reconciled with the principles of the NSDAP. In the meantime, on December 31, 1932, Robert Ley (1890–1945), Strasser's successor as director of the Reich Organization, had instructed Boehm to sever the connection to Muckermann. There were differences of opinion between Boehm and Ley as to the form of the rebuff, in which even Rudolf Heß (1894–1987), the "deputy of the Führer" intervened. Boehm won out, and sent Muckermann a formally binding letter of refusal on January 20, 1933:

> I regret to inform you that my superior holds the divergences in the opinions on certain issues, which undoubtedly do exist between you and us, and were indicated even in your letter, to be so great as to preclude a fruitful collaboration, even in loose race hygiene working groups.[40]

In the Weimar Republic Eugen Fischer and Otmar von Verschuer had always dismissed public critique of Hermann Muckermann.[41] The cooperation among the three department heads, mainly in scientific consulting on eugenics, had proceeded without a hitch. However, after the Nazis came to power, Muckermann became a liability – as the interlude of inviting and disinviting Muckermann to Boehm's working groups made apparent. Hardly a week had passed before Fischer gave in to Gütt's pressure and declared his assent to parting ways with Muckermann. Until this step had been concluded, the strategic alliance negotiated by Gütt and Fischer could not take effect. In a letter by Gütt to Rust on June 26, 1933, the interests of the two alliance partners are expressed with great clarity. Muckermann, as a Catholic priest, and because of his fundamental opposition to the National Socialist worldview, could not be tolerated as a department chief at the KWI-A. It was out of the question to use him "as a teacher of our public health officers," who were supposed to head the "later race offices." Muckermann's mere presence obstructed Gütt's plans for the KWI-A:

> The institute does not come into consideration for any cooperation as long as Mr. Muckermann is still there. [...] But because we absolutely need it for our further work in connection with race hygiene and race policy, after many discussions in my ministry Dr. Fischer accepted that the institute can no longer retain Mr. Muckermann. Although Mr. Fischer personally lobbied intensively for Mr. Muckermann, he appears to be inclined to do everything to avoid endangering a cooperation on our terms. I would thus welcome any solution we could find to use Mr. Muckermann in another position, perhaps at a Catholic university or in some other capacity, to keep from making him a martyr.[42]

---

[39] Lang to Boehm, 20/12/1932, quoted in ibid., p. 22.

[40] Quoted in ibid., p. 21.

[41] Cf., e.g. Verschuer, Eugenik und Hermann Muckermann.

[42] Quoted in Lösch, Rasse, p. 301.

Fischer and Gütt put their heads together to find the best way to achieve Muckermann's departure. Dismissal appeared inopportune because of the "Law for the Restoration of Career Civil Service" (*Gesetz zur Wiederherstellung des Berufsbeamtentums*). They sought more discrete methods to achieve a personnel change as inconspicuously as possible, without any political affront. The path that was ultimately pursued can no longer be reconstructed with any precision from the written record. Muckermann himself later spoke of a "deposition in violation of contract."[43] In 1949, he depicted the events as follows:

> A sinister twist came in the year 1933. Minister of State [sic] Frick, in the name of the National Socialist government, demanded the integration of the institute into the National Socialist state. He saw in my person an insurmountable obstacle. Therefore my immediate removal from the institute and from public life in general was resolved. [...] The document of dismissal expressly and without reserve ensures that the only reason for my dismissal was that the integration of the institute's work into the National Socialist state was otherwise impossible. The director of the institute at the time, Professor Fischer, and his later successor, Professor von Verschuer, consented to the minister's directive, while the Kaiser Wilhelm Society itself expressly ensured me that it did not approve of the minister's demands. Because I was not willing to perform bondservice for political goals, the only path left for me was to leave.[44]

Eugen Fischer, however, later portrayed the events as though Muckermann had already offered to resign in the case of political pressure, even before Fischer's visit to the Reich Ministry of the Interior, so as not to stand in the way of "the further development of the institute and its role in laying a foundation for and encouraging a biological population policy." He – Fischer – had "had to accept this magnanimous offer" and "canceled" the contract. Furthermore, according to Fischer, he and Muckermann came to an agreement to break off personal contact in order to avoid any "suspicion or reproach." In the end, the colleagues separated "in friendship."[45] Verschuer, in his memoirs, essentially upheld Fischer's version: Muckermann, when he recognized that he "was the greatest obstacle for the further work of the institute," had made the "noble decision [...] to resign from the institute voluntarily." Afterward Fischer and Muckermann became increasingly estranged from each other, while Verschuer and he at least initially remained "bound in friendship," even though Muckermann "took political paths" after leaving the institute along which Verschuer "could not follow him." Muckermann then broke off intercourse with him of his own accord, continued Verschuer, whereby what was probably decisive was "that he did not want to harm me politically."[46]

---

[43] Grosch-Obenauer, Hermann Muckermann, p. 14 (note 3).

[44] Hermann Muckermann, Aus der Chronik des Instituts, in: Studien aus dem Institut für Natur- und Geisteswissenschaftliche Anthropologie Berlin-Dahlem, 5th report, March 31, 1956, p. 374, MPG Archive, Department IX, Rep. 2, KWI-A (original emphases).

[45] Quoted in Lösch, Rasse, p. 300.

[46] Verschuer, Erbe – Umwelt – Führung, "Freier Forscher in Dahlem (1927–1935)" section, p. 21, MPG Archive, Dept. III, Rep. 86 A, No. 3–1.

Although in agreement with each other, Fischer's and Verschuer's versions are unconvincing. What is most probable is that Muckermann's contract of employment was terminated, plain and simple[47] and that Muckermann, presented with a fait accompli by Fischer, made the best of a bad situation. An ostensibly amicable departure was still the most advantageous solution for Muckermann given the circumstances: by reconciling himself to the inevitable, he did not offer the National Socialists any additional occasion for attack. For not only would an open protest against such a dismissal have had no prospect of success in the new state, it very probably would have entailed further sanctions, thus completely destroying Muckermann's remaining, albeit greatly restricted possibilities for public activity. Muckermann may also have found hope in Fischer's assurances that he would intercede on his behalf for his future professional advancement in order to avoid a public scandal. However, Gütt's recommendation that Muckermann be recommended for a chair at a Catholic university, as arranged with Fischer, did not meet with any approval in the ranks of his party comrades. And in 1934, the KWG returned the Hermann Muckermann endowment in order to provide for the material future of the former departmental director.[48]

Also to be taken into consideration in this ostensibly amicable arrangement is that Muckermann, after the turning point for race hygiene in National Socialist Germany, aligned himself "once and for all with the official 'papal line' on eugenics." This opened up to him a new field of activity under the cover of the Catholic Church, for "apparently the German episcopate could not and would not do without *the* 'people's tribunal' on eugenics, because he was needed as an advisor on eugenic issues and his services were sought in the fight against the NS compulsory sterilization law."[49] However, because the Catholic Church sought to avoid open confrontation with the National Socialist regime, in view of the *Reichskonkordat* passed by Hitler's cabinet on July 14, 1933, Muckermann could best play the role intended for him as scientific advisor to the episcopate by keeping a low profile. For this reason, too, Muckermann was better off vacating his position at the KWI-A without a fight.

Hence, all parties avoided open conflict, each for interests of its own. Observed superficially, therefore, there was indeed an amicable solution to ending Muckermann's employment. However, this was by no means a voluntary resignation, which explains Muckermann's later depiction of a "deposition in violation of contract." The fact that he broke off contact to Fischer and later to Verschuer as well presumably can be traced back primarily to Muckermann's disappointment about the all-too convenient sacrifice with which Fischer took himself, his pupil Verschuer and the institute out of the line of fire, and the complaisance with which Fischer and Verschuer made themselves the stooges of National Socialist genetic health and race policy.

---

[47] Grosch-Obenauer, Hermann Muckermann, p. 14 (note 3).

[48] Glum, Wissenschaft, p. 372. On this see also Grosch-Obenauer, Hermann Muckermann, p. 13.

[49] Richter, Katholizismus, p. 319, 320 (original emphasis).

Officially, Muckermann was "on leave until further notice."[50] Not until June 1936 did Fischer ask the General Administration of the KWG, "now that Mr. Muckermann has been retired for so long and for political reasons,"[51] to strike his name from the budget plan. Fischer's endeavors to distance him further were presumably also a consequence of the fact that Muckermann was subjected to increasing pressure from the National Socialist regime. In May 1934, he had driven the former Reich Chancellor Heinrich Brüning and his minister Gottfried Treviranus (1891–1971) to safety over the Dutch boundary in his own car, right before Hitler settled accounts with the conservative opposition in the course of the "Röhm crisis." As a consequence of his helping these targets flee, Muckermann entered the Gestapo's sights: He had to submit to several interrogations, and his passport was confiscated so that he could not take any more trips abroad. In the meantime, Muckermann had founded a "Research Center for the Shaping of Marriage and the Family" (*Forschungsstelle für die Gestaltung von Ehe und Familie*) which received active support from the Church, all the more since Konrad Graf von Preysing (1880–1950), who had been friends with Muckermann for some years, became Bishop of Berlin in 1935. Nevertheless, his possible spheres of action were visibly restricted. He was forced to resign from the Prussian State Health Council, which, however, was rapidly fading into insignificance. Starting in 1934, his publications were banned, one by one. Muckermann, who was a member of the Reich Chamber of Culture (*Reichskulturkammer*), fought against each and every ban. When the 2nd edition of his book *Grundriß der Rassenkunde* ("Outline of Race Science") was denied the required party permit for publication in 1936 by the "Official Examination Board of the Party for the Protection of NS Literature" (*Parteiamtliche Prüfungskommission zum Schutze des NS-Schrifttums*), supposedly because he had not taken the work of the "scientist of the racial soul" Ludwig Ferdinand Clauß (1892–1874) into sufficient consideration, Muckermann lodged a protest with the president of the Reich Chamber of Literature (*Reichsschrifttumskammer*). In his justification he emphasized that, especially in the final chapter of the incriminated book, he had made a contribution "to convince the readers of the importance of the race question for the shaping of the nation of the future and to induce them to understand and appreciate the biological significance of the race legislation, which was not yet promulgated when the book appeared."[52] In other words: Muckermann tendered his services for soliciting public understanding for the Nuremberg laws – an indication that Muckermann's position on race anthropology and even on National Socialist race policy was ambivalent at the least. It was all for naught, however, as ever more of his books were put on the blacklist, even those with purely religious contents. Muckermann's situation also became more precarious because his brother, the Catholic publicist Friedrich Muckermann (1883–1946), who had

---

[50] Niederschrift über die Sitzung des Kuratoriums des KWI-A am 5/7/1933, MPG Archive, Dept. I, Rep. 1 A, No. 2404, pp. 11–13, quote: p. 11.

[51] Fischer to Generalverwaltung, 14/6/1936, MPG Archive, Dept. I, Rep. 1 A, No. 2406, p. 203.

[52] Quoted in Kröner, Von der Rassenhygiene zur Humangenetik, p. 25.

criticized the National Socialist regime on several occasions, had since emigrated. In October 1936, Hermann Muckermann, who had still held a great number of well-attended lectures up to that time, was banned from speaking in the entire territory of the Reich; this ban was rescinded 2 months later (under certain conditions), but renewed in April 1937. Even the Research Center for the Shaping of Marriage and the Family had to cease its activities at this time. After that, little was heard of Hermann Muckermann. With financial support from the Catholic Church he continued to work quietly as a private scholar, assisted by Ida Frischeisen-Köhler.[53]

Frischeisen-Köhler, who worked as a doctoral student at the KWI-A in the area of genetic psychological twin research in 1929/30 and received her Ph.D. in May 1933 with her dissertation *Das persönliche Tempo* – advised by Eugen Fischer and the psychologist Wolfgang Köhler (1887–1967) as mentioned above, had already been denounced by Günther Brandt as "politically unreliable" because of her links to the Center party. In her case Fischer was not compelled to terminate her contract – he simply allowed it to expire at the end of September 1933. Frischeisen-Köhler remained Muckermann's employee until her death in 1957. Heinrich Kranz had also been classified by Brandt as "politically unreliable," and his contract, too, was not renewed in October 1933. However, it remains unclear whether this was due to political pressure alone. Heinrich Kranz – not to be confused with Heinrich Wilhelm Kranz (1897–1945), by the way, who worked as professor for genetics and race research at the university of Giessen from 1937 to 1942 and then took over as Verschuer's successor to the chair for genetic biology and race hygiene in Frankfurt – was working on a criminal biology twin study at the KWI-A. However, as criminal biology did not play any further role at the KWI-A, Kranz was left alone with his project. Thus it made sense to switch to an institute that dealt with criminal biology research. In July 1933, Fischer and the Münster psychiatrist Ferdinand Adalbert Kehrer (1883–1966), with whom he was well acquainted, endeavored to install Kranz as the director of a new department for genealogy to be founded at the Münster Psychiatric and Neuropathic Clinic, which was supposed to be expanded to a daughter institute of the KWI-A in the long term. This plan never came to fruition,[54] however, and Kranz had to see to it that he find employment in one of the existing institutes. Because his direct rival Friedrich Stumpfl was at the DFA in Munich, after leaving the KWI-A, Kranz landed more or less unavoidably as Johannes Lange's assistant at the University Neuropathic Clinic in Breslau, where he concluded his postdoctoral dissertation on the *Lebensschicksale krimineller Zwillinge* ("Life Fates of Criminal Twins") (1936). Against this background it cannot be ruled out that Kranz, as Fischer reported, did in fact leave the institute of his own accord. In any case, it must not have been very difficult for him to leave Dahlem.

[53] Grosch-Obenauer, Hermann Muckermann, pp. 16f.; Kröner, Von der Rassenhygiene zur Humangenetik, pp. 24f.

[54] Dicke, Eugenik, pp. 65f.

3.1.5 The "Faustian Bargain"

The compromise between the health leadership of the state and the KWI-A came to light of the public through two staged events. First, Arthur Gütt appeared as a "guest" to the meeting of the Board of Trustees of the KWI-A on July 5, 1933. In the discussion about Fischer's activity report, he began to speak immediately and made unmistakably clear to the leading representatives of the KWG what importance he accorded to the institute in consideration of the National Socialist genetic health and race policy, but also what expectations the new government placed in the institute – and the entire KWG – in this constellation:

> Mr. Gütt fully acknowledged the institute's important work. He requested that the institute also make its work available to the Reich government in the future. Above all he asked for assistance in realizing the law regarding sterilization as well as the prospective Reich citizenship law. Mr. Gütt also mentioned the importance of the institute in the study of hereditary diseases, especially in connection with clinical treatment. The Reich government, he asserted, places great value on the authoritative advice of the institute. Finally, Mr. Gütt proposed that the Kaiser Wilhelm Society might systematically place itself at the service of the Reich and make corresponding proposals. It would be most expedient if the Reich Ministry of the Interior were to request consultations with Mssrs. Eugen Fischer, Rüdin, Erwin Baur and Baron von Verschuer about all of the areas in question.[55]

Gütt's appeal to the KWG, to "systematically place itself at the service of the Reich with regard to race hygiene research" was attended to promptly by Max Planck (1858–1947), who had been President of the KWG since 1930, on July 14, 1933 – the very day on which Hitler's cabinet resolved the "Law for the Prevention of Genetically Diseased Offspring" (*Gesetz zur Verhütung erbkranken Nachwuchses*, GzVeN) –, by informing the Reich Ministry of the Interior that the KWG had formed a commission for this purpose, comprised of Erwin Baur, Eugen Fischer, and Ernst Rüdin, which the Reich Minister of the Interior might "dispose of in the sense suggested."[56] The fact that Fischer, and even Baur, who was also entangled in fierce exchanges with Walther Darré, were proposed for this commission in addition to Rüdin, suggests that the former two scientists were beyond the worst of it by this point in time – in contrast to Oskar Vogt, whose position at the KWI for Brain Research became increasingly untenable.[57]

For his part, Fischer sealed the compromise he had made with the National Socialist state in his first major public appearance as rector of the Friedrich Wilhelm University, with the speech he held on the occasion of Commemoration Day at the University on July 29, 1933. The address, entitled *Der Begriff des völkischen Staates, biologisch betrachtet* ("The Concept of the Volkish State, Considered Biologically") may appear ostensibly as kowtowing to the National Socialist rulers,

---

[55] Niederschrift über die Sitzung des Kuratoriums des Kaiser-Wilhelm-Instituts für Anthropologie, menschliche Erblehre und Eugenik am 5. Juli 1933, MPG Archive, Dept. I, Rep. 1 A, No. 2404, pp. 11–13, quote: pp. 11 f.

[56] Planck to Reichsinnenministerium, 14/7/1933, MPG Archive, Dept. I, Rep. 1 A, No. 2401, pp. 1.

[57] Cf. Schmuhl, Hirnforschung, pp. 562–569.

but at this juncture the interpretation by Niels C. Lösch must be emphasized vigorously: His closing ranks with the regime was intimately bound up with the conviction that what would finally be realized in the new, *volkish* state – mind you: Fischer did not speak of the *National Socialist* state – is what *he* had long demanded in vain: an energetic policy to enhance the genetic health of the German nation and a vigorous population policy to check the fall in the birth rate. In this context, the National Socialists appeared, so to speak, as the executors of an applied science whose foundations had been laid by Fischer and his school: "without the triumphant progress of genetic science, no theory of the nationalist state in eugenic-race hygiene terms." In retrospect, the grotesque distortion of the actual power relations is quite obvious – Fischer's scientocratic model of political consulting overlooked the reality of the Third Reich. Yet, Fischer's styling himself as the mastermind who paved the way for the nationalist state reflected his understanding of the societal role of science and of the scientist. In the current situation of the forced *Gleichschaltung* of his institute, it helped him save face. "The intellectually superior," he remarked, meaning himself and his colleagues, would be convinced not by the "political agenda," but by the "biological idea." "The idea of the nationalist state gains acceptance as such especially in the field of pure knowledge."

As concerned the stumbling block, his theses on miscegenation, Fischer resorted to a shabby trick. The nationalist state must, parallel to its eugenic race hygiene policy, take measures to isolate "elements alien to the nation." In Germany, this involved the Jews as the only "racially different element":

> That there are physical and intellectual differences no one can objectively deny. I am not pronouncing a value judgement when I declare this. I even go so far as to say that a nation mixed and crossed equally of Aryan and Jewish components could theoretically create a very creditable culture, but it would never be the same as one that grew on purely German national soil; it would not be a German culture, but an entirely different, half-Oriental one [...].[58]

Not *inferior*, but *different* – this formulation allowed Fischer to hold fast to the conclusions on miscegenation and "hybrid races" derived from his "bastard research" without calling into question the legitimacy of a policy of disfranchisement, economic dispossession, social exclusion, and cultural stigmatization of the Jewish minority. Fischer's shrewd argumentation bears witness, as is to be argued here in contrast to Lösch, to neither "courage" nor "remarkable naivité."[59] In the face of the Jewish university faculty dismissed in the previous months since enaction of the Law for the Restoration of the Professional Civil Service, it was nothing short of perfidy.

By fall 1933, Fischer had come to terms with the National Socialist state. On the eve of the Reichstag election of November 12, 1933, which also involved the plebiscite on withdrawal from the League of Nations, he appeared at a propaganda event in Leipzig organized by the National Socialist Association of Teachers

[58] Fischer, Begriff des völkischen Staates, pp. 6, 17, 14.

[59] Lösch, Rasse, p. 261.

(*Nationalsozialistischer Lehrerbund*) with the motto "The Day of German Science" (*Tag der Deutschen Wissenschaft*) and held a passionate speech marked by national pathos – although Lösch is correct to point out that "the withdrawal from the League of Nations [played ...] the decisive role here and [...may] not be underestimated as the catalyst of the nationalist emotions."[60] Even if the special circumstances of this speech are taken into consideration, there is still nothing to interpret about Fischer's identification with Hitler's state.

A letter by Fischer to Bernhard Rust of November 3, 1933, corroborates this. On October 28, Rust had issued an edict "for the simplification of college administration," according to which the rectors of the Prussian universities were no longer to be elected, but rather appointed, whereby the universities could submit suggestions for the position. The University of Berlin had in fact proposed Fischer, yet he feared that preference might be given to some veteran party member. He addressed the possibility as follows:

> In unreserved action for the National Socialist cause and in the will to work within it and help to integrate the universities, I believe myself to be equal to any. [...] However, the matter also concerns the cause of my science. Please do not consider me arrogant when I say here that at present there is hardly a man in Germany who can be of so much use to the entire population policy of the state government in the field of genetics and race theory, and especially as regards its assessment abroad. I place myself unconditionally at your service [...].[61]

Fischer was in fact confirmed by Bernhard Rust as rector of the University of Berlin. "The first NS campaign against him had come to an end in summer 1933 with a clear consolidation of his position."[62] Fischer and his institute had found a *modus vivendi* with the Reich Ministry of the Interior – or more precisely: with Arthur Gütt, the strong man of the state health leadership. However, in the National Socialist "double state" Fischer also had to deal with the centers of power of the "prerogative state."[63] Especially, the RuSHA under Walther Darré (1895–1953) and the Race Policy Office of the NSDAP under Walter Groß (1904–1945)[64] prepared for a new attack on Fischer and his institute in early 1934.

3.1.6 Renewed Conflicts about Eugen Fischer

At the beginning of 1934, a new National Socialist press campaign was launched against Fischer. The catalyst was another public appearance by Fischer; this time it concerned a lecture entitled *Rasse und Kultur* ("Race and Culture") held in February 1934 before the *Badische Heimat* ("Baden Homeland") association

---

[60] Ibid., p. 263.

[61] Quoted in ibid., p. 264.

[62] Ibid., p. 263.

[63] Cf. Fraenkel, Doppelstaat; Schmuhl, Rassismus; idem., Rasse.

[64] For a biography: Klee, Personenlexikon, pp. 203 f.

in Freiburg. In effect, Fischer had merely repeated his views on miscegenation once more, whereby he adhered to the prescribed terminology agreed upon in July 1933. The southern German edition of the *Völkischer Beobachter* quoted him with the words:

> What position should one take on commingling with the Jewish race? It is a matter of course that the Jewish race is not more inferior than many other races. But one thing is certain: that it is different, and this difference is also the reason why it is completely unsuitable for a crossing of cultures with the German nation.[65]

This quote was commented on critically by the nationalist magazine *Die Sonne*. Its editor Werner Kulz condemned "the demand for indiscriminate commingling that is cloaked in a scientific mantle, the so-called 'exorbitant growth of the bastards,' as a stab against the intention of the Führer and of all German race hygiene endeavors in general."[66] *Die Sonne* was not an important periodical, but because Kulz was also director of the "Supreme Inspector for Smut and Dirty Propaganda"(*Oberprüfstelle für Schund- und Schmutzpropaganda*) in the Reich Ministry for Propaganda and the Enlightenment of the Volk (*Volksaufklärung und Propaganda*), the attack made an impact. Fischer was crossed off the list of speakers at the major exhibition *Deutsches Volk – Deutsche Arbeit* ("German Nation – German Labor") in Berlin. He was outraged at this affront.

What weighed more heavily was that the Reich Ministry of the Interior apparently saw itself occasioned to reconsider the strategic alliance with the scientist so publicly attacked by the party, in order to avoid coming into the line of fire itself. In any case, in March 1934 the ministry requested the dossier about Fischer compiled by the RuSHA. The conflict had ever greater repercussions now, especially since Fischer had since been appointed to lead the German delegation to the International Congress for Anthropology and Ethnology in London, about which the British embassy had already been informed – this appointment could not be revoked without causing an international sensation. Beyond the Reich Ministry of the Interior and the RuSHA, this meant that the Foreign Office, the Ministry for Propaganda and the Reich Chancellor's office now took an interest in the "Fischer affair." Finally, the staff of the Führer's deputy took over the proceedings. "In order to effect a uniform bearing on this question," it was announced, "at present the complex of issues is under the scrutiny of the deputy of the Führer and his imminent decision will be binding for all Party offices."[67] The ministries and party offices involved were asked for their internal opinion and at the same time instructed to refrain from any further public attacks on Fischer for the time being. It did not escape his attention that a storm against him was brewing again. Seeking assistance, he turned to Arthur Gütt: "I would be greatly indebted to you for any hint about how I can defend myself effectively against attacks of this kind."[68]

---

[65] Quoted in Lösch, Rasse, pp. 244 f.

[66] Quoted in ibid., p. 245.

[67] Quoted in ibid., p. 247.

[68] Quoted in ibid.

Implicit in this request was the challenge to Gütt that it was time, now that Fischer had fulfilled the demands of the Reich Ministry of the Interior to "purge" the institute and increase its involvement in genetic health and race policy, to honor his own alliance obligations by defending Fischer against the fresh attacks. In fact, Gütt felt obliged to act, especially because the "Fischer Affair" was used by other actors in the polycratic field of powers of the "Third Reich" to break into his own area of competence. An internal debate erupted within the regime as to who was allowed to appear as a speaker for "lectures on population policy and race hygiene" and who was entitled to make the decision. On this point a meeting took place on June 17, 1934, called by Gütt and attended by such high-ranking individuals as the Surgeon General of the Reich Gerhard Wagner (1888–1939); Reich Facilitator Otto Gohdes; Falk Ruttke (1894–1955),[69] chairman of the Reich Committee for National Health in the Reich Ministry of the Interior; Walter Groß, director of the Race Policy Office of the NSDAP; as well as Bruno K. Schultz and Ernst Rüdin. Rüdin's participation proves how close he had come to the levers of power since 1933, in contrast to Fischer. The purpose of the meeting, thus, was not only to discuss how contradictory comments on the "race question" could be avoided in public, but also to examine "how it is possible to implement forces who cannot yet with certainty be designated as National Socialists, but who could perform good service to the party organizations in the areas of population policy and race hygiene." In the end, Walter Groß, with reference to the task with which the "Führer's deputy" had entrusted him, "to unify and monitor training and propaganda in the pertinent areas and also all factual issues of population and race policy,"[70] was able to claim the right to decide which speakers were admitted.

Fischer therefore had to come to an arrangement with Groß as well. Clearly, his most resolute opponents were to be found in the ranks of the RuSHA. Walter Darré had not forgotten that Fischer had panned his book *Neuadel aus Blut und Boden* ("New Nobility of Blood and Soil") (1930) in a brief, brutal review.[71] Moreover, the RuSHA had since hired Fischer's "intimate enemy" Bruno K. Schultz as well, who had machinated against Fischer at the German Society for Race Hygiene and thrown him off the editorial board of the magazine *Volk und Rasse* in November 1933.[72] Thus, to dispose of the conflict once and for all, Gütt also intervened with Darré directly. Gütt's letter makes clear in an exemplary fashion the network of mutual dependencies that emerged from the mutual exploitation of science and politics, with each using the other as a valuable resource. Gütt clarified that he, too, could not share some of Fischer's positions, especially those regarding "miscegenation," but he requested that Darré keep in mind that it would be difficult to neutralize Fischer:

[69] For a biography: Labisch/Tennstedt, Weg, vol. 2, pp. 484 f.; Klee, Personenlexikon, p. 516.

[70] Quoted in Lösch, Rasse, p. 248.

[71] Fischer, Review Walter Darré, Neuadel aus Blut und Boden.

[72] Cf. Lösch, Rasse, pp. 243 f.

> Cooperation with Professor Fischer appears unavoidable anyhow, because at this time no other institution can be placed on an equal footing with his institute; its equipment and teachers are also urgently required for the relevant training of medical officers and physicians, especially for the execution of the law on the prevention of genetically diseased offspring, etc. Further, Professor Fischer is a nationally and internationally distinguished scholar in the area of genetics and race research. [...] Discord between him and the official bureaus would very easily create, at home and abroad, the impression that Professor Fischer disapproves of the paths taken by the government to foster the race, and that thus the measures of the government must be in conflict with the findings of science.[73]

In the meantime, Fischer, presumably at Gütt's suggestion, had submitted an exposé about his position in the "race question," since the objectionable lectures had not been published. Gütt then forwarded this exposé to Darré with the request for "perusal":

> I would like to believe that you could consider the thoughts put in writing here as sufficient grounds to shelve your reservations against the person of Fischer. Incidentally, it certainly may be expected that Professor Fischer will learn from the difficulties he has accrued that henceforth he must be more cautious with remarks which could be understood as directed against the measures of the government.[74]

In his exposé, Fischer stuck to his line as far as the question of miscegenation was concerned, but he did endeavor to adapt the prescribed terminology more closely to Nazi ideology. He affirmed that he held the "Nordic race" to be the "most capable intellectually," but that the remaining "European races" were hardly inferior. The "hybridization" of these "European races" with the "Nordic race," he claimed, "under advantageous circumstances increases its capability to the highest achievement possible." The "hybridization" with "inferior," more precisely: with "races of color," however, lowered the intellectual level of performance. While Fischer did *not* count the "culture nations" of East Asia, or the Jews, among the "inferior races" – they were merely "different" – their "hybridization" with the Nordic race was nevertheless to be rejected because it changed "the spiritual orientation and the capability of the Nordic race in a non-Nordic sense."[75] Thus, Fischer held to his formulation of "not inferior, but different" in the "Jewish question," whereby the term "different" was specified here through the vague formulation "non-Nordic spiritual orientation." To see in this an "unmistakable, massive critique of Nazi race ideology,"[76] as Lösch does, means emphasizing Fischer's deviation from the party line on the "miscegenation issue" far too strongly. It cannot be overlooked that Fischer's argumentation can be used to justify all measures of National Socialist Jewish policy – and even their potential extension to "Jewish half-breeds." Moreover, Fischer's remark that the misunderstandings about his lectures can be traced back to the distortions of the "Jewish press" indicates that Fischer's scientific ideas on the "Jewish question" rested on a broad foundation of sociocultural anti-Semitism.

[73] Quoted in ibid., pp. 248 f.

[74] Quoted in ibid., p. 249.

[75] Quoted in ibid., p. 250.

[76] Ibid., p. 251.

In his exposé, Fischer once again emphasized his institute's achievements in the practical execution of genetic health and race policy and further alluded to his international contacts:

> I know that everyone abroad is critically following what I used to say about race and about the importance of the Nordic race and what I say today. I know what they would think if it came to pass that I was permitted to say *nothing* more in public.[77]

In the end, the crucial men within the National Socialist regime saw things the same way. Fischer remained the "leader" of the German delegation to the conference in London, albeit accompanied – and monitored – by Walter Groß. The relationship to Groß and his Race Policy Office was thus apparently rectified; in 1935 Groß became a member of the Board of Trustees of the KWI-A.

The relationship between Darré and Fischer eased, apparently after a clarifying conversation between the two took place. Whether the delegation of 22 physicians as trainees to the KWI-A in fall 1934 was "undoubtedly the result of the Darré's agreement with Fischer,"[78] as Lösch presumes, cannot be confirmed, but this assumption is indeed highly plausible. In any case there were no further attacks against Fischer from Darré's corner. In a file note Arthur Gütt declared the matter to be settled.

To obtain a balanced judgment about the confrontations between Fischer and the National Socialist state in the years 1933/34, it is helpful first of all to keep in mind that, as Hans-Peter Kröner remarked so accurately – in the Third Reich there was hardly a scientist of any distinction in the field of heredity and race research who "was not accused of ideological deviation at some point."[79] This point qualifies the "Fischer Affair." Kröner characterizes the political campaign against the KWI-A correctly as a "shot across the bow,"[80] intended as a drastic demonstration to Fischer that, while the new rulers were willing to base their genetic health and race policy on scientific expertise, they reserved the right to make the final decision about many a fundamental question. From the outset, they set limits to the scientocracy aspired to by Fischer. However, these limits demarcated quite a broad territory. In general the regime certainly did tend toward a scientification of policy, and that is why it sought out cooperation with Fischer and the KWI-A. Fischer, for his part, had already aspired to a politicization of science before 1933, and thus readily placed his institute at the service of the regime. The conflict during 1933/34 served merely to clarify the power relations. And here two of the National Socialist satraps, Arthur Gütt and Walter Groß, took advantage of the opportunity to draw Fischer and his institute into their sphere of influence.

---

[77] Quoted in ibid., pp. 251 f. (original emphasis).

[78] Ibid., p. 252.

[79] Kröner, Von der Rassenhygiene zur Humangenetik, p. 34.

[80] Ibid., p. 28.

3.2 Internal Structures

3.2.1 The Board of Trustees

When the Board of Trustees of the KWI-A convened on July 5, 1933 – presumably for the first time since 1928 – its ranks had grown much thinner. Of the scientists from the KWG, only Erwin Baur showed up – it was to be his last appearance on the board, for he passed away suddenly on December 2, 1933. Carl Erich Correns had already died on February 14, 1933.[81] Richard Goldschmidt, who was under increasing pressure in National Socialist Germany because of his Jewish ancestry, was still officially a member of the board, but he opted not to attend. In March 1934, as his emigration approached, he resigned from his posts as member of the Boards of Trustees of the KWI-A and the KWI for Entomology, as well as the Board of Directors of Harnack Haus, since, as he added bitterly, he did "not have the feeling that [he could] achieve useful work in these offices at the moment."[82] Ernst Rüdin had not even deemed it necessary to give notice that he would not be participating in the meeting of July 5, 1933[83] – he was moving up as a result of the National Socialist takeover, and gave the cold shoulder to the rival institute that was experiencing such turbulence.

The first chairman of the Board of Trusteees, Adolf von Harnack, had died in 1930. Max Planck had taken over his position, but showed no interest and announced that he would not be attending, so that the meeting on July 5 was chaired by Friedrich Schmidt-Ott.[84] The General Director of the KWG, Friedrich Glum, had also come, along with his deputy Max Lukas von Cranach, who had since replaced Adolf Morsbach.[85] Finally, Inspector Schröder participated for the General Administration.

---

[81] Kröner/Toellner/Weisemann, Erwin Baur, pp. 90–103; Hagemann, Erwin Baur, p. 42.

[82] Goldschmidt to Planck, 12/3/1934, MPG Archive, Dept. I, Rep. 1 A, No. 2403, pp. 69. Planck's response turned out to be rather cool: "Although I can concede that your view is justified subjectively, I do not intend to oppose your emphatically stated wishes and for the time being express to you my sincere thanks for the interest dedicated to the Entomological Institute so far. As for the rest I will reserve my opinion for later opportunities." Planck to Goldschmidt, 15/3/1934, ibid., pp. 70.

[83] File note about acceptances and declined invitations, MPG Archive, Dept. I, Rep. 1 A, No. 2399, p. 5 v.

[84] A list of invitees (MPG Archive, Dept. I, Rep. 1 A, No. 2404, p. 10) also includes the name of Karl Stuchtey, a close colleague of Schmidt-Ott's. In the Emergency Association of German Science, he had been in charge of the natural science and technical areas of specialization from 1923 to 1934 (Zierold, Forschungsförderung, p. 43). Ultimately Stuchtey did not take part in the meeting, however.

[85] Morsbach had been given leave from the KWG on March 1, 1931, to take over the direction of the Central Office of the Academic Exchange Service (*Akademischer Austauschdienst*). Therefore Cranach was appointed Deputy Director of the KWG (Henning/Kazemi, Chronik, p. 67). In his memoirs, Glum states that Morsbach was arrested in connection with the "Röhm crisis" and held in the Dachau concentration camp for around 6 weeks. He was released "a broken man" and retired in 1935. The statement by Lösch, Rasse, p. 305, that Morsbach was already being held in a camp at the time of the board meeting on July 5 is not correct.

Alfred Grotjahn had passed away in 1931. The Medical Councilor of Berlin, Wilhelm von Drigalski, had been dismissed in March 1933; his successor Wilhelm Klein (1887–1948)[86] took his place. Rudolf Hilferding had emigrated to Switzerland in March 1933. Paul Schottländer, a distinguished entrepreneur and landowner from Breslau, who had donated significant funds to the KWG, was one of the three Jewish senators who continued to be tolerated in the KWG for a time. However, after consultation with Glum he had relinquished his seat in the board of the KWI-A on July 1, 1933. Obviously, there was no desire to provoke the National Socialist government by having a Jewish member on the board of an institute that dealt with race science and eugenics.[87] Albert Vögler, finally, had announced that he would be unable to attend – he must have had more important matters to attend to in "the new Germany."

On the part of the ministerial bureaucracy, only Max Taute, Undersecretary in the Reich Ministry of the Interior, took part in the meeting of July 5, 1933. Perhaps he drew on his international reputation as a tropical physician, or possibly also from his participation in Paul von Lettow-Vorbeck's (1870–1964) campaign in German East Africa in the years from 1914 to 1918. In any case, although he was close to neither National Socialism nor eugenics, he was able to maintain his office until his death in 1934. The second representative of the Reich Ministry of the Interior, Undersecretary Max Donnevert, who played an important role in the science policy of 1933/34, did not attend. The Prussian Ministry for National Welfare had been dissolved in 1932. In its place the KWG had invited the Prussian Ministry for Science, Art, and Education to send a representative – however, to no avail. Of the representatives of the Prussian provinces on the board, the Center Party intimate Johannes Horion had just died, and Ernst Schultze had declined the invitation. Thus the provinces were represented only by the retired Undersecretary August von Schenck, who, however, was forced off the board in 1935 by the "German Council of Municipalities" (*Deutscher Gemeindetag*).[88]

The board meeting on July 5, 1933, as mentioned above, was dominated by a man who officially attended as a guest: Arthur Gütt. Presumably in preparation for the board meeting, Max Planck had endeavored to obtain Crown Prince August Wilhelm of Prussia (1882–1951) as Chairman of the Board of Trustees, who offered his support for the National Socialist regime in 1933, albeit only half-heartedly.[89]

---

[86] For a biography: Klee, Personenlexikon, p. 314.

[87] Schottländer to Glum, 1/7/1933, MPG Archive, Dept. I, Rep. 1 A, No. 2403, pp. 65a–65a v: "Since I have belonged to seven boards of trustees anyway, I would like to limit this number and restrict my activities to institutes of my actual field, biology. I would be very glad if I could continue on the boards of Rovigno, Plön, Lunz, Rossitten, namely Müncheberg; I would like to relinquish my seat in the Board of Trustees for Anthropology, which is somewhat remote for me, as well as the seat on the board of the Coal Research Institute (*Kohleforschungsinstitut*) in Breslau, since I am too far removed from the material of these two institutes [...]." Cf. Lösch, Rasse, p. 303.

[88] Schlüter to Glum, 28/5/1935, MPG Archive, Dept. I, Rep. 1 A, No. 2403, p. 84. The councilor of the German Council of Municipalities Schlüter took von Schenck's seat.

[89] Planck to August Wilhelm, Prince of Prussia, 20/7/1933, ibid., p. 67.

However, in August 1933 he declined to join the body. The Reich Ministry for Food and Agriculture (*Reichsministerium für Ernährung und Landwirtschaft*), which relayed his refusal, proposed sending Hans F. K. Günther instead – but the KWG did not acquiesce to this appointment.[90] After this the General Administration laid the matter to rest. Not until December 1934, when the next meeting of the board was settled, did Glum take the initiative again. He had already suggested to Fischer in earlier discussions that the Chairmanship of the Board of Trustees be offered to the premier of the province of Saxony, Reinhard Otto, who had been appointed senator of the KWG in 1933. Fischer, above all, in view of the fact that Otto was also chairman of the Conference of State Directors of the Prussian Provinces (*Landesdirektorenkonferenz der preußischen Provinzen*), had consented to Glum's suggestion. Now he revisited this proposal and also urged that the six free seats on the board be reassigned.[91]

In fact, the meeting of the Board of Trustees on May 10, 1935 was chaired by Premier Otto. Gütt, too, in the meantime implicitly appointed to board membership, was also invited again; this time, however, he sent a representative, the senior government official Herbert Linden (1899–1945).[92] In accordance with the previous agreement between Glum and Fischer, upon recommendation of the institute director the board resolved to recommend to the senate Fritz von Wettstein (1895–1945), Director of the KWI for Biology; Otmar von Verschuer, who had just moved to Frankfurt; Hans Reiter (1881–1969),[93] President of the Reich Health Office; and Walter Groß. The appointment of Wettstein was clearly intended to link Fischer's institute with the KWI for Biology, an idea that had been on the table even at the founding of the KWI-A, but had not been realized due to the intensive orientation toward twin research. The intention concealed behind Verschuer's appointment was also easily detected – the process of establishing him as Fischer's successor had begun. Fischer petitioned for Reiter's election with the explanation that the "ongoing expansion of his office to the field of genetic biology and population policy" makes it appear "especially important" that the president of the Reich Health Office be integrated into the Board of Trustees. Finally, the election of Groß had clearly political motives: "In the inextricable connections between the scientific research of genetic and race issues and practical population policy," Fischer argued,

---

[90] File note by Telschow of 28/8/1933, ibid., p. 68. Lösch, Rasse, p. 305, states that the crown prince had not received permission from his father to accept an office in the KWG.

[91] Glum to Fischer, 14/12/1934, MPG Archive, Dept. I, Rep. 1 A, No. 2403, pp. 61–72.

[92] For a biography: Klee, Personenlexikon, p. 373. Gütt to Otto, 26/4/1935, MPG Archive, Dept. I, Rep. 1 A, No. 2399, p. 49; Niederschrift über die Sitzung des Kuratoriums des KWI-A am 10/5/1935, MPG Archive, Dept. I, Rep. 1 A, No. 2404, pp. 48–48c. In addition to Otto and Linden, the participants were Glum, von Cranach, von Schenck, Theodor Paulstich (as a substitute for the Berlin Medical Councilor Klein), as well as Fischer and Lenz.

[93] For a biography: Maitra, "…wer imstande und gewillt ist"; Klee, Personenlexikon, p. 490. When new expert staffs were to be set up at the Reich Health Office in the course of dissolving the Reich Health Council, on June 16, 1934 Reiter proposed for the department "Population Policy, Hereditary Biology, Race Care" a list including Fischer, Lenz, Verschuer and Rüdin; for the department "Criminal Biology, Forensic Medicine, Penal Legislation," Verschuer and Rüdin. Cf. BArch. Berlin, R 1501/126.300, pp. 209–233.

the director placed "the greatest value on his [Groß's] working with us."[94] All four of the designated board members accepted their election – Groß inquired briefly as to the scope of obligations. Cranach hastened to reassure him that there were no fixed obligations besides the board meetings, which took place more or less once a year and generally lasted no longer than 2h: "The extent to which you otherwise wish to devote your interest to the institute, of course, depends entirely on your professional demands." In any case, Cranach informed him, Fischer would be very pleased "if you were to support the institute's endeavors to the extent that this is possible."[95]

By 1939, there were three new appointments. In February 1936, Glum addressed the Kaiser Wilhelm Institutes. With reference to the fact that the competencies in issues of science policy had been transferred from the Reich Ministry of the Interior to the Reich and Prussian Ministries for Science, Training, and Education of the Volk, the general director ordered that the charters of the institutes must be changed to the effect that neither representatives of the Reich Ministry of the Interior nor of the Prussian Ministry for Science, Art, and Education were to be appointed to the Boards of Trustees. Their seats were to be occupied by delegates of the Reich Ministry of Science.[96] In the case of the KWI-A this was the "military chemist" Rudolf Mentzel (1900–1987), research consultant and "strong arm" of Reich Science Minister Rust.[97] An additional change came about when Leonardo Conti, the representative whom Fischer had defied in the discussion about the Prussian draft law on eugenic sterilization in July 1932, was appointed Wilhelm Klein's successor as Berlin Medical Councilor in late 1936 and thus received by virtue of his office a seat and vote in the boards of both the KWI-A and the KWI for Brain Research.[98] In the board meeting of June 20, 1936, at Cranach's instigation, "in consideration of the recent close collaboration between the institute and the Reich War Ministry" (Fischer was on the senate of the Army Medical Corps),[99] it was resolved to propose to the senate of the KWG that General Staff Physician Anton Waldmann (1878–1941),[100] Medical Inspector of the Army, be appointed as a board member of the relevant Kaiser Wilhelm Institutes. The senate acceded to this proposal on May 29, 1937, voting Waldmann onto the boards of the KWI-A, the

[94] File note by Fischer (Anlage zum Auszug aus dem Tätigkeitsbericht, 13/5/1935), MPG Archive, Dept. I, Rep. 1 A, No. 2399, p. 57.

[95] Wettstein to Planck, 31/5/1935, MPG Archive, Dept. I, Rep. 1 A, No. 2403, p. 85; Reiter to Planck, 3/6/1935 (ibid., p. 87); Verschuer to Planck, 3/6/1935 (ibid., p. 89); Groß to Planck, 12/6/1935 (ibid., p. 91); Cranach to Groß, 14/6/1935 (ibid., p. 92, quote).

[96] Glum to all KWIs, 5/2/1936, MPG Archive, Dept. I, Rep. 1 A, No. 2402, p. 20.

[97] Übersicht über die Zusammensetzung des Kuratoriums [1940], MPG Archive, Dept. I, Rep. 1 A, No. 2403, p. 110.

[98] Planck to Conti, 1/12/1936, MPG Archive, Dept. I, Rep. 1 A, No. 2404, p. 97. Cf. also Schmuhl, Hirnforschung, p. 585.

[99] Niederschrift über die Sitzung des Kuratoriums des KWI-A am 20/6/1936, MPG Archive, Dept. I, Rep. 1 A, No. 2404, pp. 52–52e, here: p. 52a (quote), 52b.

[100] For a biography: Klee, Personenlexikon, p. 653.

KWI for Brain Research, and the KWI for Labor Physiology.[101] Waldmann, however, passed away in 1940 and never took part in a single board meeting. He was succeeded by Siegfried Handloser (1885–1954).[102]

3.2.2 *Finances and Expansion*

The budget plan for the 1933/34 fiscal year, which Friedrich Glum submitted at the board meeting on July 5, 1933, had been put together on the basis of the budget plan for the 1932/33 fiscal year, and thus persisted at the low level dictated by the global economic crisis. However, Glum added that the prospects had become "somewhat more favorable" in the meantime, so that there was the possibility of "lifting the restrictions imposed on the institute in the recent period."[103] He recommended approving the budget plan with the proviso of a later increase, which then indeed took place. In his explanation of the budget plan, the reasons with which Fischer justified the designated increase of the grant from the KWG by around 3,600 RM included the need to expand the holdings of the library, the provision of grants to print scheduled publications, and the expenditure of greater amounts for the maintenance of interior equipment, as no appreciable funds had been spent for this purpose since the institute's founding. Personnel costs also rose further, as a consequence of planned salary raises and subsidies for children. Fischer indicated that the personnel funds approved by the Emergency Association of German Science and the Rockefeller Foundation were only available for a limited time and thus an increase in the personnel budget was to be expected in the foreseeable future.[104] At the board meeting, Fischer thus stated for the record that the submitted budget plan represented "the essential minimum" and that he must reserve the right to apply for "a significant increase" at the next board meeting[105] – tentatively, he put a sum of 7,000 on the table.[106]

After the board meeting of July 5, 1933, which had developed altogether auspiciously for Fischer and his institute, the possibility arose of increasing the

[101] Auszug aus der Niederschrift der Sitzung des Senats der KWG am 29/5/1937, MPG Archive, Dept. I, Rep. 1 A, No. 2404, p. 98; Planck to Waldmann, 4/6/1937 (ibid., p. 99); Waldmann to Planck, 24/6/1937 (ibid., p. 101). Cf. Schmuhl, Hirnforschung, p. 585.

[102] For a biography: Klee, Personenlexikon, p. 223.

[103] Niederschrift über die Sitzung des Kuratoriums des KWI-A am 5/7/1933, MPG Archive, Dept. I, Rep. 1 A, No. 2404, pp. 11–13, quotes: p. 12f.

[104] File notice by Arndt of 4/7/1933, MPG Archive, Dept. I, Rep. 1 A, No. 2406, p. 123. In fact the increased funding for which Fischer applied was placed on the budget, but it was not possible to call these funds in during the 1933/34 fiscal year. Revisor Schröder, Bericht über die Prüfung des Rechnungsabschlusses zum 31/3/1934, 5/9/1934, ibid., pp. 159–162, here: p. 159.

[105] Niederschrift über die Sitzung des Kuratoriums des KWI-A am 5/7/1933, MPG Archive, Dept. I, Rep. 1 A, No. 2404, pp. 11–13, quotes: pp. 12f.

[106] File note by Arndt of 4/7/1933, MPG Archive, Dept. I, Rep. 1 A, No. 2406, p. 123.

budget to quite a different order of magnitude, however, as Arthur Gütt, with his appeal that the KWI-A might place itself in the service of the genetic health and race policy of the National Socialist regime, had set a political signal that the government would support the institute's expansion. Fischer did not hesitate to strike while the iron was hot, immediately submitting a comprehensive exposé in which he outlined the potential contribution of his institute to future genetic health and race policy:

> The Reich government needs the most intensive and accelerated [sic] elaboration of all still controversial or unclear points of human genetics (e.g. the heredity of carcinomes and other tumors, heredity of all diseases of the human constitution, heredity of normal intellectual attributes, talents, character, etc.). It further requires central processing of the results of the genetic biological and genetic pathological stock of the German nation, above all to resolve the question as to whether pathological hereditary dispositions and degenerations emerge anew. It requires exhaustive studies of genetic damage (X-rays, poisons). Added to all of this, the most copious twin studies, genealogical research, statistics and large-scale animal experiments are necessary. Also necessary is that a clinical hospital department be affiliated with the institute [...]. Additional tasks for the institute include the training of physicians in specialized human genetics, the organization of genetics instruction and race science for medical councilors, physicians, teachers, personnel in welfare institutions and sanatoriums, the clergy and judges. Teaching assistants are needed to carry out courses outside of Berlin.[107]

In order to fulfill this mission spectrum, Fischer applied for additional personnel funds amounting to 26,800 RM, for a third and fourth assistant, an additional auxiliary assistant and the institute's statistician – along with material funding of 49,000 RM. Fischer countered any potential critique of his high demands with the indication that Oskar Vogt's KWI for Brain Research was provided with an incomparably higher budget: "As significant as this brain research may be for science, I would like to contrast it with the importance of genetic research for the nation and the state."[108] Beyond the increase in operating funds, Fischer also desired the erection of an additional wing at the institute, the expansion of the animal stalls, and the construction of a garage and acquisition of an automobile – not to mention the annexation of a clinical department with a senior physician, an assistant, and corresponding nursing personnel.

These were exorbitant demands: in the fiscal year in progress, 1933/34, the grant from the KWG amounted to around 75,700 RM – meaning that Fischer demanded no less than a doubling of his current funding, not to mention significant, not yet quantified one-time grants to finance the expansion of the institute. Yet, Fischer knew how to invest his demands with force. On July 29, 1933, he circumvented official channels to send his exposé directly to the Reich Ministry of the Interior, not forwarding it to the president of the KWG until September 29 – presumably after he had received a positive signal from Arthur Gütt. Max Planck reacted with clear annoyance, warning Fischer, "in future, as I ordered in the past as well, to send all petitions to the ministries through my hands." In addition, Planck expressed his regret that Fischer had made reference in his exposé to the KWI for Brain

[107] Fischer's exposé of 29/7/1933, ibid., pp. 143–145, quote: p. 143 f.

[108] Ibid., p. 144 v.

Research "without giving me the opportunity to express an opinion on the matter."[109] Fischer's sarcastic remark about the high budget of the KWI for Brain Research was indeed a crude act just short of betrayal, as he knew very well that Oskar Vogt stood under incomparably greater political pressure than he did himself.

Despite the president's reprimand, Fischer was successful in this matter. Planck presented Fischer's petition for an increased budget during the session of the Administrative Committee on October 18, 1933, and the committee resolved to submit it to the Finance Commission.[110] Later on, oral negotiations between Fischer and Glum took place, during which an increase of the budget by 49,000 RM was agreed upon for the 1934/35 fiscal year – Fischer requested that the 26,800 RM not granted at this time be included in the budget at the next opportunity. Glum also green-lighted the preparation of the institute's expansion – the architect was to elaborate the designs to the extent that it would be possible to work out an estimate.[111] In the end this amounted to around 200,000 RM. On May 25, 1934, Fischer negotiated the further steps with Ernst Telschow (1889–1988), who had acted as the deputy managing director of the KWG since October 1933. Telschow announced that the KWG intended to demand the required funds when submitting its budget for the 1935/36 fiscal year. But Fischer had a better idea. He pointed out that the Ministry of the Interior had just obligated him to take on 20 assistants for the duration of a year "and subsequently on a regular basis for further training." "For this reason it appears to him that the present moment is suitable to submit to the Ministry of the Interior an application to approve the funds for the expansion of his institute."[112] Telschow and Glum agreed. After Fischer had estimated that the necessary increases to the budget for 1935/36, in view of the imminent expansion of the institute, amounted to 23,600 RM for personnel and 20,000 RM for material,[113] November 1, 1934, on the occasion of the inauguration of 20 scholarship recipients attended by Arthur Gütt, presented a convenient opportunity to win over the latter for the expansion plans. Following the opening of the course, Fischer guided the honored guest through the institute, as he reported to Friedrich Glum on November 6:

> During this walk through the institute, I, of course, pointed out all of the places that were too crowded. You can imagine that it was no small task to create workplaces for 20 new

[109] Planck to Fischer, 4/10/1933 (draft, according to the note here, the neat copy was sent on October 6), ibid., p. 147. The passages cited were inserted as manual corrections to the draft.

[110] Auszug aus der Niederschrift über die Sitzung des Verwaltungsausschusses am 18/10/1933, ibid., p. 148. The same is true of a corresponding petition from Ernst Rüdin. In consideration of genetic health policy, the budgets of both Fischer and Rüdin were increased at the same rate. Cf. Auszug aus der Niederschrift der Sitzung des Verwaltungsausschusses der KWG am 6/3/1934, ibid., p. 153.

[111] Fischer to Generalverwaltung, 30/11/1933, ibid., pp. 150–152. The planned enlargement of the animal stalls, as Fischer reported after consultation with the architect, was not feasible technically. Fischer suggested instead building new animal stalls in the form of barracks, which were also to house the garage for the future institute automobile. The basement rooms of the planned new wing of the institute could then be used as an air raid shelter.

[112] File note by Telschow of 25/5/1934, ibid., p. 156.

[113] Fischer to Generalverwaltung, 24/10/1934, ibid., p. 170.

> gentlemen. The doctoral students are now sitting in the basement, one secretary is in the corridor. In the course of this tour it naturally arose that I praised and vaunted the fact that the K.W.G. had already given me budget improvements and resolved to apply for funds to expand the building and increase the budget. Of course, I asked him to support this as a member of the board as well. He readily agreed and asked that he be informed about the proposed expansion so that he could advocate this project in his next interview with the finance minister. [...] He believes that there is no doubt of approval for my purposes.[114]

Glum acted immediately. On November 8, 1934, he approached Gütt with the request that he intercede with the Reich Finance Ministry to supply the construction costs of 200,000 RM as well as 43,600 RM for the expansion of operating costs.[115]

When the board of the KWI-A convened for its next meeting on May 10, 1935, there was oral assurance from the Reich Finance Ministry that the funds would be approved in short, but no written promise. The senior ministerial official Linden, who attended the meeting as Gütt's substitute, ensured that the Reich Ministry of the Interior would intercede for the approval of the funds for the new building.[116] One month later, on June 25, 1935, Glum was able to inform the senate of the KWG that the Reich government had approved the increased budget of the KWI-A "almost in its entirety," and another 170,000 RM for the construction of the new building.[117]

Thus there were no more obstacles to starting construction of the new building. Another facilitating factor was that the financial situation of the institute had improved considerably since 1933. Even the 1933/34 fiscal year concluded with open and concealed surpluses amounting to almost 14,700 RM, which came about in part because funds approved for research projects that did not advance were not expended, but also because of financial support from the Rockefeller Foundation to the tune of 11,000 RM.[118] This trend continued in the fiscal year 1934/35 – it closed with surpluses amounting to almost 22,900 RM. These were due to the "partial repose of the research activities of the institute's director"[119] as well as to the grants of around 10,000 RM each from the Emergency Association and the Rockefeller Foundation. The funding gap still existing after the approval of the 170,000 RM for the expansion – which primarily concerned the erection of a new stall building estimated to cost around 20,000 RM[120] – thus could be covered by the running

[114] Fischer to Glum, 6/11/1934, MPG Archive, Dept. I, Rep. 1 A, No. 2413, p. 10.

[115] Glum to Gütt, 8/11/1934, MPG Archive, Dept. I, Rep. 1 A, No. 2406, p. 171.

[116] Niederschrift über die Sitzung des Kuratoriums des KWI-A am 10/5/1935, MPG Archive, Dept. I, Rep. 1 A, No. 2404, pp. 48–48c, here: p. 48c.

[117] Auszug aus der Niederschrift über die Sitzung des Senats der KWG am 25/6/1935, MPG Archive, Dept. I, Rep. 1 A, No. 2413, pp. 18a–c, quote: p. 18b.

[118] The surplus from the running fiscal year amounted to around 4,000 RM; added to this were assets from previous years of around 6,500 RM along with about 4,000 RM from special grants in previous years (booked as debts in the previous year), which were "reserved for special expenditures." However, because the write-offs exceeded the surpluses, the accounts closed with an official loss of over 21,000 RM. Cf. Revisor Schröder, Bericht über die Prüfung des Rechnungsabschlusses zum 31/3/1934, 5/9/1934, MPG Archive, Dept. I, Rep. 1 A, No. 2406, pp. 159–162.

[119] Revisor Schröder, Bericht über die Prüfung des Rechnungsabschlusses zum 31/3/1935, 23/4/1935, ibid., pp. 181–184.

[120] File note by Telschow, 30/6/1936, MPG Archive, Dept. I, Rep. 1 A, No. 2413, p. 37.

operating budget. In the end the KWI-A not only contributed a grant of 22,000 RM to the expansion building in the 1935/36 fiscal year, but also furnished the new wing from its own funds, through increased expenditures for expanding the scientific inventory (around 6,500 RM) and interior furnishings (around 11,200 RM). Despite these considerable extra expenditures, the 1935/36 fiscal year still closed with a surplus of approximately 17,000 RM, from which "possible special needs arising from the occasion of the expansion of the institute could be satisfied."[121]

At the board meeting on June 20, 1936, Fischer proudly proclaimed that the renovations had been completed:

> The new building may be regarded as a complete success. Space has been created, above all, areas for the quite responsible activity of writing expert opinions; there are new laboratories, above all psychological labs with special equipment; the lecture hall is significantly improved over the old one; and, finally, the long awaited garage with an institute automobile has been obtained. The by now completely insufficient number of animal stalls was enlarged quite respectably through new constructions.[122]

Because of the institute's relaxed financial situation, until the start of World War II Fischer was able to invest heavily to furnish and equip the new building. In the 1936/37 fiscal year another 17,000 RM flowed from the operating funds into the expansion of the interior furnishings, especially into the new stall building, and yet the surpluses rose to around 23,400 RM.[123] Thus it was possible in the 1937/38 fiscal year to take 22,600 RM from the KWI-A budget in order to cover the remaining deficit in the construction budget of the KWG. For this reason the annual accounts closed with a deficit of around 10,500 RM, yet as a consequence of the amassed surplusses, the institute still had reserves of nearly 12,900 RM,[124] which actually grew again in the next fiscal year, 1938/39, to nearly 20,000 RM.[125]

The altogether positive financial development of the institute between 1933 and 1939 had several causes.[126] First, the funds allotted to the institute by the KWG increased continuously – not least as a consequence of the pressure exerted by Arthur Gütt and the Reich Ministry of the Interior: from around 75,700 RM

---

[121] Revisor Schröder, Bericht über die Prüfung des Rechnungsabschlusses zum 31/3/1936, 15/6/1936, MPG Archive, Dept. I, Rep. 1 A, No. 2406, pp. 204–207, quote: pp. 205. The institute also had another 6,000 RM at its disposal for the planned, but then delayed research trip by Fischer to Spanish Morocco.

[122] Fischer, Tätigkeitsbericht vom 1/4/1935 bis zum 31/3/1936, MPG Archive, Dept. I, Rep. 1 A, No. 2404, pp. 52b–52e, quote: p. 52e.

[123] Revisor Schröder, Bericht über die Prüfung des Rechnungsabschlusses zum 31/3/1937, 5/11/1937, MPG Archive, Dept. I, Rep. 1 A, No. 2407, pp. 14–14b.

[124] Revisor Schröder, Bericht über die Prüfung des Rechnungsabschlusses zum 31/3/1938, 3/5/1938, ibid., pp. 24a–24d. An additional KWI-A grant of 16,300 RM to the construction budget was planned for the 1938/39 fiscal year, but then was not rendered.

[125] Revisor Schröder, Bericht über die Prüfung des Rechnungsabschlusses zum 31/3/1939, 29/8/1939, ibid., pp. 48a–48d.

[126] The following numbers are given according to the overview of assets and the income and expenditure accounts of the fiscal years 1933/1934 through 1939/40 in: MPG Archive, Dept. I, Rep. 1 A, No. 2399, 2406–2410.

(1933/34) to 127,200 RM (1934/35), 139,500 RM (1935/36), 164,200 RM (1936/37) up to 168,100 RM (1937/38). Not until the end of the 1930s did they decrease slightly, to 160,500 RM (1938/39) and ultimately to 150,500 RM (1939/40). Second, Fischer was again successful in soliciting considerable third-party funds – up to the mid-1930s from the Rockefeller Foundation, and also from the German Research Association. Third, the income from expert opinions climbed steadily – from 400 RM (1934/35) to 2,300 RM (1935/36), 5,100 RM (1936/37), and finally to nearly 8,600 RM (1938/39) – advancing to become an appreciable income factor in the second half of the 1930s.[127] Fourth, until the mid-1930s the monies provided could not be expended in total because research activity did not keep pace with the approved funds. This was due above all to the fact that Fischer's research activity was at a complete standstill until late 1935, so that the practice of transmitting the funds not expended to the new household meant that considerable reserves could be accumulated. From the mid-1930s, research was in high gear again, which is reflected in such factors as the dramatic rise in expenditures for test animals and human subjects – actually, these were *one* budget line – as well as travel costs. Personnel costs climbed as a consequence of the constant increase in both scientific and nonscientific personnel, from 44,500 RM (1933/34) to 100,800 RM (1938/39). The progression of material costs from 28,600 RM (1933/34) to 95,700 RM (1937/38) was even more considerable, which can be traced back primarily to the new building, but also the constant expansion of research operations. Once the construction financing had been concluded, the material costs settled down to a level of 63,300 RM (1938/39). Thus it can be stated that the expenditures on both personnel and material nearly doubled in the first 6 years of the Third Reich.

3.2.3 *Fritz Lenz: "The Least of Evils"?*

After the forced removal of Muckermann, Fritz Lenz was appointed director of the Department for Eugenics. "Presumably," Niels C. Lösch assumes, in Lenz Fischer had chosen "the least of all evils."[128] But did Fischer see Lenz as "an evil" at all?

[127] The costs of an expert opinion were generally borne by the proband; only in cases where it was proven that the proband had no means were the costs covered by a hardship fund of the Reich Office for Genealogical Research (*Reichsstelle für Sippenforschung*). In 1936, Fischer declared himself principally against the idea of preparing expert opinions free of charge, but expressed his willingness to charge a discounted fee or to forego a fee when it was proven that the proband had no means to pay for the service. The Reich Ministry of the Interior consented to this offer, and in return left it up to the discretion of the KWI-A "to levy a surcharge to the cost of the investigation for economically better-off applicants." With prices of 50–60 RM per expert opinion the KWI-A was located in the intermediate class of the institutes entrusted with genealogical opinions. Cf. Reichs- und Preußischer Minister des Innern to Reichs- und Preußischen Minister für Wissenschaft, Erziehung und Volksbildung, 29/5/1936, MPG Archive, Dept. I, Rep. 1 A, No. 2399, pp. 80–80 v; Fischer to Generalverwaltung, 1/9/1936, ibid., pp. 86–87; Berechnung der Gebühren für Erb-, Rasse- und Vaterschaftsgutachten, ibid., p. 88; Reichs- und Preußischer Minister für Wissenschaft, Erziehung und Volksbildung to KWI-A, 20/10/1936, ibid., p. 96 (quote).

[128] Lösch, Rasse, p. 305.

Under the given circumstances, Lenz must have seemed to him the ideal candidate.[129]

Fischer had met Lenz after the latter began studying medicine at the University of Freiburg in 1906. Lenz also visited Fischer's anthropological lectures, whereby he was interested above all in their social-anthropology aspects. Inspired by his encounter with Alfred Ploetz in 1909, Lenz made race hygiene and hereditary biology his life's mission. As a logical consequence, he immediately joined the newly founded Freiburg branch of the German Society for Race Hygiene, where he worked as secretary to chairman Eugen Fischer. Since this time Fischer and Lenz had known and respected each other – in 1921, together with Erwin Baur, they wrote the fundamental German language textbook *Menschliche Erblichkeitslehre und Rassenhygiene* ("Human Genetics and Race Hygiene"). Fischer and Lenz also worked together in the race hygiene movement of the 1920s, although after moving to Munich in 1913 Lenz was oriented more toward the Munich wing, whereas Fischer was increasingly drawn to the Berlin wing of the race hygiene movement. Yet the two men's positions were not as far apart as this rough contrast might suggest: Fischer was not adverse to the Nordic idea, nor did Lenz reject the eugenic agenda of the Berlin wing. The fact that Karl Astel and Bruno K. Schultz insisted that Fritz Lenz *not* be sent as a delegate of the Munich branch to the general meeting of the German Society for Race Hygiene (Eugenics) on June 10, 1933, when Eugen Fischer was to be voted out, implies that the young National Socialist race hygienists, presumably with some justification, suspected all too little distance between the old comrades-in-arms Lenz and Fischer.

For his part, Verschuer, as mentioned above, had known Lenz since 1922, when he was studying in Munich. Lenz had been the most important mentor of the promising young scientist until 1927. He not only obtained for him the assistantship at the Medical Polyclinic of the University of Tübingen, but in 1926, he was also the one that inspired Verscheur to apply to Fischer for a post in the new KWI-A in founding in 1926. And even in the years from 1927 to 1933, Verschuer and Lenz remained in brisk scientific exchange and published together. Moreover, through his marriage to Kara von Borries in 1929 – his first wife Emmy, née Weitz, had since passed away – Lenz was related to Verschuer by marriage.[130] The relationship between Lenz and Verschuer never really became friendly – like that between Lenz and Fischer. This was due to Lenz's distanced demeanor, which made him seem cool and aloof, but even more so to his combativeness and his pedantic condescension, and his tendency toward sharp critique, which often drifted into polemics. Tactical

[129] According to his own account, Verschuer advised Fischer to bring Lenz to Dahlem. Verschuer, Erbe – Umwelt – Führung, "Freier Forscher in Dahlem (1927–1935)" section, p. 21, MPG Archive, Dept. III, Rep. 86 A, No. 3–1.

[130] Cf. Lenz to Verschuer, 20/9/1929, MPG Archive, Dept. III, Rep. 86 B, No. 1: "Now we are related by marriage, for a cousin of my wife is married to your cousin Vollprecht. I do not know whether you are aware that this relation is no coincidence. You must remember that you recommended a paper by my wife, who was completely unknown to me at the time, to Ostermann for publication in the *Zeitschrift für Volksaufartung*. I liked this paper so much that I got in touch with its author at the time (1926) [...]." Cf. also Rissom, Fritz Lenz, p. 22.

calculation was completely foreign to Lenz – if he differed on any point he did not hesitate to attack friend or foe sharply in public, without consideration for networks of supportive colleagues or political constellations. Although he did not even except himself – whenever his own standpoint was disproved, he had no problem revising it publicly – Lenz's concept of the culture of disputation did not win him any sympathy. Verschuer, for instance, reacted with great annoyance in 1932 when Lenz subjected the hypotheses put forward by Verschuer and Curtius on the heritability of twin births to a devastating critique:

> Dear Mr. Lenz, speaking quite openly: I do not understand you! You must know my opinion, you are aware of the fact that I have always retained trust and grateful reverence for you as my former teacher and friend; such a sentiment is, of course, always accompanied by great receptivity for advice and objective critique. But why the form of critique you choose here and this antagonistic tone: I have never experienced the like from scientific opponents! That is very painful, painful for me personally, but also for the cause for which, up to now, we have tended to fight on the same side.[131]

Lenz's reaction to this letter from Verschuer is typical for him. He sent a slightly milder correction proof of his critique with the remark that he had written "the manuscript 'with [too little] regard for the person' and always conscious of the fact that it would not stay in this form."[132] A few days later, he offered to let Verschuer delete from the proof everything he "sensed as invidious"; at the same time he sent a revised manuscript draft in which he responded to Verschuer's defense with renewed sharp critique: "My intention was only to convince you as drastically as possible of the untenability of your theses." In the end the adversaries agreed to delete all polemical phrases from their papers, to place them under a joint heading and to send them together as a reprint, in order to, as Lenz expressed it, make "a kind of mutual paper in this manner" and not to make "the impression of hostility."[133] In this way an open scandal was avoided, and although a clear resentment remained on Verschuer's side, the two scientists nevertheless continued their exchange. The episode shows that Lenz was an uncomfortable and unpredictable colleague, but one whose polemic was not directed toward science policy intentions of any kind. Despite bitter conflicts on the subject, Lenz never broke his connections to Dahlem, always emphasizing his willingness to participate in critical, but constructive dialog.

From the perspective of Fischer and Verschuer, Lenz had the additional advantage – in contrast to his predecessor Hermann Muckermann – of enjoying "a solid academic reputation."[134] Without doubt Lenz, whose appointment as an associate professor to the medical faculty of the University of Munich in 1923 was the first professorship for race hygiene at a German university, was considered one of the leading race hygienists in Germany. His appointment to Dahlem appreciated the institute's scientific value.

---

[131] Verschuer to Lenz, 21/11/1932, MPG Archive, Dept. III, Rep. 86 B, No. 4, pp. 29–30, quote: p. 30.

[132] Lenz to Verschuer, 21/11/1932, ibid., p. 31.

[133] Lenz to Verschuer, 8/12/1932, ibid., pp. 32–33, quotes: p. 32.

[134] Lösch, Rasse, p. 307.

- But above all, Lenz's appointment offered the inestimable advantage that the National Socialists placed no obstacles in its path. On the contrary: Lenz, who tended toward a corporatist model of society as a third way between socialism and capitalism, had already expressed his sympathies for the up and coming National Socialism back in 1931/32 – more clearly than most of the other prominent race hygienists and eugenicists – in accordance with the motto "Politically correct is the system that benefits the race."[135] In 1931, he had reviewed Hitler's *Mein Kampf* for the *Archiv für Rassen- und Gesellschaftsbiologie*. Hitler, Lenz emphasized appreciatively, was "the first politician of truly great influence to recognize race hygiene as a central task of all politics, and who intends to actively implement it for this purpose." But the ideas Hitler advocated here were not new. Rather, Lenz laid claim to the intellectual authorship for the race hygiene elements in Hitler's worldview: "Of the books actually on race hygiene, I have heard that Hitler read only the 2nd edition of 'Baur-Fischer-Lenz,' and this during his imprisonment in Landsberg. Some passages from this work are reflected in Hitler's phrases. In any case he embraced the essential ideas of race hygiene and their importance with great intellectual receptiveness and energy [...]." While he was consistent in principle with Hitler's race hygiene agenda, Lenz distanced himself carefully from some of the race anthropology, and also from the race anti-Semitic points of the agenda, about which one "sometimes shakes one's head." In the same tones as Fischer, Lenz criticized that Hitler – like Gobineau and Chamberlain – exaggerates "the detrimental consequences of miscegenation."[136] In the 3rd edition of "Baur-Fischer-Lenz," which also appeared in 1931, Lenz expressed the hope that the "affinity with the race hygiene idea" could be used in order to "win over the National Socialist movement for [...] race hygiene reforms." From his examination of the NSDAP party platform, in which race hygiene was intended not only for a "Nordization of our nation in the sense of the Nordic idea," but also for an "anti-Semitic practical application," Lenz concluded, "That National Socialism truly aspires to a recovery of the race cannot be doubted. Of course, the lopsided 'anti-Semitism' of National Socialism must be regretted."[137] In spite of his – quite reserved – critique of racial anti-Semitism, Lenz was clearly closer to the elements of National Socialist race anthropology than anyone like Eugen Fischer, which was apparent in the example that Lenz – in contrast to Fischer – reviewed Darrés *Neuadel aus Blut und Boden* in a critical, reserved tone, but quite positive in terms of content, whereby he conceived of Darré's idea of the hereditary farm (*Erbhof*) as a further development of earlier proposals of his own. Lenz hoped that the book might contribute to "directing the National Socialist movement into healthy channels."[138]

[135] Kröner, Von der Rassenhygiene zur Humangenetik, p. 34.

[136] Lenz, Stellung, p. 308, 303.

[137] Lenz, Auslese, 3rd edn., pp. 415, 418, 417.

[138] Lenz, Review R. W. Darré, Neuadel aus Blut und Boden, p. 444. Cf. also Weingart/Kroll/Bayertz, Rasse, pp. 373f., 382f.

In short: even before 1933, Lenz had made no bones about the fact that he sympathized with the National Socialists. Yes, one can even interpret his public remarks in the years 1931/32 as an offer to place his race hygiene expertise at the service of the NSDAP. Parts of National Socialism, in turn, were quite willing to entertain this offer. Since 1932 at the latest, Lenz had unofficial contacts with the *Braunes Haus* in Munich, above all to Darré.[139] Apart from Rüdin, Lenz was the most interesting cooperation partner in the race hygiene camp for National Socialist health and social policy experts, and it was no coincidence that these two scientists were the ones to take key positions in the new Expert Council for Population and Race Policy (*Sachverständigenbeirat für Bevölkerungs- und Rassenpolitik*) constituted in 1933, while Fischer and Verschuer were ignored. Thanks to Hermann Muckermann's far-reaching network of contacts the KWI-A had been able to exert direct influence on political decisions in the late Weimar Republic, at least on the Prussian level. Through the political reversal these opportunities for influence were cut off – even worse, due to its close links with the Weimar welfare state, the institute was banned to the political wilderness under the rule of the National Socialists. Lenz's appointment presented a chance to re-establish a direct line to the center of power – and this was useful to intercept the political pressure the institute was feeling. Above all, however, it corresponded with its own claim not only to perform theoretical research on hereditary biology, but also to accompany its implementation in practical genetic health and race policy in the form of scientific policy consulting.

On the other hand, Lenz was not a party member – he did not join the NSDAP until 1938, at the urging of Arthur Gütt – did not agree with the party line in all instances, and here, too, was not reluctant to express his critique of the genetic health policy of the National Socialist government public whenever he saw fit. "Similarly instructive" as he was with regard to his colleagues, "was Lenz's manner with regard to the National Socialists, here, too, underestimating the new quality of the system in reliance on the outdated status of the scientific expert."[140] As such, he had a dispute with the National Socialist genetic health policy makers in 1933 about the GzVeN – Lenz rejected compulsory sterilization. On the other hand, "the law did not go far enough for him."[141] Lenz advocated widening the spectrum of indications for sterilization on a voluntary basis beyond that of the National Socialist bill, to include "sociopaths, psychopaths and the slightly moronic."[142] The congenitally inferior tiers of society, whose share Lenz estimated to be about 10% of the population, would voluntarily make use of the possibility to sterilize themselves due to their moral instability. Against this background, the application of coercion was counterproductive. "The rejection of compulsory sterilization had its source in a purely tactical stance"[143] – this had nothing to do with the basic right to physical integrity. That Lenz was not satisfied with the

[139] Kröner, Von der Rassenhygiene zur Humangenetik, p. 35.

[140] Ibid.

[141] Ibid.

[142] Lenz, Sterilisierungsfrage, pp. 294 f.

[143] Kröner, Von der Rassenhygiene zur Humangenetik, p. 36.

National Socialist sterilization legislation was also apparent in the fact that – in contrast to Fischer and Verschuer – he did not become active as a judge in a hereditary health court, although he did produce numerous expert opinions in the framework of genetic health trials. Lenz was and remained an unconventional thinker – there was no danger of his attempting to force the institute onto the party line.

In Eugen Fischer's view, the only thing that marred the arrangement was that bringing Lenz on board meant welcoming a serious rival to Verschuer for the succession as institute director. Yet fears in this direction later proved unfounded. Lenz, who took over the direction of the Department for Eugenics at the KWI-A on October 1, 1933 – which he soon quietly renamed to "Department for Race Hygiene" – and, through the intercession of Arthur Gütt, received a professorial chair for race hygiene at the medical faculty of the University of Berlin on November 16, 1933,[144] did not develop any ambitions to take over direction of the institute, but left it up to Fischer and Verschuer to arrange the succession. His ambition was directed more toward influencing the National Socialist Regime in the sense of science policy consulting. All in all, it can be summarized, appointing Lenz was a shrewd move by Fischer to come to terms with the new rulers while maintaining a high degree of independence for the scientific work of the institute and opening up channels of influence in the arcane area of political authority.

3.2.4 *"Absolutely Acquiescent and Docile": Baron Otmar von Verschuer*

Otmar von Verschuer was regarded with suspicion by the National Socialists in 1933/34, as he – along with Eugen Fischer, Hermann Muckermann, and Arthur Ostermann – was considered to be one of the leading representatives of the Berlin wing of eugenics. Moreover, as mentioned above, toward the end of the Weimar Republic – as a counterweight to Muckermann's activities in Catholic circles – he had become involved with the Protestant Church and the Inner Mission for the cause of eugenics and genetics, and he continued this involvement in the Third Reich. As such he wrote a paper, *Die Rasse als biologische Größe* ("Race as a Biological Parameter") for the omnibus volume *Die Nation vor Gott* ("The Nation Before God") edited by Helmuth Schreiner and Walter Künneth (1901–1997) in 1933.[145] In 1937, he was supposed to travel to the World Church Conference in Oxford as an expert with the German delegation, yet German representatives of the church were forbidden to participate by the National Socialist government.[146] Verschuer's

[144] The prospect of being able to take over Benno Chajes' chair for social hygiene was probably what persuaded Lenz to accept the appointment in Dahlem – in opposition to 1927.

[145] Verschuer, Rasse als biologische Größe. Cf. Künneth, Lebensführungen, p. 96f; Künneth to Koch, 28/3/1979, MPG Archive, Dept. V, Rep. 4, Koch No. 1.

[146] Verschuer, Mein wissenschaftlicher Weg, p. 22, MPG Archive, Dept. III, Rep. 86 A, No. 3–1.

expert opinion about "Hereditary Biological Findings as Grounds for German Population and Race Policy" (*Erbbiologische Erkenntnisse zur Begründung der deutschen Bevölkerungs- und Rassenpolitik*) appeared in the book *Kirche, Volk und Staat* on the subject of the World Church Conference, edited by Eugen Gerstenmaier (1906–1986) on behalf of the Foreign Office of the Church.[147] Thus Verschuer was considered by the new rulers to be linked with the church. The fact that he was also an active member of the Dahlem parish congregation under Martin Niemöller (1892–1984), who was soon to become one of the figureheads of the "Confessing Church" (*Bekennende Kirche*) did not exactly dispel this suspicion.[148] The National Socialists scented opposition.

In fact, Verschuer did maintain a certain inner distance to the National Socialist regime and did not join the NSDAP initially. How ambivalent his political stance was, however, becomes apparent in his memoirs. He faced the change in political systems in 1933 with extremely mixed feelings. Sure enough: Verschuer had inwardly rejected the Weimar Republic. "The national crisis, the rising number of unemployed, the swelling flood of Bolshevism, the increasing corruption, all of these menacing phenomena" had ripened in him, as in many of his contemporaries, the conviction that the democratic system was ruined. Certainly he placed his hopes in a national revolution and the introduction of an authoritarian form of government. Yet: he was skeptical about Hitler and National Socialism. He designated Hitler's public advocacy for the murders of Potempa as "evil,"[149] which was also his judgment of "his tactless remarks about the venerable Hindenburg." The majority of Hitler's "closest staff" and "many of his propagandistic methods," Verschuer recalled in retrospect, had "something repulsive" for him. This is why he stayed aloof from the events of January 30 and March 21 – the "Day of Potsdam" – and May 1, 1933. Yet it was, Verschuer admitted, "a difficult internal struggle," not to join in the general enthusiasm. He attempted to "make [himself] acquainted with the National Socialist body of ideas": reading Hitler's *Mein Kampf* "with divided sensations," vehemently rejecting Alfred Rosenberg's (1893–1946) *The Myth of the Twentieth Century*, while intensely experiencing Arthur Moeller van den Bruck's (1876–1925) *The Third Reich* as enthralling reading "through a profound command of ideas and high idealism." Verschuer located himself in the camp of the "conservative revolution," whose ambivalence was expressed in an exemplary

---

[147] Verschuer, Erbbiologische Erkenntnisse. Cf. Gerstenmaier, Streit, p. 81; Gerstenmaier to Koch, 8/12/1981, MPG Archive, Dept. V, Rep. 4, Koch No. 1.

[148] Cf. Gerhard Koch, Otmar Freiherr von Verschuer und die Bekennende Kirche, Ms., ibid. Koch, Humangenetik, pp. 67–70, 77–81. Through Friedrich Wilhelm Bremer, Niemöller's brother-in-law, there was a personal connection between Verschuer and Niemöller. On the general circumstances of the church in Dahlem: Gailus, Protestantismus, pp. 306–357, esp. pp. 311 f. In 1933, Verschuer was among the signatories of an appeal by the *Jungreformatorische Bewegung* (Copy in the MPG Archive, Dept. III, Rep. 86 A, No. 12).

[149] In August 1932, five SA men had brutally murdered a Polish worker belonging to the KPD in Potempa, Upper Silesia. A special court in Beuthen sentenced the perpetrators to death on August 22, 1932, eliciting Hitler's response in a public telegram, in which he denounced the "blood sentence" and promised the sentenced that they would soon be liberated.

way in his political thought: Despite far-reaching ideological convergence and congruencies, he kept his distance from Hitler and his movement because their plebian populism and their arrant brutality repulsed him. He resolved the conflict for himself by withdrawing to a seemingly apolitical position, claiming that he regarded it as his "duty [...] to continue placing my work and my knowledge at the service of my German nation and state, free of any political obligation, and devoting myself to the hope that the men of nobility, ideals and expertise would ultimately prevail." In the end he was mistaken in this, Verschuer remarked after the war, but "over and again [he found] individual idealists," who had mobilized themselves "within the party for cleanliness, reason and objective work." In some cases they were even successful in "eliminating errors, removing objectively inappropriate personalities and thus contributing to the real work of constructing the institute."[150] With this interpretive arabesque Verschuer had no problem justifying his collaboration with parts of the National Socialist power apparatus after 1945 – all the way to complicity in the mass crimes committed in the course of heredity health and race policy, while at the same time claiming that he had *not* been part of the system, but rather located outside, even at *critical distance* to the system.

Verschuer's bridge to the Third Reich was race hygiene. Before 1933, he had been actively involved in the race hygiene movement – quasi perforce – under the auspices of the democratic constitutional and welfare state, but at the same time he had left no doubt as to his conviction that race hygiene required a different political milieu to develop. In his tract *Sozialpolitik und Rassenhygiene* ("Social Policy and Race Hygiene") (1928), Verschuer – with express reference to Othmar Spann and Bogislav von Selchow – did not conceal his opinion that race hygiene and the Weimar state were politically incongruent:

> Race hygiene can never be implemented as a means of salvation under the spiritual flag of unrestricted individualism, however. Rather, it demands a philosophy and view of life that is ready for personal sacrifice, which subjects its own welfare to the welfare of the collectivity and recognizes that the value of being a nation is higher than that of being an individual, the logical priority of the whole over the part.
>
> Race hygiene's criticism of the political, economic and societal circumstances can thus only bear fruit when certain intellectual prerequisites are fulfilled. The race hygienist therefore feels his comrades-in-arms are those who stand in the intellectual struggle of idealism against materialism, of universalism against individualism, the ideas of the "we age" against those of the "I age."[151]

In this sense Verschuer made himself available to the National Socialist state as a "genetic doctor" from 1933, supported the GzVeN in writings and lectures, made a major contribution to the training of the chairmen of the hereditary health courts and medical officers, and cooperated in the implementation of the law as a medical

[150] Verschuer, Erbe – Umwelt – Führung, "Freier Forscher in Dahlem (1927–1935)" section, pp. 18–20, MPG Archive, Dept. III, Rep. 86 A, No. 3–1.

[151] Verschuer, Sozialpolitik und Rassenhygiene, p. 32.

assessor at the Hereditary Health Court and as director of the Polyclinic for the Fostering of Genes and Race (*Poliklinik für Erb- und Rassenpflege*). The decisive National Socialist functionaries – Arthur Gütt and Walter Groß – recognized this early on. On July 14, Groß wrote to Gütt:

> Verschuer was just here and talked for an infinitely long time. He is absolutely acquiescent and docile, but not even today does he really understand deep down what the point is [...] Soon I will pay him a visit out there [in the KWI-A in Dahlem] and see what kind of propaganda we can take up with these people. However, I believe that it will be accursedly little and that one will have to advise them over and again to restrict themselves to their research.[152]

Groß's critique referred to Verschuer's commitment to keeping Muckermann at the institute, and thus was directed at his lacking loyalty. The final remark, however, implies that the scientist located outside of politics, with his theoretical research on genetic biology, could certainly be of value for the political decision makers, although it would not be advisable to deploy him at an exposed position. Gütt clarified to Rust that Verschuer, who had been nominated by the faculty for the chair for race hygiene at the University of Berlin, did not come into question for this post "because by virtue of his whole attitude he is not capable of supporting our National Socialist endeavors." But there was no reason why he could not be used "for genetics and genetic research at a less significant university."[153]

Following this line, Verschuer was appointed to the newly founded Frankfurt Institute for Hereditary Biology and Race Hygiene (*Frankfurter Institut für Erbbiologie und Rassenhygiene*) in 1935. When opposition to Verschuer's appointment arose on the part of the faculty in Frankfurt, Gütt, and Groß, the two political protectors of the KWI-A, were asked for an expert opinion. In Gütt's assessment, Verschuer was "one of the most capable scientists in the field of genetic pathology and the research of hereditary diseases and also an outstanding teacher." "As a person" he had "pleasant manners" and was "skilled at negotiation and in the practice of his profession." As to Verschuer's "attitude," Gütt appeared impressed that he stood "on the ground of the National Socialist state at this time."

> From the past he is undoubtedly encumbered by the fact that he had to work together with Father Muckermann in Professor Fischer's institute before the takeover. [...] In this context he then stood up for his employee [sic] Muckermann on many occasions, which has caused great resentment. I once spoke with him about this matter in depth, and he alleged that his interest in Muckermann was purely scientific and that he stood up for him at the time quasi as a matter of integrity. On many occasions he has assured me that he can represent the genetic biology and race legislation of the new Reich with deep inward conviction.

Gütt pointed out that he had entrusted Verschuer with the direction of the Polyclinic for the Fostering of Genes and Race in Berlin Charlottenburg. In conclusion, the expert opinion emphatically recommended utilizing Verschuer as a college professor.[154] Groß arrived at a very similar assessment:

---

[152] Groß to Gütt, 14/7/1933, quoted in Lösch, Rasse, p. 313.

[153] Quoted in ibid.

[154] Undated expert opinion by Gütt, BArch. Berlin, R 49/14.256.

> Professor von Verschuer can not be addressed as a National Socialist, not as a political man at all. From the standpoint of politics, his appointment would thus not entail a reinforcement of the National Socialist element. As far as I can judge, Professor Verschuer is completely and honestly loyal to National Socialism and endeavors to evince this posture actively even beyond the field of his science. This does not mean true inner solidarity with the idea, however, and, of course, he is also lacking any militant posture in the National Socialist sense.

Despite this political reservation Groß held an appointment "to be completely justified objectively." Verschuer's scientific achievements were indisputable, "and as a matter of course, all of Verschuer's work also contributes to the consolidation of foundations of the world of National Socialist thought." Besides, Groß added in conclusion, he could imagine "that Verschuer's objective and primarily scientific, apolitical manner can be particularly convincing for large circles of doubters, so that his appointment could also be valuable in the sense of propaganda and recruitment."[155] Gütt and Groß had recognized correctly that Verschuer was a true ally on issues of genetic health policy – and also race policy, but otherwise still persisting in his inner distance to National Socialism. However, they had also recognized how useful a seemingly politically neutral genetic scientist of Verschuer's caliber could be for the National Socialist regime, and therefore fostered his career as best they could. "In the polycratic system of rule by the National Socialists, even scientists like Verschuer consisted of a kind of reserve quantity for reinforcing their own ranks, for the ideological (scientific!) justification of its own claim," Hans-Peter Kröner aptly judges. For the scientists it was not always clear "the extent to which they were used in the gamble for power. Their very self-understanding as elites was a major obstacle to this insight." But this made it possible for them "to preserve their illusion of political independence."[156] On the other hand it may not be overlooked that the protracted battles of the political functionaries opened up free spaces, which they could exploit in turn in their maneuvering for posts and funding. And beyond the level of mutual functionalization – in the practical implementation of genetic biology and eugenic research – both sides worked together closely out of conviction.

At this juncture the thesis by Niels C. Lösch must be refuted, that Verschuer "was called away to Frankfurt according to plan"[157] in order to separate Fischer completely from all of his department heads to date, to isolate him and to bring the institute on the desirable political course. The sources Lösch gives to prove his thesis[158] are by no means convincing. The initiative to appoint Verschuer emanated

[155] Undated expert opinion by Groß, ibid. Quite similar, too, the expert opinion by the assistant professors at the University of Berlin, BArch. Berlin, BDC, DS G 117, p. 940. By the way, Lothar Loeffler, who was second to Verschuer on the list of appointees, had rejected the appointment because to him "Frankfurt appeared to conflict-laden with all of its Jews." Müller-Hill, Tödliche Wissenschaft, p. 81.

[156] Kröner, Von der Rassenhygiene zur Humangenetik, p. 31.

[157] Lösch, Rasse, p. 299.

[158] Especially the letter from Gütt to Rust cited above.

from the medical faculty of the University of Frankfurt in winter 1933/34 – above all from the radiologist (and SS member) Hans Holfelder (1891–1944).[159] Gütt and Groß supported the appointment, but there was no evidence that they did this to pursue any goals of their own connected with the institute in Dahlem. Fischer, for his part, regretted his closest colleague's departure, but supported Verschuer's appointment to Frankfurt as best he could and intervened with the Reich Ministry of Science when the proceedings were held up – not least with the ulterior motive that Verschuer's appointment to a full professorship was the prerequisite for building him up as his successor. Fischer formulated this thought in his very first reaction to the call from Frankfurt:

> For the time being I can not think of you leaving Dahlem! I do not know what I should do! But it had to be! It would be a great step for you – and in 7 or 8 years you could transfer back as director of the institute! – Let's leave the future to the future [...].[160]

Even during his time in Frankfurt, Verschuer's relationship to National Socialism was characterized by deep contradictions. On the one hand, he placed himself and his institute unreservedly in the service of National Socialist genetic health and race policy. His speech about race hygiene as science and task of the state (*Rassenhygiene als Wissenschaft und Staatsaufgabe*), which he held at the University of Frankfurt on January 30, 1936, the *Tag der nationalen Erhebung* ("Day of National Elevation"), was a single panegyrical to Adolf Hitler's state.[161] On the other hand, he was integrated in a social circle that included such individuals as the first post-war mayor of Frankfurt, Hollmann; the orthopedic surgeon Georg Hohmann (1880–1970),[162] first postwar rector of the university; and the pediatrician Bernhard de Rudder (1894–512),[163] the first postwar dean of the medical school – all of them men who kept their distance from National Socialism. Verschuer's connections to the Confessing Church were strengthened when he joined the congregation of the Trinity Church under Pastor Otto Fricke (1902–1954). At the beginning of the Third Reich, Fricke had stuck by the "German Christians," demanded the introduction of the "Führer principle" in the Protestant Church in an organ of the German Christians entitled *Fanfare*, and, in his capacity as college pastor, held the speech on the occasion of the book burning on the Römerberg on May 10, 1933. Fricke soon switched sides, however, joined the Confessing Church and found himself in increasing opposition to the National Socialist state. In August 1937, he was temporarily

[159] Verschuer, Erbe – Umwelt – Führung, "Freier Forscher in Dahlem (1927–1935)" section, p. 24, MPG Archive, Dept. III, Rep. 86 A, No. 3–1. For a biography of Holfelder: Klee, Personenlexikon, p. 267.

[160] Fischer to Verschuer, 10/9/1934, MPG Archive, Dept. III, Rep. 86 A (Münster), No. 9.

[161] Verschuer, Rassenhygiene als Wissenschaft und Staatsaufgabe. According to his own account, Verschuer expressed "fundamental thoughts and ideals of my science" in this speech. Verschuer, Erbe – Umwelt – Führung, "Professor in Frankfurt (1935–1942)" section, p. 4, MPG Archive, Dept. III, Rep. 86 A, No. 3–1.

[162] For a biography: Klee, Personenlexikon, p. 267.

[163] For a biography: ibid., p. 512.

taken into protective custody by the Gestapo.[164] Verschuer and his family, who were close friends with the Fricke family, took care of Fricke's wife during his imprisonment and later, when the rumor spread that Fricke was to be arrested again, offered the couple refuge in their apartment.[165]

With his connection to Otto Fricke and to the Confessing Church, Verschuer provoked the suspicion of the party and its organizations.[166] His relationship with the district chief of Hessen-Nassau-Süd, Jakob Sprenger (1884–1945), also remained absolutely tense – ultimately it was Sprenger who succeeded in appointing Heinrich Wilhelm Kranz to the chair in Frankfurt in 1942 against Verschuer's will. Although Verschuer finally did join the NSDAP in 1940 – more on this later – persistent doubt about his loyalty lingered in the ranks of the party. So, as late as August 1942, the Culture Policy Archive (*Kulturpolitische Archiv*) of the NSDAP responded to an inquiry as to whether Verschuer was a suitable lecturer, after consulting with the Race Policy Office:

> In the case of Professor Verschuer, who may be a party member, but does not otherwise make any political appearance and is regarded as not entirely consolidated in terms of his worldview, it must be noted that his services as a lecturer are limited to popular science and generally informative lectures. Given Professor V.'s specialized knowledge, there are no objections to lecture activities of this kind. However, it should be avoided that the announcement of his lectures, or the manner in which they are presented receive a character emphasizing politics or world view. This restriction is particularly important because in our experience aspects of politics and world view are exposed especially prominently in the advertisement of lectures about race or race hygiene.[167]

---

[164] "Alles für Deutschland," pp. 115–118.

[165] Verschuer, Erbe – Umwelt – Führung, "Professor in Frankfurt (1935–1942)" section, pp. 12–15, MPG Archive, Dept. III, Rep. 86 A, No. 3–1. Cf. also Gerhard Koch, Otmar Freiherr von Verschuer und die Bekennende Kirche, Ms., MPG Archive, Dept. V, Rep. 4, Koch No. 1.

[166] Cf., e.g. Stiftung Ahnenerbe, Forschungsstätte für Innerasien und Expeditionen to Reichsgeschäftsführer des Ahnenerbes, 3/6/1941: "In confidence I would like to add that Professor Fischer in Berlin, just like Verschuer in Frankfurt, were strongly linked with the Church. Whether they still are today is unknown to me." BArch. Berlin, BDC, DS G 117, p. 944. The suspicion against Fischer probably arose because of his daughter Gertrud, who had worked at the Fröbel Protestant School (*Evangelisches Fröbelseminar*) in Kassel since 1933 and joined the Confessing Church in 1935 (Lösch, Rasse, pp. 426 f.). Fischer himself was distant from the Church. The fact that the head of the Saarpfalz district, Josef Bürckel (1894–1944), banned a lecture by Verschuer on "Genetic Talent and Genetic Encumbrance" in Saarbrücken in 1936, by contrast, was based on a simple misunderstanding. As Bürckel reported in response to an inquiry by the Reich Minister of the Interior, he mistakenly believed the Kaiser Wilhelm Society to be one of those monarchistic "associations and little clubs from the Reich" that "abruptly fell upon the Saar region" after its reintegration. Bürckel to Reichs- und Preußischen Minister des Innern, 13/6/1936, MPG Archive, Dept. I, Rep. 1 A, No. 838, vol. 4, pp. 271 f., quote: p. 272. Many thanks to Rüdiger Hachtmann for referring me to this document. Cf. also the correspondence in BArch. Berlin, R 43 II/1227.

[167] Kulturpolitisches Archiv der NSDAP (Dr. Killer) to Deutsches Volksbildungswerk, Dept. Vortragswesen/Vortragsdienst, 21/10/1942, BArch. Berlin, NS 15/254. The assessment can be traced back to Walter Groß. Cf. Groß to Kulturpolitisches Archiv, 1/9/1942, IfZ, MA 116/16.

3.2.5 Kurt Gottschaldt

After Verschuer's departure Fischer resolved to dissolve the Department for Human Genetics and divide its tasks between the two remaining departments, the Department for Anthropology and the Department for Race Hygiene. As will be demonstrated below, this took place in the context of a deliberate division of labor between the institute in Berlin and its daughter in Frankfurt. Fischer did not want to compete with Verschuer in his traditional field of clinical genetic pathology, and thus intended to gradually discontinue this branch of research in Dahlem. This decision on research strategy also had a personnel-policy aspect: with a view to his succession as institute director, Fischer did not want to build up a potential rival to Verschuer in his own facility. Therefore, he placed a new research emphasis on genetic psychology, an area interesting for its scientific, but also political aspects, and yet not a field deemed suitable for guiding and directing the research agenda of the institute as a whole. Further, Fischer entrusted the direction of the Department for Genetic Psychology to a scientist whose career would stagnate in National Socialist Germany for political reasons.

Kurt Gottschaldt (1902–1991)[168] belonged to the Berlin school of Gestalt psychology around Max Wertheimer (1880–1943), Wolfgang Köhler and Kurt Lewin (1890–1947). From 1922 to 1926 he studied at the Psychological Institute of the University of Berlin, directed by Köhler, which was one of the leading psychological research institutions in the world during the Weimar Republic. After earning his Ph.D. in 1926, Gottschaldt worked as Köhler's assistant. In 1929, he switched to Rhine Province Children's Institution for the Mentally Abnormal (*Rheinische Provinzialkinderanstalt für seelisch Abnorme*) under the direction of Otto Löwenstein (1889–1965) founded, the "first independent stationary children's psychiatric clinic in Germany" in 1926.[169] Thus Gottschaldt was working at the top address of the still young discipline of children's and youth psychiatry in Germany and soon advanced to become a recognized expert in this field. In 1932, he earned his qualification for a professorship by working with the philosopher, cultural anthropologist and psychologist Erich Rothacker (1888–1965) at the University of Bonn. His postdoctoral thesis appeared in 1933 with the title *Der Aufbau des kindlichen Handelns* (The Structure of Child Behavior).[170]

Gottschaldt's promising career headed into a dead end as the National Socialists assumed power. Whether or not Gottschaldt actually was associated with the Communists during his years as an assistant, as his opponents asserted, can no longer be ascertained with any certainty. In any case he lost his post at the Rhine Province Children's Institution for the Mentally Abnormal in the course of a major

[168] For a biography: Stadler, Schicksal, pp. 152–154; Ash, Erbpsychologische Abteilung, pp. 120f.; Lösch, Rasse, pp. 324f.; Klee, Personenlexikon, p. 194.

[169] Waibel, Provinzialkinderanstalt, p. 67. In the account of the personnel situation (p. 84) there is no mention of Gottschaldt.

[170] Gottschaldt, Aufbau.

personnel shift because of supposed connections to the KPD. His superior Otto Löwenstein, a Protestant of Jewish extraction, fled to Switzerland in 1933; in 1939, he accepted an appointment as clinical director of neurology at the medical faculty of New York University. Gottschaldt was not only suspected for the National Socialists because of his relationship with Löwenstein, he was also doomed by the stigma of belonging to the Berlin school of Gestalt psychology, which had the reputation of being a "stronghold of Jews and Communists."[171] Gottschaldt's teacher at the academy, Max Wertheimer, emigrated to the USA in 1933, finding a position at the New School for Social Research. Kurt Lewin, upon whose methodological approach Gottschaldt based his postdoctoral thesis, also went into exile, becoming a professor for child psychology at the University of Iowa. Wolfgang Köhler, who made no secret of his rejection of the new regime, held out against the pressure of the witch hunt fomented by the National Socialists for 2 years before going into exile as well.[172] Gottschaldt found employment for a while as an assistant for Rothacker's course, until Fischer, who was personally acquainted with Rothacker, brought him to Dahlem as departmental director. At the same time Gottschaldt took over the Polyclinic for Excitable and Difficult Children and Teenagers (*Poliklinik für nervöse und schwer erziehbare Kinder und Jugendliche*) at the Berlin Children's Hospital in Wedding.[173]

In Gottschaldt Fischer acquired not only one of the most competent psychologists of children and teenagers in Germany, who – contrary to Lösch's assumption[174] – could not have been completely without proficiency on issues of genetic psychology either, as the psychiatric–genetic biology archive of the Rhine Province was attached to the Rhine Province Children's Institution for the Mentally Abnormal, where Gottschaldt had worked for 4 years.[175] At the same time, due to his precarious political position, there was little danger that Gottschaldt, who was awarded the professorial qualification for the University of Berlin in 1935 and appointed assistant professor in 1938,[176] could be called away from the institute in Dahlem or even aspire to such a move – in fact, in December 1939 he rejected a call to the chair for psychology at the University of Breslau so brusquely that he was blacklisted by the Ministry of Science.[177] On the other hand – and a tactician as shrewd as Eugen Fischer certainly would have been aware of this aspect when Gottschaldt was hired – the new department chief endangered neither his own position as director of the institute nor the ambitions of his designated successor, Verschuer. That state and party played

[171] So, for instance, Hans Preuß, director of the science office of the Berlin student union, quoted in Ash, Institut, p. 124.

[172] For details, ibid.

[173] Gottschaldt fulfilled this function well beyond the end of World War II – in 1950 a book emerged from this position, about "youth delinquency" (*Jugendverwahrlosung*). Cf. Stadler, Schicksal, p. 154.

[174] Lösch, Rasse, p. 325.

[175] Waibel, Provinzialkinderanstalt, p. 76.

[176] Cf. Copy of the certificate of appointment, MPG Archive, Dept. I, Rep. 1 A, No. 2399, p. 124.

[177] Lösch, Rasse, p. 331.

along with Fischer's game and put no obstacles in the way of his appointing a politically suspect scientist like Gottschaldt also speaks against the thesis proposed by Lösch, that the fascist rulers had attempted to separate Fischer from his politically unreliable employees and systematically infiltrate the institute. Gottschaldt had built up good contacts to Walter Groß in the late 1930s – in 1941 he supported Groß's campaign against the scientist of the "racial soul" Ludwig Ferdinand Clauss[178] – yet he did not join the NSDAP[179] and avoided any eulogies to Hitler and National Socialism in his writings.

3.2.6 Scientific Personnel

The number of "working scholars" – not counting the guest listeners from the SS – rose from 36 (1934/35) to 40 (1935/37) and finally to 50 (1938/39), before falling back to 38 (1939/40) as a consequence of the war.[180] Under the three departmental directors Fischer, Verschuer (from 1935: Gottschaldt), and Lenz, up to eight assistants, interns, or auxiliary assistants (1938/39) worked at the institute. In 1933/34, there was major turnover in the assistantships. Only Wolfgang Abel, son of the paleontologist and National Socialist rector of the University of Vienna, Othenio Abel (1875–1946), kept his post. Starting on October 1, 1933 three new assistants or auxiliaries were hired. In terms of training, all three were physicians – a first indication of the change of course toward genetic pathology after the National Socialists took power.

After his secondary school examinations in 1925, Wolfgang Lehmann (1905–1980)[181] had completed a course of study in natural science and anthropology in Königsberg, Vienna and Halle, participated in an expedition to the Lesser Sunda Islands in 1927, and then begun studying medicine, graduating with a Ph.D. in 1933 such that he could immediately take over the assistantship at the KWI-A, which he

[178] Gottschaldt wrote a devastating expert opinion about the scientific work of Clauss, who, for his part, had already delivered a devastating judgement about Gottschaldt back in 1938. Cf. Weingart, Doppelleben, pp. 81 ff., 86 f.

[179] Ash, "Positive Eugenics," p. 343.

[180] Calculated from the itemizations in the Jahresbericht 1933/34, 21/3/1934, MPG Archive, Dept. I, Rep. 3, No. 10; Jahresbericht 1934/35, 14/4/1935, MPG Archive, Dept. I, Rep. 3, No. 11; Halbjahresbericht 1935, October 1935, MPG Archive, Dept. I, Rep. 3, No. 12; Jahresbericht 1935/37, 10/4/1937, MPG Archive, Dept. I, Rep. 3, No. 13; Jahresbericht 1938/39, 16/4/1939, MPG Archive, Dept. I, Rep. 3, No. 16; Jahresbericht 1939/40, 28/3/1940, MPG Archive, Dept. I, Rep. 3, No. 17. The total numbers listed in the reports must be compared with the personnel lists accompanying them, because sometimes they count the director and the department heads, and sometimes they do not. If the SS guest listeners are included, the total for 1934/35 rises to 56, for 1935/37 to 58 scientists.

[181] For a biography: Lösch, Rasse, pp. 302 f., 569; Klee, Personenlexikon, p. 363. From October 1, 1933, Lehmann was first a "Rockefeller assistant," from April 1, 1934 an "ordinary assistant." Fischer to Generalverwaltung, 24/5/1934, MPG Archive, Dept. I, Rep. 1 A, No. 2406, p. 155.

held until 1935. He had joined the NSDAP in May 1933; later he became a member of the NSDÄB and the NS Lecturers' League (*Dozentenbund*).[182]

Martin Werner (1903–1975)[183] had run through the study of medicine in Bonn, Freiburg and Cologne from 1922 to 1927 and submitted his dissertation in 1927. He had worked as an assistant at the Medical Polyclinic of the University of Bonn and at the Medical Clinic of the University of Heidelberg. The assistantship at the KWI-A was occupied by Werner, who did not join the NSDAP, only until late 1934, but he continued to work as Verschuer's chief surgeon at the Polyclinic for the Fostering of Genes and Race in Berlin Charlottenburg.

Franz Steiner (* 1908),[184] member of the NSDAP and the SA, had studied medicine in Königsberg and Berlin and also received his Ph.D. in 1933.

A new addition in 1934 – initially as an auxiliary assistant – was the blood group specialist Engelhard Bühler (* 1908),[185] who had studied medicine in Marburg and completed his studies with a Ph.D. in 1934. Bühler, son of the painter Hans-Adolf Bühler (1877–1951) so esteemed by the National Socialists, had been a member of the NSDAP since 1932 and also joined the SA and SS. In 1934, the hitherto doctoral student Johannes Schaeuble[186] was also elevated to the rank of auxiliary assistant.

In 1935, after the departure of Lehmann, Steiner, and Werner, another changing of the guard took place at the assistant positions. Only Abel and Bühler remained. New additions were, first of all, Horst Geyer (1907–1958),[187] a specialist for neurology and psychiatry, who had worked as an assistant at the University Neurological Clinic in Kiel in 1933/34 and took part in the first course for physicians of the SS at the KWI-A in 1934/35, before becoming an assistant to Fritz Lenz. Geyer, who was already a member of the SS, joined the NSDAP as well in May 1937.[188]

Also employed at the KWI-A since 1935 was Karl-Friedrich Lüth (* 1913),[189] first as an intern, and from 1936 as an ordinary assistant to Fischer. From 1928 to 1934 he had studied medicine and natural sciences in Innsbruck, Vienna, and Greifswald and received his Ph.D. in 1934 and his Dr. med. in 1937. Since 1933, he had belonged to the SA and the NSDÄB.

---

[182] The assessment by Koch, Humangenetik, p. 39, that Lehmann had not "exposed himself [in National Socialism], neither politically nor organizationally," must be relativized accordingly. Müller-Hill, Tödliche Wissenschaft, p. 151, states that Lehmann belonged to the SS – albeit without listing any sources.

[183] For a biography: Lösch, Rasse, pp. 314–316, 576; Klee, Personenlexikon, p. 670.

[184] For a biography: Lösch, Rasse, pp. 303, 574. Steiner was initially an auxiliary assistant, and then an ordinary assistant from January 1 to March 30, 1935.

[185] For a biography: Lösch, Rasse, p. 563; Massin, Rasse und Vererbung, p. 213.

[186] For a biography: Lösch, Rasse, p. 330; Klee, Personenlexikon, p. 525.

[187] For a biography: Lösch, Rasse, pp. 565 f.; Klee, Personenlexikon, p. 182.

[188] Peter Emil Becker stated in a lecture in 1999 that Geyer was expelled from the SS because of a Jewish grandfather, a statement which cannot be verified. Klee, Personenlexikon, p. 182.

[189] For a biography: Lösch, Rasse, p. 569.

From 1935, first as an auxiliary assistant, later as an assistant to Gottschaldt, Kurt Wilde (1909–1958)[190] also worked at the KWI-A. He became a member of the NSDAP in 1937. In July 1936 Peter Emil Becker (1908–2000),[191] who was trained at the Race Biology Institute of the University of Hamburg, came as a further assistant to Lenz. Becker had been a member of the SA since 1934, and joined the NSDAP in 1940.

In 1937, Heinz Lemser (* 1913)[192] became an auxiliary assistant, and 1 year later an assistant to Fischer. Lemser had studied medicine in Giessen and Berlin and worked as an extern at Westend Hospital in Berlin before switching to the KWI-A. He had been a member of the NSDAP, where he advanced to become a political director, since November 1937.

In 1938, Herbert Grohmann (* 1908)[193] came as another assistant to Fischer. The physician, who had joined the NSDAP back in 1931, became director of the Genetic File Department (*Abteilung Erbkartei*) in the Office of Population Policy of the SS in 1936; in 1937 he worked at the RuSHA. He was one of the physicians of the SS detached to the genetic biology course at the KWI-A. During his assistantship in the years 1938/39 he processed marriage applications at the RuSHA in an honorary capacity. In September 1939, Grohmann became medical councilor and director of the Department for the Fostering of Genes and Race at the Łódž Health Office.[194]

A consequence of the fluctuation in assistantships was that the network of former employees in strategic positions became ever denser. Central figures of this network in the years from 1933 to 1939 were Friedrich Curtius, Lothar Loeffler, and Konrad Kühne, who was employed as an external scientific staff member of the KWI-A in the late 1930s, before moving to Hamburg as chief physician of the tuberculosis registration office of the social insurance carriers. As a review of the grant files of the German Research Association shows, in the 1930s, almost every research application from the KWI-A arrived via the desk of one of these three scientists. Along with Hans Weinert, who was appointed Director of the Anthropological Institute at University of Kiel in 1935 – after 1933 the "Dahlem circle" was rounded out above all by Wolfgang Lehmann, who worked as an assistant to the internist Kurt Gutzeit (1893–1957)[195] at the University of Breslau from 1935 and was appointed associate professor at the University of Strasbourg in 1943; Franz Steiner, who became an assistant at the Institute for Genetic Biology and Race Hygiene at the University of Königsberg under Lothar Loeffler in 1937; and Horst Geyer, who was Loeffler's assistant at the chair for Genetic and Race Biology in Vienna from 1943. Johannes Schaeuble, too, who switched to an assistantship at the Psychotechnical Laboratory of the Wehrmacht in 1936, received a lectureship

[190] For a biography: ibid., p. 576, Klee, Personenlexikon, p. 677.

[191] For a biography: Lösch, Rasse, p. 562; Klee, Personenlexikon, p. 35.

[192] For a biography: Lösch, Rasse, p. 569.

[193] For a biography: ibid., p. 566; Klee, Personenlexikon, p. 202.

[194] Further assistants at the KWI-A were Siegfried Tschamler, Otto Baader (* 1909) and Gerhard Philipps (* 1913), who worked as Gottschaldt's assistant in 1938.

[195] For a biography: Klee, Personenlexikon, p. 212.

for Genetic and Race Biology at the University of Freiburg in 1940 and worked with the SS Ahnenerbe from December 1942, represented an increasingly important circuit on the "Dahlem circle."

Among the scientific staff, whose number varied between 6 and 11, and the doctoral students, whose number increased from 14 (1934/35) to 24 (1938/39), there were a surprising number of foreigners. Between 1933 and 1939 guest scholars are documented from China (Wuhou Wayne King, Yun-kuei Tao), India (P. C. Biswas, Sasanka Sekhar Sarkar), Italy (Leone Franzi), Japan (Hisatoshi Mitsuda,[196] Masataka Takagi[197]), the Netherlands (Haring Tjittes Piebenga),

[196] In 1938 Hisatoshi Mitsuda († 1979) was "assistant of the Psychiatric Clinic of the Imperial University in Kyoto" (Fischer to Verschuer, 5/2/1938, MPG Archive, Dept. III, Rep. 86 A (Münster), No. 9), where he had also earned his doctorate in 1934. He maintained friendly contact with Verschuer even after World War II: "From Mitsuda, who wanted to translate my genetic pathology back then, I recently received a very nice letter. In other respects, too, some of my international relationships have started flowing again" (Verschuer to Fischer, 11/10/1951, MPG Archive, Dept. III, Rep. 86 C, No. 9, p. 17). In 1953, Mitsuda was appointed professor of psychiatry and neurology at the Osaka Medical College. Mitsuda, who also performed twin research, became known through his studies on psychiatric nosology, especially those on schizophrenia and other forms of psychosis, emerging as the leading representative of psychiatric genetics in Japan. In 1956 he founded the Japanese Society for Human Genetics, in 1963 the Japanese Genetic Society; in 1979 he was the first president of the Japanese Society for Biological Psychiatry. He translated Horst Geyer's book *Über die Dummheit* ("About Stupidity") into Japanese. Many thanks to Kazuko Kibata and Takashi Nakagawa for information about Mitsuda's biography.

[197] Masataka Takagi (1913–1962?), who earned his Ph.D. at the Department of Literature at the Imperial University at Tokyo in 1936, studied psychology, anthropology and ethnology as a Humboldt scholar at the University of Berlin from 1936. From 1938 until 1943 he was listed as a guest scholar in Gottschaldt's department. During this time he also freelanced as a translator for the Japanese Embassy. In 1944 he worked as an "animal psychologist" at the Zoological Garden in Berlin. He returned to Japan in 1948 and became a professor at the Department of Natural Sciences at Toho Daigaku University in Tokyo. His book *Pedagogical Sociopsychology* appeared in 1949. Takagi, too, frequently used the twin method.

On a trip through Germany in 1951 he visited Fischer, Verschuer, Gottschaldt and other former colleagues and staff of the institute. Upon this occasion Fischer recalled, "He is the Japanese man who traveled to Munich on every break, where he was a member of the academic section of the Alpine Association and took major mountain hikes. He usually wore a little green hat!" In fact, Takagi was better known in his homeland as an alpinist than as a scientist – in 1952 he undertook an expedition to the Himalayas. Fischer continued in 1951: "He is a professor at the U. of Tokyo and Dept. Director at the Inst. of Psycholog., whose chief pathologist he is, while Takagi runs the Dept. for Human Genetics and therefore wants to reorient himself again about the organization of twin studies and the like […]." Fischer to Verschuer, 21/9/1951, ibid., p. 14. Takagi's in-laws lived in Switzerland (Fischer to Verschuer, 13/10/1951, ibid., p. 14). When the Japanese were defeated in August 1945 he was on a mountain hike in Switzerland. Cf. also Verschuer to Fischer, 26/3/1963, MPG Archive, Dept. III, Rep. 86 C, No. 20, p. 10: "A few days ago I received a message from a Japanese colleague […] that Professor Takagi has been missing from Kobe University since last summer, when he set off on an expedition to the South Pacific. So one must assume that he met with an accident. […] Takagi was the first Japanese person to welcome us to Tokyo seven years ago, when we arrived at our hotel in the middle of the night." Takagi was registered missing from August 1962. Many thanks to Kazuko Kibata and Takashi Nakagawa for information about Takagi's biography.

Norway (Thodar Quelprud), Poland (Witold Sylwanowicz[198]), Romania (A. Hermann and Michael Fleischer, both Transylvanians), Switzerland (Erik Hug), Spain (Miguel Carmena, Luis de Lazerna, Otero Lopez, Jimena Fernández de la Vega), Czechoslovakia (Kozlik), Turkey (Seniha Tunakan[199]), and Hungary (Mihali Malán, Lajos Csik).[200]

3.3 Research Agenda and Research Praxis

The opinion occasionally advocated in the literature, that "upon the seizure of power the scientific work at the institute not only stagnated, but practically came to a standstill in many areas,"[201] because the staff was increasingly busy with activities of propaganda, training, expert opinions, and certifications in the field of race and genetic health policy, is somewhat exaggerated. However, it is true that the institute played an important role in the practice of race and genetic health policy and that its scientific work suffered under the multifarious political obligations – Fischer never missed a chance to emphasize this in his activity reports. Due to his rectorate, and a stay at the University of Santander immediately thereafter, Fischer had no time for research of his own until late 1935.[202]

[198] Witold Sylwanowicz (1901–1975) was prosector at the anatomical institute of the University of Vilnius. Cf. Fischer to KWG, 11/7/1939, MPG Archive, Dept. I, Rep. 1 A, No. 2399, p. 131.

[199] In 1942/43 the institute reported that Tunakan's dissertation, entitled *Erbpsychologische Untersuchungen über Bewegungszuordnungen* ("Genetic Psychological Studies of the Attribution of Movement") was in print. Notizen der Abteilung für Erbpsychologie für den Jahresbericht 1942/43, MPG Archive, Dept. I, Rep. 3, No. 20. Cf. also Fischer to Verschuer, 17/10/1955, MPG Archive, Dept. III, Rep. 86 C, No. 13, p. 11: "From 1938–41 I had a little Turkish woman, Miss Tunakan, as a doctoral student. By now she has become a lecturer for anthropology in Ankara and is now on vacation in Europe, for the main purpose of taking a look at the latest developments in anthropology and human genetics, literature, technology, findings, etc. [...] In Ankara she has done a great deal of publishing [...]." Among others, Tunakan visited Fischer, Verschuer, and Schaeuble.

[200] The nationality of the guest scholar Ole Pedersen is unclear. Lösch, Rasse, p. 577, lists an additional guest scholar named E. Longo (1937), whom I was unable to find. My listing is certainly incomplete, because the personnel overviews appended to the annual reports represented mere snapshots. Apparently there was also a great deal of fluctuation among the guest scholars, such that they were all not recorded by the personnel overviews. In Fischer's Tätigkeitsbericht of 1935/36, for instance, he writes that "seven foreigners (Hungary, Poland, Russia, Georgia, India, Holland)" were working at the institute during the period of the report (Fischer, Tätigkeitsbericht 1935/36, 22/6/1936, MPG Archive, Dept. I, Rep. 1 A, No. 2399, pp. 74–76 v, quote: p. 76).

[201] Lösch, Rasse, p. 299.

[202] Ibid., p. 269. His double workload as rector and institute director, which deterred him from any scientific activity, was probably the real reason for Fischer's resigning his post as Rector of the Friedrich Wilhelm University in Berlin as of April 1, 1935. The quotes Niels C. Lösch lists to show evidence of Fischer's "disenchantment" (pp. 265, 266) with National Socialist academic policy are not convincing.

Yet the scientific output of the institute remained considerable: in the period from 1934 to 1942, in addition to a flood of popular science and genetic health policy writings, 182 publications emerged from the KWI-A which were scientific in the strictest sense. Some of these publications were based on research performed before 1933, yet the practical research activity, which the institute had been forced to limit in the early 1930s due to the financial straits, started up again full scale in 1933/34.

Not least because of the increasingly close interconnections between science and policy, the emphases of research shifted to other fields. With 53 publications, genetic pathology moved far into the foreground, and genetic psychology, nudged ahead by the founding of the Department for Genetic Psychology in 1935, also experienced a clear upswing, producing 25 publications. Only 14 publications came from the research field of eugenics/race hygiene, which thus – despite the appointment of Fritz Lenz – continued to decline in importance. Twenty publications dealt with general methodological and conceptual issues. The genetics of normal attributes of humans remained the most intensively researched field, with a total of 60 publications, although it was much less predominant than it had been in the first 6 years after the institute's founding. The intersections between human genetics and anthropology continued to increase: 19 of the 60 publications on the genetics of normal attributes dealt at least in part with the aspect of race and thus can be located in the boundary region between race anthropology and human genetics. The field of geographic anthropology and paleoanthropology produced only ten publications.

3.3.1 Paleoanthropology and Geographical Anthropology

After 1933, Fischer's strategic research interest in tying anthropology, in the sense of anthropobiology, to general human genetics became even more clearly visible. Increasingly, classical physical anthropology was factored out of the institute's working plans. After Hans Weinert left the institute in 1935, the area of paleoanthropology at the KWI-A lay almost completely fallow. Gustav Perret (* 1908), who worked at the institute as a doctoral student from 1934 until 1937, studied skulls and skeletons from a neolithic burial place in Lower Hesse and – following Eugen Fischer's observation on the population of the Canary Islands – drew parallels between the prehistoric Cro-Magnon type and the recent "Phalian type."[203] Fischer himself had actually planned a "research trip to Spanish Morocco" in 1934/35 – presumably he wanted to trace the Cro-Magnon type in North Africa as well – but this trip had to be postponed "as a consequence of the burden of rectoral

[203] Perret, Cro-Magnon-Typen. Busse, Skelettreste, traced the ancient Slavic settlement of the Mark region on the basis of 27 skulls from graves in the Havel area around Potsdam, dated to between the sixth and twelfth centuries.

affairs."[204] However, on a trip to Italy, financed by the institute,[205] Fischer found a new hobbyhorse, which he introduced in a lecture to the Prussian Academy of Sciences in 1938: proceeding from the portrayals on Etruscan sarcophagi he had viewed in Italian museums, he classified the Etruscans as an "aquiline race," whose traces he believed to detect in the contemporary Italian population as well.[206] Many such individual projects existed largely unconnected with the institute's research agenda. Accordingly, the KWI-A fell hopelessly behind in the field of paleoanthropology, which was at the center of international interest at this time.[207]

Of course, the institute made itself available in a practical manner for "the prehistorical research finally enjoying its due attention in the new state," by helping recover and assess the skulls found during excavations, "as near Goslar, near Lorsch on the Rhine, near Kassel, on the island of Reichenau, now on the highway just west of Brandenburg."[208] In this context Eugen Fischer was called to Brunswick in 1935 as an anthropological expert, in order to study what was believed to be the skeleton of Henry the Lion.[209]

In the area of geographical anthropology, too, the institute developed hardly any activities worth mentioning. Wolfgang Lehmann published a short notice about his anthropological observations on the Lesser Sunda Islands, which he had journeyed through in 1927 on a joint expedition with the biologist Bernhard Rensch (1900–1990), the anthropologist Gerhard Heberer (1901–1973), and others,[210] but otherwise soon turned to genetic pathology issues. In the framework of the major project to take inventory of the German population's genetic biology, only the above-mentioned works by Hermann and Richter appeared from the KWI-A after

---

[204] Revisor Schröder, Bericht über die Prüfung des Rechnungsabschlusses zum 31/3/1935, 23/4/1935, MPG Archive, Dept. I, Rep. 1 A, No. 2406, pp. 181–184, quote: p. 182.

[205] Of the 6,000 RM earmarked for the research trip to Spanish Morocco that was not realized, Fischer received 2,000 RM for his Italian journey. Revisor Schröder, Bericht über die Prüfung des Rechnungsabschlusses zum 31/3/1938, 3/5/1938, MPG Archive, Dept. I, Rep. 1 A, No. 2407, pp. 24a–24d, here: p. 24d.

[206] Fischer, Rassenfrage der Etrusker. On Fischer's "study trip" to Italy: Fischer, Tätigkeitsbericht April 1938 – März 1939, 16/4/1939, MPG Archive, Dept. I, Rep. 3, No. 16.

[207] Cf. Massin, Rasse und Vererbung, pp. 215f.

[208] Fischer, Tätigkeitsbericht von Anfang Juli 1933 bis 1. April 1935, MPG Archive, Dept. I, Rep. 1 A, No. 2404, pp. 49–49h, quote: p. 49e (original emphasis). From a Pastor C. Gross of Chwalynsk on the Volga, who had excavated skeletal remains from the Ice Age in his district, the KWI-A received plaster casts of skull fragments in May 1934. Cf. Deutsche Botschaft Moskau to Auswärtiges Amt Berlin, 17/5/1934, MPG Archive, Dept. I, Rep. 1 A, No. 2399, no p. Skulls and skeletons from the naval expedition to Melanesia in the years 1907/1909 preserved in the institute's skull collection were loaned out to Otto Schlaginhaufen in Zurich. Cf. Preußischer Minister für Wissenschaft, Kunst und Volksbildung to Präsident der KWG, 26/10/1934, MPG Archive, Dept. I, Rep. 1 A, No. 2399, p. 32.

[209] For details on this, Lösch, Rasse, pp. 466–482. Lösch's hypothesis seems highly plausible: that Fischer was consulted in this case in order to interpret the hip luxation established on the skeleton, which – as we know today – was not congenital, but the result of a riding accident, and what is more, can probably be attributed to Henry's wife Mathilde.

[210] Lehmann, Beobachtungen. Cf. Potthast, "Rassenkreise," pp. 283f.

1933, although Fischer had proclaimed in 1933 that the surveys should be "expanded in the future with the objective of taking as complete stock as possible of all genetic lines in the nation."[211] In retrospect, Fischer attributed the failure of his own project on the Markgräfler region to the fact that, of the secondary school teachers entrusted with the field research in 1933, "very many [had been] transferred," many had also "given up the work because of political activities." He himself, as university rector, had "had neither time nor the least inclination in the subsequent two years [...] to attend to those anthropological tasks."[212] This statement confirms once more the impression that the major "German Race Science" project was not a high priority for Fischer and his institute. After the publication of the works by Hermann and Richter, only a single study along the model of cultural biological village studies was tackled in Dahlem: in 1941, Karl Heinz Schwabe (* 1917) wrote his dissertation under Fritz Lenz, a study of the race biological surveys in his hometown, Hennickendorf in Mark Brandenburg, in which he concentrated on migrations and processes of selection.[213] At the same time, Lenz spoke out critically "About the Paths and Meanderings of Race Science Studies," pointing out the methodological weaknesses of some studies in progress at the time.[214]

3.3.2 "Bastard Research"

Eugen Fischer consistently conceived of human races as "groups with the same sets of genes" – as the title of his lecture to the Prussian Academy of Sciences in 1940, *Gruppen mit gleichen Gen-Sätzen* indicates.[215] "Bastard research," which Fischer was convinced opened up a good approach to the genetic substrate of the race, was intensified after 1933. Fischer published his "New Studies of the Bastards of Rehoboth," which evaluated the material collected by Hans Lichtenecker in 1931.[216] Yun-kuei Tao concluded his dissertation about the "Chinese Male-European Female Hybrid."[217] In connection with the sterilization of the "Rhineland

[211] Fischer, Tätigkeitsbericht vom Juli 1933, Anlage zur Niederschrift über die Sitzung des Kuratoriums des KWI-A am 5. Juli 1933, MPG Archive, Dept. I, Rep. 1 A, No. 2404, pp. 14–17, quote: p. 14.

[212] Fischer to Apparate-Ausschuß der DFG, 29/4/1944, BArch. Koblenz, R 73/11.004. On December 8, 1934, Fischer informed the Westphalian Province Association that a continuation of his anthropological studies in Westphalia was not possible "because he was so occupied with the court-ordered genetic health opinions of all cases of the sterilization laws that he could no longer dedicate himself to the scientific race research in the country;" quoted in Ditt, Raum, p. 78 (note 146). Fischer proposed bringing Walter Scheidt on board.

[213] Schwabe, Erhebungen.

[214] Lenz, Wege, und Irrwege. In two lectures in the 1930s, Lenz dealt with the problem of the emergence of race. Cf. Lenz, Rassen und Rassenbildung; idem., Rasse und Klima.

[215] Fischer, Rassen als Gruppen.

[216] Fischer, Neue Rehobother Bastardstudien I und II.

[217] Tao, Chinesen-Europäerinnen-Kreuzung.

bastards," Wolfgang Abel studied 39 "half-breed children" with European mothers and Moroccan, Annamitic (Vietnamese) or Turkish fathers[218] – more on this later. Further, Abel undertook a "study trip" to Romania in 1935/36, funded by the institute and the Emergency Association, "to study Gypsies with the question of miscegenation and the environmentally determined differences between Romanian and Western European Gypsies."[219] A further "study trip" by Abel to Scotland followed in 1938/39, again "to study the Gypsies."[220]

On behalf of the institute, Johannes Schaeuble undertook a research trip to southern Chile from November 1934 until September 1935, during which he recorded the anthropological data of around 1,400 individuals – "Indians and Indio-European hybrids."[221] In 1936/37, the institute sent Rita Hauschild (* 1912), initially as a doctoral student from 1935 to 1937, then as a scientific staff member of the KWI-A, to the Caribbean to perform "bastard studies."[222] She published her anthropometric findings on "Chinese-Negro hybrids" on Trinidad and on "Chinese-Indian and Indian-Negro hybrids" in Venezuela in 1941.[223]

3.3.3 The Genetics of Normal Attributes

With a total of 60 publications, "the genetics of normal anatomical, morphological and physiological features and characteristics" of humans remained the main field of work for the KWI-A. Here, too, scientific practice for the most part remained on the trails blazed before 1933. Research on the issue of brachycephaly continued, for instance, as the works of two of Fischer's pupils looked into the connection between urbanization and skull shape.

---

[218] Abel, Europäer-Marokkaner-Kreuzungen.

[219] Fischer, Tätigkeitsbericht vom 1/4/1935 bis zum 31/3/1936, MPG Archive, Dept. I, Rep. 1 A, No. 2404, pp. 52b–52e, quote: p. 52d.

[220] Fischer, Jahresbericht April 1938 – März 1939, 16/4/1939, MPG Archive, Dept. I, Rep. 3, No. 16.

[221] Schaeuble, Beobachtungen; idem., Indianer.

[222] The trip received financial support from the Gwinner Foundation and the Culture Department of the Foreign Office. Cf. Lösch, Rasse, p. 333 (note 119).

[223] Hauschild, Bastardstudien. The research trip, which was initially restricted to anthropological studies in the German colony of Tovar in Venezuela, received financial support from the Culture Department of the Foreign Office. Additional, generous grants from the Gwinner Foundation allowed the research to be expanded to "bastard studies." On Trinidad Hauschild enjoyed significant support from the Swiss and the Chinese consuls, which was all the more important because she "worked on Trinidad at a time when the psychological prerequisites for this work were not the most auspicious. The major oil strike of summer 1937 had stirred up intense emotions. It was not only the consequence of an economic crisis, but at the same time also the expression of a race opposition between the working population of color and the ruling white upper class. Therefore it was not to be avoided that I had to leave some studies undone, although they would have been rewarding […], because the Negro parent absolutely refused to make himself the object of European research." Ibid., p. 183.

Gottfried Pessler (* 1912), in a local study on Hanover, performed anthropometric measurements on one group of parents born in the city and another born in the country, along with their school-aged children born in Hanover. This study brought him to the conclusion that the children of both groups had longer heads than their parents, whereby the parents born in the country deviated more strongly from their children than those born in the city.[224]

From January 1932 until July 1934, the secondary school teacher Walter Dornfeldt (* 1900) performed a similar study on around 900 Jewish immigrants from Eastern Europe and their 1,350 children, most of whom were born in Berlin, arriving at the conclusion that the shape of the heads of the children born in the city was narrower than that of their parents. Dornfeldt's findings, presented with emphatic objectivity, were not published until 1941, when the elimination of the European Jews was building up steam. All along the line, they confirmed the research of Franz Boas, who had taken the racial anti-Semitism based on external appearances to absurd lengths.[225]

The insights obtained by Pessler and Dornfeldt using anthropometric methods were corroborated by animal experiments performed in parallel. Continuing Fischer's early experiments on artificially influencing the form of rat skulls, the doctoral student Otto Roth (* 1899) succeeded in changing skull growth in rats with doses of pituitary and thymus hormones.[226] These projects were supplemented by Kurt Gerhardt (1912–1992), who worked at the KWI-A as a doctoral student and scientific staff member from 1935 to 1938, using the literature to study the distribution of brachycephaly in Europe, Asia, Africa, and Oceania and who also came to the conclusion that brachycephaly can be the result of many kinds of different causes, such that a classification of the human races on the sole basis of the skull index must be rejected.[227]

The dissertation published by Rita Hauschild in 1937, about race differences between Negroid and Caucasian primordial crania of the third fetal month, was considered to be a counterpart to the above-mentioned dissertation by Baeckpyeng

---

[224] Pessler, Untersuchung.

[225] Dornfeldt, Studien. Lösch, Rasse, p. 387, is correct to emphasize the lack of any anti-Semitic stereotype in Dornfeldt's work. Dornfeldt proceeded from pupils at schools run by the Jewish community, and at several youth departments of the Bar Kochba sport club and attempted to persuade the parents of the children to participate in the study as well. As he allowed, this pursuit was met in many cases with suspicion and refusal, so that, especially at the beginning, it was difficult for him to "dispel their reservations." "I then only scheduled examinations on Fridays, Saturdays and Jewish holidays after making appointments, for on many occasions I had had bad experiences on these days. Failures occurred not only because of refusals. In other cases the parents had moved to a new apartment, whose address could not be obtained. Now and then the father, or sometimes the entire family, had emigrated [...]." Dornfeldt, Studien, p. 295.

[226] Roth, Wachstumsversuche.

[227] Gerhardt, Frage, esp. pp. 465 f. The doctoral student Gerhard Schulze (* 1913), in his dissertation published in 1941, pursued the question as to the influence of rachitis on various head measurements and the length–breadth index, without obtaining any unambiguous findings. Schulze, Frage.

Kim published in 1933 about race differences in embryonic pig skulls. These two works were of particular importance for the future development of the institute, because they exhibited three elements which played a truly essential role in the refounding of the KWI-A from 1938. First, they were explicit contributions to phenogenetics; second, in that they were based on embryonic specimens, they tried out a methodological approach that was to shift further into the foreground after 1938; and third and finally, both works aspired to a race diagnostics independent of the phenotype. Fischer and Verschuer attached such importance to the work of Hauschild that they were blind to its altogether serious methodological weaknesses. For only three "Negroid" primordial crania (the first cartilaginous skeleton of the skull in an embryo is designated the "primordial cranium") were available to Hauschild, which had been entrusted to her by Adolf Hans Schultz (1891–1976), an embryologist at the Carnegie Laboratory. Because it turned out that one of the specimens could not be used, Hauschild had recourse to only two "Negroid" primordial crania, which, moreover, had been dissected into series of sections, from which three-dimensional wax models had to be produced before the measurements could be performed. Despite the very narrow material basis, and although the origin of the material was unclear, in her work Hauschild arrived at quite "far-reaching conclusions":[228] race features were already preformed in the cartilaginous skeleton of the skull; the "differences found between our Negroid primordial crania and the Caucasian ones" were to be interpreted "for the most part as racial difference."[229]

Fischer used the works by Kim and Hauschild as important building blocks in the formulation of his newly conceptualized research program in 1938 – and also the research by Konrad Kühne on the variability of the spinal column, which also had been continued after 1933. The doctoral student Maria Frede (* 1907) dissected the spinal columns of some of the rats Kühne had bred in Dahlem and believed that she could confirm the heritability of the cranial or caudal tendency in the occurrence of anomalies.[230] These studies, Fischer claimed, were "of unforeseen, fundamental importance for the understanding of the heredity of certain diseases."[231]

The variability of morphological features due to heredity and environment was also investigated further, usually using the twin method. Luise Brauns (* 1907), for instance, studied the processes of growth in 160 pairs of infant and toddler twins.[232] The maternity hospitals in Berlin ceded to the KWI-A 30 specimens of stillborn twins for scientific investigation in the first 10 years of its existence.[233] On the basis of this material, Heinz Gigas (* 1912) used the twin method to study hereditary

[228] Lösch, Rasse, p. 334.

[229] Hauschild, Rassenunterschiede, p. 274.

[230] Frede, Untersuchungen. In this project Frede also occupied herself with the arrangement of the plexus in the arm and leg and their transmission. Kühne himself published the results of his twin studies in 1936. Cf. Kühne, Zwillingswirbelsäule.

[231] Fischer, Jahresbericht des KWI-A 1933/34, 21/3/1934, MPG Archive, Dept. I, Rep. 3, No. 10.

[232] Brauns, Studien.

[233] Cf. Geyer, Hirnwindungen.

muscle varieties.[234] Various nonpathological anomalies of the body also became objects of research. In 1935 Wolfgang Lehmann, together with scientific staff member Eduard A. Witteler, described polydactyly in an identical pair of twins; in 1937 Karl-Friedrich Lüth, the increased occurrence of total albinism in one family. One year later, Lüth used the twin method to examine the occurrence of uterus myonomes.[235]

The features of the head, both those of the cranium and, above all, of the individual facial features, were analyzed by Wolfgang Abel using his own new methods of investigation – he summarized his findings in the article "Physiognomy and Facial Expressions" in the *Handbuch der Erbbiologie des Menschen* ("Manual of the Genetic Biology of Humans") edited by Günther Just. Abel further examined the teeth and jaws of 15 skulls of "Bushmen, Hottentots and Negroids and their half-breeds" from the collection of the Anthropological Institute in Vienna as well as of 19 "Tierra del Fuego Indians" – the last of these had been collected by the Austrian missionary and ethnologist Martin Gusinde (1886–1969) on his expeditions to Tierra del Fuego in 1918–1924 and made available to the institute.[236] In the context of certificates of paternity, Engelhard Bühler performed twin studies about wrinkles and furrows of the face. Also based on twin studies was his paper on the variability of the fold of the upper eyelid.[237] Thordar Quelprud continued his studies of the human auricle, processing material from 950 pairs of twins and their families.[238] The doctoral student Erich Nehse (* 1908), in part by means of the twin method, investigated the morphology, variability, and heritability of hair growth on the human head, e.g. the whorl, parting, and hairline.[239] In 1934, a paper by Heinrich Kranz was published about the hair of "East Greenlanders" and "West Greenland Eskimo and Danish half-breeds," which had been collected during the expedition by Hermann B. Peters in 1931.[240]

After 1933, the twin method was also applied systematically to normal physiological attributes of humans. Martin Werner performed physiological experiments

[234] Gigas, Untersuchungen. In 1933/34, the Spanish guest scholar Lazerna performed studies on the tissue elasticity of 50 pairs of twins (cf. Verschuer, In den Jahren 1932–34 durchgeführte, im Gange befindliche und für das nächste Jahr geplante Arbeiten, Anlage zum Schreiben Verschuers an die DFG vom 8/9/1934, BArch. Koblenz, R 73/15.341, pp. 104–108, here: p. 106). Fischer confirmed that the Spanish guest scholars Carmena and Lazerna also occupied themselves "with muscle and skin elasticity and variations of the electrical conductivity of the body" (Fischer, Tätigkeitsbericht von Anfang Juli 1933 bis 1. April 1935, MPG Archive, Dept. I, Rep. 1 A, No. 2404, pp. 49–49h, quote: p. 49e). Apparently, no publications resulted from this work.

[235] Lehmann/Witteler, Zwillingsbeobachtung; Lüth, Sippe; idem., Uterusmyom.

[236] Abel, Vererbung; idem., Physiognomik; idem., Zähne; idem., Gebiß. Cf. also idem., Vererbung normaler morphologischer Eigenschaften.

[237] Bühler, Zwillingsstudien; idem., Oberliddeckfalte.

[238] Quelprud, Erblichkeit; idem., Familienforschungen; idem., Ohrmuschel.

[239] Nehse, Beiträge.

[240] Kranz, Haare.

on pairs of twins, concentrating on the functions of the vegetative system, metabolism, and lung capacity.[241] At the 47th Congress of the Society for Internal Medicine in 1935, he thus could report on "genetic differences in a number of functions of the vegetative system according to experimental examinations on 30 pairs of twins." For these experiments, which were sponsored by funds from the Rockefeller Foundation, the subjects were admitted to the Elisabeth-Diakonissen Hospital in Berlin for a 4-day stay. Their vegetative reactions were tested after subcutaneous injections of atropine, pilocarpine, adrenaline, and histamine: pulse, blood pressure, and respiratory rate were measured at regular intervals. Further, depending on the pharmacum injected, dryness of the mouth, the secretion of sweat, heart palpitations, redness of the face or any hives occurring were observed, the level of blood sugar or amount of saliva secreted measured, or the reaction of gastric fluid checked through "fractionated siphoning off" with a tube in intervals of 10 min for 2.5 h. The experiments were quite unpleasant for the test subjects, although presumably not dangerous.[242] What was more problematic is that they took place in a gray zone of medical ethics, as some of the subjects were only 15 years old. The guidelines of the Reich Health Council of February 28, 1931 "for new kinds of therapy and for undertaking scientific experiments on humans" were clear on this point:

> Experiments on children or minors under the age of 18 are forbidden if they endanger the child or the minor in any way at all.[243]

Werner's experiments are further evidence that the guidelines of 1931 remained "without any effect"[244] in the German medical fraternity.

Blood group research at the KWI-A suffered a severe setback after the National Socialists took power. As mentioned above, Fischer had been skeptical about this field of research from the very beginning; Verschuer broke off his collaboration with Max Berliner and Fritz Schiff for reasons of political expediency. Serological research was continued by Engelhard Bühler. His work, based on the twin method, concerned the heritability of the antibodies (agglutinins) contained in the blood. Despite the assessment that the agglutinin content of the sera from persons belonging to the same blood group were very different from one another, and that even the agglutinin content in the serum of one and the same individual can be subject to strong temporal variations, Bühler posited the working hypothesis that agglutinin content could be "genetically constant and thus characteristic for an individual" within a certain range of variation. His first findings, obtained from 22 pairs of identical twins, 23 pairs of same-sex fraternal twins, and 5 pairs of different-sex fraternal twins, were presented at the Convention of the German Society for Genetics in Jena in July 1935. On the basis of the calculated rates of concordance,

---

[241] Werner, Erbunterschiede; idem., Untersuchungen; idem., Erb- und Umweltbedingtheit.

[242] Bergmann/Czarnowski/Ehmann, Menschen, p. 136, report that one of the subjects testified in an interview that "as a child he nearly died at the end of this series of experiments." However, the dosage given in Werner's essay was so low that such a reaction is quite improbable.

[243] Quoted in Winau, Versuche, p. 174.

[244] Ibid., p. 175.

Bühler concluded that "the share of heredity in the occurrence of the isoagglutinin content of serum [must be] quite considerable."[245]

Fischer immediately applied to the DFG for 4,400 RM to support the continuation of Bühler's blood group research. What is interesting is that the reasons Fischer listed, which demonstrated two practical complications above and beyond the project's importance for theoretical research on the issue of heredity vs. environment. First, it was supposed to investigate whether the "proven heritability [of the degree and the intensity of the clumping characteristics in the blood] (which I may assume after the first sample), is different according to race, which is to be tested on the Negros, Chinese, Indians living here in Berlin. (To me this question seems very problematic)." This was the first time a research application from the KWI-A explicitly mentioned the possibility of a serological race test. The second complication was that Fischer promised an improvement of immunizations against infectious diseases: "From several thousand children vaccinated against diphtheria in the district of Aachen, we sought out the twins and would like to study the question of the equivalence or difference of their immune functions. So far it is only a small, random sample. The trial, also on adults who are immunized, must be controlled by means of animal experiments."[246] The officer responsible for Fischer's application was the former assistant Curtius, who approved the application with reference to Ludwik Hirszfeld's "constitutional serology," at the same time pointing out that Fritz Schiff, who by that time was working at the Horst Wessel Hospital, was one "of the best experts in the field of constitutional serology."[247] In fact the DFG immediately improved a credit of up to 2,200 RM up to the end of the 1935/36 fiscal year.

Two further projects concerned human blood. In 1933/34, the Spanish guest scholar Jimena Fernández de la Vega (1895–1984) was occupied with measuring the size of the blood corpuscles of a large number of twin pairs – yet, although these measurements were concluded, no publication emerged.[248] Lüth investigated the blood–alcohol concentration after the ingestion of alcohol in ten pairs each of identical and fraternal twins. From the rates of concordance and discordance, he believed it was possible to conclude that hereditary factors influence the physiological processes in the decomposition of alcohol in the blood – which, as Fischer emphasized, could also be of practical interest with a view to "alcohol testing for drivers."[249] Shortly before the beginning of World War II, Fischer launched yet

[245] Bühler, Untersuchungen über die Erblichkeit des Isoagglutinationstiters, quotes: p. 464, 467. Bühler had been taught Fritz Schiff's methods; cf. ibid., p. 464 (note 1).

[246] Fischer to DFG, 9/9/1935, BArch. Koblenz, R 73/11.004.

[247] Gutachten des Fachreferenten (undated), ibid.

[248] Verschuer, In den Jahren 1932–34 durchgeführte, im Gange befindliche und für das nächste Jahr geplante Arbeiten, Anlage zum Schreiben Verschuers an die DFG vom 8/9/1934, BArch. Koblenz, R 73/15.341, pp. 104–108, here: p. 106; Verschuer, Von September 1934 bis März 1936 durchgeführte, im Gange befindliche und für das nächste Jahr geplante Zwillingsforschungen, Anlage zum Schreiben Verschuers an die DFG vom 25/3/1936, ibid., pp. 21–24, here: p. 22.

[249] Fischer, Tätigkeitsbericht vom 1. April 1935 bis 31. März 1936, June 22, 1936, MPG Archive, Dept. I, Rep. 1 A, No. 2399, pp. 74–76 v, quote: p. 75. On this: Lüth, Untersuchungen.

another project on the heritability of the normal capacity of blood vessels to dilate and narrow, cleverly making use of the National Socialist regime's antismoking campaign: "Even the receptiveness of the vessels to one of our most widespread poisons, nicotine, has been the subject of only cursory study; here, too, there is no information on the issue of heritability." The studies on the influence of cigarette smoking on blood vessels were to be performed by the newly hired assistant Otto Baader, and these studies were to use twins "so that a decision as to the involvement of a genetic disposition can be predicted with certainty. With the importance of just this issue of nicotine abuse, these studies promise to be of great importance not only scientifically, but also medically and in terms of race hygiene."[250] This argumentation fell on fertile ground, for the DFG approved a material grant for the acquisition of devices for the measurement of minimal temperature variations in the human skin, but the war prevented the experiments from being carried out.

The research begun before 1933 on the influences of heredity and environment on the age of sexual maturity was continued in 1934 by the external scientific staff member Elsa Petri using the twin method. She established that the point of time of menarche is closer on average for identical twin sisters than for fraternal ones. Lüth observed pairs of identical twins for differences in growth and in sexual maturation during puberty.[251]

3.3.4 Dermatoglyphics and Dactyloscopy

In the area of the genetics of normal attributes, the large majority of the total of 16 publications from the period from 1934 to 1942 was those which dealt with dermatoglyphics, that is, research of the pattern of epidermal ridges using the methods of dactyloscopy. After twin research, and sometimes even in connection with it, dactyloscopy constituted the second most important methodological approach to the genetic biology of humans in Dahlem in the period from 1933 to 1939, making it more important than classical craniometry, genealogy, "bastard research," or embryology. Georg Geipel worked diligently to polish the methods of dactyloscopy for the study of fingerprints, handprints, and footprints to make it useful for genetic biological research, the diagnosis of twins, and paternity testing.[252] By the end of 1934 he had analyzed the fingerprints of approximately 500 pairs of twins.[253] Verschuer evaluated these analyses, both with a view to the zygosity diagnosis and as regards issues of environmentally influenced variability and asymmetry. At the

[250] Fischer to DFG, BArch. Koblenz, R 73/11.004. Cf. in general: Proctor, Blitzkrieg, pp. 199–278.

[251] Petri, Untersuchungen; Lüth, Endokrine Störungen.

[252] Geipel, Anleitung; idem., Methode der Auswertung; idem., Formindex; idem., Gesamtzahl; idem., Verteilung; idem., Methode der Ermittlung.

[253] Verschuer, In den Jahren 1932–34 durchgeführte, im Gange befindliche und für das nächste Jahr geplante Arbeiten, Anlage zum Schreiben Verschuers an die DFG vom 8/9/1934, BArch. Koblenz, R 73/15.341, pp. 104–108, here: p. 104.

Congress of the German Society for Genetics in 1935, Geipel and Verschuer lectured about the heritability of the form index of the ridges of the fingers, emphasizing the quite high degree of concordance of the pattern of finger ridges in identical twins.[254] In 1936, the doctoral student Maria Ploetz-Radmann (* 1911) extended the dactyloscopy of twins to the lower two-finger phalanges of the human hand. A systematic analysis of the handprints of 250 pairs of twins was presented by Gisela Meyer-Heydenhagen (* 1910) in her dissertation of 1934. In 1938, Christel Steffens (* 1913) studied the toe ridges of twins. In 1937, Adolf Würth (* 1905) continued Schaeuble's studies of the hands of human embryos dissected in series of sections, with reference to the emergence of flexion creases on the palms.[255]

A whole series of individual projects that made use of dactyloscopic methods concerned race anthropology issues. The Indian guest scholar Prophulla Chandras Biswas processed fingerprints and handprints he had produced of 50 Indians; the Chinese guest scholar Wuhou Wayne King studied the fingerprints and handprints of 100 Chinese from Berlin and Paris. The doctoral student Haring Tjittes Piebenga of the Netherlands, assistant at the KWI-A from 1940, performed comparative genetic biology studies of the system of epidermal ridges of the Frisians, Flemish and Walloons; in a second paper he studied the system of epidermal ridges in the population of Urk, an island in Lake Ijssel. Wolfgang Abel, proceeding from his studies of the patterns of finger and hand lines of the Greenland Inuit, was also prominent in this field, publishing several articles in manuals and essays on methodological questions.[256]

From 1936, Abel concerned himself with a further potential application of dactyloscopy in the field of genetic pathology – he attempted to use defective patterns of papillary lines as genetic markers for physical and mental anomalies. In the investigation of fingerprints from the Police Identification Service Archive (*Erkennungsdienstliches Archiv*) of the city of Berlin, Abel was struck by a total of 31 cases characterized by major or minor defects in the pattern of papillary lines. Abel classified these defects as genetic and associated them with a supposedly criminal disposition, for, Abel reported, in contrast to what he called his "criminal material," he had not been able to find such defects in the review of 4,000 fingerprints most of which came from "German rural populations."[257] Abel even attempted to reveal connections between defective patterns of papillary lines and

[254] Verschuer/Geipel, Frage.

[255] Ploetz-Radmann, Hautleistenmuster; Meyer-Heydenhagen, Hautleisten; Steffens, Zehenleisten; Würth, Entstehung.

[256] Biswas, Hand- und Fingerleisten; King, Hautleisten; Piebenga, Untersuchungen; idem., Hautleistensystem; Abel, Frage; idem., Finger- und Handlinienmuster; idem., Verteilung der Genotypen; idem., Kritische Studien; idem., Erbanlagen der Papillarmuster. Series of fingerprints and handprints from Schaeuble's, Hauschild's and Lichtenecker's expeditions, among others, awaited analysis by Geipel. A whole series of publications emerged from these, even well after World War II.

[257] Abel, Störungen der Papillarmuster, p. 32.

certain criminal offences like murder or burglary – an undertaking that was, of course, doomed to failure.[258] Yet, the study about criminals had given him another idea:

> The dearth of these defects in the rural population on the one hand, and the not all too seldom occurrence in criminals allowed the presumption that these defects frequently might appear in connection with other (mental and physical) abnormalities, and gave occasion to the investigation of certain groups biologically more delimited, such as the feeble-minded, mentally ill, deaf-mutes, congenitally blind, mongoloid idiots, cases with physical deformities like syndactyly, polydactyly, constrictions, acrocephalus, spina bifida, etc.

Studies of this kind were "in progress." As a preliminary result, Abel stated: "In the majority of the above mentioned groups defects of the papillary lines could be found sporadically or frequently."[259] Abel introduced several cases of this kind: The volar pads on the fingers and toes, palms and soles of the feet of a 14-year-old deaf–mute boy with congenital lues show striking defects like those of an albino and several "mongoloid idiots." The provenance of these pages of fingerprints is unclear – in the diagnosis of a case of Morbus Down's syndrome patients, Hans Heinze (1895–1983),[260] at the time still director of the Brandenburg State Asylum in Potsdam, and Horst Geyer had provided assistance.[261] Abel announced that he would publish the final results of his studies of the disabled in a second part of his paper, but this second paper never appeared. However, Abel did publish a genealogical study in 1944, on a family from his "criminal material," in which he believed he could prove for the first time the heritability of defective papillary patterns. Difficulties "both in locating relatives who were still alive, and through the refusal of a number of those involved to be examined" were overcome with the help of Heinrich Himmler (1900–1945), Reichsführer of the SS. "Thus the order to the Reichskriminalamt, Department for Criminal Research, was placed by him, in consultation with Chief Detective Müller,[262] to have fingerprints taken […] of the relatives." At the end of the essay is the ominous statement, "A further study of the family has been initiated."[263]

---

[258] Ibid., p. 31.

[259] Ibid., p. 32. Also, p. 37.

[260] The institution in Potsdam was dissolved in 1938 and the majority of the children and teenagers accommodated there were moved to the state institution in Brandenburg-Görden. At the same time, Heinze took over the direction in Görden. From the very beginning, he was a member of the "euthanasia campaign" planning staff, provided "expert opinions" for children's "euthanasia" and set up the first "specialized children's department" in Görden, which served as the "Reich training station." Under Heinze Görden also functioned as a "intermediate institution" in the framework of "Aktion T4." From January 1942 on, Heinze maintained a "research department" in Görden. Around 100 children and teens from this research department were gased in the former Brandenburg jail, and their brains dissected by Julius Hallervorden at the KWI for Brain Research. Cf. Benzenhöfer, Heinze; Hanrath, "Euthanasie," esp. pp. 139 ff.; Schmuhl, Hirnforschung.

[261] Abel, Störungen der Papillarmuster, pp. 32 f.

[262] This probably refers to the chief detective and SS Obersturmbannführer Johannes Müller (1895–1961), commander of the security police and the SD in the Lublin district from July 1941 to September 1943.

[263] Abel, Fall von Vererbung, pp. 73, 85.

Although Abel's dermatoglyphic research on the disabled did not ultimately reach its conclusion, there was certainly continued interest in this cluster of issues at the KWI-A. The estate of Georg Geipel includes a folder with 93 handprint cards of mentally disabled children and teenagers, of which more than half certainly, and the rest most probably, originated from the Wittenau Sanatoriums or the City Psychiatric Clinic for Children in Wiesengrund, respectively.[264] It can be proven that a large number of these handprints were made in 1936,[265] however a number of them, as can be recognized from the admission and discharge data, certainly were taken later, with a definite accumulation in 1938/39 and 1944. Where did this material, which was never published, come from? There is an indication that it was collected by Bernhard T. Duis and sent to Geipel either by Duis himself or by Lothar Loeffler.[266] Duis, who was listed as a doctoral student of the KWI-A from 1931 to 1933,[267] had become senior physician at the Race Biology Institute of the University of Königsberg under Loeffler in 1935 and received his Ph.D. from the Medical Faculty of the University of Königsberg in 1936 with a dissertation about "Finger Ridges in Schizophrenics." Duis did not believe "to be able to diagnose

[264] Effective July 1, 1941 the "Educational Home Including a Psychiatric Clinic for Children" (*Erziehungsheim einschl. Nervenklinik für Kinder*), which had belonged to the Wittenau Sanatoriums until that time, were extracted from the entire complex of the main institution and from then on run as an independent hospital under the name *Städtische Nervenklinik für Kinder*. This new hospital contained two spatially divided areas: the educational home that continued to survive on the grounds of the Wittenau Sanatoriums, with its special school and sick bay, and the building on the grounds of the former Dalldorf farmyard, previously used as the "Wiesengrund Psychiatric Hospital for Men" (*Nervenklinik Wiesengrund für Männer*), as the actual psychiatric department. Usually reference was made simply to "Wiesengrund," although this designation did not become part of the official name until the 1950s. The extraction of Wiesengrund from the Association of Sanatoriums attended the institution of the "specialized children's department" in Wiesengrund. On the organizational structure: Beddies, Geschichte, pp. 80f.; on the "specialized children's department," see also: Krüger, Kinderfachabteilung Wiesengrund.

On the basis of the admission records of the Wittenau Sanatoriums, or of Wiesengrund, respectively, 55 of the 93 children and teenagers coud be identified definitively as patients of the Wittenau Sanatoriums/Wiesengrund. The handprint cards or corresponding file cards of fifteen of the 38 children and teens who could *not* be identified as patients of the Wittenau Sanatoriums/Wiesengrund bear the notation "*W*," which presumably means *Wittenau* (although inversely, 37 of the 55 children and teenagers who have been identified as patients of the Wittenau Sanatoriums/Wiesengrund do *not* bear the notation). Thus it can be presumed that the overwhelming majority of test subjects, if not all, came from Wittenau. A handwritten note accompanying the volume with the handprints refers to "80 feeble-minded (Wittenau)" as well as "16 children from Wittenau." According to the name lists, there is overlapping between the two groups, so that one can proceed from the assumption of around 90 test subjects from Wittenau.

[265] For 11 of the Wittenau children and teenagers the age is noted on the handprint card. Combining this information with the established birthdates, it turns out that the children's handprints must have been taken in 1936. From the perspective of the admission and discharge dates, it turns out the handprints of another 30 children could also have been taken in 1936.

[266] The folder containing the handprint cards and the corresponding file cards was originally labeled "Prof. Dr. Loeffler, Briefwechsel mit D. Duis, Univ. Königsberg, Rassenbiologisches Institut, Oberlaak 8/9."

[267] Cf. Jahresbericht 1931/32 and Jahresbericht 1932/33, MPG Archive, Dept. I, Rep. 3, No. 8.

schizophrenia from the findings about the hand and finger ridges," but found it "of anthropological interest if certain shapes of patterns were to occur more frequently for a certain selected group (like, e.g. the schizophrenics)." At the Race Biology Institute, thus, "two large control groups were set up: one was to check in an average population (East Prussian village studies) whether different papillary patterns occur in the various age groups; the other is to investigate the distribution of the papillary patterns for different groups of disorders."[268] Duis himself examined 772 schizophrenic patients from the University Psychiatric Clinic in Königsberg and from the East Prussian Sanatoriums and Homes in Kortau near Allenstein and Allenberg in the Wehlau region.

It seems plausible that the handprints were made at the Educational Home of the Wittenau Sanatoriums in the year 1936 in connection with the large-scale studies of the Race Biology Institute in Königsberg. It appears just as plausible that the contact between Duis and/or Loeffler and the Wittenau Sanatoriums continued on an ongoing basis such that material was sent from Berlin to Königsberg and Vienna. Against this background, it also becomes clearer how Loeffler came to be represented in the program of accompanying research on children's "euthanasia" with a project on "genetic biology issues according to social aspects."[269]

Duis and Loeffler – if they did not make the material available to Wolfgang Abel – may have turned to Georg Geipel for assistance in applying the particular techniques used for taking handprints, or in interpreting the prints. In any case, in his dissertation Duis referred explicitly to a technique described by Geipel for taking handprints from seriously distorted hands. Geipel received many inquiries of this kind. From his correspondence it becomes clear that he knew Duis and Loeffler well. Eugen Fischer, for his part, counted Loeffler and Duis among the "old Dahlem circle" in 1949, and in 1951 he wrote that Duis, by now an assistant to Verschuer, was among his "Dahlem pupils."[270]

Which line of attack was pursued by research on the handprints of mentally disabled children and teenagers? It may have been designed as an analog to Duis' project about schizophrenia. Thus the point was to study whether certain patterns of papillary lines occurred frequently in "the feeble minded," "morons," and "imbeciles." What is interesting is that other information is also noted on some of the handprint cards. Special attention was paid to physical conspicuities

---

[268] Duis, Fingerleisten, quotes: pp. 391 f.

[269] This emerges from a report by Ernst Wentzler (1891–1973), one of the leading pediatricians involved in children's "euthanasia," to the "Reich Committee for the Scientific Recording of Serious Genetic and Hereditary Afflications" (*Reichsausschuß zur wissenschaftlichen Erfassung erb- und anlagebedingter schwerer Leiden*) of October 17, 1942: "Since the research works already listed [...] are joined by additional works which come from outside the Reich Committee (Prof. Lothar Loeffler – genetic biological issues according to social aspects based on our file material [...]), the question of the financing of these research projects has become acute." BArch. Berlin, NS 11/94.

[270] Fischer to Magnussen, 3/10/1949; Fischer to Magnussen, 4/10/1951, quoted in Klee, Medizin, p. 365, 366.

(strabismus, nystagmus, blindness, harelip, microcephaly, cryptorchism, polydactyly), epilepsy, twin births, indications of prenatal and perinatal complications (nervous breakdown of the mother during pregnancy, difficult birth, breech delivery, brain damage from birth trauma, Lues congenitus), as well as conspicuities of character (psychopathy, *Hypothymiodismus* ["lack of emotion"]). Mental disabilities, mental disorders, social deviance, and suicide in the family were recorded painstakingly (for instance, whether the father drank, was a thief, or sex offender, whether the mother came from a home or had more than one illegitimate child). This indicates an attempt to distinguish between hereditary and nonhereditary cases of "idiocy" and to combine fingerprint patterns with other stigmata. In one case a patient was expressly noted to be "Mosaic, from Galicia" (in another case the researchers failed to note that a subject was Jewish). Duis, as becomes clear in a lecture to the Königsberg Association for Scientific Medicine (*Königsberger Verein für wissenschaftliche Heilkunde*) in 1937, was also interested in determining race differences based on the lines of the hands and fingers.

Duis' estate includes the draft of a postdoctoral dissertation entitled *Leitzeichen menschlicher Erbanlagen. Versuch einer rassenhygienischen Physiognomik unter besonderer Berücksichtigung der Lehre von den Entartungszeichen* ("Guiding Signs of Human Hereditary Dispositions. An Attempt at a Race Hygiene Physiognomy Taking Special Consideration of the Theory of the Signals of Degeneration").[271] The search for stigmata was directly connected with the National Socialist genetic health policy, to which some of the test subjects from the Wittenau Sanatoriums also fell victim: according to the Wittenau admission records, three of the teenagers whose handprints were taken in Wittenau were sterilized, and one[272] was included in the "Aktion T4."[273]

3.3.5 Genetic Pathology

Since 1933, genetic pathology had developed to become the second main working area of the KWI-A. There were various reasons for this. First, the focus of research shifted toward genetic pathological issue in keeping with the degree to which Verschuer determined the research agenda, for in contrast to Fischer, who had never practiced medicine, Verschuer had pursued marked clinical interests since his

---

[271] Klee, Medizin, p. 267.

[272] Heinz B., born 1925, was admitted to the Educational Home of the Wittenau Sanatoriums in 1933 because of mental disability. He was one of the children whose handprints were taken in 1936. According to the admission records, the "Registration Form 1" was filled out for him, but he appears to have been spared from the murder of the ill, for he was discharged "into care" in May 1944.

[273] The *Aktion T4* was a program for the gassing of around 70,000 mentally ill and mentally handicapped persons in 1940/41.

days in Tübingen. After Verschuer moved to Frankfurt, Fischer had intended to reduce the genetic pathology research in Dahlem, but there was such a backlog of genetic pathology projects that this course change in the second half of the 1930s had hardly an impact. Second, the tuberculosis research by Diehl and Verschuer, although it became the target of increasing critique in the 1930s, had provided a model of how the twin method could be applied successfully on genetic pathology issues – not least of all, with good prospects for public funding. Third and finally, and this factor had the greatest weight, a strong pull emanated from the National Socialist genetic health policy, as was reflected by the KWI-A's turn toward the neurodegenerative diseases, mental disability, epilepsy, mental illnesses, and deaf-muteness, all of which had played no role at all in the institute's research design before 1933. Here again it becomes apparent that the objects, research questions, and objectives of research changed as a consequence of the close interconnections with politics. When science serves politics as a resource, and politics serves science as a resource, then, as Mitchell Ash emphasizes correctly, the contents of research do not remain unaffected.[274] From this a change of course in terms of research strategy took place at the KWI-A, which Fischer emphasized pointedly in his 1933/34 annual report:

> The development of the genetics of humans, not least through the work of the institute, is now so far that the most urgent task is the exploration of clinical pictures with its new methods. With this the institute embarks on a fundamentally new step. […] This opens up a great new circle of challenges with much promise. The hiring of four trained physicians as auxiliary assistants is the external expression of this.[275]

From 1933 on, Fischer and his staff systematically checked all kinds of clinicial pictures for whether a genetic factor might play a role in their occurrence – and in most cases they believed to have found evidence to this end. Fischer offered the new state these findings as the foundation for the further development of genetic health policy, as proceeds from a first summary he drew in 1935:

> As was already declared as an intention and agenda item at the last board meeting, a new working area was introduced to the institute, to perform, in collaboration with clinicians, research on the heritability of certain diseases. This should not only bring knowledge about heredity and the inheritance of conditions or their dispositions which have not yet been recognized with any certainty as genetic diseases, but there is also the hope of recognizing the healthy carriers of hidden hereditary attributes, which would be of tremendous importance for the further propagation of these genes. The chain of physiological processes will also be researched with respect to their genetic foundation.[276]

As Fischer continued, the KWI-A worked closely in its genetic pathology research with the Rudolf Virchow Hospital, where the city of Berlin provided the institute with up to 365 free days of care annually, thanks to Berlin's Chief Medical

[274] Ash, Wissenschaft und Politik.

[275] Fischer, Jahresbericht des KWI-A 1933/34, 21/3/1934, MPG Archive, Dept. I, Rep. 3, No. 10.

[276] Fischer, Tätigkeitsbericht von Anfang Juli 1933 bis 1. April 1935, MPG Archive, Dept. I, Rep. 1 A, No. 2404, pp. 49–49h, quote: p. 49c.

Councilor Klein.[277] The institute accommodated ill twins there, always together with their healthy twin brothers or sisters, so that all necessary examinations could be performed on both siblings at the same time. Physiological tests were also performed at Virchow Hospital – and at Elisabeth-Diakonissen Hospital, which provided the institute with a few beds "in a small special department" for a nominal fee, thanks to the accommodation of the senior physician of the Department for Internal Medicine, Friedrich Wilhelm Bremer.[278] At the Westend Hospital in Berlin, twins were examined who suffered from diabetes or rachitis. The clinical possibilities of the KWI-A were finally expanded when Verschuer took over the direction of the Polyclinic for the Fostering of Genes and Race in Berlin Charlottenburg, founded by the Reich Ministry of the Interior in summer 1934, which was responsible for "genetic consulting, marriage consultations, genetic biology certification, taking stock of the genetic biology of Charlottenburg."[279]

Especially in the field of genetic pathology, the main method used after 1933 – besides genealogically oriented "research of descent" – was above all the twin method and this despite the fact that the project which served as a model here had clearly run up against its limits by the mid-1930s. Against Diehl and Verschuer, who had caused a furor with their monography about twin research and the genetic disposition for tuberculosis in 1933, opposition formed in the ranks of clinical tuberculosis research – and the practitioners increasingly had success in influencing health policy. Karl Diehl was certainly influenced by the harsh criticism. He expressed his doubts in a threatening letter to Verschuer of March 18, 1936. The subject was Diehl's manuscript for the second volume of the monograph on twin tuberculosis, *Der Erbeinfluß bei der Tuberkulose* ("Genetic Influence in Tubercolosis").[280] Verschuer objected to Diehl's concession to the critics, that there were "purely environmental tuberculoses." Diehl rejected his friend's arguments: The number of twins investigated was too small to be able to draw any certain conclusions at all; the discordances which also occurred between identical twins constituted a problem that could hardly be resolved; the similarities in the course of disease, which Verschuer had supposedly established in the discordant cases, could be explained with the disposition of the twin pairs' physical constitution without having to resort to a special genetic disposition for tuberculosis. Diehl completely rejected Verschuer's drawing an analogy between schizophrenia and tuberculosis. Only "after hard, internal conflicts of conscience" had he been able to "see anything more profound" at all in the few concordant pairs of twins. He held no interpretation possible other "than the one wrested from me under anguish," that

[277] Cf. also Fischer, Bericht über die klinische Erbforschung im Rudolf Virchow-Krankenhaus, 23/11/1933, MPG Archive, Dept. I, Rep. 1 A, No. 2399, pp. 16–17. Apparently this cooperation was originally supposed to take place with the Urban Hospital. Cf. Fischer, Jahresbericht des KWI-A 1933/1934, 21/3/1934, MPG Archive, Dept. I, Rep. 3, No. 10.

[278] Ibid. On Bremer, cf. Schmuhl, Krankenhäuser, p. 34.

[279] Fischer, Jahresbericht des KWI-A 1934/35, 14/4/1935, MPG Archive, Dept. I, Rep. 3, No. 11.

[280] Diehl/Verschuer, Erbeinfluß.

there was a broad continuum between purely environmental and strongly hereditary tuberculoses. Foreseeing the criticism of the tuberculosis physicians, namely Redeker, he urgently advised that the aspect of the quite strong environmental influence not be suppressed: "[…] thus everyone will see that we are not just preposterous genetic biologists, but also open for other viewpoints."[281] Diehl even asked Verschuer to consider whether the publishing date of the book, of which the galley proofs had already been produced, should be delayed for 3 years until certain material became available.[282]

What is interesting is that 4 days before, on March 14, 1936, Diehl had also turned to the President of the Reich Health Office, Hans Reiter. In his detailed letter Diehl explained that, while the "effect of a hereditary influence in tuberculosis [could be] regarded as certain on the basis of twin research," more recent research had also proven the existence of purely "environmental" tuberculoses, whereby a certain differentiation was not yet possible at the time. "These facts make taking action in terms of race hygiene considerably more difficult." Systematic family anamnesis would make a differential diagnosis possible in the majority of cases. "But on the whole I would like to advise greatest caution." At the close of his letter Diehl suggested a solution to the problem:

> The […]difficulties can only be eliminated by a systematic search for the final cause of the hereditary disposition for tuberculosis. Animal experiments on the broadest basis have been in progress for two years. Hopefully it will be possible in 2–3 years, on the basis of these studies and an even greater amount of twin material, to submit a further, hopefully final result.[283]

From the correspondence between Diehl and Verschuer it becomes apparent that Karl and Anne Diehl – the married couple worked together closely on theoretical research – in fact had been experimenting with rabbits since 1934, but had kept these experiments top secret in the early years.[284] Since Diehl hoped from his animal model to achieve a breakthrough for tuberculosis research, he did not want to attract the attention of the predominantly skeptical specialized audience of tuberculosis physicians to his experiments prematurely. Secretly he continued to breed his rabbits, always in close collaboration with Verschuer – when the latter moved

[281] Diehl to Verschuer, 18/3/1936, MPG Archive, Dept. III, Rep. 86 A (Münster), No. 7.

[282] However, the book appeared as planned. In an appendix about "Conclusions for etiological-pathogenetic tuberculosis research," the authors again discussed the issue of genes vs. environment, whereby for the most part the argumentation follows along Verschuer's lines, although Diehl's critical remarks received consideration. Diehl's thesis that there were also purely environmental tuberculoses does not appear in this text, however. Diehl/Verschuer, Erbeinfluß, pp. 150–155.

[283] Diehl to Reiter, 14/3/1936, ibid. Cf. also Diehl/Verschuer, Erbeinfluß, pp. 155–159, about "Practical Conclusions," where the authors expressly welcome the marriage ban for "the openly tuberculous" in accordance with the marriage health law, but exclude the issue of sterilization for tuberculosis sufferers.

[284] Verschuer to Diehl, 29/6/1935, ibid.: "Of course I will treat your rabbit experiments in strict confidence. I have not told anyone about them yet […]. I am completely of your opinion that one should not speak of unborn children."

to Frankfurt in 1935, Diehl researched his postdoctoral thesis there – but also in constant contact with Eugen Fischer.[285]

While twin tuberculosis research visibly was forced into the defensive, researchers at the KWI-A turned to other infectious diseases. For this purpose Bühler and Lenz methodically analyzed the Dahlem twin file catalog. Ernst-Georg Becker (* 1911), who worked as a doctoral student at the institute in 1937, studied the occurrence of pneumonias among 340 identical and 423 fraternal pairs of twins and postulated a genetic disposition or immunity, as did Lüth, who investigated the occurrence of appendicitis among 320 identical and 430 fraternal twins in 1938. In contrast, in experiments on 24 identical and 18 fraternal pairs of twins about immunization against diphtheria in 1937, Bühler was unable to establish any genetic disposition.[286]

Beyond the infectious diseases, "twin pathology" turned to internal and children's diseases as well. Franz Steiner, and above all Heinz Lemser, worked on diabetes mellitus. In 1938, Lemser put together the observations on twins up to that time, added another 27 pairs and calculated high rates of concordance between the identical twins. From the analysis of the discordant identical pairs of twins he drew conclusions on other conditioning factors and, in further publications, turned to the problem of a latent genetic disposition to diabetes and its detection. As a supplement to the twin studies, Lemser, on the basis of statistical material from all over the world, analyzed the occurrence of diabetes in various geographic regions and discussed possible causes for the unequal distribution that became apparent.[287]

In the field of diabetes research, a vehement exchange of blows between Fischer and Rüdin took place in 1937, after it became known that Hildegard Then Berg, a member of Rüdin's staff, was looking for pairs of twins in Berlin in order to clarify the genetic conditions of diabetes. Immediately the rivalry, smoldering since 1927 and further fanned by the foundation of the Department for Genetic Psychology at the KWI-A in 1935, broke out again in full force. Fischer writes a withering letter of protest:

> I am painfully surprised that you have no bones about breaking into a field in which I have been acquiring twins as material for scientific research for years now, and with great effort. It is not permissible that we enter into a kind of competition with regard to these people. The whole matter is all the more painful since it effects an intersection between your and my *scientific* areas of research. I did not know that your area of work reached so far beyond that of psychiatry. [...] Nevertheless I regard it as improper, not to mention unfriendly, to take the twin material away from me in Berlin.[288]

---

[285] In 1937, Karl and Anne Diehl received 2,000 RM for their research from the budget of the KWI-A. Fischer designated this post to the general administration as "my [sic!] tuberculosis experiments in Sommerfeld." Fischer to Schröder, 11/3/1937, MPG Archive, Dept. I, Rep. 1 A, No. 2407, p. 7.

[286] Becker, Pneumonien; Lüth, Erbliche Disposition; Bühler, Experimentelle Untersuchungen.

[287] Steiner, Untersuchungen; Lemser, Nicht manifestierte Erbanlage; idem., Kann eine Erbanlage für Diabetes latent bleiben?; idem., Erb- und Rassenpathologie; idem., Kann eine Erbanlage für Diabetes latent bleiben?, Teil II. Cf. also Bühler/Lenz, Frage; Fischer/Lemser, Frage; Störring/Lemser, Beziehungen.

[288] Fischer to Rüdin, 17/6/1937, MPIP-HA, GDA 131 (original emphasis).

Rüdin countered coolly:

> You have no right to claim a monopoly on any kind of material or a scientific research area, and I do not do the same for myself. When Mr. Kranz, who is a psychiatrist, worked at your institute on criminals, an eminent psychiatric topic (psychopaths), that is, an area which you, as it appears, regard to be my exclusive domain, I did not raise any objection. At the time, however, I assumed that you no longer placed any value on the previous agreement of a certain demarcation between our areas of research.[289]

This final remark indicates that Fischer and Rüdin had made a gentlemen's agreement to divide up their working areas after the founding of the KWI-A.[290] While Fischer had felt strong enough in the late 1920s to transgress the borderline agreed upon, with the National Socialist takeover Rüdin got the upper hand and set about to expand his area beyond the originally demarcated boundaries. He expressed this clearly in the remainder of his letter.[291]

A second emphasis of genetic research in the field of internal medicine at the KWI-A was set by Wolfgang Lehmann with his work on genetic disposition in the emergence of rachitis, which was based on observations of 134 pairs of twins with rachitis. At the Congress of the German Society for Genetics in Jena in 1935, Lehmann lectured about "dystrophic diathesis" (genetic disposition to the deficient supply of organs with nutrients) based on the study of 41 pairs of twins.[292] In several papers, the medical intern and later external scientific staff member Richard Günder (* 1912) concerned himself with hemophilia.[293] The medical intern Friedrich Hermann Haase (* 1913) attended to statistical studies of the problem of the excessive mortality for boys in infancy, which he traced back to a gender-bound disposition on a recessive gene.[294]

Genetic pathology even encroached upon dentistry and ophthalmology. The dentist Gerhard Nehls (* 1912), who spent 1938 at the institute as a doctoral student, investigated the influences of genetic disposition and environment on the emergence of caries; the external scholar Ernst Francke was interested in the connections between physique and myopia; and the doctoral student Heinz Kehl (* 1910) also discussed the heritability of nearsightedness in his dissertation.[295]

---

[289] Rüdin to Fischer, 24/6/1937, ibid.

[290] See also Lösch, Rasse, p. 323.

[291] Rüdin urged Then Berg to mention a case of discordance between identical twins documented by Lemser, in which one twin had diabetes and the other did not, "because I do not want us to suppress data that do not suit us." Then Berg to Rüdin, 9/7/1938; Rüdin to Then Berg, 12/7/1938 (quote), MPIP-HA, GDA 132. A conflict of competency between Verschuer and Rüdin had already erupted back in 1935, when Verschuer protested against Karlheinz Idelberger (1909–2003), another member of Rüdin's staff, who attempted to collect data on twins in the Rhine-Main region. Cf. Verschuer to Rüdin, 12/7/1935; Rüdin to Verschuer, 24/7/1935; Verschuer to Rüdin, 20/8/1935; Verschuer to Rüdin, 15/3/1941; Idelberger to Verschuer, 18/3/1941, ibid. Cf. also Weber, Ernst Rüdin, p. 246.

[292] Lehmann, Erbuntersuchung; idem., Zwillingspathologische Untersuchungen; idem., Bedeutung.

[293] Günder, Gerinnungsprüfungen; idem., Mitteilungen; idem., Beiträge.

[294] Haase, Übersterblichkeit.

[295] Nehls, Caries; Francke, Körperbau; Kehl, Erblichkeit.

A third emphasis of "twin pathology" in the 1930s was on neurodegenerative illnesses and mental disabilities, which had received practically no attention at the KWI-A before 1933. In 1937, Horst Geyer – together with Georg Stertz (1878–1959) – described a family with spinal ataxia.[296] In 1939 Geyer, along with the guest scholar Ole Pedersen, reported on illnesses of the central nervous system like Recklinghausen's disease and tuberous sclerosis.[297] Herbert Grohmann also provided a description of identical twin brothers, both of whom suffered from Recklinghausen's disease.[298] Geyer concerned himself with Little's disease under race hygiene aspects.[299] As Fischer proudly announced in 1936, Werner Wolfslast (* 1904), a pupil of Fritz Lenz, had discovered "spastic diplegia" (two-sided paralysis on the basis of cerebral paresis in infancy) as a "new genetic disease" and clarified its "genetic path."[300] Peter Emil Becker also worked in the area of neuropathy, studying pairs of twins suffering from sciatica.[301]

In his postdoctoral dissertation published in 1939, Horst Geyer turned his attention to "mongoloidism" (Down syndrome, trisomy 21), which he incorrectly traced back to "dysplasmatic" heredity.[302] In several overview papers between 1937, he further occupied himself with epilepsies as well as congenital "moronic conditions" and those acquired at an early age.[303] These thematic topics touched directly on National Socialist health policy, as the majority of the people subjected to compul-

---

[296] Stertz/Geyer, Erbpathologie. Spinal ataxia is caused by illnesses of the spinal cord and induces serious disturbances to muscle coordination.

[297] Geyer/Pedersen, Erblichkeit. The extremely rare Recklinghausen's disease – named after the pathologist Friedrich Daniel v. Recklinghausen (1833–1910) – is an illness of the nervous system normally characterized by the formation of numerous tumors (neural neoplasms) in the skin, nerve roots and cranial nerves. Because it is carried on a dominant gene it is particularly suitable for genetic pathology studies. Tuberous sclerosis (also known as Bourneville-Pringle syndrome) is a malformation syndrome induced by mutation, with *Adenoma sebaceum* (pasty rash) on the face, epilepsy and progressive mental disability. This syndrome attracted increased attention after 1933 because the adenoma was an externally visible stigma. It was hoped that stigmata could be found for other neurodegenerative illnesses in order to develop a method for early diagnosis. Rüdin was interested in tuberous sclerosis as well (cf. Rüdin to Verschuer, 24/7/1935, MPIP-HA, GDA 132).

[298] Grohmann, Erbpathologie.

[299] Geyer, Rassenhygiene. Cf. also Pedersen/Geyer, Auftreten von Hirntumoren. Little's disease is named after the British surgeon William John Little (1810–1894), and designates certain forms of cerebral palsy on the basis of disturbances to embryonic development or damage during early childhood to the neuronal paths for motor function in the brain. According to Fischer (Tätigkeitsbericht vom 1/4/1935 bis zum 31/3/1936, MPG Archive, Dept. I, Rep. 1 A, No. 2404, pp. 52b–52e, here: p. 52d) Geyer also performed "animal experiment studies about mutations" in 1935/36. In a letter to the KWG of March 1938, Lenz reported that Geyer was occupied – in addition to his other projects – with a paper about "Genetic Mutations in Drosophila." Lenz to KWG, 8/3/1938, MPG Archive, Dept. I, Rep. 3, No. 13.

[300] Fischer, Tätigkeitsbericht vom 1/4/1935 bis zum 31/3/1936, MPG Archive, Dept. I, Rep. 1 A, No. 2404, pp. 52b–52e, quote: p. 52d. However, Wolfslast's publication did not appear until 1943 (Cf. Wolfslast, Sippe).

[301] Becker, Erblichkeit der Ischias.

[302] Geyer, Ätiologie.

[303] Geyer, Epilepsien; idem., Erbliche Fallsucht; idem., Schwachsinnszustände.

sory sterilization in accordance with the GzVeN were done so under the indications of "congenital idiocy" and (genuine) epilepsy. With the beginning of the mass murder of the mentally ill and disabled in the years 1939/40, the differential diagnosis between congenital and acquired forms of mental disability shifted to the focus of interest – it made up a large portion of the accompanying research on "euthanasia." Here the two "research stations" of the "euthanasia" apparatus in the state sanatorium in Brandenburg-Görden and the state sanatorium and home in Wiesloch/ Heidelberg University Clinic, worked closely with the KWI for Brain Research and/or the DFA. The KWI-A no longer participated in this research complex.[304]

Independent research on mental illnesses was not performed at the KWI-A, but with two articles entitled "Who is schizophrenic?" and "Segregate the mentally ill?" Fritz Lenz participated in the debates about the heritability of endogenous psychoses,[305] about which he found himself in a vehement confrontation with Hans Luxenburger. A paper by Geyer also belongs to this field, about the opposing manifestation of "emotional disposition" based on the example of two pairs of identical twins.[306]

How new research projects proceeded from the input on National Socialist genetic health policy can be demonstrated in an exemplary fashion in the case of the GzVeN. "In the file material accruing [hereby] will be found an abundance of observations and impulses from the area of human genetics, which require a scientific evaluation. It will also be necessary to deal with some of the questions brought up by the Hereditary Health Courts scientifically." So Arthur Gütt, on June 1, 1934, justified the archiving the files of the Hereditary Health Court centrally at the Reich Health Office, and he continued: "In this manner a mutual fertilization of science and practical race fostering will result."[307] The scientific evaluation of the material was to be entrusted to Eugen Fischer, Otmar Verschuer, and Ernst Rüdin, such that Rüdin would investigate psychiatric questions; Verschuer questions from the fields of internal medicine, pediatrics, surgery, and orthopedics; Fischer and his institute with questions connected with hereditary blindness and deafness. For these purposes funds were to be provided by the Emergency Association of German Science. On the basis of the Hereditary Health Court files at the KWI-A, Herbert Grohmann did indeed set about a research project on the heredity of deaf-muteness. In this case, participation in the National Socialist sterilization legislation had provided access to new material, opened up a new research area and brought in additional financing, only – the project was located outside the fields of research worked on at the KWI-A so far. One publication emerged from the project: in 1939 Grohmann described a family in which deaf–mute parents, both of whom came from deaf–mute families, had a healthy child.[308]

---

[304] Schmuhl, Hirnforschung, pp. 594–605.

[305] Lenz, Mendeln die Geisteskrankheiten?; idem., Wer wird schizophren?

[306] Geyer, Gegensätzliche Äußerung.

[307] Quoted in Lösch, Rasse, p. 355.

[308] Grohmann, Heterogenie. Cf. also Fischer, Taubstummheit.

From early 1940, the conventional genetic pathology works based on the twin method, family research or studies of embryos and stillbirths were joined by the experimental genetic pathology of Hans Nachtsheim's department. Since this research was already intimately connected to the new paradigm of phenogenetics, it will be discussed in more detail elsewhere.

3.3.6 Genetic Psychology

Genetic psychology research had taken place in Dahlem, as already mentioned, since the late 1920s. This research continued even after 1933 and experienced a strong boom after the founding of the Department for Genetic Psychology under Kurt Gottschaldt in 1935. The new department was the result of a conscious "division of labor," which Verschuer addressed in his first report to the DFG from his new institute in Frankfurt: He himself, so Verschuer announced, would take up "genetic pathology research [...] at the Frankfurt institute with particular intensity."[309] Verschuer did not want to quit the field of genetic psychology research completely, but it was clearly recognizable that he left this field to the institute in Dahlem for the most part. Fischer, for his part, shortly before had set the new course of the KWI-A in a report to the KWG: because of the increasing fusion of human genetics with anthropology and eugenics, Verschuer's department was dissolved after his departure.

> On the other hand, in our opinion, a special part of its area has not yet, to date, been taken on by any research as consistently and urgently as might be necessary, namely the research of the heredity of mental dispositions. The research of pathological mental dispositions is served by an entire major institute, the one in Munich. When one considers that every qualitative population policy, and even more so every racial population policy, has its foundation in the fact of the heritability of mental dispositions, one understands the urgency and imperative of the task of scientifically creating an incontestable scientific foundation for this and expanding it in all its particulars. Experimental psychology can show rich results about the nature and course of mental processes, conditions, achievements. It seems high time to use their methods to research the question of the heredity of the relevant dispositions.[310]

With the establishment of the new working field of genetic psychology, Fischer was poaching – not for the first time, but now quite officially and on a large scale – on foreign territory. In board meeting of the KWI-A on May 10, 1935 the anxious question had already been raised as to whether the planned Department for Genetic Psychology could encroach on the German Research Institute for Psychiatry. Fischer emphasized in response "that a clear demarcation existed, which is also

[309] Verschuer, Von September 1934 bis März 1936 durchgeführte, im Gange befindliche und für das nächste Jahr geplante Zwillingsforschungen, Anlage zum Schreiben Verschuers an die DFG vom 25/3/1936, BArch. Koblenz, R 73/15.341.

[310] Fischer, Halbjahresbericht 1935, October 1935, MPG Archive, Dept. I, Rep. 3, No. 12.

adhered to by both sides. Psychiatric studies are performed only in Munich. The working areas certainly touch upon each other, but there was no overlapping. The collaboration between the two institutes was irreproachable."[311] Ernst Rüdin presumably had a different view. In any case, in connection with the conflict about diabetes research he vented his irritation about the founding of the Department for Genetic Psychology at the KWI-A, which he conceived of as an incursion into the area of genetic psychiatry:

> I, too, follow the maxim that the human being is a whole, and even if, in consideration of the political constellation at the time, I was forced to reconcile myself to the commandment of clearly restricting my research to the area of psychiatry, I did so only under compulsion and welcome the greater freedom in this respect today. You make use of it yourself, too, by having psychological work performed, that is, by doing something that was, up to this time, regarded as eminently appertaining to psychiatry's field of interest. But indeed, why shouldn't you, an anatomist and anthropologist, have psychological work done as well?[312]

Ultimately, however, Ernst Rüdin had to stand by powerlessly watching how Fischer expanded his research agenda toward genetic psychology, as did Wolfgang Köhler, who saw the new department of the KWI-A as competition for the Psychological Institute at the University of Berlin.[313] For all that, Köhler had his back to the wall in National Socialist Germany, and thus constituted no serious danger for Fischer, and with Köhler's emigration this adversary, too, had been driven from the field. More dangerous was Köhler's successor, Johann Baptiste Rieffert, an opportunist who had joined the NSDAP after the *Anschluss* of Austria, whose skilled intrigues were attempts to win control of Fischer's department. On the one hand, Rieffert proposed to Fischer a project on the "Psychology of Jewry". Although Fischer showed "polite interest" to keep from affronting Rieffert, he declined personal participation in the project, which did not keep Rieffert from claiming in several letters that Fischer was involved.[314] At the same time, Rieffert machinated against Fischer at the university, by arguing that Fischer's research in the field of genetic psychology was not covered by his certification to teach anthropology. Further, Rieffert revived denunciations of Gottschaldt as a Communist, with the clear intention of forcing him out of the KWI-A and replacing him with one of his own assistants. Fischer protested to the Ministry of Science, but the conflict was ended before it could escalate any further when Rieffert lost his position and was expelled from the NSDAP – he had concealed his previous membership in the SPD.

Gottschaldt found excellent working conditions in Dahlem. He was allowed one assistant position, which he gave to Kurt Wilde. He also had a pediatric nurse for the examination of twins. Best of all, however, was the state-of-the-art space of the Department for Genetic Psychology, which was housed in the new wing of

[311] Niederschrift über die Sitzung des Kuratoriums des KWI-A am 10/5/1935, MPG Archive, Dept. I, Rep. 1 A, No. 2404, pp. 48–48c, quote: p. 48b.

[312] Rüdin to Fischer, 24/6/1937, MPIP-HA, GDA 131. Cf. also Lösch, Rasse, pp. 322–324.

[313] On the following: ibid., pp. 324, 327–329; Ash, Institut, pp. 128–130.

[314] Lösch, Rasse, p. 327.

the building. It was equipped with a "psychological double laboratory," i.e. two laboratory rooms with the same furniture and equipment, separated by an observation cell. From this cell it was possible to look into both rooms through glass that was transparent only from the inside, allowing twins to be observed without their knowledge, simultaneously and independently of each other.[315] Gottschaldt also organized "genetic psychology twin camps," which served to observe twin children in as natural an environment as possible. After first attempts near Berlin, Gottschaldt's department – supported by the Youth Office of the city of Berlin[316] and the direction of the Berlin district office of the National Socialist Welfare Organization (NSV) – ran one 4-week and one 8-week time camp at the *Seehospiz* children's home on Norderney Island (48 pairs of twins and one set of triplets) and in Arendsee[317] on the Baltic (26 pairs of twins). Gottschaldt was accompanied by his assistants Wilde and Geyer, a pediatric nurse and a secretary. The researchers' goal was to create "social intimacy" and record complex patterns of behavior through participatory observation as well as comprehensive aptitude tests building on realistic everyday situations:

> The children lived in small groups. The events of the day were logged with great precision in a journal, from the time they woke up early in the morning until they went to bed, thus tracking for each pair of twins the daily rhythm of life, the basic mood, how they dealt with all of those difficulties, events and conflicts that arise in everyday life in a children's home. The material amounted to nearly 18,000 pages, and depicted for each child were more than 200 situations it faced.

This "characterological behavioral analysis" was complemented by series of psychological experiments in which "conflict situations, like those that occur in everyday life, [were] simulated experimentally":

> We attempted, among other things, to determine their attitudes toward successful and unsuccessful experiences, how they come to terms with by the social person, and their tendency to perform acts of substitution when conflictual situations arose; to record their mental needs, their range, their satisfaction and/or substitute satisfaction in experiments; and to find out their tendency toward satiation or satiableness in certain action situations and the consequent affinity to variation and such. On the social psychological area we tracked the development of the You and We relationships in new situations, incorporation into certain social communities and situations, or the imprinting through these new environments; also, the need for social recognition, the ideal of the social role which the individual aspires to play in the social circle, what is called the "persona," was investigated,

[315] Ibid., p. 326.

[316] In putting the twin catalog together, the KWI-A had worked closely with the welfare offices of the city of Berlin, which primarily recorded the poorest classes of the population. Toward the end of the Weimar Republic, three quarters of the fathers of twins were workers, the majority of them unemployed, some for longer than one year (Brauns, Studien, p. 87; cf. Massin, Mengele, p. 206). Thus most parents must have welcomed the opportunity to send their children on a free vacation to the North or Baltic Sea.

[317] The homes meant here are probably the vacation home of the Protestant Rural Youth Services (*Evangelischer Landjugenddienst*) in Bastorf near Arendsee, and the Inner Missions *Seehospiz* on Norderney.

> in part with experimental methods, in part through direct observation. In the field of intelligent activity we selected special tests that deal with inductive tasks and practical logic, the resolution of theoretical and practical problems and such.[318]

In the twin camp at Arendsee, Horst Geyer observed sleeping twins and painstakingly recorded the manner of muscle tone, the sleeping positions, the blush of the cheeks, the acts of falling asleep and waking up, as well as anomalies like talking or walking in their sleep or wetting the bed. He also checked the reflexes of sleeping twins to trace genetic differences in the reaction times of the extrapyramidal nervous system.[319]

The studies were continued in Dahlem after conclusion of the twin camps, whereby "special experiments of an inductive and affect psychology nature"[320] were on the agenda. The twin camps were repeated in 1937. During their preparation, Gottschaldt announced that for the first time "race psychology questions [were to] be considered within the racial differentiations that exist in Germany":

> The next "twin camp" is planned to consist of two large groups of pairs of twins from different racial areas. We will thus be able to compare the mental genetic structure of the people more Nordic on average with those of the people who are on average predominantly Alpine.[321]

Whether Gottschaldt actually selected the test subjects for the camp held for 96 pairs of twin at the *Seehospiz* on Norderney for 9 weeks in summer 1937 according to anthropometric attributes must remain an open question – he did not publish any race psychological results. In the framework of the 1937 twin camp Kurt Wilde addressed the issue of whether the faculty of training was heredity, which up to that time had been unquestioningly attributed to the side of environment, and attempted to support his thesis on the basis of the different "learning curves" of identical and fraternal twins. Twice a day for 35 days, Wilde asked a total of 37 pairs of twins to perform a short exercise (searching for given figures on a page), creating a competitive atmosphere by awarding prizes. He arrived at the result that training increased the difference in performance among fraternal twins, but affected an equal improvement in performance among identical twins.[322] After conclusion of the twin camp in 1936/37, Gottschaldt, as he proudly remarked, had collected material "from around 6,000 experimental trials and over 20,000 logged character and life situations."[323]

---

318 Gottschaldt, Methodik erbpsychologischer Untersuchungen, p. 522. Similarly, idem., Vererbung, pp. 6f.

319 Geyer, Schlaf; idem., Subcorticale Mechanismen. Also best located in the area of genetic psychology is the work of the guest scholar Carl R. Czapnik, who compiled the observations described in this literature in an article entitled "*Erbbedingtheit der Intersexualität*" ("The Hereditary Conditionality of Intersexuality" in 1942. Czapnik, Erbbedingtheit.

320 Gottschaldt, Methodik erbpsychologischer Untersuchungen, p. 523.

321 Ibid.

322 Wilde, Erbpsychologische Untersuchungen, esp. pp. 85–87, 105.

323 Gottschaldt, Erbe und Umwelt, p. 10.

The methods used in the twin camps picked up on Gottschaldt's previous developmental psychology studies in Bonn, which were, for their part, a first attempt to apply to children the experimental methodology designed by Wolfgang Köhler and further developed by Kurt Lewin. This approach was now combined with the twin research performed in Dahlem, whereby Gottschaldt from the outset subjected the twin method to critical reflection – more on this later. He discussed his approach in several meta-theoretical and methodologically critical papers, such as the overview article published in 1939 under the title *Erbpsychologie der Elementarfunktionen der Begabung* ("The Genetic Psychology of the Elementary Function of Aptitude") in the *Handbuch der Erbbiologie des Menschen* (Handbook of the Genetic Biology of Human Beings) edited by Günther Just. Here Gottschaldt played the role of a critic of conventional "psychotechology," which, he argued, must remain patchwork because it always measures only isolated individual functions, masks the procedural character of mental events, and thus from its very approach is not able to conceive of "totalities of action." In this critique, which was directed especially to conventional psychological aptitude testing, Gottschaldt blithely referred to – besides many other works – the emigrated Gestalt psychologists, above all Kurt Lewin.[324] However, he also based a number of his objections on the "theories of layers" to explain personality, which were gaining ground in National Socialist Germany. Aptitude testing, claimed Gottschaldt with reference to Ernst Kretschmer (1888–1964), Philipp Lersch (1898–1972), and other leading class theorists of the day,[325] remained on the surface of personality and cannot penetrate into the deeper layers of personality, which are congenital and inherited and thus cannot be changed. Gottschaldt's critique of the conventional testing methods thus cannot be assessed as simply opposition to "brown psychology" – on the contrary, especially those psychologists who proceeded from a *race* psychology approach (which, as shown above, was not entirely alien to Gottschaldt) criticized psychotechnology and demanded a holistic perspective.[326] Insofar Gottschaldt's methodological approach was in the spirit of the times, even though in his publications he *also* relied on Gestalt psychology, which had been driven into exile, and ostentatiously avoided any National Socialist phraseology.[327]

Gottschaldt proceeded from the assumption of a congenital and hereditary core personality: "In vitality, in the way successes and failures are experienced, in the pursuit of recognition, the satiation of social relations, and in the entire social attitude as a whole the genetic foundation cannot be mistaken." Genetically identical twins may demonstrate, in Gottschaldt's words, "differences in expression on the surface," but agree to a great extent in the "basic structure of personality."[328]

---

[324] Stadler, Schicksal, pp. 154 f.; Ash, Erbpsychologische Abteilung, pp. 122.

[325] Ibid., pp. 122 f.

[326] On corresponding endeavors in professional aptitude testing, cf. Schmuhl, Arbeitsmarktpolitik, pp. 248 f.

[327] Stadler, Schicksal, pp. 155 f. Einschränkend: Lösch, Rasse, pp. 330 (note 107).

[328] Gottschaldt, Erbe und Umwelt, p. 15.

Intellectual abilities in their totality, too, Gottschaldt believed to be a product of heredity and environment, whereby he assessed the weight of genetic disposition to be greater on the basis of the intelligence tests carried out at the twin camps.[329] In the twin camps a total of about 4,000 individual tests on intelligence were performed on around 70 pairs of twins, making for about 39 tests on each subject. The tests involved not only the commonly studied areas of intelligent action like language and logical thinking (vocabulary testing, conceptual mapping, clozes), but also the area of practical intelligence. The goal of the tests was "to determine the entire aptitude level in its multifarious dimensions and in its different profiles as exactly as possible, whereby the general method was to avoid the character of a testing situation, and instead using as the basis children's and teens' natural need for play and action as a dynamic incentive for the individual feats of intelligence." On the basis of the differences determined between identical and fraternal twins, Gottschaldt reached the conclusion "that in the entire area of intelligent action, the genetic influences are superior by far to the environmental moments."[330]

The Hungarian guest scholar Mihali Malán studied the capacity for spatial orientation based on 40 pairs each of identical and fraternal twins.[331] Also, three dissertations on genetic psychology topics emerged under the auspices of Gottschaldt by 1942: Karl Joachim Hene (* 1912) wrote a thesis "About the Development of the Personality of the Child," the prison pastor August Ohm (* 1894) about "The Development of the Social Personality during Remand," and Günter Stuttinger (* 1913) about "The Heritability of Optical Perception."[332]

In 1935/36, the Department for Genetic Psychology further undertook "a team project with the Psychological Department of the Reich War Ministry [...], whereby the aptitude tests of the military administration were taken as twin studies."[333] With the reintroduction of general military service in 1935 and the buildup of the Luftwaffe (now taking place openly), the Wehrmacht's interest in military psychology research increased by leaps and bounds. The Psychological Department of the Reich War Ministry sought collaboration with the new Department for Genetic Psychology at the KWI-A. The latter selected from the Dahlem twin catalog 22 identical and 18 fraternal pairs of twins in the age group obligated to perform military service, who were subjected to a military psychology examination by the army psychologists of the Reich War Ministry's Psychological Laboratory in June 1936. These tests were based on the aptitude test for officer candidates. They included an "objective thought test" (arrangement of bodies according to logical

---

[329] Gottschaldt, Methodik erbpsychologischer Untersuchungen; idem., Vererbung; idem., Umwelterscheinungen; idem., Erbpsychologie; idem., Erbe und Umwelt; idem., Methodik der Persönlichkeitsforschung; idem., Problematik.

[330] Gottschaldt, Vererbung, p. 459.

[331] Malán, Erblichkeit.

[332] Ohm, Entwicklung. The dissertations by Hene and Stuttinger are missing. Cf. Lösch, Rasse, p. 532, 534.

[333] Fischer, Tätigkeitsbericht vom 1/4/1935 bis zum 31.3.1936, MPG Archive, Dept. I, Rep. 1 A, No. 2404, pp. 52b–52e, quotes: p. 52b.

aspects), the description of a complicated technical procedure shown in a film, two essays (one on a picture selected by the candidate, one on an abstract subject) as well as practical, technical tasks. Later, the twins were ordered to report to Dahlem to undergo further special examinations. "The difficult genetic psychological elaboration of the diagnostic material supplied by the army psychologists was then the special task of the genetic psychology department, which has kept us busy for 6 months and will continue to do so for a long time to come,"[334] Kurt Wilde reported in a preliminary memorandum. He attempted to decipher the test results with regard to the interplay between heredity and environment. Wilde classified as predominantly *hereditary* the "mental pace of work," the "predominance of concrete thought," "disciplined thought," and the "agility of thoughts"; he viewed as predominantly *environmental* "independence," "practical-technical performance," "manual dexterity," and "understanding of presented components."[335] The collaboration brought with it a personnel component, for in 1936 the auxiliary assistant Johannes Schaeuble switched from the institute in Dahlem to the Psychological Laboratory of the Reich War Ministry.[336]

The other departments of the KWI-A also continued to carry out projects that were assigned to the field of physiology according to the contemporary understanding, but are more likely to be attached to psychology today. For instance, Peter Emil Becker compared the motor skills of 180 pairs of twins in psychotechnical laboratory trials. Moreover, he created a Kraepelin work curve for nine pairs of identical and nine pairs of fraternal twins – the difficult interpretation of the findings induced Becker and Lenz to present a discussion of the methodology of twin research. Becker further asked 329 identical and 410 fraternal pairs of twins about their least favorite foods – and established high rates of concordance here as well. Finally, he examined the "drawing stroke" of 50 identical and 50 fraternal pairs of twins.[337] The Spanish guest scholar Miguel Carmena tested 60 pairs of twins for their "psychogalvanic response" to investigate whether "nervousness" was a hereditary characteristic. In another project Carmena analyzed "pressure of writing" as a partial attribute of the graphological characteristics of 15 identical and 10 fraternal pairs of twins.[338] The handwriting of 150 pairs of twins collected at the KWI-A were entrusted to the philosopher and psychologist Ludwig Klages (1872–1956), who was intensively concerned with graphology and semiotics and held a "seminar for the science of expression" in Kilchberg near Zürich.[339] Eberhard Zwirner

---

[334] Wilde Intelligenzuntersuchungen, p. 513.

[335] ibid., p. 515.

[336] Lösch, Rasse, pp. 329f.

[337] Becker, Zwillingsstudien; idem., Erblichkeit der Motorik; idem./Lenz, Arbeitskurve; idem., Erbbiologie.

[338] Carmena, Affektlage; idem., Schreibdruck.

[339] Verschuer, In den Jahren 1932–34 durchgeführte, im Gange befindliche und für das nächste Jahr geplante Arbeiten, Anlage zum Schreiben Verschuers an die DFG v. 8/9/1934, BArch. Koblenz, R 73/15.341, pp. 104–108, here: p. 107; idem., Von September 1934 bis März 1936 durchgeführte, im Gange befindliche und für das nächste Jahr geplante Zwillingsforschungen, Anlage zum Schreiben Verschuers an die DFG vom 25/3/1936, ibid., pp. 21–24, here: p. 23.

(1899–1984), director of the Phonometric Department at the KWI for Brain Research, also profited from twin research at the KWI-A. In 1938, together with Fischer, he recorded "sonic films" of identical and fraternal twins "to study the hereditary roots of speech."[340]

3.3.7 Eugenics/Race Hygiene

Hardly any research was performed in the field of eugenics/race hygiene after the forced departure of Hermann Muckermann.[341] In 1936, Boeters published the study already discussed in a previous section, on the fertility numbers of German Russian peasant families.[342] In a genealogical study, the physician Johann Nepomuk Häßler (* 1898) studied the fertility ratios in a "peasant and bourgeois clan" on the Baar, a plateau between the Black Forest and the Jura district.[343] Ilse Schmidt (* 1913) looked into the connections between land flight and intelligence by evaluating the report cards of the age group born between 1889 and 1918 in a Brandenburg village near Berlin – in doing so she established that the more gifted pupils migrated to the city at much higher rates than the weaker pupils, confirming the brain drain from the country to the city so bemoaned at the time.[344] In 1941, Heinz Diedrich (* 1913) surveyed elementary school pupils in Stettin, establishing a negative correlation between school performance, aptitude and social class on the one hand and the number of children in the immediate family on the other. Pupils at schools for the disabled had considerably more siblings on average than pupils at other kinds of schools.[345] On the above-average number of children in the families of pupils with disabilities and their supposedly corruptive eugenic effects, Fritz Lenz piped up again, too.[346] Two dissertations advised by Lenz emerged in the field of race hygiene, both of them rather whimsical: in 1939, Horst Eggert held forth on the "social-biology causes" of divorce, in 1941 Karl-Georg Büscher (* 1918) looked into the "Change in the Aspects of Spousal Choice as Reflected in Private Marriage Advertisements."[347]

---

[340] Quoted in Simon/Zahn, Nahtstellen, p. 6. For a biography of Zwirner, see also: Klee, Personenlexikon, p. 699. The Phonometric Department at the KWI for Brain Research was dissolved in 1938. Cf. Schmuhl, Hirnforschung, p. 569.

[341] A series of papers by Muckermann from the year 1934 still appeared as publications of the KWI-A. Cf. Muckermann, Volkstum, Staat und Nation; idem., Leben der Ungeborenen; idem., Eugenik und Katholizismus; idem., Rassenforschung und Volk der Zukunft; idem., Eugenik.

[342] Boeters, Untersuchungen.

[343] Häßler, Untersuchungen.

[344] Schmidt, Beziehungen.

[345] Diedrich, Erhebungen.

[346] Lenz, Frage der Fortpflanzung der Hilfsschüler; idem., Gedanken zur Rassenhygiene.

[347] Eggert, Ehescheidungen; Büscher, Wandel.

3.3.8 Twin Research in Crisis

At the KWI-A twin research continued to be regarded as the king's road for genetic research on humans. Fifty-nine of the 182 scientific publications of the period from 1934 to 1942 were based on the twin method or dealt with its methodological aspects. The zygosity test was fine-tuned even further, the large-scale examination of twin placentas in connection with the zygosity diagnosis continued. The issue of asymmetry was also subjected to further study.[348] Nevertheless, after Verschuer had given an overview portrayal of the twin method in the first volume of *Zwillingstuberkulose* (Twin Tuberculosis),[349] a kind of stagnation set in with regard to its methodology. While it continued to be applied schematically to new objects, increasingly twin researchers encountered criticism.[350]

For the KWI-A this critique was all the more dangerous, because the first impulse came from its own ranks, namely from Fritz Lenz. Together with Verschuer, in 1928 Lenz had proceeded from the premise that genetic disposition and environment supplement each other *additively*, laying down the classical rule for how the quantitative shares of heredity and environment could be calculated from the variability in findings about twins:

> The differences between fraternal twins are conditioned in part by genetic material and in part by environment, whereas the differences between identical twins are conditioned exclusively by environment. Now, it may be assumed that the environmental difference between the fraternal twins is just as great as between the identical ones. Consequently the genetic difference between fraternal twins with reference to a characteristic is obtained by subtracting the difference between the identical twins from that between the fraternal ones.[351]

All twin studies were based on this rule. Disputing its validity threatened to invalidate a large share of the research findings acquired in Dahlem – among them all of the prestigious projects like the genetic tuberculosis research, the twin vertebrae study and the entire genetic psychology research. Fritz Lenz, who over and again stood out for his unconventional thinking even after 1933, and to whom every form of "esprit des corp" was alien, had no problem recanting his earlier perspective in the *Deutsche Medizinische Wochenschrift* in 1935:

> This consideration was incorrect; and through my remarks today I would like to make amends for my error. The supposition that genetic differences and environmental differences are cumulative was erroneous. In fact they combine in an entirely different manner, namely binomially.[352]

In reality the evaluation procedure subsequently proposed by Lenz, which was more or less a variance analysis, implicitly remained quite rooted in the idea of an

---

348 Lemser, Eiigkeitsdiagnose; Steiner, Nachgeburtsbefunde; Busse, Asymmetrien.

349 Diehl/Verschuer, Erbuntersuchungen.

350 On the following: Mai, Humangenetik, pp. 36–43.

351 Lenz/Verschuer, Bestimmung, pp. 425.

352 Lenz, Zwillingsbefunde, pp. 873.

additive combination of the genes combined in the genome complex along with the elements combined in the environment complex. But Lenz had drawn attention to a fundamental methodological weakness of the twin paradigm: the idea that genome and environment are two complexes effective in complete independence of each other stood on feet of clay. The critique of Johann Gottschick[353] was directed at precisely this point. In 1937 he launched a frontal attack, in the *Archiv für Rassen- und Gesellschaftsbiologie*, of all publications:

> When one keeps in mind that either genetic differences and environmental differences alone can bring about differences in attributes, [...]and that differences in attributes in human populations are most probably due to the combination of both different genetic effects and environmental effects [...], one arrives at the conviction that *findings about twins alone can never give any evidence for the heritability of an attribute.* [...] The aspiration to support heritability research must not lead to the application of unsuitable means. And the twin method is such an unsuitable means of genetic research.[354]

With his articles Gottschick triggered an avalanche of controversy,[355] in which Verschuer also found himself compelled to intervene. After the insubordinate junior scientist had failed to comply with his demand that the objectionable comments be recanted publicly, Verschuer published a rejoinder in the *Archiv für Rassen- und Gesellschaftsbiologie*. Here he had few options other than to scold the critic from the lofty tower of established normal science and threaten him subliminally:

> At many scientific institutes and clinics in nearly all cultivated countries the twin method is being worked on successfully. In view of this fact it can be presumed that the twin method meets with general trust, and that doubts as to its cogency are regarded as largely removed. A critique of the method must therefore be designated as a particularly responsible undertaking today, which requires setting a profound intellectual foundation and thorough factual justification if it is to attract attention rather than having detrimental effects for the critic.

Verschuer continued by discussing only one single, rather marginal point of Gottschick's critique of the methodology of the zygosity diagnosis, and was otherwise content with a blanket disqualification of the objections, which he claimed proceeded "from non-objective presumptions."[356] Gottschick was not intimidated, however, and reaffirmed his critique in 1939, explicitly questioning the premise that environmental influences must be regarded as equal for identical and fraternal twins, because "the genetic character usually also conditions the environmental character, that is, there exists a 'coupling' between genetic and environmental factors."[357]

---

[353] Nothing more is known about Gottschick's biography and his position in the Third Reich. In 1963 he published a psychiatry of war captivity entitled "Psychiatrie der Kriegsgefangenschaft. Dargestellt auf Grund von Beobachtungen in den USA an deutschen Kriegsgefangenen aus dem letzten Weltkrieg."

[354] Gottschick, Hauptfragen, pp. 388, 389 f. (original emphases). In 1936, the former staff member Wilhelm E. Mühlmann expressed himself similarly from a concealed position. Cf. Mühlmann, Rassen- und Völkerkunde, p. 33.

[355] Cf. Idelberger, Zwillingsforschung; Riemann, Erwiderung.

[356] Verschuer, Frage der Zwillingsdiagnose, quotes: pp. 69 f., 74.

[357] Gottschick, Zwillingsbefunde, p. 107.

While Verschuer was able to force a critic into the margins through the weight of his scientific authority, ultimately it was not possible for him to eliminate the reservations against the twin method. In the mid-1930s, Diehl and Verschuer had a difficult battle to fight on the field of tuberculosis research – the textbook example for the application of the twin method of genetic pathology was in danger.[358] Fundamental doubts on the correctness of the results obtained with the twin method and on their eugenic conclusions were raised, for instance, in 1935 by Bruno Lange (1885–1942),[359] who worked in the Epidemic Department of the Robert Koch Institute in Berlin. Lange allowed that the first reports from Diehl and Verschuer had made a great impression upon him, but after thorough consideration, he had "considerable misgivings about the cogency of Diehl's and von Verschuer's statistics." The methodological critique was focused on two points: First, Lange correctly indicated that the establishment of concordance or discordance in tuberculosis did not take place on the basis of an objectively measurable attribute, but was based "on judgments," which were "more or less subjective." Because of the "many forms of the clinical picture," a judgment about analogous or heterogeneous courses the disease can only be made on the basis of specialized medical observation and series of x-rays beginning in early childhood. Second, Lange, also completely correctly, designated as extremely problematic the assumption of equal environmental conditions upon which the twin method was based, especially in the case of tuberculosis, where the point of time when exposure and infection took place is of the essence. From the studies by Diehl and Verschuer, according to Lange, thus "a certain judgement about these extremely complex environmental conditions cannot be obtained." By no means were the findings sufficient "to justify the necessity of expanding our tuberculosis care with eugenic measures."[360] As mentioned above, in 1934/35 such doubts had induced Karl Diehl to a new start methodologically and conceptually, and to the relativization of eugenic demands. The continued critique actually had effects on the practice of tuberculosis control. The critics of a genetic biology, eugenic view of tuberculosis around Franz Redeker had dominated the Reich Tuberculosis Committee since 1933.[361] Otto Walter (1891–1964),[362] founder of the Tuberculosis Relief Organization of the NSV (*Tuberkulosehilfswerk*), Director of the Office for National Health in the Main

[358] On the following: Kelting, Tuberkuloseproblem, pp. 15–23.

[359] Lange, Bedeutung, pp. 249–253. Diehl and Verschuer responded to Lange's criticism in the second volume of their *Zwillingstuberkulose*. Diehl/Verschuer, Erbeinfluß, pp. 160–163.

[360] Ibid., pp. 250, 251, 253. Cf. also Kattentidt, Ceterum censeo, p. 256; Rößle, Verhalten, pp. 6–9. Danger for twin tuberculosis research also loomed from the theory of constitution. More and more the opinion took hold that there was a connection between leptosomatic build and disposition to tuberculosis. The supposed genetic disposition for tuberculosis thus threatened to be absorbed by constitution – which was in part determined by genetics, a typical example for the tendency of twin research to jump to conclusions in tracing a complex biological phenomenon back to a gene. Cf. Kelting, Tuberkuloseproblem, pp. 10–15.

[361] Cf. ibid., pp. 51–53.

[362] For a biography: Klee, Personenlexikon, p. 654.

Office for National Welfare in the Reich direction of the NSDAP, and later Managing Chairman of the Reich Tuberculosis Council, emphasized back in 1935 that while tuberculosis may still be conceived of as a genetic disease in large circles, for the party, the "scientific knowledge" that tuberculosis was an infectious disease was held to be "a matter of course."[363] The internist Julius Kayser-Petersen (1886–1954),[364] responsible for tuberculosis in the Department for Care of Health and the Nation (*Abteilung für Gesundheitswesen und Volkspflege*) in the Reich Ministry of the Interior, was gratified to establish in 1937 that gradually word had gotten round that tuberculosis is not a genetic disease; however, he continued, "a very special decree of a high party office" had been required "to have tuberculosis removed from the list of genetic diseases on a NSV questionnaire."[365]

Once the application of the twin method in genetic *pathology* was no longer undisputed from the mid-1930s, the field of genetic *psychology* advanced to become the new dynamo of twin research at the KWI-A. In the early 1940s, however, critique arose about genetic psychology twin research as well, and again it was Fritz Lenz who blithely sought public confrontation, without any consideration for institute politics, entangling Gottschaldt in a heated methodological debate.[366] Yet from the outset Gottschaldt had been cautious as to the cogency of twin studies:

> If the science of genetic character and genetic psychology are applied too directly to the recording of individual characteristics of the mental-emotional personality, it is all too easy to lose sight of the entire picture of the development and structure of personality, and all too frequently rash consideration of the mutual concordances and disconcordances in twins frequently leads to all too premature conclusions about the limits of educability, placticity, modifiability, etc.[367]

Quite early, Gottschaldt had recognized that heredity and environment were variables *not* independent of each other. With reference to the dissertation by August Ohm, he stressed "that even an environmental situation as coercive and impressive" as imprisonment – in Gottschaldt's words, "akin to a radical life experiment" – did not lead to a leveling out of individual personality profiles. Imprisonment, Gottschaldt deduced, is "experienced subjectively, and there is no such thing as an objective environment."[368] Gottschaldt's view of the nature–nurture problem can be characterized like Michael Stadler did: as *interactionistic* or *dialectic*.[369]

---

[363] Walter, Tuberkulosebekämpfung, p. 416.

[364] For a biography: Klee, Personenlexikon, pp. 301 f.

[365] Kayser-Petersen, Praktischer Arzt, p. 13.

[366] Lenz, Problematik der psychologischen Erbforschung; Gottschaldt, Problematik; Lenz, Problematik der psychologischen Erbforschung und der Lehre vom Schichtenbau der Seele; Gottschaldt, Bemerkung.

[367] Gottschaldt, Vererbung, p. 4.

[368] Gottschaldt, Umwelterscheinungen, pp. 433 f.

[369] Stadler, Schicksal, p. 156 (original emphases).

Methodological caution was also exerted by Gottschaldt's assistant Kurt Wilde, who responded to Lenz's critique in 1941 by presenting a comprehensive methodological analysis, in which he discussed problems like the correct selection of twins (minimum number, age, sex, creation of a series without selection bias, possibilities of mathematical standardization in a nonhomogeneous study sample), the differentiated scaling of results from psychotechnical tests and the calculation of measurement errors. In the decisive point – the nature-nurture problem – Wilde subscribed to Lenz's view that "genetic and environmental forces are not added, but combined."[370]

In summary it must be stated that twin research was able to defend its dominant position in human genetic research into the period of World War II, but came under increasing pressure. With regard to the refined evaluation procedure proposed by Lenz, Hans Luxenburger confirmed in his 1940 *Handbuch der Erbbiologie des Menschen* that the application possibilities of the twin method were "quite limited."[371] Against this background Eugen Fischer, as will be demonstrated, started in 1938 to seriously consider a paradigm shift.

3.4 The Kaiser Wilhelm Institute for Anthropology, Human Heredity and Eugenics and National Socialist Genetic Health and Race Policy

From the beginning, the KWI-A made a significant contribution to the practical implementation of National Socialist genetic health and race policy. Fischer and his staff influenced the political decision makers as members of expert commissions. In lectures and popular science publications, they made propaganda in the German public for the National Socialist genetic health and race policy and gave it a legitimate scientific basis. With their expert opinions they took over a key role in the recording and selection of the "genetically ill," "Rhineland bastards," Jews, Sinti, and Roma. They were also substantially involved in the training and continuing education of the specialized personnel who executed the genetic health and race policy.

3.4.1 Training and Continuing Education, Lecturing Activity, Preparation of Expert Opinions and Certificates

Popular science lectures and publications had occupied a major role in the work of the institute from the outset. From 1928 on paternity certificates were issued for the courts, from 1930 the institute held continuing education courses for medical

[370] Wilde, Meß- und Auswertungsmethoden, quote: p. 51.

[371] Luxenburger, Zwillingsforschung, p. 233.

officers, and since the beginning of the 1930s the institute's leading scientists were increasingly integrated into political decision making through their scientific consulting function. Nevertheless, the year 1933 marked a break: The practical activity advanced not only into a new *quantitative* dimension, which led to a detraction from its scientific work, but it also made a *qualitative* leap, because from this point on coercive measures were supported, some of them well outside any legal framework.

The continuing education courses for medical officers were continued seamlessly. In connection with the State Medical Academy (*Staatsmedizinische Akademie*) and on behalf of the Prussian, later the Reich Ministry of the Interior, between 1933 and 1935 eight 1-week continuing education courses for medical officers about genetic biology and race hygiene were held at the KWI-A, for between 50 and 185 participants. In total Fischer estimated that around 1,100 physicians had been trained.[372] In retrospect Verschuer asserted that "by the time the GzVeN took effect on January 1, 1934, nearly all public health officers in Germany had been to our school."[373] In subsequent years, too, continuing education courses for public health officers and heads of medical departments took place at the KWI-A, and also monthly courses for physicians of the SS, arranged by the Reich Ministry of the Interior.[374]

In collaboration with the State Medical Academy, in 1933/34 the KWI-A also carried out a semi-weekly continuing education course for judges, biology teachers, and pastors.[375] Of particular importance here was the "course for jurists," which was held from February 12 to 14 in the auditorium of Harnack Haus. The Prussian Minister of Justice delegated the chairmen of the Hereditary Health Courts and Appellate Hereditary Health Courts to attend this course. In so doing he obligated the course participants "to make records of what they hear and see and impart the essential results to their deputies or even a larger circle of interested judicial officers."[376] The course agenda listed the following speakers and topics:

Monday, 12/2/1934

- Senior Secretary and Councillor Dr. Conti, Race Hygiene in the Volkish State
- Professor Dr. Eugen Fischer, Genetic Research and its Application to Human Beings
- Professor Dr. Lenz, Causes of Degeneration

[372] Tätigkeitsbericht 1933/35, MPG Archive, Dept. I, Rep. 1 A, No. 2404, pp. 49–49h, here: pp. 49–49a. On the sequence of these course dates: Lösch, Rasse, p. 319.

[373] Verschuer, Erbe – Umwelt – Führung, "Freier Forscher in Dahlem (1927–1935)" section, p. 22, MPG Archive, Dept. III, Rep. 86 A, No. 3–1.

[374] Tätigkeitsbericht 1935/36, 22/6/1936, MPG Archive, Dept. I, Rep. 1 A, No. 2399, pp. 74–76 v, here: p. 74; Jahresbericht 1935/37, 10/4/1937, MPG Archive, Dept. I, Rep. 3, No. 13.

[375] Jahresbericht 1934/35, 14/4/1935, MPG Archive, Dept. I, Rep. 3, No. 11. The activity report of 1933/35, MPG Archive, Dept. I, Rep. 1 A, No. 2400, pp. 49–49h speaks of a "course for jurists," a "course for biologists" and a "course for pastors."

[376] Preußischer Justizminister to Oberlandesgerichtspräsidenten and Landgerichtspräsidenten, 3/2/1934, Landesarchiv Greifswald, Rep. 76 G, No. 536, pp. 16–16 v.

- Medical Councilor Dr. med. Klein, Blood Groups
- Professor Dr. Baron von Verschuer, Paternity Certification

Tuesday, 13/2/1934

- Professor Dr. Fischer, Genetic Research and its Application to Human Beings
- Professor Dr. von Verschuer, Practical Genetic Prognostification for Sterilization
- Senior Physician Dr. Pohlisch (Psychiatric Clinic of the Charité), Exhibition of Genetically Conditioned Mentally Ill Patients

Wednesday, 14/2/1934

- Professor Dr. (Martin) Staemmler,[377] Population Movement
- Professor Dr. Lenz, Race Hygiene and Population Policy
- Medical Director Dr. med. Neupert, Sterilization Technology
- Undersecretary Dr. med. Gütt, The Law on the Prevention of Genetically Diseased Offspring[378]

The scientists of the KWI-A were further involved as lecturers in numerous other courses, including those organized by the Reich Committee for the Continuing Education of Physicians (*Reichsausschuß für das ärztliche Fortbildungswesen*), by the Central Institute for Training and Instruction (*Zentralinstitut für Erziehung und Unterricht*) and by the State Academy of the Public Health Service.[379] Abel, Bühler and Geyer also taught as instructors at the "German Academy for Politics" (*Deutsche Hochschule for Politik*), a pet organization of Reich Propaganda Minister Joseph Goebbels (1897–1945) – where Abel took over direction of the Department for the Fostering of Race in 1934 – and Fischer also appeared as a speaker occasionally. Fischer and his staff held a number of lectures and lecture series, for instance at the State Medical Academy (*Staatsmedizinische Akademie*), the Academy for Administration (*Verwaltungsakademie*), at the National Socialist Teachers' League, at the NSDÄB, and at the Hanns-Kerrl-Gemeinschaftslager für Justizreferendare, a camp for judiciary trainees in Jüterbog.[380] These activities were supplemented by numerous tours of the institute for physicians, biologists, teachers, "stewards of might," caregivers, and the like.[381]

---

[377] For a biography: Labisch/Tennstedt, Weg, vol. 2, p. 502; Klee, Medizin; idem., Personenlexikon, p. 594.

[378] Landesarchiv Greifswald, Rep. 76 G, No. 536, p. 17. The participants had to submit a declaration regarding their "Aryan ancestry." Ibid., p. 20 v.

[379] Jahresbericht 1933/34, 21/3/1934, MPG Archive, Dept. I, Rep. 3, No. 10.

[380] Ibid.; Tätigkeitsbericht 1933/35, MPG Archive, Dept. I, Rep. 1 A, No. 2404, pp. 49–49h (here Fischer, Lenz, Verschuer, Abel, Bühler, Werner, Steiner and Lehmann are listed as lecturers); Tätigkeitsbericht 1935/36, 22/6/1936, MPG Archive, Dept. I, Rep. 1 A, No. 2399, pp. 74–76 v (here Fischer, Lenz, Abel, Bühler, Geyer, Lehmann, Steiner and Werner are mentioned as lecturers); Jahresbericht 1938/39, 16/4/1939, MPG Archive, Dept. I, Rep. 3, No. 16. Cf. also Lösch, Rasse, p. 320.

[381] Tätigkeitsbericht 1933/35, MPG Archive, Dept. I, Rep. 1 A, No. 2404, pp. 49–49h. On December 13, 1933 a group of diplomats toured the institute. Cf. the list of participants, MPG Archive, Dept. I, Rep. 1 A, No. 2399, p. 24.

On November 1, 1934, on the basis of an agreement with the Reich Ministry of the Interior, the KWI-A welcomed 20 physicians from the SS for a 9-month training course. The participants of the course were selected by the Race Policy Office of the NSDAP and by the RuSHA, whereby Fischer was granted a voice in the decision. Women, nonphysicians and nonmembers of the party were excluded from participation, and the majority of the hand-picked participants ended up coming from the SS medical corps. The trainees were housed in a "community camp" in the Stubenrauch Hospital in Berlin-Lichterfelde. The Reich Committee for National Health Service in the Reich Ministry of the Interior bore the costs. The trainees were to be "trained in genetic theory, race science and race hygiene" in Dahlem. Fischer, Lenz, and Verschuer "instructed them daily in lectures, seminars and colloquia, in addition they participated in all scientific investigations at the institute." The course served the dual objective of educating junior scientists in the material while indoctrinating them ideologically. The "political education"[382] of the trainees was the responsibility of Falk Ruttke. Graduates of the course, which ended on July 1, 1935, "for the most part [entered] the newly created health offices."[383] In 1936 another such course took place for 18 physicians from the SS.[384]

As Engelhard Bühler later recalled, the SS trainees brought an "atmosphere of misgivings"[385] into the KWI-A. On the other hand, the courses led to a progressive interconnection between the SS and the KWI-A. First, they affirmed the presence of the SS at the institute, even in the long term: of the 20 trainees in the first course, two became members of the scientific staff at the KWI-A – Horst Geyer, as already mentioned, entered the institute as an assistant immediately following the course, Heinrich Schade (1907–1989)[386] first went to Frankfurt as Verschuer's assistant and then returned with him to the KWI-A in 1942. Second, the institute's influence on the regime's genetic health and race policy was propagated through the physicians of the SS who were trained there, as many of the trainees rose to occupy important positions. Thus, for instance, Helmut Poppendick (* 1902),[387] a participant of the first course. After absolving the course, the internist, previously senior physician at Virchow Hospital in Berlin, became first Arthur Gütt's adjutant at the Reich Ministry of the Interior and Chief of Staff in the SS Office for Population Policy and Fostering of Genetic Health, which was absorbed by RuSHA in 1937. In 1943, Poppendick was promoted chief of the personal staff of the Reich Physician SS Ernst Grawitz (1899–1945) and in this function was involved in a number of medical experiments on concentration camp prisoners. Another graduate was Georg Renno

[382] Tätigkeitsbericht 1933/35, MPG Archive, Dept. I, Rep. 1 A, No. 2404, pp. 49–49h, quotes: p. 49a.

[383] Halbjahresbericht 1935, October 1935, MPG Archive, Dept. I, Rep. 3, No. 12.

[384] Jahresbericht 1935/37, 10/4/1937, MPG Archive, Dept. I, Rep. 3, No. 13.

[385] Quoted in Müller-Hill, Tödliche Wissenschaft, p. 149.

[386] For a biography: Lösch, Rasse, p. 573; Klee, Personenlexikon, p. 522.

[387] For a biography: Ebbinghaus/Dörner (ed.), Vernichten, p. 638; Klee, Personenlexikon, pp. 469f.

(1907–1997),[388] who worked as an evaluator in the framework of "Aktion T4" from 1940 and turned on the gas at the "euthanasia" institute in Hartheim.

Fischer and his staff served National Socialist genetic health and race policy not only as speakers and teachers, but also as experts and evaluators. Fischer, Verschuer, and Lenz – the last as member of the Board of Experts for Population and Race Policy (*Sachverständigenbeirat für Bevölkerungs- und Rassenpolitik*) – were often called in by the Reich Ministry of the Interior for consultations; the Reich Health Office also availed itself of the three department directors' expertise. What is more, Fischer was a medical expert and medical assessor on the Hereditary Health Appellate Court in Berlin, and Verschuer on the Hereditary Health Court in Berlin-Charlottenburg. In 1935/36 Fischer also became a senator of the Army Medical Corps.

In the end the consultation and expertise activities took on a substantial volume. In the 1934/35 annual report Fischer explained:

> Quite numerous are the expert opinions demanded by all, especially the race opinions for the Reich Office for Genealogical Research of the Reich Ministry of the Interior, expert opinions about the academic expansion of our subjects and junior scholars, by various state offices, and finally consultations on prehistorical excavations at numerous locations. The Film Censorship Office lays claim to our services for expert opinions on certain films, the Reichswehr for anthropological advice in hiring, the Superior Court for determinations of paternity.[389]

For the 1935/36 fiscal year, Fischer gave concrete data about the scope of the expertise activities:

> The number of expert opinions demanded rose constantly: For the Reich Office for Genealogical Research over 60 expert opinions about racial purity were provided (Fischer, Abel); for Superior Courts, 28 opinions about paternity (Fischer, Bühler); for Hereditary Health Courts and Appelate Courts, around 20 on sterilization cases (Fischer, Lenz); for the Emergency Association, 41 on research grants (Fischer).[390]

In all reports Fischer emphasized his institute's achievements in the theoretical underpinning and practical implementation of genetic health and race policy, and he never missed a chance to mention that the scientific work suffered under the multifarious new obligations. "Allow me to say honestly that we all sigh sometimes at having to [take] so much time and performing difficult and responsible work that departs from scientific research," he complained in the annual report of 1934/35, for instance. But, so he continued, "it is necessary these days and we will never disappoint in performing it."[391] In 1935, he explained to the Board that the institute

[388] For a biography: Klee, Personenlexikon, pp. 491 f. Only the names of the institutions of the course participants are listed. In some cases clear identification is thus impossible with the present state of knowledge.

[389] Jahresbericht 1934/35, 14/4/1935, MPG Archive, Dept. I, Rep. 3, No. 11. Similar: Tätigkeitsbericht 1933/35, MPG Archive, Dept. I, Rep. 1 A, No. 2404, pp. 49–49h, here: p. 49b.

[390] Tätigkeitsbericht 1935/36, 22/6/1936, MPG Archive, Dept. I, Rep. 1 A, No. 2399, pp. 74–76 v, quote: pp. 74–74 v.

[391] Jahresbericht 1934/35, 14/4/1935, MPG Archive, Dept. I, Rep. 3, No. 11. Cf. also Auszug aus dem Tätigkeitsbericht, 13/5/1935, MPG Archive, Dept. I, Rep. 1 A, No. 2399, pp. 52 f.

had been force to restrict or even discontinue some scientific work, "as in these times of rebuilding our volkish state, this task, which could be performed by no other institution in this way, is of the highest priority."[392] Of course it must be taken into consideration that tactical calculations always played a role in these reports as well – Fischer emphasized to Gütt and Groß how indispensable his institute was in the realization of genetic health and race policy and how willingly it made itself available for this task. The allusion to neglected research projects was also suited to justifying an increase in funding. Nevertheless, the sheer volume of obligations in the areas of lecturing, instruction, expert opinions, and evaluations makes it seem quite likely that research work actually did suffer. And it is altogether believable when Fischer writes that he shared the burden of restricting research out of conviction. The political commitment of the KWI-A cannot be explained by political pressure alone.

3.4.2 "The Sword of Our Science": Eugen Fischer and Otmar von Verschuer on the International Stage

Since National Socialist genetic health and race policy met with rejection in many states, under the aspects of foreign policy, heightened attention was paid to the participation of German population scientists, physicians, geneticists, anthropologists, and race hygienists in international congresses, as such events could – with skilled direction behind the wings – be used to disperse skepticism and critique of the sterilization legislation and Jewish policy abroad. Indeed, the appearance of the German delegations, as Sheila F. Weiss portrayed convincingly, was painstakingly set in scene by the Foreign Office, the Reich Ministry of Education, the Reich Ministry for Propaganda and the Enlightenment of the Nation, along with the German Congress Headquarters and party offices such as the Race Policy Office of the NSDAP and the Nazi organization of instructors (*NS-Dozentenschaft*), in order to exploit science as a foreign policy resource.[393] For the scientists this meant that, if they wanted to visit international congresses, it was up to the benevolence of the responsible science policymakers whether or not they were included in a delegation, and that they had to submit to the delegation's discipline when abroad. They were expected to work as propagandists of National Socialist genetic health and race policy. Regardless of how closely they were bound to state and party, most of the scientists had no problem taking on this role. In his speech on the occasion of the *Tag der nationalen Erhebung* on January 30, 1936 at the University of Frankfurt, Verschuer's martial vocabulary swore an oath to the alliance of science and politics:

---

[392] Tätigkeitsbericht 1933/1935, MPG Archive, Dept. I, Rep. 1 A, No. 2404, pp. 49–49h, quote: p. 49.

[393] Weiss, Sword; pp. 3–6 with interesting indications to the guidelines of the German Congress Headquarters for Delegation Leaders (*Deutsche Kongreß-Zentrale für Delegationsführer*).

> The battle of opinions between the nations is an especially vehement one as regards questions of genetic biology and race hygiene; there are many endeavors under way to attack through the scientific route the fostering of genes and race in National Socialist Germany – the sword of our science must therefore be sharpened and well handled![394]

In July 1934, Eugen Fischer traveled to London as the "leader" of the German delegation to the International Congress for Anthropology and Ethnology.[395] His first major appearance on the international stage after the National Socialists had taken power thus took place at a point in time when his relationship to state and party had yet to be clarified. Although he had already come to an agreement with the state health policy leadership, the conflict with Walter Darré and the RuSHA had not yet been resolved. There were thus major reservations in party circles about sending Fischer abroad in such a responsible position. However, due to foreign policy considerations it did not seem opportune for the Foreign Office to rescind Fischer's appointment as delegation leader. As already discussed, the problem was resolved by having Walter Groß accompany Fischer – and keep him under surveillance. Fischer, who also functioned as the vice president of the congress, fulfilled the expectations of the National Socialist regime. He unprotestingly accepted that the Race Policy Office forbid that the anthropologist Karl Saller – a former staff member of the KWI-A, after all – attend the conference because of his unorthodox views on the "race question."[396] Attempts by the American anthropologist Franz Boas and his Czech colleague Ignaz Zollschan (1877–1948) to pass an "anti-racist position paper" failed due to the "disinterest" of the participants and the "opposition of the organizers of the congress."[397] Fischer did his part to ensure that no critique was voiced: According to his own claims, he intervened with the congress directors to have a "tactless" remark directed at National Socialist Germany stricken from the official conference record.[398] Fischer had passed the test administered by the National Socialists with flying colors, which must have played a significant role in the fact that the political campaign directed against him died down.

Also in 1934 was the Congress of the International Federation of Eugenic Organizations in Zurich. The German race hygienists and their foreign sympathizers had a much easier time here from the outset. In 1932 Ernst Rüdin had been elected president of the IFEO, after Eugen Fischer had refused to run due to his excessive workload.[399] Rüdin also led the 10-man German delegation to Zurich, whose members had been painstakingly selected in advance. What is

[394] Verschuer, Rassenhygiene als Wissenschaft und Staatsaufgabe, pp. 8 f.

[395] Weiss, Sword, pp. 6 f.; Lösch, Rasse, p. 269.

[396] Weiss, Sword, p. 7.

[397] Kühl, Internationale, p. 147. Cf. Barkan, Retreat, pp. 308 f.

[398] Weiss, Sword, p. 7.

[399] Weber, Ernst Rüdin, p. 169.

notable is that Otmar von Verschuer also belonged to this hand-picked circle.[400] Although individual critical voices were audible from the ranks of the French and Dutch delegations in Zurich, the assembly passed a resolution that can be conceived of as a "scientific blank check"[401] for National Socialist genetic health and race policy. In the subsequent period a cartel of German race hygienists, "orthodox eugenicists of the first generation" like Harry Hamilton Laughlin (1880–1943), Davenport, Lundborg, and Mjöen, and "younger sympathizers of National Socialist Germany" such as the Englishman George Pitt-Rivers (1890–1966) and the Swede Torsten Sjögren (1896–1974) succeeded in "bringing [the IFEO] largely under their control."[402] Rüdin took pains to integrate into the IFEO not only Fischer and Ploetz, but also younger German scholars: in 1935/36 Verschuer, Kurt Pohlisch as representative of the Genetic Biology Institute at the University of Bonn and Ernst Rodenwaldt from the Hygienic Institute of the University of Heidelberg joined the organization.[403]

With their first appearances on the international stage, Fischer and Verschuer had commended themselves for additional tasks as ambassadors of German science and politics. They celebrated their greatest propaganda success with the International Demographics Conference in Berlin from August 26 to September 1, 1935.[404] Back in 1931, the International Union for the Scientific Investigation of Population Problems had agreed to hold its second international congress in the capital of the German Reich. After the National Socialist takeover, doubts arose, particularly among the American demographers, who feared that the National Socialist regime would use the congress as a political demonstration, but the board of directors of the IUSIPP staunchly held fast to Berlin as the conference location – not least because of Eugen Fischer's scientific reputation. The US American demographer and eugenicist Raymond Pearl (1895–1940), for instance, proved optimistic, writing the IUSIPP president, Charles Close that, as president of the conference, such an "established and far-sighted scientist"[405] as Fischer offered the guarantee for upholding the scientific standard and political neutrality. However, this expectation was not fulfilled. As chairman of the organizational committee, Fischer excluded the Board of Directors of the IUSIPP from the congress planning, which was financed in large part by the National Socialist government and placed under the "honorary presidency" of Reich Minister of the Interior Wilhelm Frick. As president

[400] Other members of the delegation were Alfred Ploetz, Ernst Rodenwaldt, Lothar Gottlieb Tirala (1886–1974), Heinz Kürten (1891–1966), Lothar Loeffler, Walter Groß, Falk Ruttke and Karl Astel. On the other hand, Rüdin prevented a long-planned lecture by the Dresden hygienist Rainer Fetscher (1895–1945), who was considered politically unreliable. Kühl, Internationale, p. 128. Cf. Weindling, Health, pp. 504 f.

[401] Kühl, Internationale, p. 129.

[402] Ibid., pp. 137 f.

[403] Ibid., p. 138.

[404] On the following, ibid., pp. 131–137; Ferdinand, Bevölkerungswissenschaft, pp. 74–83; Weiss, Sword, pp. 8–10; Lösch, Rasse, pp. 268 f.

[405] Pearl to Close, 17/12/1934, quoted in Kühl, Internationale, p. 131.

of the congress, Fischer was bound to political neutrality, but by interpreting the National Socialist population, genetic health and social policy as the practical implementation of genetic biology and race hygiene research, and portraying a scientifically grounded race policy as a constitutive element of population policy, it required no effort for him to have his inaugural address – just a few weeks before the "Nuremberg Laws" were decreed – culminate in a eulogy to National Socialist race policy and its creator Adolf Hitler:

> Many nations have burning race issues among their population problems. Only the geneticist can help the demographic statistician in resolving these, with advice and assistance. It is the right and the holy duty of each and every nation to keep its national stock as racially pure as it was in the days of our great-grandfathers [...]. And we are full of proud joy to be able to enjoy the uplifting feeling [...] that the government of our Reich, and above all our Führer and Reich Chancellor Adolf Hitler, clearly recognized this deepest and gravest meaning of demographics and has the will to draw the consequences. Today, therefore, at the start of our work we may, with thankful heart, think of this man whose strong hand has the will and, God willing, the strength, to avert from the German volk the fate of populations which led bygone cultures and nations to their demise. And I hope and wish the same for those in control of the states and governments of all other nations and peoples. In this wish for all we greet deferentially and are mindful, when we assemble for work here on German soil and in the capital of the Reich, of the Führer and Reich Chancellor of the German nation, and I ask you to greet him with me: The Führer and Reich Chancellor *Adolf Hitler* Sieg Heil![406]

The "International Congress for Demographics," to which 500 scientist from 38 countries, appeared in Berlin, 180 of them from abroad, was "for a complete success the National Socialists."[407] Even among the foreign specialists there emerged broad acceptance of the National Socialist genetic health policy. Only one single participant, Jean Dalsace (1893–1970), a French eugenicist linked with the Communists, openly condemned the GzVeN as reversion to barbarism, but his critique was rejected decisively, even by foreign participants – Alfred Mjöen and Cora Hodson, the secretary of the IFEO. Verschuer's employee Heinrich Schade reported in *Erbarzt* that the congress had demonstrated that the "most renowned scientists of all countries [had recognized] the path taken by the German National Socialist government to be expedient and promising."[408] This statement certainly contained a grain of truth, even though American, British, and Dutch demographers criticized the politicization of the Berlin congress after the fact. In public, however, the impression dominated that the scientific community of demographers assembled in Berlin had given the genetic health and race policy of the "new Germany" a positive reception. The National Socialist propaganda machine swung into full gear to exploit this circumstance.

[406] Fischer, Introductory Lecture, pp. 42 f. (original emphasis).

[407] Kühl, Internationale, p. 135. Verschuer was also represented, with a lecture about "Genetic Biology as the Foundation of Population Policy."

[408] Schade, Internationaler Kongreß für Bevölkerungswissenschaften, p. 140. Cf. also idem., Stimmen (where the former guest scholar Fritz Schrijver is also cited).

In the second half of the 1930s, too, the network of German scientists around Rüdin, Fischer and Verschuer were eager to silence opponents of National Socialist genetic health and race policy at international conferences for eugenics, demographics, anthropology, and ethnology as well as genetics. This task was easiest within the international eugenics movement. When fears arose during preparations for the international conference of the IFEO, which took place in Scheveningen in July 1936, that the Dutch eugenicists Gerrit Pieter Frets and Jacob Sander could use this opportunity to attack Nazi Jewish policy publicly, Rüdin undertook to put together a "quantitatively and qualitatively outstanding delegation of German scholars"[409] – to which, of course, Verschuer belonged. In contrast, Rüdin obstructed an appearance by Hermann Muckermann, who had been asked by the hosts in the Netherlands to advocate the standpoint of Catholic eugenics.[410] The German delegation succeeded in preventing the conference from adopting the sharp criticism of the National Socialist mass sterilizations voiced by the Dutch eugenicists around Frets – with a view to sterilization legislation in their own countries, the eugenicists from the USA and Scandinavia were not willing to resolve on a protest against the GzVeN. In the selection of a successor for Rüdin as president of the IFEO the German delegation also negotiated successfully – according to Verschuer, one could be "satisfied"[411] with the election of Torsten Sjögren, Director of the Psychiatric Institute in Lillhagen near Göteborg, for Sjögren, a pupil of Lundborg, sympathized with National Socialist Germany and was a passionate advocate for organizing a major race hygiene congress in Berlin under the auspices of the IFEO. After considerable to and fro behind the wings, the plan took shape to hold the Fourth International Conference for Eugenics and Race Hygiene in Vienna in August 1940,[412] but the beginning of the war prevented the plan's realization. The attempt by Sjögren and Rüdin to make the IFEO a "marionette" of the National Socialist regime led to a "splintering" of the international eugenic movement in the second half of the 1930s. The most important group to secede was the International Human Heredity Committee around Frets, Gunnar Dahlberg (1893–1956), and Otto Louis Mohr (1886–1967), but since Verschuer succeeded in infiltrating this grouping as one of its vice presidents, it was paralyzed politically.[413]

On the other hand, stubborn resistance within the IUSIPP took shape after the Second International Demographics Conference in 1935. During preparations for the third conference, which took place in 1937 as part of the Paris World Exposition, open confrontation seemed unavoidable. The French group *Races et Racisme*, an initiative against National Socialist race policy launched by French scientists

[409] Kühl, Internationale, p. 139.

[410] Weber, Ernst Rüdin, p. 232.

[411] Verschuer to Fischer, 24/7/1936, MPG Archive, Dept. III, Rep. 86 A (Münster), No. 9. Verschuer added: "Unfortunately neither side of the discussion was free of political admixtures and some unwise words were spoken."

[412] Kühl, Internationale, pp. 141–143. Members of the already formed preparatory committee were Rüdin, Astel, Burgdörfer, Fischer, Lenz, Linden, Loeffler, Reiter, Ruttke, and Wettstein.

[413] Ibid., p. 144.

around Henri Laugier (1888–1973), Célestin Bouglé (1870–1940), and Paul Rivet (1876–1958), had held a conference about racism during the expo and then excluded itself from participation in the International Demographics Conference. Together with Boas and Zollschan, it constituted a counterweight to the German delegation, which was originally to be headed by Eugen Fischer. Yet, Fischer was aware of the imminent conflict and attempted to put together a top-class scientific delegation for the trip to Paris. In the first instance he endeavored – with success – to win over his friend Verschuer:

> It is imperative that the Paris congress be visited by a few renowned German scholars. It is well and good that Ruttke and Groß and Linden and Reiter or their representatives attend, but in my opinion, the main thing is that they must be accompanied by representatives of scientific researchers [sic] who may be regarded as politicians. Burgdörfer and Zahn and several statisticians have announced lecture. I myself intend (presumably as leader of the delegation) to hold a lecture or two from the area of miscegenation or the emergence of race. I also want to ask Loeffler, Staemmler and Astel. Rüdin will come and bring two or three pupils.[414]

In the end Fischer had to hand over the direction of the delegation to Rüdin due to illness. The preparatory committee around the French Minister of Labor Adolphe Landry (1874–1952) took various measures to prevent a politicization of the conference, whereby the preliminary survey of the lectures, performed for the first time here, was directed toward the opponents of German policy. All the same, the critique could not be silenced. Boas, Zollschan, and the Czech eugenicist Maximilian Beck directed sharp attacks against German race hygiene and race anthropology, to which the German side, extremely irritated, responded with malicious anti-Semitic invective. The scale of critique was great that it could no longer be concealed from the public. Verschuer reported that the critiques had not been able to prevail "in the face of German scientific diligence."[415] Yet it could not be overlooked that the climate on the international level had cooled significantly since the congresses in London, Zurich, Berlin, and Scheveningen.[416]

The German delegation to the Second International Congress for Anthropology and Ethnology in Copenhagen in 1937 was conscious of headwind as well.[417] With his paper about "Race and the Heredity of Mental Characteristics,"[418] Eugen Fischer found himself in the line of fire, with Melville J. Herskovits (1895–1963) leading the charge. Herskovits replied by pointing out to Fischer the study by his thesis

---

[414] Fischer to Verschuer, 6/2/1937, MPG Archive, Dept. III, Rep. 86 A (Münster), No. 9. Verschuer initially had not accepted because he intended to travel to the World Church Conference in Oxford.

[415] Quoted in Kühl, Internationale, p. 152.

[416] The tension between the German race hygienists and the eugenicists from the Netherlands propagated all the way to the executive committee of the IUSIPP, resulting in this organization's almost complete lack of activity after 1937. Upon Rüdin's suggestion, Fischer had been elected as one of the vice presidents in 1937. Ibid., pp. 151 f.

[417] Kaufmann, "Rasse und Kultur," pp. 323 f.

[418] Fischer, Rasse und Vererbung geistiger Eigenschaften.

advisor Franz Boas, about "Culture and Race," which provoked polemic invective against the latter from Fischer: "The views of Mr. Boas are in part quite ingenious, but in the field of heredity Mr. Boas is by no means competent [...]." He saw "not the slightest refutation"[419] of his remarks. This was indeed a clear case of *sacrificium intellectus*, as a great number of research projects at the KWI-A which had picked up on Boas' studies about immigrants in New York had confirmed his findings – including the study by Walter Dornfeldt about Eastern European Jews in Berlin. Fischer resorted to polemic simply because he had no arguments to counter the Boasians' critique.

In the scientific community of geneticists, too, increasing opposition to National Socialist genetic health and race policy took form, especially on the international level. After the Seventh International Congress for Genetics, which originally was to take place in Moscow in 1937, was first postponed indefinitely and then cancelled by the Soviet government, much to the relief of National Socialist makers of science policy, the standing committee of the International Genetics Congress decided to hold the event in Edinburgh from August 22 to 31, 1939. The value attached to this event in National Socialist Germany becomes apparent in a letter by the head of the German delegation, Wettstein, to the German Congress Headquarters:

> The International Genetics Congress is doubtlessly the most important congress to which we send representatives this year. There the general questions of heredity in plants and animals are discussed, it covers the problems of plant cultivation and animal breeding, and, finally, all questions of heredity in humans, genetic hygiene and race hygiene are also discussed. Since much of our state and political life, a large share of our legislation today is built on the application of genetic research, these questions' treatment before an international forum has a very special meaning, even general importance for us.
>
> Having attached such value to the application of genetic research, it is imperative to prove at the congress, in front of the whole world, that we are also leading the way in genetic research. [...] After all, the application to humans has been achieved by us through the legislation of genetic hygiene. These measures are discussed frequently abroad and will also be the subject of conversation at this congress. Therefore it is most urgently imperative that our best experts are on hand for this event.[420]

At this opportunity, Wettstein reported that 130 registration forms had been submitted, of which 40 had been initially selected. Fischer had registered, along with his assistants Bühler, Geyer, and Lemser, Verschuer with his chief physician Ferdinand Claußen (1899–1971)[421] and assistants Heinrich Schade and Josef Mengele (1911–1979). After all, Fischer had been the dominant voice in the preselection process. On March 9, 1939, he had reported to Verschuer that 43 registration forms had been submitted for the area of human genetics alone:

[419] Fischer's discussion paper, ibid., pp. 185 f.

[420] Wettstein to Deutsche Kongreß-Zentrale, 28/4/1939, BArch. Berlin, R 4901/3016.

[421] For a biography: Klee, Personenlexikon, pp. 94 f.

> Recently Wettstein, Lenz and I had a meeting at which we classified the registered colleagues in order, according to the degree of our approval. We classified 21, leaving out the rest entirely. The first 12 are those who are absolutely indispensible, and, of course, you and I belong to this group! Claußen is also among these divine 12!! Schade is in the next group; Mengele is no longer under consideration, however.[422]

The 32-man German group of scientists ultimately included Fischer, Lenz, and Lemser for the KWI-A, plus Hans Nachtsheim, who became director of the Department for Experimental Genetic Pathology shortly thereafter, as well as Verschuer and Schade for the Frankfurt Institute for Genetic Biology and Race Hygiene.[423] After the USA, the German Reich constituted the second largest delegation at the congress attended by around 400 participants.

The looming scientific debate about National Socialist genetic health and race policy did not take place in the end, however, as the congress was already overshadowed by the threat of war breaking out. The delegation from the USSR had to cancel at short notice by command of the Soviet government. The German delegation did arrive, but after the signing of the German–Soviet nonagression pact on August 23, 1939, the Foreign Office ordered the German scientists to depart abruptly. Most of the other continental European scientists returned to their home countries as well,[424] so that the British and American geneticists were more or less alone forthwith. In this situation Hermann J. Muller succeeded in convincing a portion of the most renowned geneticists to sign a *Geneticist's Manifesto* he had written, in which the racism of the National Socialist state was condemned from the perspective of a socialist eugenics. In Germany this manifesto could no longer have any influence. After the beginning of the war, international academic relations increasingly came to a standstill, and National Socialist science policy did its best to create its own empire of science in the occupied and allied states.

[422] Fischer to Verschuer, 9/3/1939, MPG Archive, Dept. III, Rep. 86 A (Münster), No. 9. Verschuer had intervened energetically with the conference director Gunnar Dahlberg, who had urged that the German lecturers and papers be named. Verschuer to Fischer, 13/9/1938, ibid.

[423] Nikolaj V. Timoféeff-Ressovsky represented the KWI for Brain Research, Hans Luxenburger and Friedrich Stumpfl the DFA, while the applications of Rüdin and seven other employees of the DFA were decided in the negative. Cf. the correspondence in BArch. Berlin, R 4901/3016.

[424] Bericht Verschuers an das Wissenschaftsministerium, 4/9/1939, ibid.; Verschuer, Erbe – Umwelt – Führung, "Professor in Frankfurt (1935–1942)" section, pp. 20–25, MPG Archive, Dept. III, Rep. 86 A, No. 3–1. Aside from the congresses mentioned, in September 1935 Verschuer traveled to scientific lectures in Riga, Dorpat and Reval, in October 1936 to Hermannstadt (advanced training course for German physicians in Romania), in May 1937 to Cracow, in November 1937 to Budapest, in April 1938 to Vienna – immediately after the "Anschluss," in June 1939 to London and in February 1940 to Rome. Cf. ibid., pp. 3, 4 f., 8 f., 11, 12, 17–20, 21.

3.4.3 The Kaiser Wilhelm Institute for Anthropology, Human Heredity and Eugenics and the Law on the Prevention of Genetically Deficient Progeny

National Socialist genetic health policy started off with a bang. On July 14, 1933 Hitler's cabinet passed the "Law on the Prevention of Genetically Deficient Progeny," large passages of which were based on the Prussian draft law of 1932, but opened up possibilities for the state to apply force that would have not been politically enforceable under the presidential governments at the end of the Weimar Republic.[425] Eugenic sterilization legislation was already in power in a number of the states of the USA (since 1907), in the Swiss canton Waadt (1928), in Denmark (1929), in the Canadian provinces Alberta (1928) and British Columbia (1933), and in the Mexican state of Vera Cruz (1932)[426] – and most of these laws also permitted the application of force. The National Socialist sterilization program, however, because it legalized open violence beyond the circle of institutionalized psychiatric patients – in contrast to the sterilization legislation of other states – exploded all dimensions known at the time. Between January 1, 1934, when the law was enacted, and May 8, 1945, around 360,000 people on the territory of the German Reich in the borders of 1937 were sterilized in accordance with the GzVeN, most of them by force – nearly 1% of the population between the ages of 16 and 50. If the areas annexed after 1937 are taken into consideration as well, the total number of race hygiene sterilizations in National Socialist Germany must amount to approximately 400,000.[427] In comparison, the number of eugenic sterilizations in the USA between 1907 and 1945 amounted to a total of around 45,000. In other words, in the German Reich nearly ten times as many people were sterilized between 1934 and 1945 than in the USA between 1907 and 1945. Based on the population, more than 30 times more sterilizations took place in the Third Reich than in the USA.[428] Because of the tremendous force, National Socialist genetic health policy had calculated in an increased mortality risk from the outset. Indeed, in the sterilizations according to the GzVeN, around 5,000–6,000 women and about 600 men lost their lives.

---

[425] On the basics: Bock, Zwangssterilisation.

[426] By 1937 the number of US states with sterilization laws rose to 32. Cf. Hoffmann, Rassenhygiene; Laughlin, Sterilization; Popenoe, Rassenhygiene; Blasbalg, Gesetze; Steinwallner, Gesetzgebung; Haller, Eugenics; Reilly, Solution; Trombley, Right; Kline, Race. On Switzerland: Keller, Schädelvermesser, pp. 152 ff.; Müller, Sterilisation, pp. 37 ff.; Huonker, Anstaltseinweisungen; Ritter, "Winde." On Denmark: Hansen, Something Rotten. On Canada: McLaren, Master Race; Dowbiggin, Keeping America Sane. On Mexico: Stepan, Hour, pp. 130–133. After the GzVeN was passed, several states followd in Germany's footsteps: Norway (1934), Finland (1935), Sweden (1935), Estonia (1936), Latvia (1937), and Iceland (1938). The texts of most of the laws are in Binswanger/Zurukzoglu (ed.), Verhütung.

[427] Bock, Zwangssterilisation, pp. 237 f.

[428] Ibid., p. 242.

According to the GzVeN, a person could be sterilized "if in accordance with the experience of medical science it was to be expected with high probability [...] that his offspring [would] suffer from serious physical or mental genetic defects."[429] This rubber formulation was an expression of the genetic diagnostics dilemma. While psychiatric genetic research was certain that the genetic factor played a decisive role in the emergence of the illnesses and disabilities named in the law, clarity about the exact heredity existed only in the rarest of cases. Only in the case of Huntington's chorea did it appear that heredity in accordance with the Mendelian rules could be demonstrated; otherwise scientists had to resort to indirect methods like twin research, family research, or the method of "empirical genetic prognosis" developed by Ernst Rüdin and his staff, which worked with statistical probabilities.[430]

Considered as "genetic illnesses" in the sense of the law were "congenital feeble-mindedness," schizophrenia, "manic-depressive insanity," hereditary epilepsy, "hereditary St. Vitus's dance" (Huntington's chorea), hereditary blindness, hereditary deafness, "serious hereditary physical deformities," and "serious alcoholism." Of particular importance was the group of "feeble-minded," who made up around two thirds of those sterilized. Since the heritability of the various forms and degrees of "feeble-mindedness" – at the time the terms were "idiocy," "imbecility," and "debility" – was difficult to prove in the individual case, the legislator replaced the concept of "heredity" with that of "congenital" feeble-mindedness, which covered all cases in which no exogenous cause could be demonstrated. The burden of proof thus lay on the person to be sterilized. The situation was similar for schizophrenia. While the possibility of heredity was the subject of vehement controversy in the psychiatric research of the day, the legislator determined that, in cases of doubt, the decision was to fall *against* the person to be sterilized – an exogenous cause for schizophrenia had to made plausible by the afflicted themselves. Their objection that the illness had never before occurred in the family was rejected, on occasion with reference to the "concealed heredity of schizophrenia."[431] Similarly, genuine epilepsy was presumed in all cases in which no injuries to the skull or brain, or no advanced stage of syphilis could be proved.[432]

Sterilization proceedings mixed medical and social diagnostics. In addition to "serious alcoholism," it was above all the diagnosis of "congenital feeble-mindedness" that offered the possibility of access to asocial psychopaths, who were accused of "moral feeble-mindedness" – independent of their intelligence. In this manner there was the danger of compulsory sterilization for people who did not finish high school, who had an illegitimate child, who did not have a regular job, who had

[429] GzVeN, § 1, quoted in Kersting/Schmuhl (ed.), Quellen, p. 444.

[430] Bock, Zwangssterilisation, pp. 329f.

[431] In the scientific discourse, the opinions about the heredity of schizophrenia diverged widely at this time. Cf. Gerrens, Medizinisches Ethos, p. 32. On the interesting case of Hans Luxenburger: Kaufmann, Eugenische Utopie.

[432] Even an obviously symptomatic epilepsy was unquestioningly declared to be genuine epilepsy and thus an indication for sterilization.

no proof of fixed residence, or who had come into conflict with the law for petty crimes. This dimension of social diagnostics is tangible in practice. In many sterilization proceedings, widely divergent norms and values, manners and styles of speech collided: judges and doctors were confronted with lifeworlds completely alien to them, those of farmers and farmhands, female factory workers and house maids, tradesmen, and minor employees – and because the educated bourgeoisie, which sat here in to judge "simple people," did not understand their system of values and norms, they did not hesitate to deny that they possessed any "higher powers of reason" or any powers of critique or discrimination. Under these conditions, communication was impossible from the very start. Everything the people to be sterilized, their family members and guardians brought forward followed common sense, rather than adhering to the strict scientific logic of psychiatry, and as a consequence was dismissed by the court as immaterial. This contained no little potential for conflict – in the population, far beyond the circle of the directly affected, the GzVeN met with increasing rejection and distrust. In the power struggle with Arthur Gütt and the Ministerial Department of the Reich Ministry of the Interior, Chief Reich Physician Gerhard Wagner picked up on the population's displeasure and demanded that the criterion of "the preservation of life" be weighted more strongly than the medical genetic prognosis in the sterilization proceedings.[433]

According to the GzVeN, applications for sterilization could be submitted, first, by the person to be sterilized or his/her legal representative. Also authorized to make applications, however, were medical officials, that is, the relevant public health officers or district medical officers and their representatives, and further the directors or directing physicians of institutions of healing and of care, and of hospitals and penal institutions, who were authorized to apply for the sterilization of their wards. According to the implementation ordinance of December 5, 1933, furthermore, all persons involved in treatment, especially licensed physicians, had the duty to register patients who met the conditions for mandatory sterilization with the responsible public health offer. The applications for sterilization were evaluated by the 205 Hereditary Health Courts (*Erbgesundheitsgerichte*, EGG), which were affiliated with the district courts, and the 31 Hereditary Health Appellate Courts (*Erbgesundheitsobergerichte*, EGOG), which were associated with the appellate courts.[434] The EGGs were comprised of a district judge as chairman, a medical official, and another physician, who was supposed to be familiar with genetics and the care of genetic health. The EGGs were authorized to order inquiries, hear witnesses and expert witnesses, exhibit the person to be sterilized and perform a medical examination. Patient/physician confidentiality was not respected in the framework of sterilization proceedings. The decisions of the EEGs could be

[433] Cf. Bock, Zwangssterilisation, pp. 341–344; Schmuhl, Rassenhygiene, Nationalsozialismus, Euthanasie, pp. 164–168.

[434] The numbers refer to the year 1934. Due to restructuring and the founding of new courts, the number of EGGs and EGOG changed many times up to the end of the Third Reich.

appealed, but the EGOG made the final decision on such appeals. Once a conclusive sterilization decision had been made, the operation was supposed to be performed within 2 weeks, by a licensed physician at one of the institutions authorized to do so by the state authorities – if a sterilization candidate refused, then under application of physical force.

The execution of the National Socialist sterilization program was decisively dependent on the involvement of the sciences, in no less than six aspects: First, the National Socialist regime depended on human geneticists, genetic pathologists, genetic psychiatrists, and eugenicists bringing their expertise into the legislation in the context of scientific policy consulting. Second, in view of the prevalent and deep distrust of the German population, it was important for politics that renowned scientists threw their authority into the ring as specialists to give the race hygiene propaganda, which was supposed to prepare the soil for mass sterilizations, the veneer of scientific seriousness. Third, it was important from the foreign policy perspective that high-ranking German scientists offered resistance to the criticism of National Socialist genetic health policy from abroad on the international stage. Fourth, if the sterilization legislation was to be implemented effectively in practice, the medical officers, judges delegated to the EGGs, the physicians in institutions of healing and care, hospitals and prisons, the social workers in the ambulant psychiatric care stations, etc., had to be trained in genetic biology and race hygiene – and specialized knowledge was required for this as well. Fifth, genetic pathology and psychiatry were called upon to solidify the still shaky scientific foundations of genetic health policy through increased research efforts to illuminate the heredity of illnesses and disabilities. Feedback effects were hoped for here: The files of the sterilization proceedings were to be evaluated scientifically in order to acquire more detailed explanations for the genetic foundation of the illnesses and disabilities in question. Sixth, and finally, the active collaboration of geneticists was of strategic interest, be it in writing expert opinions or serving as medical assessors on the EGGs and especially on the EGOGs, the second instance, which often deliberated on questions of fundamental importance.

Placing the KWI-A in the proper position of this analytic framework, it becomes apparent that Eugen Fischer, Otmar von Verschuer, Fritz Lenz, and their institute played quite an important role in terms of the scientific foundation, justification, implementation, and accompaniment of the National Socialist sterilization program, surpassed only by that of Ernst Rüdin and the German Research Institute for Psychiatry. Certainly: in the formulation of the GzVeN Fischer and Verschuer made *no* contribution – in contrast to the Prussian draft law of 1932, because they were not involved in the "Expert Council for Population and Race Policy" (*Sachverständigenbeirat für Bevölkerungs- und Rassenpolitik*) put together by Arthur Gütt, which convened for its first session on June 28, 1933 to discuss the draft of the law. However, it must be taken into consideration that the GzVeN was an only slightly, "though portentously"[435] edited version of the draft law of 1932,

[435] Weingart/Kroll/Bayertz, Rasse, p. 463.

the essentials of which, as already explained elsewhere, bore the signatures of Fischer, Verschuer, and Muckermann. Nonetheless, the essential part of the new law – the catalog of indications – was the work of the DFA, and especially its director Ernst Rüdin, who, as a key figure in the Working Group II (Race Hygiene and Race Policy) of the Expert Council for Population and Race Policy, had a decisive influence on the wording, and above all as co-author of the official commentary, on the further interpretation of the law as well.[436] Not until the appointment of Fritz Lenz to Director of the Department for Eugenics was the KWI-A to gain access to the Expert Council – Lenz belonged to both Working Group II, which was directly responsible for the sterilization legislation, and Working Group I (Tax and Finance Legislation, Statistics, Social Policy [*Steuer- und Finanzgesetzgebung, Statistik, Sozialpolitik und Siedlungswesen*]).[437] Yet the influence of Lenz, whose rejection of *compulsory* sterilization had maneuvered him into the political wilderness for the time being, must not be overstated. Fischer proved more flexible. As early as May 1933, he publicly advocated the application of *indirect* force: "The genetically ill" who eluded voluntary sterilization should lose their right to welfare support and be taken into custody. Fischer implied that a corresponding draft law, to which he had made no small contribution, was in preparation[438] – thus conveying to his audience the incorrect impression that he was involved in the current legislation process. His advocation of indirect force, which went considerably further than his positions expressed up to this time, can by all means be conceived of as a concession to the new rulers,[439] yes, even as an attempt at placating the leadership so that he could be reintegrated into scientific policy consulting – for naught.

After the GzVeN was enacted, the institute in Dahlem placed itself unconditionally on the ground of the law. All of the misgivings brought forward during the Weimar Republic were thrown overboard; in the sterilization issue the institute switched, with flying colors, to the camp of the radical race hygiene in the National Socialist mold. A comment by Verschuer – printed in the *Deutsches Adelsblatt* ("Nobility Gazette"), and thus not directed to a specialized audience – is characteristic of this shift:

> The beneficial experiences, above all those made on a larger scale in California, allow us to expect that the law will be of great benefit for Germany as well. This may be expected all the more because, on the one hand, the law corresponds completely to the results and demands of genetic theory and fulfills the wishes of the leading race hygienists, and on the other hand, the preparation of the medical associations and public opinion for the acceptance of such a law in Germany has made particular progress. While people with a liberalistic and individualistic bent repeatedly expressed misgivings, these reservations have been completely eliminated thanks to the breakthrough of the National Socialist world view.

---

[436] Gütt/Rüdin/Ruttke, Gesetz.

[437] Aufgabengebiete der Arbeitsgemeinschaften des Sachverständigenbeirats, MPG Archive, Dept. I, Rep. 1 A, No. 2399, pp. 20–22. Also on this: Weingart/Kroll/Bayertz, Rasse, pp. 460–463; Lösch, Rasse, pp. 309–312.

[438] Fischer, Fortschritte der menschlichen Erblehre.

[439] So Kröner, Von der Rassenhygiene zur Humangenetik, p. 27.

Today it is a matter of course that the individual must make sacrifices for the welfare of the whole, his volk. From the genetically ill, who may be certain that the healthy will love and care for them, we must demand the sacrifice of doing without offspring, because this is a necessary requirement in order to achieve the goal of improving the genetic health of our volk.[440]

At the beginning of the Third Reich, Verschuer – even more than Fischer – played an important role as propagandist of National Socialist genetic health policy. First, by continuing to win over people in the periphery of the Inner Mission for the ideas of eugenics,[441] he contributed to the GzVeN's encountering a high degree of acceptance in the Protestant Church and its welfare organizations. Second, as a clinical genetic pathologist, Verschuer exerted a direct influence on his fellow physicians. At the beginning of his textbook for physicians on genetic pathology, published in 1934, he placed a chapter entitled *Der Erbarzt im völkischen Staate* ("The Genetic Physician in the Volkish State"), which linked a new professional ethic for physicians in the age of race hygiene with a eulogy to Adolf Hitler:

The mission of the physician for state medicine is completely new and much greater in today's *volkish, National Socialist state*. Under volk today we understand a spiritual and a biological unit. For the fact that we are a *spiritually unified volk* again today we are indebted to our Führer *Adolf Hitler*. [...] The *new mission of state medicine* today is: *Cultivation of the body of the volk* through the preservation and encouragement of healthy genetic material, through the elimination of pathologic genetic material, and through the maintenance of the racial character of our nation – through the *fostering of genes* and *the fostering of race*.[442]

By this time Verschuer, with the help of the National Socialist health leadership, had created a publicity platform upon which he could propagate among German physicians the professional image – and term – he had invented of the "genetic physician". The first issue of the journal he founded, *Der Erbarzt* (The Genetic Physician), appeared on June 16, 1934 as an insert to the *Deutsches Ärzteblatt*, the weekly journal of the German Medical association. The new journal's mission was, as Arthur Gütt wrote in his foreword, "to unite the German medical community in the National Socialist spirit and henceforth introduce German physicians to new trains of thought."[443] *Der Erbarzt*, as Klaus-Dieter Thomann emphasizes, was "the race hygiene periodical most widely propagated within the medical community."[444] Ludger Weß even holds *Der Erbarzt* to be "probably the most influential race

[440] Verschuer, Gesetz zur Verhütung erbkranken Nachwuchses.

[441] Cf., e.g. Verschuer's article about "Eugenic Questions" in the monthly newsletter of the Soziale Arbeitsgemeinschaft evangelischer Männer und Frauen Thüringens of August/September 1933, which emphasized above all positive eugenics (education and training, tax legislation, control of immigration and emigration, genetic biological stock-taking, marriage counseling), but also indirectly demanded bars to marriage for "those of alien ancestry," "ill and deformed persons and the genetically ill from encumbered families." (p. 44), but mentioned sterilization only in passing.

[442] Verschuer, Erbpathologie, pp. 1 f. (original emphases).

[443] Gütt, Geleit, p. 1.

[444] Thomann, Otmar Freiherr von Verschuer, p. 45.

hygiene journal" in the Third Reich, "whose scientific papers even [met with] interest and acknowledgment abroad."[445] On the whole, Fischer, Verschuer and the KWI-A were eager to help National Socialist genetic health policy acquire a scientific foundation and justification in the international arena.

With the genetic biology and race hygiene courses in the years 1933/34, the KWI-A made quite a decisive contribution to the fact that the sterilization proceedings were set in motion at unparalleled speed after the GzVeN was enacted on January 1, 1934. It is striking that there was a kind of division of labor between the KWI-A and the DFA in terms of training and further education: while the KWI-A trained the public health officers, who played a key position in the application process, and the presiding judges of the EGGs and EGOGs, the DFA brought the psychiatrists working in the treatment and care facilities of the German Reich in line with a genetic biology-race hygiene course, which took place under Rüdin's direction from January 8 to 16, 1934.[446] In these courses guidelines for diagnostics were issued, which were to provide direction in the filing, evaluation of and decisions on sterilization indications and applications – and, as could be demonstrated on the basis of example cases, did indeed play a leading role in such procedures. Moreover, the KWI-A and the DFA worked together on the scientific evaluation of the file material accumulated from sterilization practice. Beyond Fischer's and Verschuer's concrete research tasks, the stronger orientation of the institute to genetic pathology, especially to neurodegenerative illnesses, epilepsy, and mental disability, was intimately linked with sterilization practice.

Both Eugen Fischer and Otmar von Verschuer were directly involved in sterilization proceedings starting in 1934, where they served as medical assessors. Fischer's appointment as judge of the EGOG Berlin presumably occurred at the suggestion of Karl Bonhoeffer, the director of the Psychiatric and Neuropathic Hospital of the Charité in Berlin. In any case, in a letter to the Dean of the Medical Faculty on February 17, 1934, Bonhoeffer wrote:

> I – like Mr. Lenz – am of the opinion that, for the Hereditary Health Appellate Court, above all psychiatrists would be suitable, since the salient point is certainly the securing the diagnosis of the case in question. I would nominate Messrs. Fischer and Verschuer, and, as a psychiatrist, I would be willing to enter myself. As deputy I nominate the private physician Dr. [Kurt] Pohlisch.[447]

Bonhoeffer, who was known in the Weimar Republic as a cautious advocate of voluntary[448] eugenic sterilization on the basis of established genetic science, supported the GzVeN as a scientific expert and medical assessor, but – corresponding to his primary scientific interest in organic mental dysfunctions – advocated a narrow

[445] Weß, Humangenetik, p. 169.

[446] Rüdin (ed.), Erblehre.

[447] Quoted in Neumärker/Seidel, Karl Bonhoeffer, p. 274.

[448] In connection with the Prussian draft law of 1932, Bonhoeffer had violated this guideline, of course, when he encouraged in a written position paper to delete § 6, which excluded the application of coercion in principle, because he figured that the mentally ill might "rescind consent previously granted in the fear of the moment during preparations for the operation (shaving, disinfection, etc.)." Quoted in Gerrens, Medizinisches Ethos, p. 85.

interpretation of the indications listed in the GzVeN.[449] In his memoirs published after 1945, Bonhoeffer explained that the reason for his willingness to make himself available to the EGOG Berlin as a medical assessor was that he hoped to bring differential diagnostic aspects to bear in the EGG's practice of judgment. Indeed, the two courses on genetic biology which Bonhoeffer held at the Charité in 1934/36 furnish evidence that he was at pains to emphasize exogenous factors in the emergence of mental illnesses, epilepsy, and mental disability, to draw attention to diagnostic uncertainties and in this manner to prevent the concept of genetic illness from being stretched too far.[450] It is interesting to note that Bonhoeffer thought he recognized in Fischer and Verschuer allies to this cause – in the Weimar Repubic both had kept a low profile on the sterilization issue and were still considered moderates in the field of eugenics.

Bonhoeffer's later assessment, that due to his influence "the diagnostic opinions were handled cautiously in Berlin and even in the provinces,"[451] must, as statistics from the Reich Ministry of Justice for the period from 1934 to 1936 evaluated by Uwe Gerrens demonstrate, be regarded skeptically: While the number of sterilization proceedings pending and the number of sterilization decisions per capita in the territory covered by the EGOG Berlin did lie *under* the average for the Reich, and the rate of rejection in 1935–1946 was *above* average, the differences are not very great.[452] The moderating influence on the EGOG Berlin claimed by Bonhoeffer cannot be proved.

Fischer played no significant role on the EGOG Berlin. The two judge positions for physicians were occupied by three men each. The position of the official medical assessor was held by Leonardo Conti, represented by Franz Redeker, by now the Head of the Medical Department of Police Headquarters in Berlin, and Eugen Fischer. The position of the second medical assessor, who was required to be familiar with genetic theory and the fostering of genetic health, was occupied by Karl Bonhoeffer, represented by the senior physician Thiele from the Wittenau Sanatoriums and the assistant professor (and later "euthanasia" expert) Kurt Pohlisch, at the time a senior physician in Bonhoeffer's department.[453] Evaluating the files of the EGOG Berlin from the years 1935 to 1940, Niels C. Lösch comes to the conclusion that Fischer was involved in only 67 of a total of 1,016 proceedings – his numerous obligations as rector and institute director must have left him little time to take part in the sessions of the EGOG. What is decisive, however, is that in taking his office at the court he had set a clear signal that he and his institute placed themselves at the service of the practical implementation of the sterilization legislation.

---

[449] In this sense, Gerrens, Medizinisches Ethos, against Bock, Zwangssterilisation, p. 292; Damm/Emmerich, Irrenanstalt Dalldorf-Wittenau; Grell, Karl Bonhoeffer; Neumärker/Seidel, Karl Bonhoeffer.

[450] Bonhoeffer (ed.), Aufgaben; idem. (ed.), Erbkrankheiten.

[451] Bonhoeffer, Lebenserinnerungen, p. 102.

[452] Gerrens, Medizinisches Ethos, pp. 189–191.

[453] Lösch, Rasse, p. 350.

Bonhoeffer's proposal to appoint Verschuer to the EGOG as well was not adopted. Instead, Verschuer became the second medical assessor on the EGG of the district of Berlin-Charlottenburg and was appointed director of the "Polyclinic for the Fostering of Genes and Race," which was opened in October 1934 at the instigation of Arthur Gütt, in a barracks on the grounds of Kaiserin Auguste Viktoria Haus of the Imperial Institute for the Fight against Infant and Child Mortality (*Reichsanstalt zur Bekämpfung der Säuglings- und Kindersterblichkeit*) in Charlottenburg. In addition to Verschuer and his assistant Martin Werner, other physicians working at the polyclinic were Dr. Bockmann and the intern Dr. Pfotenhauer, both as scholars of the Emergency Association of German Science – they received two of the five scholarships Gütt had requested from the Emergency Association for assisting Fischer, Verschuer, and Rüdin in evaluating the sterilization files. Aside from this fixed personnel, the assistants at the KWI-A and also the physicians of the SS delegated to the institute were on rotation for shifts at the clinic. The point of this exercise is revealed by remarks from the first activity report:

> The Polyclinic for the Fostering of Genes and Race is obliged for its creation to the consideration at authoritative levels, that the scientist in the field of genetics must be placed in close touch with living human beings. From the creation of such positions then proceed, on the one hand, the possibility for the researcher to acquire new human material for his studies, and on the other, the result for the national community that the measures of a race-fostering and population-policy nature regarded as necessary in the conceptual work of the scientific institutes can be tried out and implemented in practice.[454]

The clinic thus was conceived as "a kind of superior office of expertise for public health officers;" the state health offices being erected were to obtain standards for "marriage consulting and certification for marriageability and for sterilization."[455] Apparently, the SS physicians delegated to the KWI-A were to pass on the experiences accumulated in Charlottenburg to health offices everywhere in the German Reich. The polyclinic was also directly integrated into sterilization practice, however, in the first 6 months of its existence, 290 potential sterilization cases were reported in accordance with the GzVeN; over 1,717 genetic biology opinions were written about people to be sterilized for the Hereditary Health Courts.[456]

What he had begun in Charlottenburg, Verschuer continued in the second half of the 1930s in Frankfurt: after assuming his post there, he succeeded in convincing the Reich Ministry of the Interior to allow his institute to perform free clinical examinations and set up a genetic health information center, whose catchment area soon covered the entire Frankfurt metropolitan area. Parallel to this Verschuer launched a large-scale genetic biology inventory project – in 1938 data on around half of the population of Frankfurt had been recorded in the institute's files.[457]

---

[454] Quoted in ibid., p. 315.

[455] Verschuer, Erbe – Umwelt – Führung, "Freier Forscher in Dahlem (1927–1935)" section, p. 22, MPG Archive, Dept. III, Rep. 86 A, No. 3–1.

[456] Lösch, Rasse, p. 315.

[457] Cf. Daum/Deppe, Zwangssterilisation, pp. 68–74; Stuchlik, Frankfurter Institut, pp. 166f.

Verschuer vigorously demanded observing the obligation to report sterilization candidates stipulated in the GzVeN – just as exactly as the obligation to report infectious diseases – even in "suspected cases." He criticized that the medical records of the general hospitals frequently did not mention "feeble-mindedness," and recommended "obligating assistants to conscientious examination with regard to the issue of feeble-mindedness."[458] Against the opposition of the City Health Office of Frankfurt, Verschuer succeeded in receiving permission that expert opinions for sterilization proceedings could be produced at his institute.[459]

Not only Fischer, but also Verschuer and Lenz produced expert opinions for the Hereditary Health Courts – the last although he did not act as a medical assessor because of his criticism of the GzVeN.

3.4.4 The Sterilization of the "Rhineland Bastards"

In the sterilization of the "Rhineland bastards," as the children conceived by nonwhite soldiers of the French, Belgian, and American occupation troops after World War I and German women were disparagingly called, the KWI-A placed a key role.[460]

Back in 1927, discussions had taken place in the Reich Ministry for the Occupied Areas and in the Reich Health Office about the sterilization of the "Rhineland bastards," the oldest of whom had just reached the age of 8. Such a measure was not permitted by law at the time, but apparently the consideration was less about changing the law than about violating it.[461] After the National Socialists came to power, demands in this direction became increasingly vocal. Hans Macco, for once, dedicated an entire paragraph to the "Rhineland bastards" in his work *Rassenprobleme im Dritten Reich* (Race Problems in the Third Reich), published in late 1933:

> Another significant reason for the deterioration of our race lies in commingling with races alien to us. The first task is to excise the remnants of the Black Disgrace on the Rhine. These mulatto children either originated in violence or the white mother was a prostitute. In either cases there is no moral obligation whatsoever to this progeny of an alien race. Now almost fourteen years have passed; any of these mulattos still alive is now reaching sexual maturity, so there is no longer much time for long discussions. Let France and other states deal with their race issues as they will, for us there is only one solution: the destruction of everything alien, especially in these ravages originating in brutal violence and immorality. As a Rhineland man I thus demand: The sterilization of all mulattos left behind by the Black Disgrace on the Rhine! This measure must be carried out within the next two years; otherwise it will be too late and this race degeneration will continue to advance in centuries to come.[462]

---

[458] Verschuer, Mitarbeiter, pp. 156f.

[459] Thomann, Otmar Freiherr von Verschuer, p. 46.

[460] Pommerin, "Sterilisierung der Rheinlandbastarde"; Lilienthal, "Rheinlandbastarde"; Lebzelter, "Schwarze Schmach"; Lösch, Rasse, pp. 344–348.

[461] Pommerin, "Sterilisierung der Rheinlandbastarde," pp. 30–32.

[462] Macco, Rassenprobleme, pp. 13f.

What the author did not know is that the new rulers had long since set about preparing for the "excision" of the "mulatto children." The first step was the collection of statistics, for up to that time there was only incomplete and unreliable information about the number of these children. On April 13, 1933, because Franz von Papen had been dismissed as Commissioner of the Reich in Prussia just 6 days earlier, the Prussian Minister of the Interior, Hermann Göring (1893–1946), asked the presidents of the governments in Düsseldorf, Cologne, Koblenz, Aachen and Wiesbaden to put together statistics about the number and ages of these children. On February 28, 1934 the Prussian Ministry of the Interior reported the result of this survey: According to these records, in Prussia there were 145 children fathered by non-white soldiers, and they were born between 1919 and 1930, that is, between 4 and 15 years of age. The district of Düsseldorf registered four, Aachen six, Trier 16, Cologne six, Koblenz 24 and Wiesbaden 89 "mulatto children."[463] The Ministry estimated the actual number of mixed-race children at around 500–600. To supplement the first Prussian survey, in April 1934 the Reich Ministry of the Interior directed the state governments in Baden, Bavaria, Hesse, Oldenburg and Prussia to put together lists of the mixed-race children.[464] In the course of this listing, other nonwhites of German citizenship well beyond the circle of the "Rhineland bastards," such as circus artists and musicians – including those from the former German colonies in Africa – were also registered with the authorities in greater numbers.[465]

In parallel with the recording of statistics, a race anthropology study was launched. On April 19, 1933 the Undersecretary in the Prussian Ministry of the Interior, Arthur Ostermann, informed the president of the government in Wiesbaden that a young anthropologist would be sent to Wiesbaden for this purpose – Ostermann, who, as mentioned on several occasions, had occupied a leading role in the eugenics movement at the time of the Weimar Republic, was retained in office by the National Socialists despite his associations with the Catholic Center Party,[466]

---

[463] Pommerin, "Sterilisierung der Rheinlandbastarde," pp. 44f.

[464] Ibid., p. 58. In the Bavarian Palatinate, Karl Heinrich Roth-Lutra (1900–1984) began with the anthropological-genetic biological examination of the "bastards" in Kaiserslautern in late 1936. In spring 1937 he set about expanding his examinations to the entire Palatinate. They probably served as the foundation for the recording and compulsory sterilization of the "Rhineland bastards" in the Palatinate. In performing his studies, Roth-Lutra was in direct contact with the KWI-A, namely with Eugen Fischer and Georg Geipel. (On this, Wolfgang Freund, Volk, Reich und Westgrenze. Wissenschaften und Politik in der Pfalz, im Saarland und im annektierten Lothringen 1925–1945, Diss. Universität Saarbrücken 2002. I thank Wolfgang Freund for referring me to Roth-Lutra.) Later Roth-Lutra worked at the Ethnological Museum in Dahlem, in order to prepare an anthropological expedition to central Brazil, a project that was never realized. During this time Roth-Lutra was a frequent visitor to the laboratory course at the KWI-A. Cf. Fischer to Verschuer, 9/3/1939, MPG Archive, Dept. III, Rep. 86 A (Münster), No. 9.

[465] Pommerin, "Sterilisierung der Rheinlandbastarde," p. 60, 63.

[466] Weindling, Health, p. 455, designates Ostermann incorrectly as a "socialist doctor." Weingart/Kroll/Bayertz, Rasse, p. 253, and Schwartz, Eugenik, p. 318 (note 186), proceed from the assumption that Ostermann was a member of the Catholic Center Party. Cf. also Richter, Katholizismus, p. 104 (note 18).

until he went into retirement for reasons of poor health in fall 1933. Due to the extremely close connection between Ostermann and Fischer in the Weimar Republic, it was obvious that the Prussian Ministry of the Interior would turn to the KWI-A in connection with the race anthropological study of the "Rhineland bastards," especially since Fischer was considered *the* expert on the subject of "bastard biology."[467] Fischer entrusted his assistant Wolfgang Abel with this mission. Because Abel fell ill, the study did not take place until July 1933. Abel published the results in 1937 in the *Zeitschrift für Morphologie und Anthropologie*. Because statistical record of the "Rhineland bastards" was not yet complete in 1933, Abel restricted his study to the Wiesbaden district, where a larger number of "mixed-race children" were already known by name. Abel examined 39 children from Wiesbaden und Biebrich, supposedly fathered by nonwhite occupation soldiers, of whom he categorized 27 as "Moroccan bastards" and 6 as "Annamese bastards."[468] He photographed all of the children, took anthropometric measurements, and evaluated the health forms kept on them by the school doctor and their report cards. Physical attributes like color of skin, eyes, and hair, shape of nose, lips, and head were examined. He found the children's state of health to be "very poor" on the whole. In more than half of the cases he diagnosed rachitic changes to the rib cage or knock-knees, further, on several occasions a high palate and bad teeth, swelling of the glands, flaccid posture, wryneck, skew or flat feet. Abel further emphasized the increased occurrence of tuberculosis, especially among the "Moroccan bastards," whereby he left open the extent to which this could be traced back to the circumstance that the Moroccans were seldom exposed to an infection in their home country and therefore had not built up any "antibodies," or to the fact that under these conditions "no selection of the resistant and elimination of the less resistant individuals" occurred. Either way, Abel presumed, "in the paternal genes of the Moroccans, the average resistance to Tbc [was] lower." Abel further thought he could establish an "accumulation of early psychopathic stigmata," among which he counted "crying out at night, nail-biting, eyelid flutter, spreach impediments" and nervous excitability.[469] Even though it could be assumed, in Abel's estimation "that the majority of the mothers of the bastards do not exactly embody the best heritage," nevertheless "the accumulation of early psychopathies was very striking."[470] However, Abel was not able to completely exclude the possibility that "the social conditions [might]

[467] Back in 1925, when the Swedish pastor Martin Liljeblad (1877–1950), who had launched a protest movement against the "black disgrace," wanted to organize an international congress in Sweden about race questions, which was to concern above all the "Rhineland bastards," the Foreign Office proposed sending Rudolf Martin (1864–1925) from the University of Munich, but also Eugen Fischer, at the time still an assistant professor in Freiburg. Pommerin, "Sterilisierung der Rheinlandbastarde," p. 27.

[468] In two cases the fathers were Turks, in one case each a Scotsman, a "white Frenchman" and a "mixed Englishman-Indian," in one case the father was unknown. In one case Abel classified the mother as "Fx mulatto." Abel, Europäer-Marokkaner-Kreuzungen, p. 312.

[469] Ibid., p. 321. In this connection Abel referred to the work by Tao, Chinesen-Europäerinnen-Kreuzung, which had arrived at similar results.

[470] Abel, Europäer-Marokkaner-Kreuzungen, p. 324. The indication "of the frequent psychopathies of the Jews" could not be left out here.

also have been an influence": "Some of the bastards are supported by the mothers alone, who have a hard time getting through life as household maids or minor employees, and thus have little time to raise the children."[471] In a number of cases the later husband of the children's mothers had refused to accept the child in his household, so that it found refuge with foster parents. In other cases, however, the later husband had adopted the child.

In evaluating 1,500 grade cards from 993 children, Abel arrived at the result that the average scholastic achievement of the "bastard children" was only 86.9% of the class average. With reference to Ernst Rodenwaldt, Abel also pursued the question as to "whether the achievements of the bastard children show differences, that is, worsening, before and after 1933," as it could be presumed "that through the broader propagation of general race science knowledge, these children themselves became aware of their bastard nature and differentness, and that this consciousness, in turn, influenced their achievements."[472] Abel disposed of this objection however, remarking that the scholastic achievements of the "bastard children" he studied had slightly *improved* since 1933 – the examiner had not the slightest grasp of the gauntlet these children were subjected to in National Socialist Germany. For him, the poor grades at school – and also the circumstance that the "bastard children" were flunked approximately seven–eight times more frequently than their classmates – were proof of their mental inferiority. According to the testimony of the teachers and social workers "and according to my own impressions," the "Moroccan bastards" were distinguished above all by a "stubborn nature," "difficultness," indeed even "ineducability," "disobedience, slovenliness, predilection for life on the street, great excitability and irascibility," while the "Annamese bastards" were easier to control.[473] In summary Abel came to the conclusion that "the main cause of the adverse condition of the Rhineland bastards within our population is found in the mixture of Caucasian with Negroid and Mongoloid races."[474]

Abel's study constituted the foundation for the following decision processes. He made his own opinion known in an article entitled *Bastarde am Rhein* ("Bastards on the Rhine"), which he published in the February 1934 edition of the journal *Neues Volk* put out by the Race Policy Office of the NSDAP. Without any mention of his study, Abel unfurled an infamous agitation against the nonwhite occupation children, which was directed expressly to the political decision makers "in whose hand it lies to prevent the propagation of suffering."[475] With this he recommended in barely veiled form the sterilization of all of these "half-breeds" – although such a measure was not covered by existing legislation.

A special law on the sterilization of "bastard children" was out of the question due to foreign policy considerations – here the Foreign Office put up resistance,

---

[471] Ibid., p. 325.

[472] Ibid., p. 326. Abel referred to Rodenwaldt, Seelenkonflikt.

[473] Abel, Europäer-Marokkaner-Kreuzungen, p. 327.

[474] Ibid., p. 329.

[475] Abel, Bastarde am Rhein.

repeatedly urging to restrict the race legislation to the Jewish minority in Germany in the interest of relations with Japan, China, India, Turkey and South America.[476] As such, the sterilization of the "bastard children" remained irreconcilable with the legal situation even after the GzVeN was enacted on January 1, 1934. However, the GzVeN opened up a loophole in the form of indication of "congenital feeble-mindedness," to which the Prussian Minister of the Interior directly referred the presidents of the relevant governments: The hereditary health courts could also order the sterilization of men and women to whom "moral feeble-mindedness" was imputed because of their way of life. As Abel's findings provided evidence for the "inferior, mental and spiritual disposition" of the "bastard children" (the Ministry enclosed Abel's polemic article in *Neues Volk* with this letter) in many cases it was possible to obtain sterilization in accordance with the GzVeN. The responsible government offices thus were to be instructed "to direct their particular attention to these half-breeds" in the execution of the GzVeN.[477]

On March 11, 1935 a meeting of the Working Group II "Race Hygiene and Race Policy" of the expert council for population and race policy took place. Among the agenda items was a speech by the Director of the Race Policy Office of the NSDAP, Walter Groß, about "Paths to Resolving the Bastard Question," which was to be followed by a discussion. The text of Groß's speech is not included, but an extremely lengthy protocol of the session provides detailed information.[478] Groß stated that a total of 385 "Negro bastards" had been recorded, and that he estimated the actual number to be 500–800. They were not under the special protection of the League of Nations, as illegitimate children they had the citizenship of the mother according to German law, but upon recognition of paternity by a French occupation soldier, according to French law they would have French citizenship in addition to German. Because the German Reich had complete legal latitude in the treatment of the "bastard question," it could proceed according to political expediency. The reproduction of "bastard children" was undesirable and had to be prevented. There were basically two paths to achieve this: "removing" them from "German *Lebensraum*" or sterilization. Groß emphasized once more that current law did not offer any foothold for the sterilization of the "bastard children." One could, first, in a tacit agreement between public medical officers and hereditary health courts, in "awareness of the higher purpose [and] against one's better judgment," stretch the provisions of the GzVeN so far that the "bastard children" were covered as well. Second, one could create a new legal regulation. Third and finally, one could have the sterilization performed *illegally* by party organs especially empowered for this purpose – independent of the usual Hereditary Health Court proceedings. Groß excluded the first possibility on the grounds "that Germany today does not yet have the discreet and reliable apparatus at its disposal in order to commit violations of the law in such special cases, silently and out of a sense of responsibility to the

[476] Pommerin, "Sterilisierung der Rheinlandbastarde," pp. 61–71.

[477] Quoted in ibid., p. 57.

[478] Ibid., pp. 71–77.

volk, without being noticed." The second option – the creation of a new legal foundation – was to be welcomed as such, as the "open and honest path,"[479] but was out of the question for reasons of foreign policy. Thus Groß ultimately argued for the third possibility, illegal sterilization according to the model of abortions for eugenic indications, which Reich Physician Führer Gerhard Wagner had just launched with Hitler's support.[480] However, such an illegal sterilization was only possible on a voluntary basis, with the written consent of the mothers. Before initiating sterilization, the expert opinion of an anthropologically trained physician was to be obtained and the paternity of a nonwhite occupation soldier to be established beyond doubt. Groß emphasized the important role of the office to be consulted for an expert opinion. One mistake that met with opposition and attracted attention could foil the entire project so inconspicuously set in motion.

In the ensuing discussion, the assembled experts wavered. Lenz stated for the record that the difficulties of the material had not been clear to him until Groß's speech. At the moment he could see no feasible solution, although in his opinion the option of "exporting"[481] the children upon the parents' approval seemed more suitable than the others. Rüdin wanted to offer the "bastards" the alternative of voluntary sterilization or leaving Germany. Herbert Linden, with reference to Abel's study, argued for sterilizing the "Moroccan bastards" on the basis of the GzVeN. For the "Annamese bastards," and for other "half-breeds" living in the German Reich, another solution must be found. The assembly ultimately reached a consensus to aspire to either compulsory or voluntary sterilization in accordance with the GzVeN, or illegal sterilization on a voluntary basis, but not to pursue the emigration plan any further, in part due to financial considerations. The final decision was left to Hitler.

The remaining course of the decision-making process can no longer be reconstructed from the archives. What is clear is that the "bastard question" was ultimately "resolved" in the manner proposed by Walter Groß, through illegal sterilization, whereby the principle of voluntariness he admonished was ignored. In all probability one must presume that Hitler – as in the case of eugenic abortions – issued an oral mandate. The execution of the order was entrusted to the Gestapo. To this end, in spring 1937 a "Special Commission 3" was set up in the Secret State Police Office in Berlin. The Reich Ministry of the Interior made available the collected documentation material; the Special Commission requested supplementary documents, e.g. files on guardianship, from the responsible offices.

Three commissions were formed in the territory of the Reich,[482] which also included anthropological specialists provided by the KWI-A: Wolfgang Abel,

---

[479] Quoted in ibid., p. 72.

[480] Cf. Schmuhl, Rassenhygiene, Nationalsozialismus, Euthanasie, pp. 161–164.

[481] Quoted in Pommerin, "Sterilisierung der Rheinlandbastarde," p. 75.

[482] Among the members of Commission I in Wiesbaden was Heinrich Schade, who had participated in the first annual physicians' course at the KWI-A and was an assistant and chief physician of Verschuer's from 1935 on. In 1942 he followed his boss to Dahlem. For a biography: Klee, Personenlexikon, p. 522.

Engelhard Bühler, Herbert Göllner, and in one case, even Eugen Fischer himself.[483] On the basis of the race anthropology expert opinions, 385 were, in fact, forcibly sterilized in 1937. Eugen Fischer, Fritz Lenz, Wolfgang Abel, and their institute had contributed their scientific expertise to an undertaking that was clearly illegal according to the existing legal situation. Walter Groß had emphasized how important the selection of the scientific institution to be entrusted with the expert opinions was in view of this illegal character – the institute in Dahlem fulfilled expectations. It participated actively and with initiative, even proposing itself that race anthropological sterilizations be expanded: The doctoral student Yun-kuei Tao, as Fischer reported to the board on May 10, 1935, had found "much to our surprise" several "Chinese-European bastards [...]not only in Paris and London, but also here in Berlin," which was "not only of great interest scientifically, but above all nationally."[484]

3.4.5 *The Kaiser Wilhelm Institute for Anthropology, Human Heredity and Eugenics and the "Jewish Question"*

Having offended the party ideologues repeatedly at the beginning of the Third Reich with his remarks on the "Jewish question," Eugen Fischer was conspicuously reserved on this subject until 1938, and even resisted the temptation to orient his institute toward National Socialist Jewish policy. Fischer's scientific views on the "Jewish question" had not changed in essence since 1913, as is clearly recognizable in the protocol notes on a lecture about "The Race of the Jews" Fischer held for the "Mittwochs-Gesellschaft" on June 7, 1933. According to this talk, the region of the Near East had originally been settled by a single race. This "Near Eastern race" was considered the "anthropological foundation" of the later processes of miscegenation:

> Through Semitization, from the fourth, more from the third century on, comes the miscegenation with the "Oriental" race. Certain immigrations, like those of the Hethites and later the people known as the Thraco-Phrygians contribute (in mixtures) "Nordic" elements; others, like the Philistines, "Mediterranean" elements. [...] The later Jews are thus primarily a Near Eastern-Oriental crossbred with the elements mentioned above. – According to Harnack's estimation, around the year of Christ's birth there were around ½ million Jews in Palestine, 1 million in Syria, 1 in Egypt and 2 in the remainder of the Roman Empire. (4 ½ million). There was certainly assimilation of much alien blood. The Sephardim (Africa and Southern Europe) adhered to the Oriental type, the Achkenazim (Eastern European Jews) assimilated very large amounts of alien races. So the empire of the Chasans crossed over to Judaism in the 8th century, which thus assimilated the Mongolian-Alpine European race. – The adherence of the race attributes corresponds entirely to the Mendelian laws of heredity, this is true for all races – but in the Jews the attributes are merely strange to us and thus preserved conspicuously![485]

---

[483] Pommerin, "Sterilisierung der Rheinlandbastarde," p. 78; Lösch, Rasse, p. 347.

[484] Fischer, Tätigkeitsbericht von Anfang Juli 1933 bis 1. April 1935, MPG Archive, Dept. I, Rep. 1 A, No. 2404, pp. 49–49h, quote: p. 49e.

[485] Quoted in Scholder, Mittwochs-Gesellschaft, pp. 69 f.

This argumentation was not particularly original; it corresponded for the most part to the paleoanthropological analysis of Felix von Luschan, which had become generally accepted in the field of anthropology back around the turn of the century. Here it becomes apparent why Fischer refused to speak of *the* "Jewish race," but had no problem talking about the "Jewish *races*." He regarded Jewry as a "race mixture," whereby the "ratio of mixture" could vary widely – which thus explains the deliberate contrast between Western and Eastern Jews. Fischer's hesitation to categorize the "Jewish races" across the board as "inferior" on the race anthropology scale also originates here – after all, he presumed from the assumption that the genetic material of some Jews even contained "Nordic parts." The thesis of the Jews as a "race mixture" with "Nordic elements," connected with the insights on the "exuberant growth of the bastards" in the "bastards of Reheboth" studies, had been the basis of Fischer's implicit critique of a policy of strict apartheid with regard to the Jewish minority in the years 1933/34. Once he had recognized how politically explosive his conclusions were, Fischer followed Gütt's advice to be more cautious with public statements that could be understood as directed against the party, and did his best to avoid making any comments on the "Jewish question" in public. This probably became most obviously evident in the 4th edition of "Baur-Fischer-Lenz" published in 1936 under the changed title *Menschliche Erblehre und Rassenhygiene* ("Human Genetics and Race Hygiene"), where he abruptly left out the description of the individual races. Niels C. Lösch's interpretation that this happened under the pressure of "NS censorship" is misleading – in the Third Reich there was a multitude of censorship offices, blacklists, a system of banning writers, not to mention innumerable raids, banned books, and newspapers, there was *no* regular system of *preliminary* censorship. Fischer certainly could have published his deviant views on the "Jewish question" – providing he found a publisher who would print it, which could not have been all too difficult. Fischer's caution was a consequence of mental cutting – he subjected himself to *self*-censorship to keep from risking his altogether advantageous arrangement with the regime. Lösch's judgment that Eugen Fischer was not an anti-Semite cannot be endorsed in this form, either. Correctly, Lösch admits that Fischer was caught "in the cultural code of anti-Semitism" (Shulamit Volkov), but did not share the anti-Semitism of the National Socialists, yet because of his being formed by this "cultural code" he was not able "to decipher the message of this new anti-Semitism."[486] What is certain is that Fischer's inventory of ideas – and this even before 1933, between 1933 and 1945, and also after 1945 – included all of the dregs of anti-Jewish stereotypes that had developed since the Kaiser's day. The occasional anti-Semitic asides in his private correspondence provide eloquent evidence of this. However, Fischer was also a *racial* anti-Semite in the fundamental sense that he regarded the "Jewish question" as a "question of race" – only that his answer to this "question of race" turned out to be more differentiated than that of the party ideologues. Insofar Lösch's dictum that Fischer was "no anti-Semite," but rather a "racist," misses the point.

---

[486] Lösch, Rasse, p. 296, with reference to Volkov, Antisemitismus.

Also interesting in this context is a look at the library of the institute in Dahlem.[487] The largest department by far (A VI) was a collection of works on ethnology and geographic anthropology. This department was subdivided into six divisions, of which four followed a geographical order (A VI a: Arctic, North, and South America; A VI b: Africa; A VI c: Europe; A VI d: Asia; A VI e: Australia and Oceania), while the sixth (A VI f) was dedicated exclusively to Jewry. Thus the very arrangement of the library – which certainly can be traced back to Eugen Fischer – reflected a tendency to exclude the Jews from the community of human races and peoples, for in terms of pure logic the literature on Jewry would certainly have had to be included in the divisions for Asia or Europe. Thirty-seven titles were associated with the division A VI f, the bibliographies of which could be determined for the great majority. Among these were, first, books from the periphery of Zionism like *Das Rassenproblem unter besonderer Berücksichtigung der theoretischen Grundlagen der jüdischen Rassenfrage* (The Race Problem under Special Consideration of the Theoretical Foundations of the Jewish Race Question) (1910) by the anthropologist Ignaz Zollschan, *Der Untergang der deutschen Juden. Eine volkswirtschaftliche Studie* (The Decline of the German Jews. An Economic Study) (1911) by the sexologist Felix A. Theilhaber (1884–1956) and *Les juifs dans le monde moderne* (The Jews in the Modern World) (1934) by Arthur Ruppin (1876–1943), Director of the Palestine Office in Jaffa from 1908 on. The second group of works were written in the period before 1933 which dealt seriously with the history, sociology and anthropology of the Jews, such as the *Geschichte des jüdischen Volkes seit der Zerstörung Jerusalems* (History of the Jewish People since the Destruction of Jerusalem) (1908), written by the "Christian Zionist" Carl Friedrich Heman (1839–1919), *Die Juden und das Wirtschaftsleben* (The Jews and Economic Life) (1911) by the economist Werner Sombart (1863–1941), *Die Rassenmerkmale der Juden. Eine Einführung in ihre Anthropologie* (The Race Attributes of the Jews. An Introduction to their Anthropology) (1913), the study by the American anthropologist Maurice Fishberg which first took the myth of the "Jewish nose" to absurd lengths, and *Das Judentum als landschaftskundlich-ethnologisches Problem* (Jewry as a Problem for Landscape-Geography and Ethnology) (1929) by the geographer Siegfried Passarge (1867–1958). In addition, however, this department of the library in Dahlem also had a third division, containing the nastiest anti-Semitic concotions of the Third Reich, like *Antisemitismus der Welt in Wort und Bild* (The Anti-Semitism of the World in Words and Pictures) (1936), *Judas Kampf und Niederlage in Deutschland* (Judas' Battle and Defeat in

[487] In taking stock of the KWI-A's inventory ultimately moved to Frankfurt/Main, a detailed list was prepared (MPG Archive, Dept. I, Rep. 1 A, No. 3034), where the books of the library are listed in the order in which they were taken out of the moving boxes, with call number and short title. In the framework of this study these data were sorted by call number, so that the original structure of the library becomes visible again. For the Sub-Department A VI f, then, on the basis of the short titles, the complete literature references were reconstructed as well as possible.

Germany) (1937), *Der Jude als Verbrecher* (The Jew as Criminal) (1937), *Das Judentum, das wahre Gesicht der Sowjets* (Jewry, the True Face of the Soviets) (1941) and the omnibus volume coedited by Julius Streicher (1885–1946) about the "court Jews" (1939).[488] And, of course, the anti-Semitic classics like the *Handbuch der Judenfrage* (Handbook of the Jewish Question) (49th edn. 1943) by Theodor Fritsch, the *Semigothaische genealogische Taschenbuch aristokratisch-jüdischer Heiraten* (Semigothaic Genealogical Pocket Book of Aristocratic-Jewish Marrying) (1912) and Hans F.K. Günther's *Rassenkunde des jüdischen Volkes* (Race Science of the German People) (2nd edn. 1930). It is presumably not amiss to deduce from this selection of literature how Fischer stood on the "Jewish question" because of the simple fact that most of the library was put together from Fischer's own private volumes, and did not become institute property until Fischer retired.

The fact that his personal conduct with Jewish employees and colleagues was on the whole "irreproachable"[489] did not change the overall picture – all in all, in this point, too, the total impression is quite ambivalent: On the one hand, in 1931 Fischer protected his doctoral student Harry Conitzer against excessive criticism apparently motivated by anti-Semitism. On the other hand, in a letter of defense directed to Arthur Gütt in 1933, he defended himself from the accusation that he advised too many Jewish doctoral students, pointing out that he could not refuse to advise Jewish doctoral students as a university professor, but that he had never considered a Jew when selecting his assistants. As chairman of the Berlin Society for Anthropology, Ethnology, and Early History (*Berliner Gesellschaft für Anthropologie, Ethnologie und Urgeschichte*) up to his resignation in late 1937, Fischer did not undertake any steps to exclude "non-Aryan" members like Franz Boas.

Fischer's abandoning his cautious stance in 1938 can be attributed to Verschuer's urging. The latter portrayed his position on the "Jewish question" at length in his manuscript "Mein wissenschaftlicher Weg"("My Scientific Path"), written after World War II:

> I looked into the Jewish question intensively. For those of us who joined the [youth organization] Wandervogel it was a matter of course, on the quest for the sources of a healthy life and from the endeavor to keep the German type pure, to reject everything unnatural (like, for instance, life in the city) and everything alien. Preserving the purity of the race was thus one of the basic life principles. I was raised by my parents with the same standpoint, which was self-evident and not linked to any deprecation of other races. On the contrary: Through the concern with the colonial question so active in Germany at the time, the interest in strange races was quite vibrant. The Jewish question was in practice the most pressing race

[488] Pugel/Körber, Antisemitismus; Reventlow, Judas Kampf; Keller/Andersen, Jude; Deeg/Streicher (ed.), Hofjuden; Agthe/Poehl, Judentum. From the period of the Third Reich, also: Eichstädt, Bibliographie; Fried, Aufstieg; Günther, Judentum in Mainfranken; Hinkel, Judenviertel; Kessler, Familiennamen; further, the first six volumes of *Forschungen zur Judenfrage*. The title "Rassensieg in Wien, der Grenzfeste des Reiches" could not be determined using bibliographic methods.

[489] Lösch, Rasse, p. 295. On the following: ibid., pp. 294 f.

> question for the German nation, because the Jews were the only people of a strange race who lived within our volk in large numbers. The awareness of a separation in terms of blood, however, did not by any means exclude personal association with individual Jews. [...] Even within the race hygiene movement I had sustained friendly relations in Berlin to the Jewish ophthamologist Dr. Czellitzer[490] and undertook a joint scientific project with the Jewish serologist Schiff. This did not deter me from seeing the dangers that threatened the German volk as a whole from Jewry. I received the most flagrant impressions of this from my years in Berlin from 1927–1933, where one encountered Jewry at every turn in the economy and in cultural institutions and could always feel its close link with the rampant corruption prevalent at the time. The volkish and racial separation between Germans and Jews thus seemed to me a necessary demand to resolve the emerging difficulties for both sides. Through my scientific work I was at pains to work out the foundations to resolve this issue.[491]

This text is a perfect example of the "cultural anti-Semitism" that developed in the "republic of outsiders,"[492] especially in Berlin, in three aspects: in equating large cities, modernity, and Jewry; in the differentiation between *the* Jew, who was, so to speak, a foreign body in the German volk, and *the individual Jews*, with whom one had social or professional contact; and finally, in the demand for social segregation – whatever this was supposed to look like. In the 1920s Verschuer had still pled for the standpoint that the "fight against Jewry" was above all a "volkish-political battle," and therefore had not gone into the "Jewish question" in any detail in his scientific writings. In his political writings, though, even back in the mid-1920s he advocated checking the "massive immigration"[493] of Poles and Eastern European Jews, to stem an excessive foreign influence. After the Nazis came to power, anti-Semitic thought increasingly played a role in Verschuer's scientific work as well. For instance, the key word "Jews" (*Juden*) appeared in the index of the second edition of his textbook *Erbpathologie* (Genetic Pathology), which came on the market in 1937; an entry that was not included in the first edition published in 1934.

By this time the KWI-A had been integrated into Nazi Jewish policy on the practical level as well. A number of staff members produced "genetic and race science certificates of descent" (*erb- und rassenkundliche Abstammungsgutachten*) for the "Official Expert for Race Research" (*Sachverständiger für Rassenforschung*), Achim Gercke, and from March 1935 on for the "Reich Office for Genealogical Research (*Reichsstelle für Sippenforschung*), renamed in November 1940 to the

---

[490] Arthur Czellitzer (1871–1945) worked as an ophthamologist in Berlin from 1900 until 1938. He was the founder and president of the Society for Jewish Family Research (*Gellschaft für jüdische Familienforschung*) and a member of the Central Office for German Personal and Family History (*Zentralstelle für deutsche Personen- und Familiengeschichte*). In the Berlin Society for Race Hygiene he was active as the second deputy secretary. In 1938 he was forced to emigrate to the Netherlands, in 1943 he was confined in Breda, deported to Poland and murdered. For a biography: Labisch/Tennstedt, Weg, vol. 2, p. 395.

[491] Verschuer, Mein wissenschaftlicher Weg, pp. 22–24, MPG Archive, Dept. III, Rep. 86 A, No. 3–1.

[492] Unparalleled on this: Gay, Weimar Culture.

[493] Verschuer, Auslese und Rassenhygiene, p. 750.

"Reich Ancestry Office" (*Reichssippenamt*), in the Reich Ministry of the Interior.[494] In the 1935/36 fiscal year, as mentioned above, Fischer and Abel supplied over 60 of such certificates; in total, Wolfgang Abel estimated in retrospect, around 800 such expert opinions were prepared for the Reich Office for Ancestral Research.[495]

With the "Aryan Clause" in the "Law for the Restoration of Career Civil Service" (*Gesetz zur Wiederherstellung des Berufsbeamtentums*) of April 7, 1933 and with the "Nuremberg Laws" of September 15, 1935,[496] the proof of descent – the "Small Certificate of Descent" (*Kleiner Abstammungsnachweis*) went back to the grandparents, the "Large Certificate of Descent" (*Großer Abstammungsnachweis*) traced the ancestors back to 1800 – took on ever greater importance. If it was not possible to reconstruct the complete line of descent through genealogical research, the Reich Office for Ancestral Research[497] could request a genetic and race science expert opinion from one of the authorized scientific institutes. The KWI-A played a key role in this from the very beginning, having concerned itself intensively with similarity diagnostics in the context of twin research and already having been entrusted with paternity opinions by the courts of the Weimar Republic. A race science opinion on descent involved performing four studies: determination of the blood group (Engelhard Bühler was the KWI-A specialist responsible for this), determination of the patterns of papillary lines (Georg Geipel's specialty), anthropometric measurements (mainly by Fischer and Abel), and photographic documentation according to anthropometric aspects.[498]

The number of expert opinions requested increased so dramatically that Fischer pulled the emergency brake in November 1938. The flood of requests had let to a "double emergency." For the people affected "the long waiting is not only emotional torment," it also entailed "most serious economic consequences." The institute, for its part, suffered from a "workload [...] that was impossible to bear" unless two or three new assistant positions were created.

> Under the present conditions I have no choice but to submit the request that radical redress be provided. It is impossible for me to take on new expert opinions, as painful as it is for me to leave the Reich Office for Ancestral Research, with which we have always enjoyed working and never had any problems, in an embarrassing position.[499]

---

[494] Seidler/Rett, Reichssippenamt; Kröner, Vaterschaftsbestimmung; Lilienthal, Anthropologie; Schulle, Reichssippenamt; Geisenhainer, "Rasse," pp. 236–239. Cf. in general: Meyer, "Jüdische Mischlinge."

[495] Müller-Hill, Tödliche Wissenschaft, p. 138.

[496] They were acknowledged positively by Fischer before the theological faculty of the University of Berlin, because they made it possible for "genetic researchers [...] to make their research results of practical service to the volk as a whole." Quoted in Müller-Hill, Tödliche Wissenschaft, p. 37.

[497] From 1938, in certain cases prosecutors as well.

[498] Cf. Lösch, Rasse, pp. 341 f.

[499] Quoted in ibid., p. 343.

The KWI-A continued to deliver such expert opinions in the subsequent period, although with decreasing frequency. What is more: it was at the center of the system of expert opinions. In 1943, the 22 institutes authorized to prepare race science certificates of descent included not only the KWI-A, but also the Institute for Race Biology and the Institute for Race Hygiene at the University of Berlin under Wolfgang Abel and Fritz Lenz. Six of the other institutes were also headed by scientists who can be reckoned among the "Dahlem circle" (the Genetic Pathology Department of the First Medical Clinic of the Charité Berlin under Friedrich Curtius, the Anthropological Institute at the University of Kiel under Hans Weinert,[500] the Institute for Genetic Biology and Race Hygiene in Cologne under Ferdinand Claußen, the Race Biology Institute at the University of Königsberg under Bernhard T. Duis, the Institute for Race Biology at the Reich University in Strasbourg under Wolfgang Lehmann, and the Race Biology Institute at the University of Vienna under Lothar Loeffler). Of the 10 remaining individual experts authorized, one, Johannes Schaeuble, Director of the Department for Genetic and Race Science of the Anatomy Institute at the University of Freiburg, was a former employee of the KWI-A.[501]

In his endeavor to place Jewish policy on a strict scientific foundation, Verschuer did not balk at offering his expertise to the regime. In 1936, he readily accepted the invitation to participate as an official expert in the newly founded "Jewish Question Research Department" (*Forschungsabteilung Judenfrage*), part of the "Reich Institute for the History of the New Germany" (*Reichsinstitut für Geschichte des neuen Deutschlands*).[502] At the department's second workshop from May 12 to 14, 1937 Verschuer held a lecture entitled "What Can the Historian, the Genealogist and the Statistician Contribute to the Research of the Biological Problem of the Jewish Question?"[503] In this talk Verschuer urgently advocated the scientific underpinning of every form of race policy. He left no doubt as to the objective of such policy: "Cross breeding of a foreign race into a nation" – be it by "Negroes, Gypsies, Mongols, South Sea islanders or Jews" – always led to a change in the "biological character of the volk" and its "characteristic culture."[504] His remarks, Verschuer reported in a letter to Fischer, were "received with applause, even though the extreme anti-Semites certainly could not have been satisfied." He further writes that he was asked to present "in a short exposé, proposals for the practical race biological registration of the Jews and Jewish half-breeds in Germany." Before submitting this exposé to the Reich Institute, Verschuer wanted to discuss it with Fischer, "because in all of these matters I would like to go hand in hand with you all the way." Verschuer also hinted at the direction of his thoughts in this letter:

---

[500] For details on this: Meyer, "Jüdische Mischlinge."

[501] Seidler/Rett, Reichssippenamt, p. 162; Lilienthal, Anthropologie, pp. 80 ff.

[502] Cf. Weinreich, Hitler's Professors, pp. 51, 55f.; Heiber, Walter Frank, p. 421; Lösch, Rasse, pp. 286–292.

[503] Verschuer, Historiker.

[504] Ibid., p. 222.

> In addition to what I already hinted at in my lecture, I believe that another promising path would be to organize the physical examinations for military service, such that anthropological investigation is part of the examination of Jews (in accordance with the Law of the Reich Citizenship [*Reichsbürgergesetz*]) an anthropological investigation is performed as well. It is true that only Germans and half-breeds are drafted into military service, but certainly there must be the possibility of a special regulation for Jews with which such an examination could be combined. The Reich Institute for the History of the New Germany should then use its influence at the highest offices to arrange for such dispositions.[505]

These plans came to nothing, but Verschuer's association with the Jewish Question Research Department of the Reich Institute for the History of the New Germany remained intact. In October 1937 he accepted the invitation of Wilhelm Grau (1910–2000),[506] the director of the Jewish Question Research Department, to hold a lecture on the "Race Biology of the Jews" at the next meeting from July 5 to 7, 1938. Upon this opportunity it was proposed – it remains unclear whether by Verschuer or Grau – to invite Eugen Fischer as a speaker as well, to discuss the emergence of the Jewish race. However, Fischer hesitated – not only was he once bit, twice shy, but he probably also feared for his scientific reputation, especially abroad. Verschuer urged Fischer to accept, whereby the reasons he offered are extraordinarily interesting:

> The lectures attract great attention and thus present a special opportunity to contribute to a calmer perception of the matter. This duty appears to justify this action to me. On the other hand, I believe that we have no reason to fear that such a lecture could be injurious to our scientific prestige. International Jewry is completely clear about which side we are on; our participation in such a meeting plays no role. However, it is important that our race policy – in the Jewish question, too – receives an objective scientific background that is also recognized by other circles.[507]

In actuality, both Verschuer and Fischer appeared as speakers at the third "Workshop on the Jewish Question" of the Reich Institute for the History of the New Germany in Munich in July 1938. In his talk about the "Race Biology of the Jews" Verschuer contradicted the generally accepted idea that Jews could be recognized by the shape of their nose or their blood group. Instead he referred to the emerging science of comparative race pathology. A number of illnesses and disorders occurred more frequently in Jews than among the non-Jewish population: diabetes, neuroses, flat feet, myonomes, xeroderma pigmentosum, hemophilia, and deaf-muteness. "Amaurotic idiocy" (Tay-Sachs syndrome) and torsion dystonia are particularly prevalent among the Eastern European Jews. To explain this phenomenon Verschuer advanced population-genetic argumentation: Through conscious segregation from their "host volk" the Jews had " 'bred' their race themselves" in genetic isolation.[508] Verschuer's typical example was tuberculosis. Because the Jews "have lived predominantly

---

[505] Verschuer to Fischer, 20/5(1937, MPG Archive, Dept. III, Rep. 86 A (Münster), No. 9.

[506] For a biography: Heiber, Walter Frank, passim; Klee, Personenlexikon, p. 197.

[507] Verschuer to Fischer, 5/11/1937, MPG Archive, Dept. III, Rep. 86 A (Münster), No. 9.

[508] Verschuer, Rassenbiologie der Juden, pp. 148 f.

in urban environments longer than all other peoples and were thus exposed more frequently to infection by tuberculosis," the individuals with a disposition for this disease had been "eliminated," "whereby the average resistance to tuberculosis infection rises from generation to generation."[509]

In his speech "The Emergence of Race and Ancient Race History of the Hebrews," Fischer – after long-winded remarks on the domestication of animals and genetic theory – repeated his old familiar interpretation of "race relations" in the Near East in prehistory and Antiquity, based on Felix von Luschan, whereby – presumably in consideration for his audience – he attempted to underplay the skandalon of "Nordic race elements":

> Oriental race as the basic stock, Mediterranean race admixed, Near Eastern as the second basic stock and Nordic elements occasionally interbred. That the Nordic element is mentioned repeatedly in this depiction must not create the impression that its impact was particularly great! The Nordic race simply had (and has!) an enormous migratory power and role as conqueror [...].[510]

Anti-Semitic echoes are most apparent in Fischer's short remarks about the mental-spiritual characteristics of the ancient Hebrews. So "even in the early history of the Jewish nation, the passion, the hate and the cruelty [can be seen] that so often spent itself in bloodthirstiness," according to Fischer, a racial attribute of the "herdsman of the Oriental race," along with the "cleverness, adaptability, cunning and lust for power of the city founders of the Near Eastern race."[511] Several anti-Semitic stereotypes are assembled in this statement, but Fischer hastily added that he did not intend to portray the physical nor the mental characteristics of today's Jews. He thus stuck to the cautious line he had followed since 1933. For all that, with his final sentences, he legitimated National Socialist Jewish policy including the "Nuremberg Laws":

> But the Jews – the Oriental-Near Eastern amalgamated race is unique, one of a kind – a foreign species to us Europeans, different in body and above all in soul – but most foreign and opposite with regard to the Nordic race, whose emergence took place under conditions contrary in every way – and we perceive this instinctively even today.[512]

After July 1938 Fischer did not take part in any more of the Reich Institute's workshops. In February 1941, however, he was appointed as the last scientist to the Academic Council of the Reich Institute, now totaling 68 members (among them Verschuer), but he did not become involved in any further activities on this body.[513]

---

[509] Verschuer, Erbpathologie, 2nd edn., pp. 123f.

[510] Fischer Rassenentstehung, p. 133.

[511] Ibid., p. 135.

[512] Ibid., p. 136.

[513] Heiber, Walter Frank, pp. 609f. The late appointment of Fischer, according to Helmut Heiber, should be seen as a reaction to his participation in the opening event of the Frankfurt Institute for the Research of the Jewish Question (*Frankfurter Institut zur Erforschung der Judenfrage*) on March 27/28, 1941 – more on this later. In general, Heiber advocates the view that the council was concerned above all "with the collection of decorative names and positions." Ibid., p. 596.

At the workshop in Munich Fischer had met the Viennese theologist Gerhard Kittel (1888–1948),[514] a dyed-in-the-wool anti-Semite who had specialized in the history of the Jews in Antiquity. In 1944, Kittel and Fischer published a jointly authored book about "World Jewry in Antiquity" – more on this later. That Fischer did not even back off from his positions on the "Jewish question" after the November pogrom, but rather, as regarded the political implications, used even more radical formulations in his public appearances, was demonstrated in a lecture of June 20, 1939 for the Association for Mining Interests (*Verein für bergbauliche Interessen*) in Essen:

> The Jew is [...] of a different kind and therefore to be fended off if he wants to invade. It is self-defense. In this I do not designate Judaism as a whole as inferior – like Negroes, for instance, and I do not underestimate the great enemy we have to combat. But I reject him openly and with all means, to protect the genetic material of my volk.[515]

The same is true for Otmar von Verschuer. After the November pogrom, which filled him with "horror and disgust," as he asserted after the war, Verschuer no longer complied with further requests "to be scientifically active in the field on the Jewish question," and retreated "from the race question thus bespattered and burdened with guilt."[516] But there is conclusive evidence to the contrary. As early as January 1939 – that is, just 2 months after the November pogrom – he held another speech about "The Physical Race Attributes of Jewry" in the framework of a lecture series called "Jewry and the Jewish Question" organized by the Reich Institute for the History of the New Germany in the main auditorium of the University of Berlin. According to the report in the *Frankfurter Zeitung* he described attributes that were supposedly typical for the Jewish population, including dispositions for and resistance against certain illnesses, emphasized the "Jew's need for doctors and fear of illness" and even referred to allegedly specific forms of criminality: "[...] the Jews showed a reduced rate of crime for assault and theft, but penalties were considerably above average for defamation, fraud and forgery [...]." As an explanation, Verschuer, picking up on Fischer, alleged "that the typical characteristics for today's Jew are mainly derived from the Near Eastern-Oriental basic stock," which had "experienced a certain loosening up" through miscegenation.

> On the other hand, a consolidation and standardization also occurred in the genetic material of the Jewish race through the breeding influence (selection preferences), to which the Jews have been subject during the last two millennia: far away from the natural bonds to a certain landscape, adapted to the living conditions of the city, with a preference for working in commercial profession, constrained by Talmudic education to purely formal logical intellectual activity and in the religious-volkish belief that the Jews are the "chosen people," Jewry was able to preserve itself as a racial type and reshape itself.[517]

---

[514] For a biography: Klee, Personenlexikon, pp. 311 f.

[515] Fischer, Erbe als Schicksal, lecture, held at the 8th Technische Tagung des Vereins für bergbauliche Interessen in Essen on June 20, 1939, quoted in Müller-Hill, Tödliche Wissenschaft, p. 40.

[516] Verschuer, Mein wissenschaftlicher Weg, p. 25, MPG Archive, Dept. III, Rep. 86 A, No. 3–1.

[517] "Die körperlichen Rassenmerkmale des Judentums." Ein Vortrag an der Berliner Universität, in: Frankfurter Zeitung, 17/1/1939, clipping in BArch. Berlin, BDC, DS B 43. Cf. also Heiber, Walter Frank, p. 627.

Striking is that Verschuer applied to the "Jewish question" a dynamic concept of race founded in Darwinism and oriented toward population genetics, reminiscent of Walter Scheidt's "breeds," while Fischer proceeded from a long outdated, static Gobinist concept of race when discussing the Jews. In this, Verschuer not only adapted racial anti-Semitism to the modern science of genetics, but at the same time made it possible to apply the concept to National Socialist Jewish policy. Verschuer made a practical contribution to this in the years 1943–1945, when he worked intensively on a research project to develop a serological race test – and this was of fundamental importance for National Socialist race and Jewish policy. When Verschuer designated a "new total solution of the Jewish problem" as "a political demand of the present" in the second edition of his *Leitfaden der Rassenhygiene* ("Manual of Race Hygiene") published in 1944,[518] this project was in full gear –Verschuer still hoped to be able to lay the scientific foundations for this "final solution."

[518] Verschuer, Leitfaden, 2nd edn., pp. 138 f.

Chapter 4
In the Realm of Opportunity: The Kaiser Wilhelm Institute for Anthropology, Human Heredity and Eugenics during World War II, 1938/42–1945

4.1 The "Reorganization" of the Institute under the Banner of Phenogenetics, 1938–1942

4.1.1 Preliminary Considerations in the Years 1938/42

On March 8, 1940, Eugen Fischer wrote a long, confidential letter to Otmar von Verschuer, director of the Institute for Genetic Biology and Race Hygiene at the University of Frankfurt at that time. In this letter Fischer expressed critique – and certainly also self-critique – about the scientific development of his institute since the mid-1930s.

At first glance this critical assessment seems surprising. The KWI-A had profited considerably from the genetic and race policy of the National Socialist regime. Money flowed abundantly, research projects received generous support, the expansion of the institute proceeded. As the deputy chairman of the Medical Biology Section of the "Academic Council" of the KWG, and even more so as a member of the Expert Committee for "Anthropology and Ethnology" of the Emergency Association of German Science and the Reich Research Council, respectively, the position Fischer held within his area of expertise was central in terms of research strategy.[1] The political prestige and social recognition of the KWI-A, and the scientists working there increased constantly. Fischer himself received a number of honors and accolades in the Third Reich, of which his election to membership in the Prussian Academy of Sciences in 1937 was the most important. Just a year earlier, in 1939, Fischer had been awarded the Goethe Medal for Art and Science.[2]

Yet in March 1940 Fischer was not satisfied with the development of the institute. The "Faustian bargain" he had entered into with the National Socialists entailed numerous additional duties for Fischer and his staff, as experts in political bodies, as assessors on the hereditary health courts, as evaluators, teachers, popular

[1] Lösch, Rasse, p. 273.

[2] Cf. the "Ehrenliste" ibid., p. 277.

speakers for public lectures, as representatives of the new Germany at international congresses, all of which were performed at the cost of the scientific work. More serious was that the emphases of research had shifted as a consequence of the interconnections with politics – and not necessarily in the direction Fischer would have wished. In 1933, under pressure from Arthur Gütt and with a view to the genetic health policy of the new rulers, Fischer had placed the stress on genetic *pathology*. With Verschuer's departure in 1935, Fischer had ceased forcing genetic pathology research, although the projects in progress were continued, and the emphasis was shifted to strengthening genetic *psychology* instead. The dual course shift had the result that the research program was visibly fragmenting into unrelated, individual projects. Fischer recognized that the institute was in danger of losing its scientific focus. As he wrote in his letter to Verschuer of March 1940:

> I have been very concerned recently, even as long ago as a year or two before the war, that the institute is working, so to speak, "without a plan." That was not always the case. When you were here, our first major task was to elaborate and test the twin method, and with this method to set human genetics properly on its feet. And I believe we can say that we managed to do this. In *this* field only touching up should be necessary. *Also* quite important. The point is to fill in the quite significant gaps and deepen our knowledge. But some areas, for instance, normal morphological attributes, but also numerous diseases, have pretty much been exhausted for twin research.[3]

Here it becomes apparent once again that from the founding of the KWI-A into the second half of the 1930s, twin research had the function of a scientific paradigm for the institute, providing an ensemble of axioms and premises, concepts, theories and models, methods, tools, and model examplars, symbolic generalizations, and implicit valuations to guide research practice. That this paradigm no longer carried the research in Dahlem – as Fischer expressed clearly – had to do with more than just Verschuer's departure. Rather, the paradigm had exhausted itself. According to Thomas S. Kuhn, a paradigm shift occurs when the anomalies within a disciplinary system accumulate, that is, when the number of riddles that cannot be solved using the means of the normal science guided by the paradigm predominant up to that time increases at such a rate that the disciplinary system as a whole enters a state of crisis.[4] And indeed, in the 1930s, twin research, which for several years had been considered the king's road of human genetics, as discussed above, reached its conceptual and methodological limits and became the target of criticism. What was sought was a new paradigm that opened up the perspective to solve problems the old one could not. Where should new impulses come from now? Fischer did not believe that the orientation of Verschuer's institute in Frankfurt could serve as a model for Dahlem:

> You work more clinically in Frankfurt. I don't want that here. And perhaps it is correct that a pure research institute like the one here leaves the actual clinical matters to the university institutes, those for genetic biology and truly clinical ones.[5]

[3] Fischer to Verschuer, 8/3/1940, MPG Archive, Dept. III, Rep. 86 A (Münster), No. 9. Quotes and emphases added by hand to the original.

[4] Kuhn, Structure of Scientific Revolutions.

[5] Fischer to Verschuer, 8/3/1940, MPG Archive, Dept. III, Rep. 86 A (Münster), No. 9.

Aside from the problem of demarcation from the university clinics and institutes working in the area of genetic pathology, Fischer probably foresaw that a clinical orientation of his institute would cause conflicts of competence with two other Kaiser Wilhelm Institutes: the Kaiser Wilhelm Institute for Brain Research, which, under the direction of Hugo Spatz since 1937, had turned increasingly to questions of genetic pathology in neurological disorders, mental illnesses and mental disability, and above all with the German Research Institute for Psychiatry in Munich under Rüdin. In Munich the very founding of the Department for Genetic Psychology had been taken as an affront, and the vehement exchange of blows concerning diabetes research in 1937 had shown that the two institutes were bound to get in each other's way in the field of genetic pathology. Against this background it seemed to Fischer, who had never worked purely clinically anyway, that a one-sided orientation of research at the KWI-A toward genetic pathology – while neglecting Fischer's own original research field, the heredity of "normal" (not pathological) attributes – to be a strategic mistake, although he certainly held genetic pathology to be a constitutive element of his further research strategy.

But as Fischer established self-critically, his own research in the field of physical anthropology did not have the potential to constitute a new paradigm either:

> For my own (personal) work, of course, I have plans that include my doctoral students. The one project is, as for many years, the research of the conditions for the shape of the skull; the other the bastardization problem. But these two topics cannot fill an entire institute, and assistants from the field of *medicine*, who want to pursue practical activity later, cannot, or at least not exclusively, be set to work on such subjects. And since you have been gone I feel both a great void and a sense of being orphaned.[6]

The Department for Human Genetics had been dissolved in 1935, as mentioned above. In addition to the Department for Anthropology led by Fischer himself, the only pillar of the institute that remained was the Department for Race Hygiene headed by Fritz Lenz. However, Fischer lamented, it could hardly be expected that new impulses would come from there.

> For Lenz does not take care of any of this. He works without planning, assigns themes to pupils without a plan. These themes are individual questions of a genetic pathology or race hygiene biological nature, which occur to him while editing Baur-Fischer-Lenz or during his critical perusal of the literature.[7]

Such occasionally arising, isolated themes were to be pursued in the future as well, and the work of the Department for Genetic Psychology under Kurt Gottschaldt was to be continued according to Fischer's wishes.

> But as to the main point, such an institute needs an ambitious plan. And because I can not receive it from Lenz's sphere of interest, let alone through his initiative, I do it alone. But such a plan is conceived for a number of years, and certainly – not only presumably – longer than I will be in office here. Because I do not doubt in the least that you will be my successor someday, in truth I would like to begin with a long-term plan only if I have the hope that you like it enough to pursue it further.

---

[6] Ibid. In parentheses: handwritten edition. Emphases added by hand to the original.

[7] Ibid.

> I once told you in passing that I would like to set up a collection of work for the genetic biology of humans. I will do this in any case. But what I would most like to do now, or at least right after the war, is to begin work in this sense as well. I am thus thinking of an ambitious phenogenetics.[8]

With this the decisive catchword had fallen. It is important to keep in mind that the idea was not new and that this was not the first time Fischer had discussed it with Verschuer. Rather, phenogenetics was already structured within the concept of anthropobiology, which had been one of the institute's founding fields. From 1938 on, Fischer – in lively intellectual exchange with Verschuer – worked intensively to elaborate a concept of phenogenetics. The impetus had come from the zoologist Alfred Kühn (1885–1968), since 1937 Deputy Director and then Director of the Department for Animal Genetics at the Kaiser Wilhelm Institute for Biology. Kühn had turned to Fischer in February 1938 in his capacity as chairman of the German Society for Genetics, in order to discuss with him the thematic conception of a congress to be held in fall 1938. While the zoology and botany sections were to deal with topics from the area of "Chromosome Structure, Genes and Effective Agents," Kühn proposed the general topic "Phenogenetics" for the human genetics section. In the conversation between Kühn and Fischer, the initial idea was to deal with the topic in three separate talks about "normal," "pathologic" and "psychopathic" attributes, whereby Kühn placed particular value on supplementary demonstration cases. Fischer warned "that human genetics in most cases [does not yield] results as clear [...] as the experimenters are accustomed to from their material."[9] Verschuer, with whom Fischer consulted immediately, confirmed this:

> We are, of course, still far from being able to perform phenogenetics of the kind Kühn conducts on the flour moth. The objective lies clearly before us, and we are seeking to move forward by combining research on the history of development and pathogenetics with pure genetic analyses. At first we must be content to perform causal analysis of the phenotypical manifestations of variation. If phenogenetics is thus conceived in this more humble sense, I consent to the topic. I also think it is the most current one.[10]

Verschuer also agreed with the "tripartition" of contents. It was out of the question that Fischer himself would give the talk about normal morphological and physiological attributes – Kühn had also advocated this solution, solely by virtue of the works on the spinal column he had cowritten or initiated. Supplementary demonstrations, Verschuer wrote, could be presented by Fischer's staff members Wolfgang Abel and Engelhardt Bühler. The second talk about "pathologic morphological and physiological attributes, that is, the entire area of pathology with the exception of psychiatry," Verschuer allowed, would "certainly appeal" to him, yet due to his workload he could not take it on. In his stead he proposed his chief physician Ferdinand Claußen [11] and offered to "advise [him] in every way in the selection of

---

[8] Ibid.

[9] Fischer to Verschuer, 8/2/1938, ibid.

[10] Verschuer to Fischer, 10/2/1938, ibid. The following quotes also come from this letter.

[11] As an alternative to Claußen, Verschuer suggested Friedrich Curtius. "The pupils of Rüdin seem to me to be somewhat one-sided specialists."

examples and in working through the material, especially with regard to the clinical side." He was also willing to add on to Claußen's presentation "a demonstration or two. We have abundant material at our disposal, and during the editing of the chapter 'Anomalies of the Shape of the Body' for 'Baur-Fischer-Lenz' the scientific problems just keep coming." The third talk about "normal psychology and psychopathy including psychiatry" was to be held by Kurt Gottschaldt, director of the Genetic Psychology Department at the KWI-A.[12] Verschuer's proposals were implemented, and so the human genetics section at the Congress of the German Society for Genetics, which finally took place in Würzburg in September 1938, was firmly in the hands of the KWI-A and its daughter institute in Frankfurt – Verschuer had been responsible for directing the section.[13] The three talks by Fischer, Claußen and Gottschaldt were published immediately – in greatly expanded versions – in the *Zeitschrift für induktive Abstammungs- und Vererbungslehre*, the journal founded by Erwin Baur in 1908 which was thus the oldest journal in the world on the field of genetics.[14] Above all Fischer's "Attempt for a Phenogenetics of the Normal Attributes of Humans," one of the few publications by Fischer from the second half of the 1930s that was scientific in the strict sense, and the most extensive by far, constituted the conceptual basis for the restructuring of the KWI-A at the start of World War II.

4.1.2 Phenogenetics: A New Paradigm

What did the term *phenogenetics* mean to Fischer, and where did the term and its content come from? As related in detail in previous sections, in the 1920s, German genetics – in the sense of the *developmental genetics* – had opened up to evolutionary biology, developmental physiology, and embryology. One of the earliest attempts to close the gap between genetics and developmental physiology was the "phenogenetics" introduced by the zoologist Valentin Haecker (1864–1927) of Halle in 1918.[15] In contrast to experimental embryology, which proceeded from the

[12] Verschuer suggested calling upon Luxenburger, Stumpfl, and Conrad of the German Research Institute for Psychiatry to present supplementary demonstrations.

[13] In his memoirs, Verschuer emphasized the scientific yield of this conference less than its special atmosphere: it was completely dominated by the effect of the "Sudetenland crisis": "The leaden weight of an impending storm lay on everyone's mood. One listened to the scientific lectures only as if from a great distance. In one session dealing with the topic of the phenogenetics of humans [...] I had to preside. Every movement was mechanical, as if the scientific thoughts moved themselves along in the same old groove, while emotional life was occupied completely by the impending war. I had to leave the congress early and return to Frankfurt to stand ready for soldierly disposal." Verschuer, Erbe – Umwelt – Führung, "Professor in Frankfurt (1935–1942)" section, p. 16, MPG Archive, Dept. III, Rep. 86 A, No. 3–1. Cf. Geyer, Würzburger Vererbungskongreß.

[14] Fischer, Phänogenetik; Claußen, Phänogenetik; Gottschaldt, Phänogenetische Fragestellungen.

[15] Haecker, Eigenschaftsanalyse; idem., Aufgaben der Phänogenetik; idem., Aufgaben und Ergebnisse (here, p. 100, Haecker indicated that – aside from his own doctoral students – 17

fertilized egg to follow the further development of an organism, this "History of Development Analysis of Attributes" (*Entwicklungsgeschichtliche Eigenschaftsanalyse*) started with the finished phenotype and drew conclusions about the effects of the genotype. Still quite bound up in the descriptive methodology of the nineteenth century, Haecker's phenogenetics did not survive much past the early death of its founder in 1927. In the late 1930s it could be regarded as out of date, and the question arises as to why Fischer referred explicitly to Haecker in 1938. Upon closer observation, however, it becomes apparent that Fischer took over from Haecker little more than the term phenogenetics. In content he referred instead to Richard Goldschmidt, who had submitted a far more modern version of a genetics based on the history of development in his work *Physiologische Theorie der Vererbung* ("The Physiological Theory of Genetics") published in 1927.[16] Fischer referred to this work explicitly, albeit in a rather concealed location. It appears that in the conceptual-programmatic sections of his talk, especially the formation of concepts, he was reluctant to use Goldschmidt as a basis, for Goldschmidt, who had been a Deputy Director and the head of the Department for Animal Genetics of the Kaiser Wilhelm Institute for Biology (and, as mentioned above, also a member of the Board of Trustees of the KWI-A), had been forced into emigration. In fact, Goldschmidt's successor Kühn had urged Fischer in this direction when he suggested the framework topic for the human genetics section.

The scientific object of the newly sketched field of research was *phenogenesis*. Under this, Fischer understood the development "between the genome generated upon the fertilization of the egg and the complete phenome." Instead of the conceptual pair "genotype/phenotype" introduced by Wilhelm Johannsen (1857–1927), Fischer proposed using the terms "genome" and "phenome" to differentiate between the genetic type and the way in which it is manifested. The process of phenogenesis, Fischer continued, was influenced on the one hand by the effects and reciprocal actions of the genes, on the other hand by "a multiplicity of environmental and influencing conditions," which he designated with the term "peristasis." Peristasis was, as Fischer emphasized expressly, more broadly conceived than the concept of "environment," which generally only covered the influences exerted on an organism *after* its birth, like nutrition, light, chemical substances, trauma, movement, and rest. Fischer also counted the various developmental stages of an organism like youth, maturity and age, then pregnancy and illness, and finally "the mental" as environmental factors. According to Fischer these factors mainly affected the phenogenesis of attributes that do not manifest themselves until later in life. Peristasis included not only these factors, but all prenatal influences, such as the metabolism between the fetus and the maternal organism through the placenta, but also pressure, tension, or swelling caused by the position of the unborn child which can influence its development, as well as the regulation processes effective in the development of the embryo. Peristasis was thus a collective

---

predominantly German-speaking biologists had taken over his concept and usually the term phenogenetics as well); idem., Bestrebungen. Cf. Harwood, Styles, pp. 52–55.

[16] Goldschmidt, Physiologie.

term for all factors affecting phenogenesis that did not lie directly in the genes. Fischer emphasized that the "series of forces" triggered by genome and peristasis do not simply complement each other cumulatively, but are related in a very complex system of interdependence and synergy that is subject to constant change. The task of phenogenetics was thus to disentangle the networks of effects exerted by genes and peristatis for analytical purposes and to pursue their effects and interactions all the way up to the complete phenome. Methodologically this task was to be approached through a combination of classical genetics, embryology and developmental mechanics, anthropometry and clinical diagnostics, whereby Fischer repeatedly emphasized the utility of combining the animal model with observation of humans.

At this juncture it is worth taking another look across the Atlantic. Since the early 1930s the interest of geneticists there, too, encouraged in part by the Rockefeller Foundation, shifted increasingly to physiological genetics. In sharp competition with the research group around Kühn, the geneticist George Beadle (1903–1989) and the embryologist Boris Ephrussi (1901–1979) worked in Pasadena to bridge the gap between classical genetics and developmental physiology. They may have lost the race to identify kynurenin, but in the early 1940s Beadle and Edward Lawrie Tatum (1909–1975) succeeded in finding evidence for the "one-gene-one-enzyme" hypothesis they had advanced during experiments on the neurospora mold – a pioneering success, for which they were awarded the Nobel Prize in 1958. While this had directed attention to the effects of genes, the embryologist, geneticist, and evolutionary biologist Conrad H. Waddington broadened the horizon in 1942 with his concept of "epigenetics," in which he – in nearly the same words as Eugen Fischer 4 years previously, and also with reference to Valentin Haecker's phenogenetics – shifted the complex developmental processes from the genotype to the phenotype into the focus of interest:

> For the purposes of a study of inheritance, the relation between phenotypes and genotypes can be left comparatively uninvestigated; we need merely to assume that changes in the genotype produce correlated changes in the adult phenotype, but the mechanism of this correlation need not concern us. Yet this question is, from a wider biological point of view, of crucial importance, since it is the kernel of the whole problem of development. Many geneticists have recognized this and attempted to discover the processes involved in the mechanism by which the genes of the genotype bring about phenotypic effects. The first step in such an enterprise is – or rather should be, since it is often omitted by those with an undue respect for the powers of reason – to describe what can be seen of the developmental processes. For inquiries of this kind, the word "phenogenetics" was coined by Haecker.[17] The second and more important part of the task is to discover the causal mechanisms at work, and to relate them as far as possible to what experimental embryology has already revealed of the mechanisms of development. We might use the name "epigenetics" for such studies [...].[18]

[17] In this Waddington referred to Haecker, Eigenschaftsanalyse.

[18] Waddington, Epigenotype, quote: pp. 18f.

The concept of epigenetics has since undergone a series of transformations, but has managed to persist – off the track of mainstream genetics. Indeed, in the most recent bioethical debates many hopes are pinned on the concepts of epigenesis and epigenetics, as they promise to break open the "reductionist approach" of classical genetics.

From Fischer's perspective, the new paradigm of phenogenetics offered several advantages. First, it permitted a whole series of projects that had been performed at the KWI-A and its periphery in the 1930s, which were rather loosely connected, to be related to each other under a complex of issues that applied to all of them. This was true first of all for a great number of works by Wolfgang Abel, Georg Geipel, Bernhard Duis, and others on the methodology of dermatoglyphics, that is, the genetics of epidermal patterns, on the increased frequency of characteristic epidermal ridge patterns for certain human races, and on the connections between defective epidermal ridge patterns and physical disability or mental diseases. Fischer attributed a central position in the field of phenogenetics to the works of his earlier pupil Konrad Kühne on the genetics of the variations of the spinal column, which had been continued at the KWI-A in Maria Frede's work on rats. Great value was also ascribed by Fischer to the embryological studies by Rita Hauschild on the skulls of Negroid and Caucasion fetuses and by Baeckyang Kim about race differences in embryonic pig skulls – even at the time, both studies were understood explicitly as contributions to phenogenetics.[19] According to Fischer, various other works originating from the KWI-A on morphology and the genetics of human hair growth, the auricle, on asymmetries in body structure, on the heritability of stature, on miscegenation and on genetic pathology could also be classified under the umbrella of phenogenetics.

On the other hand, second, the paradigm of phenogenetics demarcated a broad research area that the existing works had barely begun to cover, and which was underdeveloped in terms of both breadth and depth. At the congress in Würzburg, Fischer established retrospectively in 1940, it became apparent that the "true course of development" was only really known for coloboma, the congenital gap in the eye area due to the insufficient closure of the fetal eye cleft, on the iris, choroid, lens, or lid.

> Just as we have a history of development of every normal organ, we should have an exact history of development of every hereditary disease. Everything has yet to be done here. Much is also missing on the heredity of normal things.[20]

The paradigm of phenogenetics was thus open enough to provide the foundation for a comprehensive research program with questions covering all of its areas.

A further advantage of turning toward phenogenetics was that, third, the research focused on a form of human genetics that was compatible with the orientation of

---

[19] "We are currently working on breeding embryonic material for the entire heredity of vertebral varieties. I had the idea for this and took the first steps back in 1933, when Miss Frede was working on the first rat embryos, then later, when Kim examined the embryonic pig skulls and Hauschild the Negro skulls. That was true phenogenetics. But that is really everything." Fischer to Verschuer, 8/3/1940, MPG Archive, Dept. III, Rep. 86 A (Münster), No. 9.

[20] Ibid.

developmental genetics predominant in German animal and plant genetics at the time.[21] In his lecture in Würzburg, Fischer referred to the work of the developmental physiologist Hans Spemann (1869–1941), who had performed experiments on amphibian embryos in the 1920s, which proved that some parts of the embryo, such as the primitive roof of the mouth and the eye socket, act as "organizers" to "induce" the formation of other structures in the embryo. Spemann had not attempted to explain the inductive effect of the organizers by investigating the genes. Alfred Kühn made more progress on this with his experiments on the flour moth *Ephestia kühniella*, to which Fischer referred several times in his comments. Kühn had found a mutant Ephestia with red eyes rather than the usual black ones, and his doctoral student Ernst Wolfgang Caspari (* 1909) was able to prove that, by injecting tissue from wild moths into the larvae of the red-eyed mutants, the eye color of the mutant could be adapted to that of the wild type. A substance missing in the mutants was apparently added through the injection. Genes, it was concluded, obviously work through enzymes. If an enzyme is lost through mutation, this can block the transformation of a certain substance into another. Through artificial implementation of the missing enzyme – in the case of the light-eyed Ephestia this was the tryptophan derivate kynurenin, as two assistants to Adolf Butenandt (1903–1995), Erich Becker and Wolfhard Weidel, were able to prove – it was possible to generate a "phenocopy" of the wild type, an idea that apparently fascinated Eugen Fischer. Yet, even more often than he referred to Kühn, Fischer brought up Hans Nachtsheim's work on the genetic pathology of rabbits.

Consequently, in March 1940 Fischer presented to Verschuer the idea of bringing Nachtsheim to the KWI-A as director of a new Department for Experimental Genetic Pathology, in order to supplement his studies on the phenogenesis of genetic illnesses of the rabbit with "parallel studies of a clinical nature on humans" as a way of connecting animal and human genetics. Fischer's remark, "I have no idea whether he [Nachtsheim] would want to," indicates that Fischer had not yet negotiated with Nachtsheim at this time. He first wanted to await his designated successor's opinion of this plan, and further of the plan associated with it – quite explosive in terms of institute politics – to "completely dismantle" the Department for Race Hygiene headed by Fritz Lenz, as "race hygiene could then be taken care of in the university institute." In other words, the KWI-A was supposed to give up race hygiene as a field of research, and Fritz Lenz gradually be forced to the margins.

Fischer closed his letter to Verschuer with the request that he not answer in writing, for he hoped that there would be opportunity at Easter 1940 to discuss the complex of topics in person. Hence we know nothing about Verschuer's immediate reaction. Yet the further course of events suggests that Verschuer fully agreed with

[21] "Drosophila genetics has been our pacemaker until now. This appears to be nearly over; it no longer teaches us anything new." (ibid.). This can be interpreted as a renunciation of the drosophila genetics at the Genetic Department of the KWI for Brain Research under Nikolaj V. and Elena A. Timoféeff-Ressovsky. Of course, it must be remembered that the Timoféeff-Ressovskys, before turning to the genetics of mutations and populations, made significant contributions to developmental genetics. Cf. Harwood, Styles, pp. 55 f.

the research plan developed by Fischer. The two most important conceptual works by Verschuer from the year 1939 – his lecture about "The Genotype of Humans" to the main assembly of the Kaiser Wilhelm Society in Breslau on May 24, 1939 and his presentation "On the Genetic Analysis of Humans" for the 7th International Congress for Genetics in Edinburgh[22] (which he was not able to give himself due to the early departure of the German delegation, but was read to the audience on August 28, 1939[23]) – first of all show that Verschuer picked up on Fischer's impulses immediately, and also illuminate the background against which Fischer's and Verschuer's new conception must be viewed: over the course of the 1930s, classical Mendelian genetics was undergoing a dramatic and extensive process of transformation. The idea generally accepted up to that time, that every attribute was simply transmitted as dominant or recessive, monofactorial genetic information, did not hold up to the results of mutation research, population genetics and developmental physiology. Thus Mendelian genetics was giving way, in the words of the day, to "higher Mendelism,[24] which presumed much more complicated mechanisms of heredity. It became generally accepted that genes could not be observed in isolation from each other, but only in the context of the genotypical setting – the effect of one gene was always influenced by other genes, and even by the genome as a whole. It was acknowledged that the genes on the chromosomes are not just pearls strung on a string in any order, but that the effect of each gene depends on its position in the genome. With increasing clarity it became apparent that these mutual effects within the genome, but also prenatal influences on the intra-uterine environment during maturation of the embryo, and even influences from the external environment, had modifying effects on the way genes were manifested in the process of phenogenesis. The phenomenon of "weak genes" made its first appearance. The team of the KWI for Brain Research around Nikolaj Vladimirovich Timoféeff-Ressovsky and Elena Aleksandrovna Timoféeff-Ressovsky had attempted to grasp the phenomenon of the variations in how such genes were manifested in terms of the three concepts "penetrance" (the frequency with which a genetically conditioned attribute develops in the phenotype), "expressivity" (the degree to which it develops) and "specificity" (the nature of its development depending on the part of the body the gene must affect for this development) – a terminology that was picked up everywhere, including by Fischer and Verschuer. It was acknowledged that in many cases a single gene is involved in the development of several attributes (pleiotropy) and, inversely that the development of a single attribute can be influenced by multiple genes (polygeny). Furthermore, the advances in differential diagnostics showed that one and the same clinical picture can be caused by both genetics and environment (heterogeny). Finally, the results from radiation and

[22] Verschuer, Erbbild vom Menschen; idem., Bemerkungen zur Genanalyse.

[23] Cf. Roth, Schöner neuer Mensch, pp. 11–13.

[24] The term was coined in 1934, probably by Günther Just, and picked up on immediately by Verschuer. Cf. Just, Probleme des höheren Mendelismus; Verschuer, Genetic pathology, 2nd edn., p. 7. Cf. Weß, Humangenetik, pp. 173–176.

population genetics suggested that the rates of mutation were higher than originally presumed, but that the heterozygotic mutants did not become visible because the gene did not necessarily develop in their phenotype.[25]

Verschuer, as his lecture texts indicate, was completely up to date in the contemporary specialized discourse; his institute in Frankfurt had even made a significant contribution to theoretical research in human genetics in 1938, when one of Verschuer's staff members, Bruno Rath, on the basis of a family study of a "bleeder clan," succeeded in finding the first proof of a crossing-over (exchange of genes or gene sections through the recombination of chromosome fragments) in humans.[26] Against the background of higher Mendelism, Verschuer was fully aware that the previous conception of human genetics required greater differentiation, an expanded catalog of questions, and a larger arsenal of methods. Fischer's suggestions could hardly have come at a better time.

In his Breslau lecture Verschuer first related the success story of human genetics. By that time around 1,000 of the estimated 30,000–60,000 genes in humans were known, along with several hundred hereditary diseases, for most of which the heredity had been illuminated. "The human being is an object of the human sciences that has been examined in manifold ways [...]." Yet there remained much to be done. Genetic analysis could no longer be content with using the methods of family and twin research to reveal the dominant or recessive mode of inheritance of a gene. On the contrary, a whole bundle of new questions had to be posed:

> What meaning does a gene have for development? At what point in time and at what location does it become visible? What changes in the tempo or in the chronological sequence of certain developmental processes does it cause? Is there the possibility of preventing pathological consequences? How do the individual genes work together? Do certain genes reveal peculiarities according to the race or constitution of the individual human being?[27]

In summary it can be stated that the cognitive advances in the field of human genetics in the 1930s practically forced an expansion of classical Mendelian genetics. Theoretically, expansion was conceivable in various directions – from mutation research, to the "synthetic theory of evolution," all the way to molecular genetics. However, considering its own resources, and also with a view to the orientations of competing research institutions, the paradigm of phenogenetics seemed most promising for the KWI-A. Moreover, it was quite advantageous for Verschuer, because his personal research interest in clinical genetic pathology dovetailed perfectly with the new paradigm. As such – another point that must not be overlooked – it was practically tailored to Verschuer and provided him a weighty advantage over Lenz as a potential rival for Fischer's succession.

---

[25] This final aspect is placed all too strongly in the foreground by Roth, Schöner neuer Mensch, esp. pp. 22, 25, 39 f., 54–56.

[26] Rath, Rotgrünblindheit. Verschuer reported about this before the fact in January 1938, to the Frankfurt Medical Society. Cf. Verschuer, Erster Nachweis von Faktorenaustausch; idem., Frage des Faktorenaustausches.

[27] Verschuer, Erbbild vom Menschen, pp. 5 f.

4.1.3 The Succession Issue

At this juncture our discussion turns to the thesis by Niels C. Lösch, who goes into great detail about the changes that took place under the banner of phenogenetics at the KWI-A in the years 1938–1940. However, because he is fixed too one-sidedly on personnel-policy strategies, he interprets the development as a kind of "false label" designed to "prepare the ground for Verschuer."[28] There is no question that Fischer, who celebrated his 65th birthday on June 5, 1939, had been building up his pupil and friend Verschuer as his successor for a long time already, and that the establishment of the new paradigm of phenogenetics was linked intimately with Verschuer's person. It is also indisputable that Fischer had long since begun taking precautions to nip in the bud any aspirations Fritz Lenz might have for the post of the institute director – although, it must be added, there are no indications that Lenz pursued ambitions in this direction.

In September 1934, when Fischer, much to his dismay, learned of the plans to call Verschuer to Frankfurt, he immediately began considering whether the chair in Frankfurt could serve Verschuer as a stepping stone on the path to succeeding Fischer. In November 1938 – that is, while the Würzburg lecture was being printed – Fischer then set the course for the future development of the institute in a talk with Ernst Telschow, General Secretary of the Kaiser Wilhelm Society: "For the case of his departure upon reaching the age limit in 3–4 years, Professor Fischer nominated Professor Verschuer of Frankfurt as his successor."[29] In March 1939 Fischer informed Verschuer of this conversation, after congratulating him for the most recent evidence of the Kaiser Wilhelm Society's favor – "exchange professor in London! Speaker at the general meeting [...] Some time ago Mr. Telschow and I had a long talk, during which we also discussed you in great detail. He is informed for now and for the future and was entirely of my opinion."[30] In his exchange of opinions with Telschow, Fischer had indicated that he rejected the plan to set up a major institute for anthropology at the Friedrich Wilhelm University in Berlin, advising Telschow "to pare this institute down considerably and plan it [...] merely as an institute for race hygiene. As director Prof. Lenz would then be suitable on a full-time basis."[31] This is yet another indication of Fischer's strategy of strengthening Lenz's role at the university and pushing him to the margins of the KWI-A, even though at this point in time he was still advocating Lenz's appointment as

[28] Lösch, Rasse, p. 374.

[29] File note by Telschow about a meeting with Fischer on 18/11/1938, 21/11/1938, MPG Archive, Dept. I, Rep. 1 A, No. 2399, p. 130.

[30] Fischer to Verschuer, 9/3/1939, MPG Archive, Dept. III, Rep. 86 A (Münster), No. 9. The KWG had appointed Verschuer for an exchange of professors with the Royal Society in London, which took place in June 1939. Cf. Verschuer, Erbe – Umwelt, Führung, "Professor in Frankfurt (1935–1942)" section, pp. 17–20, MPG Archive, Dept. III, Rep. 86 A, No. 3–1.

[31] File note by Telschow about a meeting with Fischer on 18/11/1938, 21/11/1938, MPG Archive, Dept. I, Rep. 1 A, No. 2399, p. 130.

deputy director of the institute. In July 1940 – by this time Fischer had coordinated his plans for reorganizing the institute with Verschuer – the departing director expressed himself more clearly to Telschow:

> In repetition of earlier conversations, Prof. Eugen Fischer designated Prof. von Verschuer in Frankfurt as a suitable successor. It would then be appropriate to grant Prof. Lenz the title of "Director" because of his age, without entrusting him with the direction of the institute. Prof. Fischer held it even more appropriate to transfer Prof. Lenz to the Institute for Race Hygiene at the University of Berlin, which – at present consisting of two rooms in the Hygiene Institute – would have to be expanded.[32]

Then, in October 1940 Fischer got down to brass tacks:

> Within his institute Prof. Fischer wants to have a special Institute for Race Hygiene under Prof. Lenz, who thus would receive the title of "director," as it were, but without becoming Prof. Fischer's deputy. On the contrary, Prof. Fischer wants to prevent this, in consideration of the proposed succession to his position by Prof. von Verschuer.[33]

At the same time Fischer conveyed his intention to rename his institute. In the future it was to be called the "Kaiser Wilhelm Institute for Genetic and Race Science" (*Kaiser-Wilhelm-Institut für Erb- und Rassenkunde*) and thus in its very name express the demarcation from race hygiene.[34]

Fischer's attitude toward race hygiene was expressed quite clearly 1 year later, in October 1941, when he argued – much more aggressively than Lenz, who lacked the requisite tact – for the expansion of the university's Race Hygiene Institute:

> [...] today race hygiene [has] become a state policy, it no longer requires propaganda. Race hygiene is a required lecture and examination subject for medical students. I can no longer recognize race hygiene as such as a research subject; rather, the research subject is its substrata, first of all human genetics and then demographics. [...] For these reasons I hold the expansion of a university institute for race hygiene at the greatest German university to be a quite self-evident necessity. [...] Because of the auspicious historical development in the Third Reich, at the Kaiser Wilhelm Institute this race hygiene department must be dismantled rather than expanded.[35]

Race hygiene, as the theory of the practical implementation of the knowledge in human genetics and demography, one could summarize Fischer's argument, had a right to exist at the KWI-A during the Weimar Republic, but in the Third Reich race hygiene seemed to him unnecessary baggage that distracted the institute from its theoretical research. By no means did Fischer want to force Lenz out of the institute entirely – as an astute critic he made a major contribution to the conceptual foundations

---

[32] File note by Telschow about a meeting with Fischer, 24/7/1940, ibid., p. 141.

[33] File note by Telschow about a meeting with Fischer, 18/10/1938, ibid., p. 142.

[34] Cf. also the undated (written around April 1940) paper by Fischer, "*Der Name des Institutes*" ("The Name of the Institute"), MPG Archive, Dept. III, Rep. 86 A (Münster), No. 9 (here Fischer advocated renaming the institute to "*Kaiser-Wilhelm-Institut für menschliche Erb- und Rassenforschung und Institut für Rassenhygiene*" ("Kaiser Wilhelm Institute for Human Genetic and Race Research and Institute for Race Hygiene").

[35] Fischer to Reichsminister für Wissenschaft, Erziehung und Volksbildung, 20/10/1941, MPG Archive, Dept. I, Rep. 1 A, No. 2400, pp. 218–219 v, quote: pp. 218 v-219.

of the research at the KWI-A. The fact that he did not want to see Lenz in the director's post concerned not only his lack of qualities as a science manager, and was not founded only in personal animosities between the two scientists – that, too – but above all Lenz was far too much a proponent of classical race hygiene, which Fischer did not believe had much potential for innovation.

The result is indisputable: Fischer machinated behind the wings in order to guarantee that Verschuer would be appointed director of the KWI-A and to prevent Lenz from offering himself as an alternative candidate. In contrast to Niels Lösch, however, this author advocates the thesis that the paradigm of phenogenetics was *also*, but by no means *only* a means to an end in order to prejudice an impending personal-policy decision. The realization of the newly developed research conception presupposed a kind of package solution: the new orientation of research in progress; opening up new areas of work, but also relinquishing areas that could not be fit into the new paradigm in a meaningful way; integrating scientists who fit into the new research profile; changes to the internal structure of the institute; the creation of an infrastructure to implement new methods in practice, and finally the solicitation of the additional financing these tasks would require. In the years 1938–1940 Fischer resolutely pushed ahead in all of these directions, but initially he met with considerable resistance.

4.1.4 The Alliance Between Eugen Fischer and Leonardo Conti and the Decisive Board Meeting in 1941

The first step turned out to be quite easy. In November 1938 Fischer submitted to the General Administration his plan to set up an external department of the KWI-A for tuberculosis research, in the TB hospital "Waldhaus Charlottenburg" in Sommerfeld near Beetz in the eastern Havel region, under direction of that institution's Directing Physician of the Surgical Department, Karl Diehl. Telschow indicated that there would be no problem providing the 3,000 RM needed to set up the outpost in the coming fiscal year,[36] so that from 1939 on the KWI-A officially comprised four departments. Thus, the first phase of reorganization went off without a hitch.

But in July 1940, when Fischer submitted to Telschow his plan to found a Department for Experimental Genetic Pathology and asked for a budget increase of 10,000 RM for this purpose, the general secretary was less forthcoming. An increase in financing, he replied to Fischer, probably would be impossible "as long as the war lasts." Apparently Telschow believed that the war would end in 1941, but he did not want to raise Fischer's hopes for more funding in that year, either. All the same, he announced that he would attempt to get Max Planck interested in the plan and encourage the provision of support from the Planck fund – a plan Planck rejected immediately upon Telschow's proposal.[37]

[36] File note by Telschow about a meeting with Fischer on 18/11/1938, 21/11/1938, MPG Archive, Dept. I, Rep. 1 A, No. 2399, p. 130.

[37] File note by Telschow, 24/7/1940, ibid., p. 141.

This setback opened up a precarious phase in the reorganization of the institute, for the new Department for Experimental Genetic Pathology was to become one of the pillars supporting phenogenetic research, if not *the* supporting pillar. Fischer needed political protection to succeed in implementing his plan by circumventing the General Administration – but this strategy was also shaky, as Hans Nachtsheim was not exactly considered a convinced National Socialist. Nevertheless: Fischer sought and found the necessary patronage of a functionary located high in the machinery of the National Socialist regime: Leonardo Conti.

It was convenient that a connection to Conti already had been established, as mentioned above, albeit a loose one. In December 1936 Conti had taken over the duties of the Medical Councilor of Berlin. In this capacity he was entitled to a seat and vote on the boards of the KWI for Brain Research and the KWI-A.[38] The KWI for Brain Research must have interested him less – in any case in November 1937 he appointed one of his closest staff members, Director of the Department for the Care of Genes and Race in the Main Health Office of Berlin, Dr. Theodor Paulstich (* 1891),[39] as his permanent representative on the board of this institute, although he reserved the right to participate in future board meetings himself.[40] At first Conti did not have anything to do with the KWI-A, either – in the years from 1937 to 1940 no board meetings took place. Conti had since moved up to the pinnacle of civilian health care: In April 1939 Hitler had appointed him as Director of the Main Office for National Health (*Hauptamt für Volksgesundheit*) and "Führer of Physicians of the Reich (*Reichsärzteführer*) and awarded him the title of *Reichsgesundheitsführer* ("Reich Health Leader"). In August 1939 Conti was also appointed State Secretary for Health Care in the Reich Ministry of the Interior and thus held all of the reins to steer the health matters of the state and the party.[41]

When Fischer endeavored to call a Board Meeting in January 1940,[42] he discovered that the previous chairman of the board, the premier of Saxony, Landeshauptmann Richard Otto, who had resigned his office as senator of the Kaiser Wilhelm Society in 1937 "in quite an abrupt manner,"[43] no longer considered himself to be in office.[44] Because the General Administration – after consulting with Ministry Director Mentzel – did not regard the option of convincing Otto to remain on the

---

[38] Planck to Conti, 1/12/1936, MPG Archive, Dept. I, Rep. 1 A, No. 2403, p. 97. Max Planck solicited Conti's interest insistently: "Because the next sessions of the two boards will probably not be held until the coming spring, perhaps you might first find an opportunity to tour the two institutes."

[39] For a biography: Klee, Personenlexikon, p. 452.

[40] Cf. Schmuhl, Hirnforschung, p. 585.

[41] Kater, Conti; Labisch/Tennstedt, Weg, vol. 2, pp. 393–395.

[42] Fischer to Otto, 10/1/1940, MPG Archive, Dept. I, Rep. 1 A, No. 2403, pp. 108–109. Justifying the long interruption in board meetings, Fischer stated that there had "never occurred anything in particular and on the other hand the years were so eventful politically that one wanted to dispose of the time of such very busy men as sparingly as possible."

[43] Note by Telschow, 13/3/1940, on a letter by Fischer to Telschow, 30/1/1940, ibid., p. 102.

[44] Otto to Fischer, 23/1/1940, ibid., p. 107.

institute's board to be opportune,[45] the question of a successor was raised, an issue Fischer and Telschow discussed in a meeting on March 17, 1940. Apparently Telschow's first suggestion here was Leonardo Conti, followed by Walter Groß. On the following day Fischer expressed his opinion on these suggestions in writing, declaring himself

> completely agreeable to State Secretary Dr. Conti. I find this proposal of yours especially good. Of course, I would have nothing against Dr. Groß either; on the contrary, I would be pleased. But here my good personal relationship with Groß should not be the crucial factor. As a responsible representative of race policy, Groß is not as professionally close to the objectives of my institute as Conti, the Director of the Medical and Race Hygiene Department of the Reich Ministry of the Interior. The connection to him would presumably be more important to the institute; in any case, I already have a connection with Mr. Groß.[46]

It would soon become apparent that this was quite a clever move, especially since Fischer's strongest ally in the Nazi health leadership up to that point, Arthur Gütt, had been ousted by an intrigue in 1938, clearing the way for Conti.[47] But Fischer needed strong political protection to realize his ambitious – and exceedingly costly – plans for the reorganization of the institute under the banner of phenogenetics. On October 18, 1940 the General Administration of the Kaiser Wilhelm Society officially filed Fischer's proposal to offer Conti the Chairmanship of the Board of Trustees of the KWI-A. Fischer, it was recorded there, wanted "first to personally approach [Conti] on this matter."[48] This personal meeting between Fischer and Conti took place on November 12, 1940. It can be presumed that Fischer took this opportunity to relate his plans for reorganizing the institute to the new strong man in the health policy leadership and acquire his support. In any case Conti declared himself willing to accept the Chairmanship of the Board of Trustees and call a board meeting immediately, which was initially scheduled for December 11, 1940, but then postponed to January 9, 1941 due to conflicts with Conti's schedule.[49]

---

[45] Telschow to Fischer, 13/3/1940, ibid., p. 105.

[46] Fischer to Telschow, 18/3/1940, ibid., p. 106.

[47] The situation was all the more piquant because Gütt remained a member of the board. Conti proved magnanimous. He informed Fischer that he woud find it "especially nice if Gütt were retained on the board without further ado." Fischer, as he let Telschow know, had "the sense that he, too, wanted to avoid the appearance of having forced him [Gütt] out." Fischer himself spoke for Gütt's remaining on the board: "Since Mr. Gütt always had especially friendly interest in the institute and did much for it, I, too, would be very pleased if he remained on the board. Of course, this is only possible if, first, the number of members would not be raised to beyond that allowed by any existing regulation, and second, if Mr. Gütt is not expressly nominated as a representative of his ministry." (Fischer to Telschow, 13/11/1940, ibid., pp. 112–112 v.) Both were not the case, and consequently Gütt remained a member of the board. Telschow to Conti, 15/11/1940, ibid., p. 114.

[48] File note by Telschow, 18/10/1940, ibid., p. 111.

[49] Fischer to Telschow, 13/11/1940, ibid., pp. 112–112 v; file note by Miss Reinold, 14/11/1940, ibid., p. 113; Reichsgesundheitsführer, Verbindungsstelle Berlin to Geschäftsführenden Vorstand der Kaiser-Wilhelm-Gesellschaft, 22/11/1940, MPG Archive, Dept. I, Rep. 1 A, No. 2399, p. 143; invitation to board meeting on 11/12/1940, 28/11/1940, ibid., p. 145; telegram from Conti to the KWG, 7/12/1940, MPG Archive, Dept. I, Rep. 1 A, No. 2400, p. 171; Telschow to the members of the Board of Trustees of the KWI-A, 17/12/1940, ibid., p. 175.

Attending this decisive meeting were – besides Fischer and Conti – from the side of state and party, Walter Groß and Hans Reiter, further the Medical Councilor of Berlin, Theobald Sütterlin (* 1893);[50] and – as representative of the German Council of Municipalities – from Kiel, Dr. Klose; then the Inspector of the Army Medical Corps, General and Chief Staff Physician Siegfried Handloser; from the side of the Kaiser Wilhelm Society, General Secretary Ernst Telschow and Friedrich Schmitt-Ott; as representatives of science, finally, Fritz von Wettstein, director of the KWI for Biology, Otmar von Verschuer, and – as a guest – Fritz Lenz.[51] The new chairman of the board set a political signal right in his welcome message, by pointing out

> that activity and research of the Kaiser Wilhelm Institute for Anthropology, Human Heredity and Eugenics is of particularly great importance for the state and that it would be wrong if – as it sometimes seems – interest in the meaning of issues of the heredity and race of our nation were to decline. The new Greater Germany needs such knowledge urgently, the next generation of scholars in this area must be provided for.

The KWI-A as the "first and most outstanding" in this area must "serve and influence other institutes as a model."[52] For this reason he, Conti, had accepted the Chairmanship of the Board.

Telschow's comments about the institute's budget plans from 1937 to 1940 turned out to be considerably more sober. Cuts of 20,000 RM from the regular budget had been necessary. The Reich Education Ministry was not able to refrain from this cut, "although the other ministries relevant for the Kaiser Wilhelm institutes had not made such cuts in consideration of the institute's acknowledged status as essential for the war." On this subject, Fischer elaborated that the institute had been able to "get over" the reduction due to the decline in the personnel budget, which had been eased as staff members were called up for military service, and thanks to savings in the material budget achieved by the "restriction of experiments" – although, Fischer emphasized, at "the detriment to scientific achivement." However, a glance at the revenue and expenditure accounts of the institute and the auditing reports of the KWG for the fiscal years 1939 and 1940 indicate that Telschow and Fischer painted an exaggeratedly gloomy picture of the institute's financial situation, which did not correspond to reality – more on this later.

The fifth agenda item, the "Director's Report about the Erection of a New Department for Experimental Genetic Pathology" was the sensation. In a speech explaining the entire framework of his proposal, Fischer submitted to the board his plans to reorganize the institute under the paradigm of phenogenetics. He started by

---

[50] For a biography: Klee, Personenlexikon, p. 616.

[51] The invitation was declined by Arthur Gütt, who had been promoted to State Secretary after his resignation; General Director Vögler, Senator of the KWG; and Ministry Secretary Rudolf Mentzel of the Reich and Prussian Ministry for Science, Training and Education.. Anlage 1 zum Entwurf des Sitzungsprotokolls, n.d., MPG Archive, Dept. I, Rep. 1 A, No. 2400, p. 185.

[52] Niederschrift über die Sitzung des Kuratoriums des KWI-A am 9/1/1941, ibid., p. 186 and 195, quotes: p. 186.

providing detailed reasons why, in the middle of the war, he was submitting a research plan that pointed so far into the future. "The impending victorious conclusion of the war and the vast expansion of the Greater German Reich," Fischer claimed, would also pose "great and new challenges" to the research institutes. While until now the institutes directly important for the war had stood at the foreground, like those in the areas of physics, chemistry, and technology, in the "near future" all institutes that dealt with "questions of genetic health, race, human selection, environmental influences" would become more important, as these were "of consequence for leadership." One could never know "how pure scientific research, often of a seemingly completely theoretical nature, will work out in practice in the future." And thus, Fischer added somewhat less than humbly, there had been no way of knowing that his bastard studies of 1908 "one day could lay a foundation for race legislation." Until 1933 his institute had transformed "the young field of human genetics into a securely founded, widely developed theory, [...] which measured up to all demands of practical application in genetic consulting, genetic legislation, and as a basis of race theory and race legislation."[53] By this time human genetic research was so far, Fischer proclaimed boastfully, that the genes for all essential normal and pathological attributes were known "in principle," the external phenotype could be related to these genes and "to some extent [...] the approximate scope of the environmental effects" was known.[54] Then came the transition to Fischer's project of phenogenetics:

> But one large area here is still quite dark. This is the question of *how* a given genetic disposition actually develops, how it works, how the gene "does its thing" (metaphorically speaking) to obtain the external appearance it is due. The path from the finished genetic *disposition* to the completely developed genetic *attribute* is still unknown.[55]

To legitimate the new research program, Fischer's argumentation stressed applicability. Phenogenetics was not only of "greatest scientific interest;" beyond this it promised "practical medical utility, the direction of which I can only hint at: differentiability of genetic conditions, prophylaxis for the genetically encumbered and corresponding marriage consulting, treatment of symptoms."[56]

Embedded in this context of justifications, Fischer concretized his ideas for reorganizing the institute: First he emphatically championed the hiring of Hans Nachtsheim. For because "human embryonic material with *certain* pathologically determined genetic dispositions [could be] received only in *very* restricted amounts," one had to rely on "model experiments" on animals – and Nachtsheim's rabbit breeding was the most suitable model by far. The study of genetic conditions of the rabbit must be linked closely with clinical research.

---

[53] Anlage 2 zur Niederschrift über die Sitzung des Kuratoriums des KWI-A am 9/1/1941: Bericht über die Neueinrichtung einer Abteilung für experimentelle Erbpathologie, erstattet vom Direktor, ibid., pp. 187–193, quotes: p. 187.

[54] Ibid., p. 189.

[55] Ibid., p. 189f. (original emphases).

[56] Ibid., p. 191.

In order to establish a connection with embryology, too, Fischer demanded, second, the erection of a "Central Genetic Biology Collection," which was to include fetuses, miscarriages, and organ specimens from humans and animals, especially from twins, with a view to race attributes, genetic illnesses, and deformities.

Central Genetic Biology Collection
– Phenogenetics of Humans and Mammals –

I. Twins

1. Fetuses and newborn bodies of identical twins (IT) and fraternal twins (FT)
2. Organs of child and adult IT and FT
3. Dual deformities of all kinds
4. Animal multiple births and dual deformities

II. European races: fetuses, newborns, and organs

1. Belonging to the races of the German nation
2. Belonging to other nations of Europe
3. Jews

III. Non-European races: as above

1. Asia
2. Africa
3. South Sea and Australia
4. America
5. Arctic region

IV. Genetic illnesses: fetuses, newborns, and organs from families with certain genetic pathological dispositions (later sorting by illnesses)
V. Domestic animal races: fetuses and organs
VI. Animal genetic illnesses: fetuses, newborns, and organs from breeds with certain genetic pathological dispositions[57]

Lösch advances the thesis that this central collection was "a new label for the institute's already existing, extensive collection of specimens."[58] This is a misinterpretation, however – there had not been an embryologically oriented collection of fetuses, premature births and stillborn children at the institute before this time. Back in 1939 Fischer had placed appeals in the *Deutsche Medizinische Wochenschrift* and the *Wiener Klinische Wochenschrift* ("Vienna Clinical Weekly"), asking practical physicians to supply the institute with such material.[59] Moreover, Verschuer

[57] Anlage zur Niederschrift über die Sitzung des Kuratoriums des KWI-A am 9/1/1941, MPG Archive, Dept. I, Rep. 1 A, No. 2400, p. 194.

[58] Lösch, Rasse, p. 373.

[59] Fischer, Menschliche Erblehre (Deutsche Medizinische Wochenschrift); idem., sic (Wiener Klinische Wochenschrift). On this also the undated draft for this appeal in the MPG Archive, Dept. III, Rep. 86 A (Münster), No. 9.

pursued his predecessor's ambitious goals. A note (undated) in Verschuer's handwriting, presumably made in preparation for the negotiations with the KWG about his appointment, outlining his ideas about the future development of the institute, includes mention of "studies of human embryos from genetically diseased families" and the "collection of all cases of embryos of women whose pregnancies were terminated."[60] In November 1942, Verschuer announced that "from inside the institute" he would "set in motion an organization according to which all women's clinics in Germany that perform abortions on genetically ill women would collect the embryos and deliver them to us."[61] It was emphasized in particular that this could only be realized in collaboration with the "Reich Committee for the Scientific Recording of Serious Genetic and Genetically Disposed Conditions" (*Reichsausschuß zur wissenschaftlichen Erfassung schwerer erb- und anlagebedingter Leiden*), that is, with the steering apparatus of the children's "euthanasia" that decided when pregnancies should be terminated on the basis of a eugenic, race, or ethical indication.[62]

Third, Fischer proposed a reclassification of the institute, which was to comprise five departments in the future:

- Department for Human Genetics (Fischer)
- Department for Genetic Psychology (Gottschaldt)
- Department for Race Science (Abel)
- Department for Experimental Genetic Pathology (Nachtsheim)
- External Department for Tuberculosis Research (Diehl)
- Central Genetic Biology Collection (Fischer)

The Department for Human Genetics dissolved in 1935 thus was to be restored, whereby Fischer – so to speak as a placeholder for his successor Verschuer – was to take over direction, just as he intended to take care of the Central Genetic Biology Collection himself. Fischer's own Department for Anthropology, on the other hand – renamed the Department for Race Science – was to be handed over to his pupil Wolfgang Abel, who then was also to be appointed Fischer's successor to the chair for anthropology at the University of Berlin. Further additions were those conceptualized in the context of the phenogenetic project: the Department for Experimental Genetic Pathology under Hans Nachtsheim, and the External Department for Tuberculosis Research under Karl Diehl. The Department for Genetic Psychology under Kurt Gottschaldt was to survive, whereby Fischer certainly assumed that it would be integrated into the program of phenogenetics as envisioned in Gottschaldt's talk at the Würzburg congress in 1938.

[60] Note by Verschuer, undated, MPG Archive, Dept. III, Rep. 86 B, No. 35.

[61] Verschuer to Stadtmüller, 16/11/1942, MPG Archive, Dept. III, Rep. 86 B, No. 36, pp. 26–29, quote: p. 27. Shortly thereafter Verschuer wrote that it was his desire "that material of human embryos collected according to this plan, which came from families with certain genetic illnesses (in particular physical deformities), should be studied by a specialized embryologist here at the institute." Verschuer to Starck, 30/11/1942, ibid., pp. 38–40, quote: p. 38.

[62] Schmuhl, Rassenhygiene, Nationalsozialismus, Euthanasie, p. 226.

The only problem left was what should become of Fritz Lenz and his Department for Race Hygiene. This was to "remain linked securely to the whole,"[63] but granted autonomy as an "Institute for Race Hygiene" under "Director" Fritz Lenz. To the Board of Trustees, Fischer sang Lenz's praises as a race hygiene pioneer. His strength lay in "positive suggestions, consulting with the responsible offices and oral and written instruction for students, physicians and the general public." Due to his "unique character," however, he could "not be considered [...] for the organization of the institute – a Fritz Lenz needs and has received unreserved relief and liberation from simple administrative and other institute activities in the interest of his theoretical work."[64] This was an extremely elegant formulation to express that Fischer held his department head to be unsuited for the post of director. The fact that Lenz attended the meeting and did not contribute to the discussion again confirms the impression that Lenz was altogether satisfied with the solution of an "institute in the institute" to which Fischer aspired. As the negotiations concerning the extension of the Race Hygiene Institute commencing later that year showed, Lenz pinned his hopes on his institute in the institute receiving its own budget and the right to hire its own staff, and, if this would be guaranteed, was even willing to relinquish any claim to a larger Institute for Race Hygiene at the planned University Clinic.[65]

In closing Fischer addressed the delicate issue of financing. He offered to finance the equipment and furnishings of the new Department for Experimental Genetic Pathology from institute funds, since the budget offered some latitude as a consequence of the restrictions to its work necessitated by the war. But additional finances were required for future operating costs, of which the personnel costs of 23,800 RM comprised the lion's share, as Fischer intended to hire not only the department head, but also an assistant with experience in anatomy, pathology, and histology, a technical assistant, and an animal keeper. Fischer estimated the additional material costs incurred by keeping animals at 10,000 RM, so that the future additional requirements amounted to 33,800 RM annually, and Fischer wanted this sum in the form of a regular budget increase rather than as a special allocation. Finally, new land was also required for the construction of stalls for the rabbits, as no more room was available on the grounds of the institute.

Lösch presumes that such a comprehensive concept for the reorganization of the institute "was expected by hardly any of those attending," and that it was "new in this dimension"[66] even for Telschow. However, this is not the case. In fact, Fischer had sent a written draft of his talk to both Telschow and Conti back on December 3,

---

[63] Appendix 2 to the protocol of the meeting of the board of the KWI-A on 9/1/1941: Bericht über die Neueinrichtung einer Abteilung für experimentelle Erbpathologie, erstattet vom Direktor, MPG Archive, Dept. I, Rep. 1 A, No. 2400, pp. 187–193, quote: p. 192.

[64] Ibid., p. 189.

[65] Lenz to Reichsministerium für Wissenschaft, Erziehung und Unterricht, 31/7/1941, MPG Archive, Dept. I, Rep. 1 A, No. 2400, pp. 212–213 v.

[66] Lösch, Rasse, p. 371.

1940.[67] Telschow thus would not have felt "affronted," as Lösch presumes; rather, his comments about necessary budget cuts simply made clear once and for all whence the funds for the institute's modernization would *not* come. In so doing he had hit the ball back into the politician's court, where it was readily received. The game was rigged. For back on October 18, 1940, in the very same conversation between Fischer and Telschow in which Fischer officially made the proposal to give Conti the chairmanship of the Board of Trustees, and offered "to personally approach [the *Reichgesundheitsführer*] on this matter," Fischer had laid out to the General Secretary of the KWG his plan to equip the Department for Experimental Genetic Pathology, applied for a 20,000 RM increase in the personnel budget and 10,000 RM in the material budget, and for an investment of 7,000 RM for the rabbit hutches. Fischer even brought with him to this meeting Hans Nachtsheim, who used the opportunity to negotiate with Telschow about his future salary. His appointment was slated for January 1, 1941. At the same time, according to a file note by Telschow, agreement was reached that the additional funds would "of course not be demanded in Professor Fischer's budget request until after conclusion of the war."[68] Since, as we showed above, Telschow did not believe that the war would be over within the year 1940, only one conclusion is possible: Fischer and Telschow had agreed to ask Conti for help in procuring the missing money from other sources for the time being. The course of the board meeting makes unmistakably clear that Fischer had done precisely this in his meeting with Conti on November 12, 1940, and that Conti had pledged his support.

Under Conti's direction, the board thus recorded in the protocol that the discussion had reiterated for the record "the special importance of the new department"; the provision of 33,800 RM was also "designated as urgently necessary." As regards the purchase or leasing of property for the rabbit hutches, the Reichgesundheitsführer pledged his "active support." Conti's confidant Sütterlin seconded the motion, signalizing the interest of Berlin's City Medical Administration and promising its support as well. At the same time, Sütterlin stated for the record "his satisfaction with the cooperation with the institute achieved in the working group with Dr. Diehl in Sommerfeld." Even should the hospital change leadership, Sütterlin ensured, Diehl's research could be continued without restriction. Finally, Army Medical Inspector Handloser also wished to have "his special interest in the work on rabbit tuberculosis" written in the protocol.[69] Yet again it must be emphasized that Fischer's push surprised neither Conti and his right hand Sütterlin nor Telschow, nor Verschuer and Lenz – in a sense, the roles had already been distributed in the preliminary talks, and the course and result of the consultations set beforehand. The entire meeting was completed in just 90 min, and the society retired to Harnack House for a snack.

---

[67] This proceeds from the handwritten marginals on this draft by Telschow, dated to 4/12/1940. MPG Archive, Dept. I, Rep. 1 A, No. 2400, p. 159.

[68] File note by Telschow about a meeting with Fischer, 18/10/1940, MPG Archive, Dept. I, Rep. 1 A, No. 2409, p. 81.

[69] Niederschrift über die Sitzung des Kuratoriums des KWI-A am 9.1.1941, MPG Archive, Dept. I, Rep. 1 A, No. 2400, pp. 186 and 195, quotes: p. 195.

4.1.5 The Conversion Takes Shape

On the very day after the board meeting, Conti turned to Telschow to coordinate the next steps. The General Secretary of the KWG drew up two letters in the name of the Reichgesundheitsführer, which he forwarded for Conti's signature.[70] One was directed to Rudolf Mentzel, the president of the German Research Association. It contained a request to the DFG to approve Fischer's application for a research grant of 40,000 RM for the 1941/42 fiscal year to finance the Department for Experimental Genetic Pathology[71] – the sum had increased over Fischer's original estimate, as Fischer now wanted to hire a clinical physician for the department as well. To the KWG Fischer justified this decision with the Board of Trustee's express wish "to bring the experimental [...] results in as rapid and lively connection with the human-clinical questions as possible." "From the close cooperation between the zoologist and theoretical genetic researcher Nachtsheim and a clinical physician [he hoped for] an acceleration of the results and and adaptation of the formulated questions to the burning questions of medicine."[72] Fischer submitted the application heralded by Conti to the German Research Association on March 13, 1941, and the grant was issued by the Reich Research Council on March 26 without further ado.[73] "The influence of Conti," Lösch established correctly, "was worth its weight in gold [...]."[74]

But Conti's patronage did more than make the money sources gush forth: The second letter Telschow prepared for Conti in January 1941 was directed to the responsible District Economic Office (*Bezirkswirtschaftsamt*) and applied that the KWI-A be recognized as strategically important for the war because of the research to be performed at the new Department for Experimental Genetic Pathology in the course of formation. Since the beginning of the war it had been a formidable obstacle to the work of the institute that it was the only one of the total of 14 Kaiser Wilhelm Institutes (including the General Administration of the KWG) in the region of the Mark Brandenburg province not to be classified as a "W" concern (for *Wehrwirtschaft*, army economy).[75] Here, too, Conti sought to remedy the problem,

[70] Telschow to Conti, 24/1/1941, MPG Archive, Dept. I, Rep. 1 A, No. 2399, pp. 199–200. In this Telschow complied with a request by Conti. Conti to Telschow, 10/1/1941, MPG Archive, Dept. I, Rep. 1 A, No. 2413, p. 64.

[71] Conti to Mentzel (draft), 23/1/1941, MPG Archive, Dept. I, Rep. 1 A, No. 2399, p. 201. The letter was actually sent in this form. Cf. Fischer to Deutsche Forschungsgemeinschaft, 13/3/1941, BArch. Koblenz, R 73/11.004.

[72] Fischer to KWG, 21/1/1941, MPG Archive, Dept. I, Rep. 1 A, No. 2413, pp. 65–65 v, Quotes: p. 65 v. On the very same day, 20/1/1941 Fischer had submitted the calculation without the clinical physician. Fischer to KWG (with handwritten supplement), 20/1/1941, MPG Archive, Dept. I, Rep. 1 A, No. 2409, pp. 90a–90a v.

[73] Fischer to DFG, 13/3/1941; Reichsforschungsrat to Fischer, 26/3/1941, BArch. Koblenz, R 73/11.004.

[74] Lösch, Rasse, p. 375.

[75] Conti to Bezirkswirtschaftsamt für den Wehrwirtschaftsbezirk III (draft), n.d., MPG Archive, Dept. I, Rep. 1 A, No. 2399, pp. 203–204.

just as he also supported Telschow's request for the classifying the pathologist Otto Baader as "indispensable" with the responsible Military District Command (*Wehrkreiskommando*) after such an application by Fischer had been rejected.[76]

Conti supported the undertaking as best he could, even after Fischer had finally secured the appointment of Abel as his successor to the professorship for anthropology, and Verschuer as his successor for the directorship of the institute in winter 1941/42. Verschuer came to Berlin on May 5, 1942 and put his ideas and demands on record. In the case of his appointment, he guaranteed, he would continue the research under the banner of phenogenetics according to the wishes of Eugen Fischer, but in doing so would stick to his own research profile, shifting the emphasis to genetic pathology. Further, he and his pupils would continue the twin and family studies already begun. Accordingly, a small polyclinical and a small clinical department were to be created at the institute, which was to employ two national social workers (*Volkspflegerinnen*) and two nurses. His own Department for Human Genetics, for which Verschuer requested two further assistant positions, would continue to work closely with the Department for Experimental Genetic Pathology. "As a central, connecting node between these, a new Department for Embryology should be set up." The planned changes, so Verschuer calculated, necessitated an increase of 38,000 RM in the personnel budget and 23,000 in the material budget. Besides this, Verschuer's plan earmarked non-recurring expenditures – for a new stall building for Nachtsheim's rabbit breeding, a laboratory for the animal breeding, equipment of the clinical and polyclinical departments, etc. – totaling 106,000 RM. Verschuer also requested, if possible, a full professorship at the Medical Faculty of the University of Berlin.[77] Despite the strained financial situation, the KWG accepted Verschuer's ambitious plans surprisingly readily – after Verschuer had conducted a conversation with the president of the KWG on May 8, 1942, the fulfillment of his demands was approved, initially orally; and this approval – upon his express wish – was confirmed in writing shortly thereafter.[78] There is no indication that the KWG, and be it "even only pro forma," had been on the lookout for another candidate for the post of director. Correctly, Lösch assesses: "Fischer had been successful with his tactics of making Verschuer out to be the only sufficiently qualified candidate."[79]

In July 1942 Fischer received the message that his son Hermann had been killed in action on the Eastern Front – he lost any interest in the work of the institute and

---

[76] Telschow to Conti, 24/1/1941, MPG Archive, Dept. I, Rep. 1 A, No. 2400, pp. 199–200.

[77] Niederschrift vom 5.5.1942, ibid., pp. 225–227.

[78] Verschuer to Telschow, 10/6/1942, MPG Archive, Dept. I, Rep. 1 A, No. 2409, p. 125; Telschow to Verschuer, 13/7/1942 (transcript of excerpts), ibid., pp. 126–126 v. The request for a written confirmation, Telschow wrote, seemed to him "indeed somewhat unusual, since you know the Kaiser Wilhelm Society and therefore also know that oral approvals from the direction of the society are always observed and viewed as binding." The budget increase would presumably "not come into question until after the end of the war […], because you could not use the increased budget now due to a lack of personnel." (ibid.).

[79] Lösch, Rasse, p. 390.

moved to Freiburg in August. Due to an illness he was not able to resume the business of the institute. Since Verschuer was not able to take over direction of the institute until October 1, 1942, in September – irony of fate – Fritz Lenz was appointed interim director.

On October 28, 1942 Verschuer continued his negotiations with the KWG. During these negotiations he appeared full of self-confidence and demanded that the grants from the Reich and the Prussian state be increased considerably in the next fiscal year.[80] Yet finances were not the decisive problem – although the increase in public grants demanded by Verschuer was rejected, he ultimately received the money from the DFG and from the "Sponsorship Association of German Industry" (*Förderergemeinschaft der deutschen Industrie*). More difficult to master, as the negotiations on October 28, 1942 evince, were the restrictions on facilities and personnel due to the war. Here, too, Verschuer pinned his hopes on Conti. With the assistance of the Reichgesundheitsführer, the new director hoped to win back the lower rooms of the institute, which had been used as sanitary facilities up to that time. Further, Conti was to procure the construction permit for extending the attic – until then the skull collection had been kept there, which now was to be transferred to the university – into a sickroom and rooms for the nurses. Further, Verschuer hoped to achieve with Conti's help that Gottschaldt and his colleague from Frankfurt Hans Grebe (1913–1999) be classified as "indispensable."[81] Originally Conti was to be addressed in a board meeting, but since this never took place, Verschuer and Walter Forstmann (1900–1956) from the General Administration visited Conti at his office on November 24, 1942. The Reichgesundheitsführer willingly pledged his support on all points, inquired as to the works in progress and promised to tour the institute over the course of the next 6 months.

4.1.6 Conti's Interests: Tuberculosis Research and Population Policy in the East

Whence the interest of the Reichgesundheitsführer? Why did he regard the research which the newly oriented institute intended to take on as a resource for his political ambitions? In any case the tuberculosis research performed by Diehl and Verschuer was highly interesting for the Reichgesundheitsführer. In late 1939 Conti still struck a positive balance: tuberculosis may have increased, but only because of improved diagnostics; the tuberculosis mortality, by contrast, had diminished.[82] In a lecture – not intended for the public – about the "Health Balance in the Second Year of the

[80] File note by Telschow about a meeting with Verschuer on 28/10/1942 (transcript of excerpts). MPG Archive, Dept. I, Rep. 1 A, No. 2409, p. 136.

[81] File note by Telschow about a meeting with Verschuer on 28/10/1942, MPG Archive, Dept. I, Rep. 1 A, No. 2400, pp. 239–239 v.

[82] Typed manuscript, based on the stenographic transcript of a lecture held by Conti in late 1939 in Münster. Estate of Leonardo Conti, private collection.

War," which Conti held at the Humboldt Club in Berlin on August 7, 1941 for the editors of German journals, he expressly designated tuberculosis as

> [L]east advantageous point [...]. Tuberculosis increased, especially in the areas with endangered air like Hamburg and Kiel. [...] I presumably do not need to indicate especially that we have not yet been able to implement a major social program. We have had no time for this since 1933. Before 1933 the German Volk was standing at the brink. Then came the political revolution. Many retrenchments had to be undertaken for the fortification of the German Volk.[83]

Under these circumstances the genetic research of tuberculosis had to be of immense interest for the Reichgesundheitsführer, especially as it can certainly be presumed that Fischer played this card in the decisive conversation with Conti. Back at the beginning of World War II he had justified his application to the General Administration for feed for the 200 rabbits in Beetz by claiming that there was "no doubt that these studies from the area of one of the worst national epidemics promise to be of great importance for fighting human tuberculosis"[84] – a justification that was forwarded by the General Administration to the responsible Food Office (*Ernährungsamt*) almost word for word.[85]

That Conti was familiar with Diehl's and Verschuer's tuberculosis research can be proved on the basis of an (undated) typed lecture manuscript on the subject of "Genes and Performance" (*Erbgut und Leistungsfähigkeit*):

> The genetic disposition also plays a role in infectious diseases. The views about heritability have oscillated extremely. First it was observed that tuberculosis occurred in certain families, then the pathogen was discovered and the way it befalls the diseased, namely in earliest childhood; the disease is then carried forth and does not break out until puberty and professional life and even later: at that point no one considered that the disease in question might have been acquired in childhood. It was twin research that illuminated us to the fact that this congenital inferiority is important in tuberculosis. It was possible to establish that identical twins who grew up separately nevertheless got tuberculosis. If a person is resistant he will not become ill if he is only susceptible. It is clear that someone who may be absolutely resistant, but becomes a tuberculosis doctor or nurse, ultimately does take in the bacillum, which then spreads in the body. A doctor who may come from a tuberculous family, but so far has remained entirely healthy, may not become a tuberculosis doctor, for the risk of infection is too great. In other respects the environmental influences are important in fighting tuberculosis; reasonable living conditions must be created.[86]

---

[83] Lecture by Reichgesundheitsführer State Councilor Dr. Conti, "Die Gesundheitsbilanz im zweiten Kriegsjahr," held on August 7, 1941, in the Humboldtclub Berlin for editors of the journal press, typed manuscript, pp. 15–16. Estate of Leonardo Conti. In the discussion Conti responded to an inquiry from the audience: "The possibilities of accommodating those who have just fallen ill are exhausted. It is a great effort for me to create new tuberculosis sanctuaries. All of the homes and sanatoriums I have today have been taken for resettlement. The SS, police and HJ and other institutions require rooms for their purposes. A vast lack of tuberculosis beds exists. I turned to the head of the district to request that beds be made available to us. The serial tuberculosis study can no longer be expanded, and because it also can no longer examine all suspects. Thus the series study must mark time. For the future it will be carried out without a doubt." Ibid., pp. 17–18.

[84] Fischer to Telschow, 8/9/1939, MPG Archive, Dept. I, Rep. 1 A, No. 2399, p. 136.

[85] Telschow to Ernährungsamt Berlin-Zehlendorf, 14/9/1939, ibid., p. 136 a.

[86] "Vortrag: Erbgut und Leistungsfähigkeit," estate of Leonardo Conti.

If any further proof is needed that tuberculosis research was irresistible bait for the Reichgesundheitsführer, it is provided by a letter which Verschuer wrote on January 27, 1941 – that is, just 3 weeks after the decisive board meeting – to his friend Karl Diehl. The latter had asked for advice about whether he should set about expanding the rabbit hutches in Sommerfeld. Verschuer's advice ran as follows:

> I would undertake absolutely everything that is at all possible. So build with all of the money and material you get! Since your research activity was acknowledged at the Board Meeting of the institute in Dahlem by the relevant people, above all by the City Medical Councilor and Reichgesundheitsführer, and its continuation declared to be urgently necessary, you need not have any concern about your future. Your tendency toward moving forward is thus altogether correct.[87]

In the end, Verschuer's assessment turned out to be right. Diehl's project enjoyed high priority up to the end of the Third Reich. Upon Verschuer's application, the Reich Research Council classified Diehl's research on tuberculosis as "important for war and state" and issued a corresponding research contract on August 18, 1943. Achim Trunk is correct to emphasize that Diehl's project was the only one of all of the research projects being conducted at the KWI-A in 1944 to be granted the higher priority of "SS."[88] Since the costs of the project ultimately consumed a large portion of the institute budget, in February 1944 Verschuer submitted an application for funding of 10,000 RM to the Reich Research Council, which was also approved without a hitch.[89] "It is truly unpleasant for me to be the greatest consumer of the institute's funds," Diehl commented about the application. "Couldn't money be saved? But where? Everything I have is still so meager and yet so much money. It embarrasses me. And if anything is to come of it, this is only the beginning."[90]

Conti's interest in tuberculosis research is easy to understand. But here the thesis will be advanced that Conti was interested by no means only in genetic pathology research, but also in research under the banner of phenogenetics. This thesis is supported by a source from the estate of Leonardo Conti, which indicates that, at the time when the negotiations about the reorganization of the KWI-A were under way, the Reichgesundheitsführer was fervently interested in issues of "ethnic cleansing" in occupied Poland, the "Germanization" of Poland and the resettlement of German nationals. Conti was concerned with this complex of subjects because of the danger of epidemics associated with the resettlement of German nationals. At Himmler's request, in December 1939 the RuSHA had presented the draft of a "Selection System for the Settlement of the New Reich Districts" (*Ausleseordnung für die Besiedlung der neuen Reichsgaue*), which also entailed the participation of the Reichsgesundheitsführer.[91] In this context Conti,

[87] Verschuer to Diehl, 27/1/1941, MPG Archive, Dept. III, Rep. 86 A (Münster), No. 7.

[88] Trunk, Zweihundert Blutproben, pp. 44 f.

[89] Verschuer to Diehl, 25/2/1944, MPG Archive, Dept. III, Rep. 86 A (Münster), No. 7.

[90] Diehl to Verschuer, 29/2/1944, MPG Archive, Dept. III, Rep. 86 A, No. 32.

[91] Heinemann, "Rasse," pp. 233 f.

accompanied by the internist Heinz Kalk (1895–1973)[92] flew to Przemysl on January 13, 1940, in order to personally witness the arrival of a trek of German nationals from Volhynia in the Ukraine.[93] On February 29, 1940 Conti held a lecture for the SS Reichsführer to convey the impressions from his trip and to present his proposals for "ethnic cleansing." A handwritten sheet upon which Conti noted talking points for this meeting provides information about the Reichgesundheitsführer's ideas for the new population order:

East Prussia – Silesia – Danzig- W[est Prussia] – Warthegau[:] 26 million people, half of them German, 7½ million Poles, ½ million Jews. Government: 14 million [,] of whom 2½ million Jews. *Problem* must stop. Different types.

Nordic type who profess to be Polish, does not submit [.] Pure Huns, all variants of cross-breeds, made nations through our blood. Führer even solicited, for no order possible otherwise. First separation, then interbreeding, colonists fetched later. Power of the Reich diminishing, increasing. Ebb and flow, at times colonists fetched again. Language is accepted. […] The Germans by blood then became the best Polish soldiers, always the bravest opponents. Endangers only our own blood. Let nothing more flow away, get it all back. Rigor in the goal, adaptability in the method. Elimination of leading personalities required.

Race and nationality mixed up [?]

Congress Poles in part better racially than Poles from Poznan, and Silesia Poles in part better than German nationals.

Volhynia and Galicia 135,000 (110,000 peasant families[,] rest tradesmen) Nat. Russia around 20,000[.] Southern Tyrol 230,000.

Germans elsewhere in the world is a question which may not be touched on. Likewise 40,000 Germans in Lithuania.

General government: Training [residential?] area: Polish self-administration required. 40,000 German nationals must be returned from Lublin and Chelm.

Polish workers marked. Strict segregation. Business, shopping for Poles only on certain day. Polish workers hanged for Rassen[schande?] ("race disgrace" – interracial intercourse). Polish women available for the Poles. German women to concentration camp.

Baltics: take luggage, all want to go to Poznan.

Volhynia Germans: fabulous. Believe in the Führer. Surrendered gold and food. Bought horses and brought with them. Wanted to surrender wedding rings.

Now there is no purchase of land, no application for settlement, etc. Those who have moved in may not have the feeling of discrimination and must not be discriminated against.

Improve climate in the East by planting.

Settle border zone and build bridges. Split up settlements as they were split up militarily.

Settle mixture of Volhynia Germans, German nationals and Reich Germans in village.

Inbreeding of German nationals ceases. Political fertilization.

Führer after the war: Off to the East.

Merge estates in the former Reich fragmented by distribution among heirs.

Racially and politically good people to the East.[94]

[92] For a biography: Klee, Personenlexikon, p. 296.

[93] On this the estate of Leonardo Conti includes a photo album compiled by Heinz Kalk, who accompanied Conti on this flight.

[94] Note by Conti, "Himmler, 29/2/1940," estate of Leonardo Conti (original emphases).

In reality, Conti's influence on "Germanization policy" in the occupied areas remained marginal[95] – but that could not be foreseen in 1940. One can certainly presume that Fischer, in their conversation on November 12, 1940, drew the Reichgesundheitsführer's attention to the fact that Fritz Lenz had turned to the "burning issues of resettlement and race hygienic population policy"[96] in 1939/40. Similarly, Fischer must have referred Conti to the KWI-A scientists' activity preparing evaluations for the Reich Genealogical Office. In his meeting with Conti Fischer probably also addressed the research on embryonic animal and human skulls carried out at the institute, which had the objective of determining race differences by means of embryological methods. Presumably he also mentioned the studies in progress on race dermatoglyphics, which were based on the serial anthropological studies Fischer had initiated in the Łódž ghetto. Finally, it cannot be excluded that Fischer presented in his talk with Conti considerations on a serological race diagnostics, like the one developed in connection with Engelhardt Bühler's project on the heritability of agglutinines in 1935. In any case, placing the KWI-A under his protection appeared to offer the Reichgesundheitsführer the chance to secure the political monopoly on potentially groundbreaking methods of race diagnostics, which would be far superior to the anthropometric methods applied in occupied Poland.

4.2 Internal Structures

4.2.1 Finances

The reorganization under the banner of phenogenetics had significant consequences for the institute budget. The size of the budget grew continuously in the war years. The sum of revenues and expenditures shown in the yearly accounts in the 1943/44 fiscal year was 208,000 RM, clearly higher than the 144,000 RM in the 1940/41 fiscal year – compared to the 1933/34 fiscal year, when the revenues and expenditures

[95] In the subsequent period Conti rarely succeeded in getting through to Himmler. His suggestion to resettle entire villages of German nationals from Bessarabia as communities was the subject of consultations between Himmler and SS Gruppenführer Ulrich Greifelt (1896–1949), the chief of the Office of the Reich Commissioner for the Consolidation of German Nationhood (*Dienststelle des Reichskommissars für die Festigung deutschen Volkstums*) in May 1942. Conti himself was no longer consulted. Cf. Witte et al., Dienstkalender Heinrich Himmlers, p. 432, note 66.

[96] Fischer, Tätigkeitsbericht 1939/1940, MPG Archive, Dept. I, Rep. 3, No. 17. In his activity report for 1940/41 Fischer wrote that Lenz had written several exposés on "The Population Policy of the Peasantry," on "Resettlement" and on "The Methodology of Race Research," the first two of which were not intended for publication. In the draft of the report he wrote: "Mentioned in particular are certain works by Mr. Lenz that are not intended for publication, on issues of resettlement and the assessment of demographic and population policy works and issues." Fischer, Tätigkeitsbericht 1940/1941 (draft, clean copy, resp.), MPG Archive, Dept. I, Rep. 3, No. 18.

amounted to 104,700 RM, the total balance had nearly doubled. On the expenditure side, the personnel costs exploded as a result of the new hirings over the course of setting up new departments or reorganizing old ones. They climbed from around 76,500 RM (1940/41) to 131,500 RM (1942/43), while the material costs, aside from a temporary rise through new acquisitions in the 1942/43 fiscal year, persisted at around 50,000 RM. On the revenue side, the latitude for an increase in the grants from the Reich and Prussia was limited. This had not yet been a problem in the Fischer era, since the institute had significant reserves at its disposal, with which the rising costs could be defrayed initially. In the Verschuer era, by contrast, this financial padding dwindled rapidly, as the running expenditures peaked their pinnacle. Since the inflexible personnel costs made up the lion's share, savings measures were practically impossible without reducing personnel. From the 1942/43 fiscal year on, the institute lived beyond its means. In order to be able to continue working in the same order of magnitude as it had until then, it was dependent on the constant flow of third-party funds of considerable scope. This, in turn, had effects on the research program and practice.

The initial financial situation when Fischer set about to reorganize his institute was not as bad as he had portrayed it to the Board of Trustees. It had been possible to stop the gap left by the cuts at the start of World War II through the significant surpluses amassed in the years from 1933 to 1939. In September 1939 the subsidies from the Reich and Prussia planned for 1939/40 were cut to 150,500 RM, despite Fischer's vehement protest[97] – the shortfall of nearly 10,000 RM resulting from this cut ate up around half of the credit balance from the previous years.[98] In the 1940/41 fiscal year the subsidies from the Reich and from Prussia were cut by another 10,000 RM to 140,500 RM, but the expenditures dropped so sharply due to the drafting of nearly all scientific staff and "the cessation of research works associated therewith"[99] that a surplus of nearly 17,000 RM remained at the close of the fiscal year. This, together with the remaining surpluses from the previous years, yielded a credit balance of almost 27,000 RM, which was transferred to the new budget in view of the research projects the institute had been forced to defer because of the war. Added to this was a travel fund of 10,000 RM, leaving the

---

[97] Cf. Fischer to Generalverwaltung, 23/9/1939, MPG Archive, Dept. I, Rep. 1 A, No. 2409, pp. 52–52 v: "A certain reduction as a result of the war situation is certainly understandable. But the overall situation is not such that the institute will be closed. […] All of our operations, which, of course, were greatly restricted in the past, first weeks of the war, are coming back into gear. It would be entirely wrong to perform exclusively chemical and physical science because these can be put directly in the service of military economy. Our studies about hereditary diseases are at least as important for the *Volk*." Thus Fischer rejected – successfully – *huge* cuts to his budget.

[98] Schröder, Bericht über die Prüfung des Rechnungsabschlusses zum 31/3/1940, 5/9/1940, ibid., pp. 74–76; Vermögensübersicht zum 31/3/1940, ibid., p. 77; Einnahmen- und Ausgabenrechnung für das Rechnungsjahr 1939, ibid., p. 78.

[99] Schröder, Bericht über die Prüfung des Rechnungsabschlusses zum 31/3/1941, 28/4/1941, ibid., pp. 99–101, quote: p. 99; Vermögensübersicht zum 31/3/1941, ibid., p. 102; Einnahmen- und Ausgabenrechnung für das Rechnungsjahr 1940, ibid., p. 103.

KWI-A with "secret reserves" of around 37,000 RM in April 1941. In other words: The balances could barely conceal that Fischer was again hoarding money in his institute for future research projects.

In comparison to the balances for the 1940/41 and 1941/42 fiscal years, however, it becomes apparent that funds flowed even faster now. The subsidies from the Reich and Prussia increased by about 16,500 RM. They reached the level of 157,000 RM and thus more or less that of the late 1930s. Despite the dramatically increased personnel costs – a consequence of founding the Department for Experimental Genetic Pathology – at the end of the fiscal year a new surplus of over 5,000 RM remained, so that the surpluses, including the full-to-bursting travel fund, totaled over 42,000 RM – and this although of the 40,000 RM earmarked for Nachtsheim's department from the German Research Association, only 22,000 RM were called in right away. And because this subsidy could not be spent in any reasonable way, with the consent of the General Administration it was used for the purchase of the library and collection of specimens from the private property of the departing director – the money thus flowed into Fischer's pockets. The rest could be transferred to the next accounting year.[100] In other words: At the start of the 1942 budget year the institute had "silent reserves" of 70,000 RM at its disposal, more than a third of the entire KWI-A budget.

Not until the 1942/43 fiscal year did the unchecked expansion thrust result in a hefty deficit. The subsidies from the Reich and Prussia diminished to 140,000 RM, and although the oustanding payment of 18,000 RM from the DFG balanced this out, the revenues were not sufficient to cover the dramatically increased expenditures. The personnel costs were the largest post – in this fiscal year alone, three new scientific assistants and eight technical and administrative employees were hired. The deficit ultimately amounted to 29,000 RM, through which the accumulated reserves dwindled to just under 3,000 RM; however, this did not include the travel fund of over 10,000 RM, which remained untouched.[101]

The new director Otmar von Verschuer was faced with a weighty problem. The personnel costs had exploded so greatly as a consequence of creating the Departments for Human Genetics and Experimental Genetic Pathology that they far surpassed the level of the usual grants. In the negotiations with the General Administration about his appointment on October 28, 1942 Verschuer thus submitted a cost estimate of 234,000 RM for the 1943/44 fiscal year. In oral negotiations Telschow made clear that, while an increase had been requested from the Reich Education Ministry, it could not be expected in such an order of magnitude.[102] This assessment was to prove correct: The allocations from the Reich and the Prussian state did

[100] Schröder, Bericht über die Prüfung des Rechnungsabschlusses des KWI-A zum 31/3/1942, 27/7/1942, ibid., pp. 127–129; Vermögensübersicht zum 31/3/1942, ibid., p. 130; Einnahmen- und Ausgabenrechnung für das Rechnungsjahr 1941, ibid., p. 131.

[101] Schröder, Bericht über die Prüfung des Rechnungsabschlusses des KWI-A zum 31/3/1943, 27/4/1943, ibid., p. 146 f.; Vermögensübersicht zum 31/3/1943, ibid., p. 148; Einnahmen- und Ausgabenrechnung für das Rechnungsjahr 1942, ibid., p. 149.

[102] File note by Telschow, 28/10/1942 together with the cost estimate for 1943/44, ibid., p. 135 f.

increase back to 156,000 RM in the 1943/44 fiscal year, however, under the condition that the remaining funding gap of 77,000 RM be covered by another source. In this situation, Alfred Kühn, deputy director of the KWI for Biology, who had followed the reorganization of the KWI-A around phenogenetics since 1938 with interest, leapt into the breach. In his capacity as Chairman of the Biology and Medicine Section of the Academic Council of the KWG, he took part in the meeting of the KWI-A Board of Trustees on May 4, 1943, which was dominated by the financial crisis. Kühn suggested turning to the Association of Sponsors of German Science (*Stifterverband der Deutschen Wissenschaft*), which had free funds at its disposal at the time. To this effect Friedrich Schmidt-Ott, the Chairman of the Association of Sponsors, was to be addressed, who also belonged to the Board of Trustees, but had not attended the meeting on May 7. Also missing was Reichsgesundheitsführer Leonardo Conti, who had supported all of the institute's financial requests so effectively before. Conti's star was waning by this time, and it is striking that Fischer, Verschuer, Telschow, and Kühn, who were alone at the meeting on May 7, 1943, no longer included Conti in their calculations, but rather decided to arouse the interest of Conti's former rival, Karl Brandt (1904–1948),[103] who had since overtaken Conti in importance as the "accompanying physician" of the Führer, one of the two figures responsible for the "euthanasia" program, and since July 1942 also Hitler's authorized representative for the Medical and Health Service, in the institute's work.[104]

Telschow took immediate action. Just 1 day after the board meeting, on May 8, 1943, he addressed Schmidt-Ott – with express reference to Kühn.[105] Since the Association of Sponsors no longer had such a high sum at its disposal, Schmidt-Ott forwarded the letter from Telschow to Albert Vögler, president of the KWG since 1941, who suggested directing a petition to the "Sponsorship Association of German Industry" (*Förder[er]gemeinschaft der deutschen Industrie*), to request a nonrecurring grant of 100,000 RM.[106] Verschuer kept this possibility under his hat for the moment. For in the meantime, on May 24, 1943, the German Research Association – in response to an application by Verschuer on March 23, 1943 – had approved 40,000 RM for the institute in Dahlem, for "studies in the area of comparative genetic pathology."[107] In June 1943 Verschuer reported that Kühn was

---

[103] Süß, Aufstieg; idem., "Volkskörper"; Klee, Personenlexikon, pp. 70 f.

[104] Niederschrift über die Sitzung des Kuratoriums des KWI-A am 7/5/1943, MPG Archive, Dept. I, Rep. 1 A, No. 2404, p. 68 f.

[105] Telschow to Schmidt-Ott, 8/5/1943, MPG Archive, Dept. I, Rep. 1 A, No. 2409, pp. 152–152 v: "Prof. Kühn is well disposed to instruct Your Excellency personally about the areas of work currently being worked on at the institute, especially about those we have just begun to study."

[106] Vögler to Telschow, 21/5/1943, ibid., p. 158. Schmidt-Ott belonged to the Administrative Council, Vögler to the Board of Trustees of the Sponsorship Association. Cf. Schulze, Stifterverband, pp. 91 f.

[107] Verschuer to Präsident des Reichsforschungsrates, 23/3/1943, BArch. Koblenz, R 73/15.342, pp. 97–98; Reichsforschungsrat to Verschuer, 24/5/1943, ibid., p. 96. Verschuer's application had been supported by Reichsgesundheitsführer Conti. Cf. Reichsgesundheitsführung, Verbindungsstelle

negotiating with Schmidt-Ott about the remaining deficit of 37,000 RM.[108] These negotiations ultimately resulted in resorting to Vögler's offer: on September 7, 1943 the Association of Sponsors of German Industry approved a 3-year research grant of 47,000 RM annually for the KWI-A.[109]

The Association of Sponsors of German Industry had been founded officially in November 1941. The motives that led to the founding of this organization were located on two levels: For one, in view of the profit restrictions imposed upon business by the National Socialist regime, sponsoring research was simply a possibility for "investing the considerable war profits, when the traditional possibilities for reducing profits, that is, increasing share capital and increasing capacity, no longer appeared interesting." Second, leaders in industrial circles were concerned about theoretical research and the sponsorship of young scientists – at a point in time that coincided with the "first disillusionment about Germany's chances of military success" and in view of "the future existence of business and research in a postwar period." Extremely interesting – and until now disregarded – is that the economic leaders assembled in the Association of Sponsors accorded such great importance to phenogenetic research in Dahlem in this context that they approved quite a considerable amount for the KWI-A. By way of comparison: In spring 1943 the Association of Sponsors had an endowment of 22 million RM, of which a total of around 800,000 RM in interest yields were available for distribution.[110]

In the 1943/44 fiscal year, besides the subsidies from the Reich and Prussia, the KWI-A received third-party funds from the Association of Sponsors and the DFG amounting to 87,000 RM. In the 1944/45 fiscal year this total even increased, to 97,000 RM, as the DFG not only renewed its grant of 40,000 RM,[111] but also, as mentioned above, responded to Verschuer's petition by providing an additional 10,000 RM for Diehl's tuberculosis research.[112] In August 1944 Verschuer was able to state with satisfaction, in a letter to his friend Bernhard de Rudder:

> Surprisingly, the cutbacks I expected in my institute have not come to pass; on the contrary, great value is placed on continuing the research important to the war. And so the cogs remaining in my institute machine are turning at full speed, as if the entire machine were still running. But I am glad that so much remains in operation, and that thus still quite a bit of productive work can be performed.[113]

Berlin, to Geschäftsführender Beirat des Reichsforschungsrates, 24/5/1943, ibid., p. 95; Geschäftsführender Beirat des Reichsforschungsrates to Reichsgesundheitsführung, Verbindungsstelle Berlin, 2/6/1943, ibid., p. 94.

[108] Note for the file by Reinold, 8/6/1943, MPG Archive, Dept. I, Rep. 1 A, No. 2409, p. 158.

[109] Arndt to Deutsche Industriebank, 10/2/1944, ibid., p. 169.

[110] Schulze, Stifterverband, pp. 89–94, quotes: p. 90.

[111] Verschuer to Präsident des Reichsforschungsrates, 20/3/1944, BArch. Koblenz, R 73/15.342, pp. 78–78 v; Reichsforschungsrat to Verschuer, 16/5/1944, ibid., p. 77.

[112] Verschuer to Präsident des Reichsforschungsrates, 25/2/1944, ibid., pp. 92–92 v; Reichsforschungsrat to Verschuer, 6/4/1944, ibid., p. 91.

[113] Verschuer to de Rudder, 31/8/1944, MPG Archive, Dept. III, Rep. 86 A (Münster), No. 8.

Since the subsidies from the Reich and Prussia had been fixed at 156,000 RM, the ratio of public subsidies to third-party funds was 3:2 in the final budget year.[114] In other words: The subsidies from the German Research Association and the Association of Sponsors of German Industry were of vital importance for the institute from 1943 on.

4.2.2 Hans Nachtsheim, Director of the Department for Experimental Genetic Pathology

The new Department for Experimental Genetic Pathology was tailored precisely to the zoologist Hans Nachtsheim. He studied zoology in Bonn and medicine in Munich from 1909 to 1912.[115] In 1913 he received his Ph.D. under Richard Goldschmidt at the University of Munich, where he became an intern at the Zoological Institute. In 1914 he moved to the Zoological Institute at the University of Freiburg under Erwin Baur, where he also met Eugen Fischer. In 1915 he was drafted into military service, where he spent most of his duty working as a censor at the military surveillance posts in Karlsruhe, Freiburg, and Munich. In summer 1919 he was a member of the Epp Freikorps. After working as an assistant to the zoologist Richard Hertwig (1850–1937) at the University of Munich from 1916 to 1921 and receiving his qualification as a professor there, in 1921 he joined Erwin Baur at the Institute for Genetic Research at the Agricultural Academy of Berlin in Dahlem, as Director of the Zoological Department. Nachtsheim spent the years 1926/27 as a Rockefeller Foundation scholar in the United States, where his experiences included sitting in at the laboratory of the drosophila geneticist Thomas Hunt Morgan at Columbia University in New York. Increasingly, Nachtsheim turned to the genetics of domesticated animals and began systematically breeding strains of rabbits with pathological attributes.

At the Congress of the German Society for Genetics in 1937, which took place at Verschuer's Institute for Genetic Biology and Race Hygiene in Frankfurt, Nachtsheim introduced his breeding experiments to the genetic community, and also aroused the attention of those geneticists working on the genetic pathology of humans. After the war, Fischer admitted that the idea of winning Nachtsheim for his institute occurred to him at this congress.

When Fischer approached Nachtsheim in September 1940,[116] the latter did not hesitate. In the very next month he gave notice that he would be leaving his senior

[114] "Zusammenstellung über Einnahmen und Ausgaben 1937–1946" (MPG Archive, Dept. I, Rep. 1 A, No. 3025) lists the current *private* revenues of the institute: 42,430 RM (1941/42), 52,100 RM (1942/43), 82,000 RM (1943/44), 99,000 RM (1944/45), 13,700 RM (1945/46, actual amount) and 661 RM (1946/47, actual amount). This compilation does not reveal how the revenue post is put together.

[115] For a biography: Schwerin, Experimentalisierung, esp. pp. 338–341; Lösch, Rasse, p. 368.

[116] Fischer to Verschuer, 20/9/1940, MPG Archive, Dept. III, Rep. 86 C, No. 1: "Nachtsheim accepted in principle and very happily; next week we negotiate with Telschow."

assistantship. His professional status was precarious.[117] Since 1935, when the Institute for Genetic Research, along with the Agricultural Academy, had been integrated into the Agricultural Faculty of the Friedrich Wilhelm University, his room for maneuver had been severely restricted. Although Nachtsheim had been appointed associate professor of the university in 1939, there was no prospect of a regular professorship, especially since he was considered suspect in party circles. He never joined the NSDAP.[118] In 1933 he had been dismissed as chairman of the Reich League of German Rabbit Breeders. For Nachtsheim entirely new possibilities for continuing to advance his research on comparative genetic pathology opened up with the switch to the KWI-A, on a secure material foundation, shielded by the Kaiser Wilhelm Society. The new orientation presented no difficulty for him, since his research on the genetic pathology of mammals had been conceived from the outset as a model for human genetics. In 1940 Nachtsheim then also switched from the agriculture to the mathematical-natural sciences faculty of the University of Berlin – he could not bring himself to decide to switch to the medical faculty.

As in the cases of Kurt Gottschaldt and Karl Diehl,[119] in Hans Nachtsheim, too, Fischer opted for a scientist whose career seemed to have hit a dead end in the Third Reich – and who was rather distanced from National Socialism. This proved to be a skillful move, for Nachtsheim, too, justified the trust placed in him and built the new Department for Experimental Genetic Pathology into a supporting pillar of the KWI-A in a very short time.

4.2.3 A Director for the Department of Embryology

In Fischer's and Verschuer's plans for reorganizing the institute in Dahlem, the triangle of clinical genetic pathology of humans, the animal model and embryology was assigned decisive importance in terms of research strategy. The high value placed on embryology in this concept is often overlooked, because the planned Department for Embryology was never founded due to the war. This was not for Verschuer's lack of trying. In June 1942 – that is, a full three months before he took over as director of the institute in Dahlem – he began asking around in his circle of colleagues in order to find candidates for the position of director of the new "Department for Embryology or Genetic Developmental Physiology" in planning. Since the new department was not only to study animal embryos from Nachtsheim's Department for Experimental Genetic Pathology, but also "to build the bridge […] to humans" and to work "on human material" as well, Verschuer elaborated to

[117] Schwerin, Experimentalisierung, pp. 248 f.

[118] He did become member of the NS Dozentenbund, however. Kröner, Von der Rassenhygiene zur Humangenetik, p. 38 (note 105).

[119] Gottschaldt was not a member of the NSDAP, but had joined the NS League of Teachers in 1933 and left it in 1934. In 1936 he joined the NS Dozentenbund. Diehl did not join the NSDAP until 1937. Kröner, Von der Rassenhygiene zur Humangenetik, p. 38.

inquiries that no zoologist, but only an "embryologist coming from the field of anatomy"[120] came in question. Preferably, Verschuer was searching for a young scientist who was nevertheless well-versed in embryological methods – postdoctoral qualifications were not required. The survey produced very few indications of any utility. An acute lack of young anatomists was a problem at the time, and most of the few younger scientists in this area had been drafted, so that ultimately only one of the candidate's names seemed at all suitable to Verschuer: the university lecturer Dietrich Starck (* 1908), prosector and senior assistant at the Anatomical Institute of the University of Cologne. Verschuer entered into detailed negotiations with Starck and his superior Franz Stadtmüller, the Director of the Anatomical Institute. Starck indicated that he was interested, but expressed from the very beginning reservations because he was "an anatomist, body and soul,"[121] and could find that a move to Dahlem could "'sideline' [his chances as] *an anatomist*"[122] and end up doing himself out of a chair in anatomy. In early 1943 Verschuer and Starck agreed to put the negotiations about the appointment on ice for the time being. In May 1943, Verschuer invited the still hesitant Starck to hold a lecture at one of the upcoming "biological evenings"[123] in Harnack House. These evenings were presided over by Alfred Kühn – the invitation to Starck underlines an earlier indication by Verschuer that "through the close cooperation with the neighboring Kaiser Wilhelm Institutes, above all with von Wettstein, Kühn and Butenandt, quite special working possibilities are presented" for the new Department for Embryology, especially "in joint colloquia and team projects."[124]

As becomes apparent in the correspondence between Verschuer and his friend de Rudder, in summer semester 1943 Starck actually did appear "at a Dahlem biological evening" and held a talk about "The Importance of Developmental Physiology for Comparative Anatomy, Explained on the Example of the Head of Vertebrates," which, according to Verschuer, was "outstanding in form and content." Personally, too, Starck had "made the best impression." Ultimately he rejected the appointment to Dahlem, however, because he did not wish to "endanger his anatomical career." "Despite the high qualification of Mr. Starck," Verschuer continued, he was "not unhappy about the rejection," as he had since believed to have found "another and [...] apparently more suitable candidate for the position of department director." The person in question here was the university lecturer Wouter Frans Hendrik Ströer, prosector at the Anatomical-Embryological Institute of the University of Groningen, who had worked as a guest scholar at the KWI-A for several months in 1943. "Ströer is a Dutchman, but entirely on our side." He was "an outstanding researcher personality" and "decidedly the best man

[120] Verschuer to Hermann Bautzmann, 2/7/1942, MPG Archive, Dept. III, Rep. 86 B, No. 36, pp. 8–9, quotes: p. 8.

[121] Starck to Verschuer, 9/12/1942, ibid., p. 41.

[122] Stadtmüller to Verschuer, 20/6/1942, ibid., pp. 24–25, quote: p. 24 (original emphasis).

[123] Verschuer to Starck, 7/5/1943, ibid., pp. 88–49, quote: p. 48.

[124] Verschuer to Stadtmüller, 16/11/1942, ibid., pp. 26–29, quotes: p. 28.

I could think of for my institute." Nevertheless it was open to question whether his move to Dahlem would take place. Ströer himself had "doubtlessly the greatest inclination." However, "by order of the Reich Commissioner for the Netherlands," he was supposed "to take over a professorship at the new Reich University in Groningen."[125] The decision was still open.

Verschuer had informed the Reich Education Ministry of his intention to appoint Ströer to departmental director on July 12, 1943 – mediated by the General Administration of the KWG:

> Dr. Ströer is a scientist known for his superior research work in the field of developmental history and genetic pathology [...]. He has been occupied with phenogenetic studies as a guest assistant at my institute for some time [...]. Politically I hold him to be altogether reliable and pro-German – he is a storm-trooper of the Germanic SS in the Netherlands.[126]

In the end Verschuer was not able to get his way. As late as September 1944 he reported to de Rudder that Ströer was "still being held back by the Reich University of Groningen (by now one must write 'former'!) for the time being," but his wife and three small children had been "sent here into my protection, as their lives were threatened directly by their fellow countrymen." "Emergency quarters" had been set up for them at the institute.[127] In a further letter by Verschuer written a short time later, this time to Fischer, he stated that Ströer was "stationed with the SS in Arnheim" and "certainly took part in the heavy fighting there."[128]

4.2.4 Scientific and Non-scientific Personnel

The total number of "working scholars" remained – on paper – nearly unchanged during World War II: From 37 (1939/49) it fell slightly to 35 (1940/41) and finally to 33 (1941/42), and then rose again over the course of Verschuer's takeover, to 38 (1942/43), then 39 (1943/44).[129] Yet this impression of relative stability is misleading,

[125] Verschuer to de Rudder, 7/11!1943, MPG Archive, Dept. III, Rep. 86 A (Münster), No. 8. In January 1943 Verschuer had proclaimed to Fischer: "As of 1/4/1943 Ströer is coming to the institute as a guest assistant. Perhaps he will be our future embryologist someday [...]." Verschuer to Fischer, 7/1/1943, MPG Archive, Dept. III, Rep. 86 A (Münster), No. 9.

[126] Verschuer to Reichserziehungsministerium, 12/7/1943, MPG Archive, Dept. I, Rep. 1 A, No. 2400, p. 249.

[127] Verschuer to de Rudder, 16/9/1944, MPG Archive, Dept. III, Rep. 86 A (Münster), No. 8.

[128] Verschuer to Fischer, 29/9/1944, MPG Archive, Dept. III, Rep. 86 A (Münster), No. 9. In an earlier letter to Eugen Fischer, Verschuer mentioned that Ströer had "integrated himself into the German Wehrmacht." Verschuer to Fischer, 9/9/1944, ibid. – After World War II Ströer lived in the Netherlands, undisturbed. Cf. Verschuer to Fischer, 5/2/1953, MPG Archive, Dept. III, Rep. 86 C, No. 11, p. 4f.

[129] Calculated according to the statements in the annual reports 1939/40, MPG Archive, Dept. I, Rep. 3, No. 17; 1940/41, MPG Archive, Dept. I, Rep. 3, No. 18; 1941/42, MPG Archive, Dept. I, Rep. 3, No. 19; 1942/43, MPG Archive, Dept. I, Rep. 3, No. 20; 1943/44, MPG Archive, Dept. I, Rep. 3, No. 21.

for the numerous drafts into the Wehrmacht thinned out the scientific personnel extremely. As Fischer's activity report for the 1939/40 fiscal year shows, this began as early as spring 1939:

> The activity of the institute was restricted, for even in the first five months of the year covered by this report, which were before the outbreak of the war, all assistants but one had been drafted for military drills, sometimes alternately, sometimes simultaneously.[130]

At the beginning of World War II, in addition to the Department Director Kurt Gottschaldt, *all* assistants of the KWI-A and the majority of the male doctoral students were drafted. Until late 1942 three assistant positions remained unfilled, and the remaining assistants – Wolfgang Abel, Otto Baader, Heinz Lemser, and Siegfried Tschamler[131] – were in the Wehrmacht and held contact with the institute only sporadically. Thus, the "central block" was lost, so to speak, which not only had the consequence that all of the assistants' research projects lay idle. The supervision of the foreign guest scholars and the remaining doctoral students suffered as well. The fact that the assistants were drafted also meant that Eugen Fischer and Fritz Lenz had to take on more duties in academic instruction – in winter semester 1941/42, Fischer himself had to hold the practical course in anthropology at his university chair, which he had been able to load off to his assistant Abel until that time.[132] Demands on Lenz's time were made by academic instruction duties, but primarily through the supervision of a great number of dissertations. Further, due to the loss of their assistants, Fischer and Lenz had to take on an even higher degree of activities in producing expert opinions and evaluations.

The situation remained unchanged in the 1940/41 fiscal year – despite Leonardo Conti's intervention there was no success "in freeing up even a single assistant from military service, through which the scientific activity of the institute is greatly

[130] Tätigkeitsbericht 1939/40, MPG Archive, Dept. I, Rep. 3, No. 17. The text here continues: "Through this not only their work, but also that of the doctoral students and guests was impeded severely, especially since those who were not drafted had to be drawn upon more frequently than usual to the extensive activity of producing expert opinions that is such a burden for the institute."

[131] Tschamler fell in action on October 1, 1943. Cf. "Meldung für die Ehrentafel," MPG Archive, Dept. I, Rep. 3, No. 21.

[132] "When the semester began here, our personal scientific work practically came to a halt. For all that, I have six hours per week, have to take the subway into town twice, around one hour there and one hour back each trip. These two mornings are completely unavailable. For the major practical course I have to prepare myself first, because I have not held it for ten years now, but then have to prepare it technically as well, blood groups, order in twins, etc." Fischer to Verschuer, 2/7/1941, MPG Archive, Dept. III, Rep. 86 A (Münster), No. 9. On summer semester 1942 is written: "From April to July, increasing activity by his colleague took many claims on Professor Fischer's time, because the younger lecturers of the faculty were unavailable." Tätigkeitsbericht by Verschuer 1942/43, MPG Archive, Dept. I, Rep. 3, No. 20. – Considering that Fischer had gone to Rome for three months as an exchange professor after the decisive board meeting on January 9, 1941, and further that he lost any interest in the institute after the death of his son in July 1942, moved to Freiburg in August 1942, took ill there and never returned to Berlin, it becomes clear that Fischer hardly worked as a scientist in the last 2 years of his tenure as Director of the KWI-A. Cf. Lösch, Rasse, p. 389, 392.

limited,"[133] Fischer lamented. In the first draft of his activity report the final clause read as follows: "[...] since even among the doctoral students only one foreigner and two ladies remained, the scientific activity of the institute, aside from Professor Lenz, Professor Gottschaldt and the institute director, was completely extinguished."[134] Fischer's final activity report as Director of the KWI-A, which referred to the 1941/42 fiscal year, began with the resigned observation:

> Through the further duration of army service of all assistants, one departmental director [Abel] and at times a second [Gottschaldt], through the lack of nearly all male doctoral students, the institute has not been able to carry out scientific activities on a larger scale.[135]

Not until the change in institute leadership did the personnel situation improve. Wolfgang Abel, by now director of the Department for Race Science, was finally classified as "indispensable" in October 1942,[136] as was Kurt Gottschaldt, so that all departmental director positions were filled. In the first round of negotiations with the KWG in May 1942, Verschuer had also, as mentioned above, managed to acquire two further assistantships for his own Department of Human Genetics to be reestablished, which he wanted to occupy with his closest colleagues from Frankfurt. In July 1942 Verschuer reported to the race biologist Wolfgang Lehmann of Strasbourg, a member of the "Dahlem circle": "I will take almost all of my staff from here, first of all [Heinrich] Schade and [Hans] Grebe, later [Josef] Mengele and Fromme [...]."[137] In November 1942 Verschuer then was able to report to Fischer that starting on December 1 he would have "in addition to my Dr. Grebe from Frankfurt, further a Dr. [Siegfried] Liebau as assistant, whom the SS has commanded to us for training. Thus some operations should be able to get back into gear."[138]

From December 1942 two assistants were thus on location in Dahlem again: Hans Grebe and Siegfried Liebau. The physician Hans Grebe had closed ranks with National Socialism at an early date, joining the NS Student League in 1931, the NSDAP and SA in 1933, and the NSDÄB in 1937. He took his Ph.D. under Verschuer with a dissertation about genetical and nongenetical blindness. After that he worked as an assistant at the Horst Wessel Hospital in Berlin, before becoming Verschuer's assistant at the Frankfurt Institute for Genetic Biology and Race Hygiene in 1937. Released from the Wehrmacht after he was seriously wounded, Grebe qualified as a professor under Verschuer in 1942 with a postdoctoral dissertation about chondrodysplasia. He moved to Berlin with his mentor, becoming an assistant at the KWI-A and lecturer at the university. From December 1942 on he was among the most active scientists at the KWI-A, until October 1944, when he was appointed at the age of just 31 to associate professor and Director of the

---

[133] Fischer's Tätigkeitsbericht 1940/41, MPG Archive, Dept. I, Rep. 3, No. 18.

[134] Fischer's Tätigkeitsbericht 1940/41 (draft), ibid. This passage was deleted from the final version.

[135] Tätigkeitsbericht 1941/42, MPG Archive, Dept. I, Rep. 3, No. 19.

[136] Verschuer to Fischer, 22/10/1942, MPG Archive, Dept. III, Rep. 86 A (Münster), No. 9.

[137] Verschuer to Lehmann, 11/6/1942, MPG Archive, Dept. III, Rep. 86 A (Münster), No. 5.

[138] Verschuer to Fischer, 23/11/1942, MPG Archive, Dept. III, Rep. 86 A (Münster), No. 9.

newly founded Institute for Genetic Biology and Race Hygiene at the University of Rostock.[139]

Siegfried Liebau (* 1911)[140] had worked at the RuSHA and as an adjutant of the SS Medical Academy of Berlin since 1938, from May 1940 to September 1942 he was a personal consultant in the SS Paramedical Office in Berlin. From December 1942 to October 1943 he was detached to the KWI "for professional training in the areas of anthropology, human genetics and race hygiene."[141] The posting of Liebau, whose wife Ingeborg, née von Ekesparre, was a close friend of the Verschuer family,[142] apparently can be traced back to a request by Verschuer on November 10, 1942. As will be shown below, Liebau carried out twin studies in Auschwitz during his time at the institute. In the further course of the war he became Chief Physician for the Higher SS and Police Leadership of the Adriatic coastal region and Italy.

The two assistants Grebe and Liebau were joined in the 1942/43 fiscal year by two auxiliary assistants: Karl Joachim Hene, who had entered Gottschaldt's Department for Genetic Psychology as an auxiliary assistant in 1939 and taken his Ph.D. in 1940 with a genetic psychology dissertation about twins in early childhood, returned from military service. Further, the teacher Hans Ritter (* 1903),[143] who had begun a second university degree in zoology, anthropology and psychology in 1941, but then had been drafted into military service, started work as an auxiliary assistant in Abel's Department for Anthropology, where he dedicated himself to "Gypsy twin research."

Additional reinforcements came in the course of 1943: Karin Magnussen (1908–1997), working at the KWI-A on a scholarship since 1941, was promoted to an assistantship – during the war period she was the only woman to hold this status. Finally, Heinrich Schade was also hired. Schade, member of the NSDAP and SA since 1931, had taken part, as already mentioned elsewhere, in the first yearly course held at the KWI-A for physicians from the SS in 1934/35. In 1935 he collaborated in the sterilization of the "Rhineland bastards." In the same year he started at the Frankfurt institute as Verschuer's assistant and senior physician. In 1939 he submitted his postdoctoral dissertation about the genetic biological inventory of the population of the Schwalm region, located between Treysa and Alsfeld in Hesse. In December 1942 he moved – on paper – to the KWI-A as Verschuer's

139 For a biography: Lösch, Rasse, p. 566; Klee, Medizin, pp. 265–267; idem., Personenlexikon, p. 198. Cf. Reichsminister für Wissenschaft, Erziehung und Volksbildung to Grebe, 20/1/1945, MPG Archive, Dept. I, Rep. 1 A, No. 2400, p. 264.

140 For a biography: ibid., p. 371.

141 SS-Führungshauptamt to KWI-A, 12/11/1942, BArch. Berlin, BDC, RuSHA, Liebau's personal file.

142 Cf. Massin, Mengele, p. 226 (note 93). On this also Liebau to Reichsführer-SS, 1/12/1935, BArch. Berlin, BDC, RuSHA, Liebau's personnel file (here Verschuer is named as a guarantor for Ingeborg von Ekesparre in the application for permission to engage and marry). The formulation in Verschuer's letter to Fischer of November 1942 (see note 139), however, indicates that Verschuer had not made Liebau's personal acquaintance at the time he was requested.

143 For a biography: Lösch, Rasse, p. 573.

senior physician. However, because he had been drafted into the Wehrmacht, he was not able to start his new position right away. Not until the turn of the year 1943/44 did Schade come to Berlin, in the course of a military command,[144] where at times he was able to continue his work of evaluating the genetic biological inventory of the Schwalm region. He must have been detached to the front again later, for he experienced the end of the war as a Yugoslavian prisoner of war.[145]

Of the veteran scientific staff, only Georg Geipel remained at the institute over the entire period of the war.[146] Otherwise German scientists could only be recruited sporadically, like the "lateral hire" Karin Magnussen, and – as a convalescent – the physician Gerhard Koch (1913–1999).[147] The ranks were filled instead with foreign guest scholars from neutral or allied states.[148] Their number initially dropped from five (1939/40) to three (1940/41), but then rose back up to eight (1942/43) and remained at this level to the end. The staff was joined by several foreign doctoral students. In the period from 1939 to 1945, guest scholars from Bulgaria (Nicolaus Ilkow), Finland (Magister Longfors), India (Sasankar Sekhar Sarkar), Japan (Masataka Takagi, Masaji Kamitake[149]), Croatia (Franjo Ivanicek),[150] the Netherlands (Haring T. Piebenga, Wouter Ströer, Hendrik Scalogne),[151] Norway (Thordar Quelprud), Portugal (José Ayres de Azevedo), Romania (Marius Sulica), Switzerland

---

[144] Verschuer, Tätigkeitsbericht 1943/44, MPG Archive, Dept. I, Rep. 3, No. 22.

[145] Sparing, Rassenhygiene; Lösch, Rasse, p. 573; Klee, Medizin, p. 267; idem., Personenlexikon, p. 522. – Also appointed to an assistantship in 1943 was the physician Walter Beck, but since he had been called up to the Wehrmacht he was hardly present at the institute.

[146] Max Fischer died in July 1940. Konrad Kühne was listed on paper as a scientific staff member until 1945.

[147] For a biography: Lösch, Rasse, p. 568; Klee, Personenlexikon, p. 323. – Koch joined the Hitler Youth in 1930, the Bund der Artamanen in 1932, then the NS Student League, the NSDAP, the SA and the SS. On October 26, 1942 Verschuer had asked the office of the Army Medical Inspector to grant Koch a 3-month working vacation at the KWI-A. Verschuer to Heeressanitätsinspektion Berlin, 26/10/1942, MPG Archive, Dept. I, Rep. 1 A, No. 2400, p. 238.

[148] Cf. also Fischer's Tätigkeitsbericht 1941/42, MPG Archive, Dept. I, Rep. 3, No. 19.

[149] Masaji Kamitake (1909–1983) had taken his Ph.D. in psychology at the Tokyo Bunri Daigaku University. He studied in Germany as a Humboldt scholar. From 1942 to 1945 he worked as a Japanese editor. After the war he became a professor for psychology at the Kyoiku Daigaku (Teacher's College) in Tokyo. In 1963 he published a work about "New Developmental Psychology," in 1971 one about "Heredity and Environment in their Effect on the Psychological Functions. Studies of Twins." Kamitake worked predominantly with the twin method. Warm thanks to Kazuko Kibata and Yasushi Maruyama for information on Kamitake's biography.

[150] Ivanicek had come to Dahlem as a doctoral student under Fischer. In October 1942 he conceived a plan to acquire his doctorate in medicine with the work on "Australian skulls" entrusted to him by Fischer. In this matter he intended to consult with the anatomist Hermann Stieve (1886–1952) in Berlin. "Apparently Mr. Ivanicek received funds from Croatia to extend his period of study." Verschuer to Fischer, 22/10/1942, MPG Archive, Dept. III, Rep. 86 A (Münster), No. 9. In February 1943 Verschuer wrote: "Ivanicek was absent here for several weeks, now he has turned up again. I think, however, only to conclude his work." Verschuer to Fischer, 9/2/1943, ibid. Cf. also Ivanicek, Beiträge.

[151] Piebenga, Ströer and Scalogne were the editors of the Netherlands Journal for Race Hygiene. Piebenga was supposed to become director of an institute "for the execution of certificates of ancestry and race" in 1942. To make inquiries about him, L. ten Cate, a consultant for questions

(Erik Hug),[152] Spain (Jésus Cabeza), Turkey (Senhia Tunakan) and Hungary (Mihali Malán, Lajos Csik, Anton Steif, Ladislaus Apor).[153]

The number of doctoral students at the KWI-A oscillated between 21 (1940/41) and eleven (1943/44), whereby the number of those who worked under Fritz Lenz – 14 in 1940/41 – is not included. The comparatively high numbers are deceiving in this case, too, however. Numerous doctoral students had been called up to the Wehrmacht – those who were able to work at the institute with any continuity were generally only the foreign doctoral students and the female doctoral students, whose number oscillated between two and four.[154]

Finally, a glance at the nonscientific personnel, which also grew considerably in the course of reorganizing and expanding the institute. At the beginning of World War II four technical assistants, eight secretaries, one nurse, and five "wage earners" (gardener, driver, keeper, cleaning ladies) had worked at the KWI-A; in 1943/44 there were five technical assistants, 14 secretaries, one photographer, one laboratory technician, one nurse, one auxiliary technical assistant, one caretaker, plus the married couple who worked as caretakers in the External Department for Tuberculosis Research in Sommerfeld, as well as four "wage earners" (keepers, cleaning ladies) and several "temps."[155]

Despite Verschuer's fears to the contrary,[156] the institute was able to maintain all of its personnel in the second half of 1944 as well, as impending drafts were postponed for the time being:

> Apparently it is primarily thanks to the vigorous action of Prof. Osenberg [Werner Osenberg (1900-1974)] of the Reich Research Council that research is so protected at the moment and should be continued to its full extent. Thus I have increased only the working

---

of ancestry at the Ministry of the Interior of the Netherlands, paid a visit to the KWI-A in October 1942. Cf. Verschuer to Fischer, 22/10/1942, ibid.

[152] Hug apparently had a scholarship from the Alexander von Humboldt Foundation. Therefore, so Verschuer to Fischer in October 1942, he would "have to be retained as a guest for a while longer." Verschuer to Fischer, 22/10/1942, ibid. In February 1943 Verschuer then reported: "Hug did not receive a visa for Germany. He asked me to use my influence with the political offices on his behalf, but I was not willing to do so, since I do not know him and, according to everything I have heard about him, cannot take any responsibility for him." Verschuer to Fischer, 9/2/1943, ibid.

[153] According to Lösch, Rasse, p. 577, the guest scholar Ernst Wiedemann was also a foreigner, whereby his nationality is unclear. The doctoral student Martin Haetinger (* 1915) came from Brazil; it is unclear whether or not he was a German citizen. Cf. idem., Stellung.

[154] These were Lieselotte Block (* 1918), Eva Justin (1909–1966), Ruth Rohloff (* 1920), Clärchen Steer and Senhia Tunakan. Inez de Beauclair, who was counted as a doctoral student for the entire period of the war, was on a research trip to China. Lösch, Rasse, p. 571, also lists the doctoral student Margot Irene Oetting in 1943/44 (Cf. Oetting, Haut- und Fingerleisten). – Not taken into consideration here are the female doctoral students working under Lenz, like the social worker Gertrud Maas (* 1894), the physicians Hildegard Schwarz and Johanna Schötzau (* 1916). Cf. Lösch, Rasse, pp. 563, 570, 574.

[155] Aufstellung über die Personalzusammensetzung 1939/40 and 1943/44, respectively, MPG Archive, Dept. I, Rep. 3, Nos. 17 and. 21, respectively. Also, the salary cards and lists, MPG Archive, Dept. I, Rep. 3, Nos. 2 and 3.

[156] On August 11, 1944 Verschuer still expected that Abel and Gottschaldt would be called up again, and perhaps Nachtsheim as well, who had been ordered for a physical examination.

> hours of the institute – in keeping with the times – but only to such a degree that overstraining is avoided [...].[157]

Verschuer's private household had initially employed foreign civilian workers. In the move from Frankfurt to Berlin in November 1942, the Verschuer family had a Croatian maid.[158] At the beginning of 1944 the family appears to have employed an additional female "Eastern worker." In late February, Verschuer reported in a letter to Bernhard de Rudder, that there had been "all kinds of sagas with our Russian East worker." "It turns out there had been minor thievery, with which she provided provisions for all kinds of male compatriots [...]." Yet she had been "kept again on probation."[159] Shortly thereafter Verschuer again complained of "troubles at home with our Russian (Bolshevik!)."[160] Besides this, since 1943 at the latest, an additional female "Eastern worker" was working at the neighboring institute. In August 1944, after the renewed proclamation of "total war," Verschuer feared that he would probably "have to give up the two Eastern workers from the house and the institute."[161] Yet it never came to this. In late September 1944 Verschuer wrote to Fischer in Freiburg that the "Russian woman" in service in his household had "run away" – as once before in 1943, and this time "the Russian woman from the institute [...] ran away with her."[162]

4.3 Research Agenda and Research Praxis

Between September 1939 and November 1942 most of the scientific work at the KWI-A – aside from the External Department for Tuberculosis Research in Sommerfeld and (from 1941) the Department for Experimental Genetic Pathology

---

"And for me, too, it will be a matter of course to put on the gray uniform again when the call is issued to me." Verschuer to Fischer, 11/8/1944, MPG Archive, Dept. III, Rep. 86 A (Münster), No. 9.

[157] Verschuer to Fischer, 31/8/1944, ibid.

[158] Verschuer to de Rudder, 4/11/1942, MPG Archive, Dept. III, Rep. 86 A (Münster), No. 8. On the basis of a recruting agreement Croatian workers came to work in the German Reich *voluntarily*, and thus were *not* among the compulsory laborers brought in from abroad. On the demarcation: Schmuhl, Zwangsarbeit.

[159] Verschuer to de Rudder, 29/2/1944, ibid. "You must be stricter with your Eastern girl," de Rudder responded. They were "on average poor things indeed," but one must not let them get away with anything. "Is she useful otherwise? Now your wife must learn Russian as well as Croatian! By the way, there are nice, simple little dictionaries for household purposes, I picked one up recently." De Rudder to Verschuer, 6/3/1944, ibid. On the "female Eastern household workers," cf. Winkler, "Hauswirtschaftliche Ostarbeiterinnen." For a definition of "Eastern worker," cf. Schmuhl, Zwangsarbeit.

[160] Verschuer to de Rudder, 10/3/1944, MPG Archive, Dept. III, Rep. 86 A (Münster), No. 8.

[161] Verschuer to Fischer, 11/8/1944, MPG Archive, Dept. III, Rep. 86 A (Münster), No. 9.

[162] Verschuer to Fischer, 29/9/1944, ibid. Foreign compulsory laborers were apparently also in service on the Verschuer family estate in Solz. In any case Verschuer reports in his memoirs about the end of the war at Easter 1945 in Solz: "One danger was the many foreign workers, who looted

– came to a standstill due to the fact that so many departmental directors and assistants had been drafted. This was not immediately apparent to the outside world, however. Between 1940 and 1942 the institute still published 43 scientific papers; however, this was the result of a "publication backlog." Most of the publications from this period – to the extent that they were not simply overview papers, conceptual or methodological discussion papers – resulted from research projects that had been concluded before the outbreak of the war. Only very few papers, such as a paper by Karl Diehl and Eugen Fischer about the tuberculosis experiments on rabbits in Sommerfeld and the papers by Hans Nachtsheim about "The State of Convulsion Readiness and Genotype," referred to current projects.

The change came with the new director. Verschuer was successful in getting the departmental director Kurt Gottschaldt and Wolfgang Abel classified as "indispensable," filled the ranks of the assistants with Hans Grebe, Siegfried Liebau, Hans Ritter, and Karin Magnussen and obtained a larger number of foreign guest scholars. And – not to be forgotten – Verschuer achieved a budget hike and solicited considerable third-party funds. Thus research resumed on a large scale from December 1942 on. Of course, this was not immediately reflected in the publication lists. Even so, between 1943 and 1945 the institute in Dahlem produced another 46 publications, whereby – in addition to the general intensification of the war situation – it must be taken into consideration that nearly all of the publications that had accumulated in the prewar period were published at this time. The papers published in the last two years of the war were almost without exception minor works presenting intermediate findings from projects in progress, and some of them were based on material that had been collected previously. A number of publications that had been available in manuscript form or were even in print were lost in the chaos of the war's end; others were not completed before the collapse of the Third Reich. Some of these papers were still published after World War II, for others this did not seem opportune because they were all too closely associated with the state crimes of National Socialist Germany.

In the final 2 years of the war, the shift in emphases between the fields of research of the institute in Dahlem, observable since 1933, continued at a faster pace. Genetic pathology moved to center stage with 19 publications – and this was, so to speak, only the tip of iceberg, as several large-scale projects in the field of genetic pathology never found their way to publication. This dominance of genetic pathology had various reasons: First, the two departments that had been able to keep up their operations in the first war years, that is, the External Department for Tuberculosis Research and the Department for Experimental Genetic Pathology after 1941 – both worked in this area. Second, the focus of interest of the new director, Verschuer, remained on genetic pathology. Third, in bringing Hans Grebe from

---

and took revenge. Fortunately our farmer Cornelius had always treated them well, so that nothing bad happened on our estate." Verschuer, Erbe – Umwelt – Führung, "Direktor des Kaiser-Wilhelm-Institutes für Anthropologie, menschliche Erblehre und Eugenik (seit 1942)" section, p. 16, MPG Archive, Dept. III, Rep. 86 A, Nos. 3–1. Cf. on compulsory workers in the service of the KWG Strebel/Wagner, Zwangsarbeit.

Frankfurt to Dahlem, Verschuer had an assistant who, because of a project for the collection of stillborn fetuses in progress since the prewar period, and because of his postdoctoral dissertation about chondrodysplasia, had a rich fund of pathological material at his disposal, which could be evaluated without any great cost. Fourth, through the activities of preparing assessments and evaluations, individual cases of genetic pathological interest (including all of the important genealogical information required for their genetic pathological evaluation) came to the attention of the researchers in Dahlem. Fifth and finally, the findings of genetic pathology promised a direct practical utility with regard to the measures of both genetic health policy, as well as eugenic sterilization, marriage bans in accordance with the "marriage health law," the allocation of "matrimony loans" and so on. The KWI-A extolled this practical aspect of genetic pathology research quite audibly, which was evident in the mere fact that the research application which covered the major portion of the work in this area bore the keyword "race hygiene."[163] In total it can be asserted that the research field of genetic pathology increasingly pushed its way into race hygiene over the course of World War II: there was hardly a genetic pathology study that was not oriented to genetic health policy, and hardly a race hygiene paper without clear references to genetic pathology.[164] The trend toward specialization observed in the final years before World War II, which inclined to lead race hygiene and genetic pathology (and race biology) away from each other,[165] was reversed at the KWI-A from 1943 on.

For a concrete example, in the years 1943–1945 there were nine research contracts with the keywords:

- "genetic pathology research" (Hans Nachtsheim)
- "tuberculosis" (Karl Diehl)
- "specific proteins" (Otmar von Verschuer)
- "eye color" (Karin Magnussen)
- "twin camps" (Kurt Gottschaldt)
- "race hygiene" (Hans Grebe et al.)
- "genetic biological inventory" (Heinrich Schade)
- "stillborn fetuses" (Grebe)
- "pneumoconiosis" (Grebe)[166]

---

[163] Cf. Bericht über das Forschungsprojekt "Rassenhygiene," n.d. [March 1944], BArch. Koblenz, R 73/15.342,p. 67; ibid. [October 1944], ibid., p. 41 (a dissertation by Klaus Gnirke, "Kasuistischer Beitrag zur Klinik und erbpathologie der Myotonia congenita" ["Casuistic Contribution to the Clinical and Genetic Pathology of Myotonia congenita"] is also announced here).

[164] As, e.g. Maas, Kinderzahl.

[165] Weingart/Kroll/Bayertz, Rasse, pp. 436 f.

[166] Wehrmacht contract numbers: pp. 4891–5376 (1591/10) – III/43 ("genetic pathology research"), SS 4891–5377 (1592/10) – III/43 ("tuberculosis"), S 4891–5378 (1593/10) – III/43 ("specific proteins"), K RO/RFR-0295/1594/10 – III/43 ("eye color"), K RO/RFR-0296/1595/10 – III/43 ("twin camps"), K RO/RFR-0297/1596/10 – III/43 ("race hygiene"), K RO/RFR-0298/1597/10 – III/43 ("genetic biological inventory"), K RO/RFR-0299/1598/10 – III/43 ("stillborn fetuses") and K RO/RFR-0300/1599/10 – III/43 ("pneumoconiosis").

Five of these nine research contracts – "genetic pathology research," "tuberculosis," "race hygiene," "stillborn fetuses" and "pneumoconiosis" – were located directly in the field of genetic pathology, one further – "genetic biological inventory" – had strong bearings on genetic pathology. This illustrates the dominance of genetic pathology even more strongly than the analysis of the publication list. The genetics of normal attributes, even and especially under the aspect of race was relegated down to second place, with two research contracts – "specific proteins" and "eye color" – and a total of 13 publications.[167]

Weighting the individual research projects according to their financial, political, and research-strategic value, it becomes apparent that from 1943 on, four areas were of fundamental importance for the future of the institute: comparative genetic pathology (Nachtsheim), research on the heredity of tuberculosis (Diehl), the project on the phenogenetics of eye color (Magnussen), and the project to develop a serological race test (Verschuer). Their progress determined whether the budget could be fixed at a high level, whether sources of financing outside of the regular budget kept flowing, and whether research operations could be maintained in their entirety. They decided whether the project of phenogenetics, above all its integration into general genetics and biology, would succeed. And they were eminently important for genetic health and race policy.

4.3.1 Genetic Pathology and Race Hygiene

Verschuer energetically pushed ahead with the concept of phenogenetics developed by Fischer, but placed the emphasis on genetic pathology research,[168] whereby, of course, he consequently conceived of genetic pathology as "medical genetics," thus embedding it in general human genetics. Besides, Verschuer understood genetic pathology as a principle encompassing and penetrating all subdisciplines of medicine and urged – in keeping with his concept of the "genetic doctor" – that it be indulged generously in both specialized and general medical practice. With his attempts to influence the medical students' conditions of study and examinations, his house journal *Der Erbarzt*, and – since 1937 – *Fortschritte auf dem Gebiet der Erbpathologie, Rassenhygiene und ihrer Grenzgebiete* ("Advances in the Field of Genetic Pathology, Race Hygiene and Their Boundary Areas"),[169] and finally with

[167] The remaining areas had nearly no importance at all: In the area of geographic and paleoanthropology six papers appeared (mainly connected with colonial science research on "White Africa"); four papers dealt with subjects that were decidedly race hygiene; four works were dedicated to conceptual and methodological issues. No papers appeared on genetic psychology.

[168] Cf. e.g. Verschuer to Stadtmüller, 15/11/1942, MPG Archive, Dept. V a, Rep. 16, No. 2.

[169] Verschuer coedited this journal with the psychiatrist Johannes Schottky (* 1902), director of the Hildburghausen Sanatorium in Thuringia. Cf. Klee, Personenlexikon, p. 558.

his *Leitfaden der Rassenhygiene* ("Manual of Race Hygiene"),[170] Verschuer contributed to the process of making the findings of genetic pathology research flow into practice.

The orientation on genetic pathology was also expressed in the erection of a genetic biological examination office in the attic of the KWI-A, which was to be expanded to a "clinical and polyclinical station [...] in order to be able to continue the activities of consulting and producing expert opinions and also genetic clinical and genetic pathology research."[171] Verschuer had already operated such an office in Frankfurt – the model for it had been the "Polyclinic for the Care of Genes and Race" in Berlin-Charlottenburg, which Verschuer had run in 1934/35. From Frankfurt he brought nurse Emmi Nierhaus (* 1880) from the Protestant Social Services Association (*Evangelischer Diakonieverein*), who not only took over the administration of the institute as his "right hand," but also provided nursing care for the examination office.[172]

"For as intensive specialized study of the research material as possible" Verschuer further founded a "Genetic Pathology Working Group," to which he invited – besides the staff of the institute – prominent representatives of "pathological anatomy, radiology and all clinical specialities." This working group, which convened for the first time in March 1943, was also supposed to "discuss difficult questions in the practical care of genes and race and prepare the basic decisions for the state offices."[173] In his journal *Erbarzt*, too, Verschuer emphasized the Genetic Pathology Working Group's orientation to practice: Over and again he was "enlisted for genetic medical consultations and evaluations, by the Health Offices as a genetic biology consultant, and by the Hereditary Health Courts and Appellate

[170] The 2nd edn. of *Leitfaden der Rassenhygiene* appeared in 1944. In 1943 Verschuer reported that a French edition was in printing, and a Portuguese one in preparation. Cf. Verschuer, Tätigkeitsbericht 1942/43, MPG Archive, Dept. I, Rep. 3, No. 20.

[171] Verschuer to the Evangelischer Diakonieverein, 22/5/1942, Archiv des Evangelischen Diakonievereins Zehlendorf, W 3848 (pre-archive). Cf. also Verschuer, Tätigkeitsbericht 1942/43, MPG Archive, Dept. I, Rep. 3, No. 20.

[172] Emmi Nierhaus started at Verschuer's institute in Frankfurt in September 1937, followed him to Dahlem in December 1942 and from there to the lay-by in Solm. After the war Nierhaus continued to work as Verschuer's assistant, from July 1947 on, officially in the service of Protestant Social Services (*Evangelisches Hilfswerk*). After a short interruption in 1951/52 she joined Verschuer at the University of Münster. Her responsibilities proceed from a letter by Verschuer from the year 1952: "I would like to assign nurse Emmi the same group of duties she used to perform for me in Frankfurt and then in Berlin in such an excellent manner: it means a great deal to me that those people who come to use for scientific examination (e.g. twins), for evaluation (e.g. paternity certificates) or for their own consulting and examinations (e.g. marriage counseling), enjoy nursing care. The help of a nurse during the examinations currently in progress at the institute would thus have the highest priority. Added to this would be the economic direction and administration of the institute, along with the many individual tasks associated with these duties, in which nurse Emmi has proved particularly invaluable in the past." Verschuer to Oberin Sprenger, Evangelischer Diakonieverein, 26/2/1952, Archiv des Evangelischen Diakonievereins Zehlendorf, W 3848 (pre-archive).

[173] Verschuer, Tätigkeitsbericht 1942/43, MPG Archive, Dept. I, Rep. 3, No. 20.

Courts and other offices contracted to carry out race hygiene measures, as a chief evaluator."[174] In Frankfurt, whenever a specialized medical examination became necessary, he had turned to his specialist friends and their clinics – the working group in Dahlem was supposed to serve an equivalent function. In his report about the 1943/44 fiscal year Verschuer reported that the Genetic Pathology Working Group had held "several sessions [...] at which not only scientific cases from the field of genetic pathology were presented and discussed, but also practical issues of the care of genes and race debated, in order to provide to the Reich Ministry of the Interior and the Hereditary Health Appellate Courts a position on evaluations."[175] The *Erbarzt* published the protocols of two meetings of the working group, those held on March 17 and May 19.[176] In his memoirs, published in 1993, Gerhard Koch stated that he attended a further meeting of the Genetic Pathology Working Group in July or August 1943, in which the subjects were hip luxation and club foot and whose participants included the internist Friedrich Wilhelm Bremer, the orthopedic surgeon Lothar Kreuz (1888–1969),[177] director of Oskar Helene Heim and the Orthopedic Clinic of the University of Berlin, and the pathologist Robert Roessle (1876–1956).[178] On this occasion Kreuz claimed he advocated the elimination of these two congenital disabilities from the catalog of indications in the GzVeN; his proposal had been agreed to, even "by the high-ranking medical officers of the army and Waffen-SS attending this session, whose names were not known to me." Koch presumes that the publication of the protocols of the meeting was "suppressed by the censors."[179] This could in fact be the reason why – in contrast to the original proclamation – after the first two, no further protocols of meetings by the Genetic Pathology Working Group were published. However, it must be taken into consideration that Koch's account is not confirmed by any other source and that Koch has a tendency to exaggerate the frictions between genetic pathology research at the institute in Dahlem and NS genetic health policy, not to mention the importance of censorship.

A key role in the area of genetic pathology was played by Hans Grebe, who came to Berlin from Frankfurt with Verschuer. It is essentially due to his influence that a new emphasis on the field of the differential diagnosis of congenital defects developed at the KWI-A from 1943 on. In 1938 Grebe had begun with comprehensive studies on chondrodysplasia (hereditary disproportional dwarfism). He wrote a circular to 98 health offices in southern, western, and northwestern Germany,

---

[174] Verschuer, Erbpathologische Arbeitsgemeinschaft, p. 91.

[175] Verschuer, Tätigkeitsbericht 1943/44, MPG Archive, Dept. I, Rep. 3, No. 22.

[176] Grebe, Hydrophthalmus; idem., Erbpathologische Arbeitsgemeinschaft.

[177] For a biography: Fuchs, "Körperbehinderte," pp. 126–130; Klee, Personenlexikon, p. 340.

[178] For a biography: ibid., p. 503.

[179] Koch, Humangenetik, p. 115. At Verschuer's institute in Frankfurt intensive work on hip luxation and club foot had been performed (cf. e.g. Dönges, Fragen; Schwarzweller, Beitrag), and at the DFA in Munich, too, these topics were dealt with during the war (cf. Idelberger, Frage der anlagemäßigen Entstehung; idem., Frage der exogenen Entstehung).

with which the institute in Frankfurt had already been in contact regarding further professional training for medical officers. Eighty-five health offices responded to this survey and reported a total of 115 people with "dwarfism," nearly all of whom Grebe visited personally and subjected to a thorough clinical and radiological examination, together with the members of their families (parents, siblings, children, uncles, aunts, nieces, nephews, and cousins). Family tables were produced on the basis of registry office and church records, whereby particular attention was paid to the question of whether the parents were related by blood. For the purpose of comparison, Grebe consulted the specimens of miscarried and stillborn chrondodysplastic fetuses that had been dissected at the Pathological Institute of the University of Frankfurt in the years from 1936 to 1939.[180] Grebe had to discontinue his work after the beginning of the war because he was called up for military service. In summer 1942 – as previously mentioned, Grebe had been discharged from the Wehrmacht because he was seriously wounded – the "main part" of the work performed at the University of Frankfurt was submitted as his postdoctoral thesis. The manuscript was sent to Thieme-Verlag in Leipzig for publication, but the proofs were destroyed there by a bombing – not until 1955 was the work published in *Analecta Genetica*, largely unchanged, by Luigi Gedda (1902–2000), the founder and director of the Mendel Institute in Rome.[181]

Grebe had examined a total of 118 families with around 9,350 persons,[182] frequently against the bitter opposition of the subjects. One of the probands, who had been sterilized at the age of 41 in 1938, as Grebe reports casually in 1955, had reacted to "a clinical and radiological examination and especially the production of photographs [...] with the greatest resistance." Among these probands, he continued, there was the "highest degree of mental sensitivity, which was also shared by most of the members of the family."[183] Of one 43-year-old subject he writes that she was "very sensitive mentally" and seemed "decidedly depressive. For instance, during the examination, against which she put up vehement resistance, she began to cry. During a later visit, too, her mental behavior seemed melancholy."[184] Only in the case of a 17-year-old girl, who had been sterilized in spite of an appeal to the Hereditary Health Court, did Grebe express a degree of sympathy: "The resistance brought against our examination was particularly great under these circumstances."[185] The boundaries between voluntariness and compulsion were blurred to the extent that some health offices used Grebe's survey to request an opinion as to whether a marriageability certificate could be issued for certain probands.[186] Occasionally a Hereditary Health

---

[180] Grebe, Chondrodysplasie, pp. 53–55.

[181] Ibid., pp. VII–VIII.

[182] Ibid., pp. 355–366.

[183] Ibid., p. 79. Grebe did not even hesitate to secretly take a picture of a female subject whose behavior was guarded and suspicious. Ibid., p. 69.

[184] Ibid., p. 116.

[185] Ibid., p. 202. Further indications of resistance on, e.g., pp. 86, 98, 100 p. 114, 119, 123.

[186] Cf. e.g. ibid., p. 105: "With the negative family finding and the particular professional prowess of the proband, who also successfully graduated from a rural vocational school," in this case

Court requested that the Frankfurt Institute for Genetic Biology and Race Hygiene provide an evaluation in accordance with the GzVeN – in these cases it was not possible for the subjects to refuse an examination. As Grebe adhered strictly to his analysis of the conditions of heredity, he sometimes took a position *against* sterilization,[187] yet this did not change the fact that in this situation he confronted his subjects in compulsory proceedings as an officially empowered evaluator with comprehensive powers of attorney. This constituted a transgression of the boundaries of scientific ethics of major importance, both potential and in principle.

The material Grebe had collected in the course of his study of chondrodysplasia constituted the basis for a series of publications in the years 1942–1945, as he had run across an abundance of additional physical defects, mental disabilities, and mental disorders in his comprehensive genealogical studies.[188] This material increased when, after completing his postdoctoral dissertation, Grebe set about recording miscarried and stillborn fetuses on a large scale. By March 1943 he had examined nearly 100 families who had produced a stillborn child with a serious defect.[189] Verschuer's activity report for the 1943/44 fiscal year relates:

> Grebe concluded a major family research project using a non-selected series of deformed stillborn fetuses. Generally speaking he was able to prove that heritability plays a much greater role in the problem of stillbirths than was previously presumed. For the first time he was able to prove that certain forms of congenital defects are hereditary.[190]

Thus Grebe, proceeding from his collection of stillbirths, described three families in which multiple intestinal deformities had occurred (stenoses, atresias, ventricles, cysts).[191] In 1944 he published a paper about the problem of a genetic disposition for hernias (inguinal and umbilical), based on observations of twins and families.[192] Grebe attempted to prove that a hereditary factor was involved in the etiology of both cases. In other cases he endeavored to illuminate the hereditary precisely. In 1944, for instance, he published an essay that proceeded from the "Stillbirths" project, on the emergence of arhinencephaly (absence of olfactory tract, olfactory bulbs, and frequently the frontal lobe of the brain), whereby he presumed an

---

Grebe took a position *for* issuing a marriageability certificate. A case with similar circumstances is depicted on p. 182.

[187] Cf., for instance, ibid., pp. 104, 139. Perusing the book, indications of more than twenty sterilization trials are found, whereby in one case (p. 186) the application was supposed to have been submitted by the subject herself.

[188] Cf. Grebe, Fistula; idem., Struma; idem., Erblichkeit; idem., Akrocephalosyndaktylie; idem., Lipomatosis.

[189] Verschuer, Bericht über das Projekt "Totgeburten" für die Zeit vom 1/10/1943 bis zum 31/3/1944, 17/9/1944, BArch. Koblenz, R 73/15.342, p. 57.

[190] Verschuer, Tätigkeitsbericht 1943/44, MPG Archive, Dept. I, Rep. 3, No. 22. Cf. also Grebe, Todesursache.

[191] Grebe, Erblichkeit.

[192] Grebe, Hernien, pp. 66f. This study was based on a collection of material Grebe had collected during his previous work at the Horst Wessel Hospital and on cases from the twin files in Dahlem. It continued earlier works by Weitz and Verschuer.

"irregularly dominant genetic disposition."[193] In the same year he published a major paper on acrocephalosyndactylia (a syndrome characterized by skull deformation and webbing of the fingers). With reference to the etiology of this syndrome he rejected all "exogenous attempts at explanation (above all deficient amnion, lues and hypophysis damage)." Grebe assumed a specific mutation and concluded that "no race hygiene measures proceed from acrocephalosyndactylia at this time."[194]

In all of his research Grebe endeavored to make as precise a differential diagnosis as possible.[195] He assumed that one congenital defect could have very different genetic or even exogenous causes (heterogeny). Verschuer emphasized the importance of Grebe's research in this direction in his activity report for the 1942/43 fiscal year:

> In the area of the typical clinical picture [of chondrodysplasia] it was possible to establish several gene types that could be differentiated clinically and genetically. On the margins of the typical complex of symptoms there are many other genetic conditions of the cartilage-skeletal system, some of which could be observed and described for the first time. The project thus yielded a very far-reaching heterogeny, which is of fundamental importance.[196]

This finding made it seem very important to demarcate the different clinical occurrences as precisely as possible, to explain the genes responsible in each of the hereditary forms, and to reveal genes that were manifested to a hardly perceptible degree or not at all. In the context of his study about chondrodysplasia Grebe published a family study in which he pursued the question as to whether the heterozygotic carriers of the recessive gene for chondrodysplasia could be recognized on the basis of minor, nonpathological varieties. In x-rays he established that the heterozygotic family members showed minimal changes in the bone structure of the hands and feet. "But should it not be possible," Grebe asked at the close of his article, "to find a way to recognize the heterozygotes in the future, for other recessive genetic conditions as well?" For the "practical care of genes and race" the importance of this question "could not be underestimated."[197]

Yet another study by Grebe of the year 1943 must be viewed against the background of his search for stigmata. This particular work dealt with a family with an increased frequency of lipomatosis (painless symmetrical diffuse deposits of fat), but also "mental anomalies (schizophrenia, schizoid psychopathy, feeble-mindedness to greater or lesser degree, suicide, epileptic-type fits, melancholy)" as well as physical deformities (chondrohypoplasy, microcephaly, wryneck, hernias). In this case, however Grebe discarded the hypothesis of a genetic connection. Rather, he traced the coincidence of the various anomalies back to "sifting by mating." Moreover, it was possible "that the effect of one or more pathological genes on the manifestation of other genes resulted from the particular frequency of anomalies in the family described."[198]

---

[193] Grebe, Ätiologie, p. 145.

[194] Grebe, Akrocephalosyndaktylie, pp. 259 f. Cf. idem., Untersuchungen, which deals with disorders of the papillary lines in cases of syndactylia.

[195] Cf. also Grebe, Differentialdiagnose; idem./Weisswange, Chondrodysplasie.

[196] Verschuer, Tätigkeitsbericht 1942/43, MPG Archive, Dept. I, Rep. 3, No. 20.

[197] Grebe, Nachweis, p. 9.

[198] Grebe, Lipomatosis, p. 62.

In a further essay entirely tailored to practical race hygiene, in 1943 Grebe discussed the question of how high the risk should be estimated that a mother who already had experienced a miscarriage or stillbirth would give birth to yet another child with defects. In the case of very serious defects that made survival impossible, Grebe summarized his considerations, "more or less complete destruction" resulted on its own. However, a "complete elimination" of all genes responsible for defects was not possible, "first of all because only some of the carriers of very frequent, irregularly dominant and recessive genes become phenically ill, and further, because constant new generation through mutation is possible." After all in many cases the probability that a further deformed child would be born after the birth of a non-viable, seriously deformed child was so low that there was no need to advise against a new pregnancy. And even for minor changes "that can hardly be addressed as defects" there was no need to take action. "On the other hand, great misgivings about the conception of additional children must be expressed in cases of defects which allow the affected child to survive and reproduce, but reduce to some degree the capability of the adult to work or perform military service." In every consultation, however, "the total value of the given family [must be] considered."[199]

Methodologically speaking, genetic pathology research at the KWI-A was committed to higher Mendelism at the time of the World War II, and its objectives thus differed from those of practical race hygiene. In terms of contents a clear emphasis on the area of physical defects emerged, due above all to Grebe's research interests. In the Department for Race Hygiene the research "about the heritability of deaf-muteness and the race hygiene prospects of its prevention"[200] begun in the prewar period was continued. Hereditary blindness, too, remained an object of interest.[201] Finally, these were joined by the research on epilepsy in the Department for Experimental Genetic Pathology. This constitution of emphases entailed a clear division of labor with the German Research Institute for Psychiatry and the Kaiser Wilhelm Institute for Brain Research in Berlin-Buch, both of which, closely connected with the Nazi "euthanasia," were concerned with the differential diagnostics of the various forms of mental disability, schizophrenia, and neurodegenerative diseases at this time. This area played no further role at the KWI-A during World War II.[202]

[199] Grebe, Mißbildung, pp. 488 f.

[200] Fischer, Tätigkeitsbericht 1939/40, MPG Archive, Dept. I, Rep. 3, No. 17.

[201] Cf. Grebe, Häufigkeit; idem., Hydrophthalmus.

[202] However, the typed list of publications for the year 1943/44 lists an essay by Hans Grebe about "*Eine Sonderform der Athetose mit Hörstörung und Schwachsinn*" ("A Special Form of Athetosis with Hearing Defects and Feeble-Mindedness)", which was to be published in the *Psychiatrisch-Neurologische Wochenschrift*, but had not been published by the end of the war. Gertrud Veit was listed as coauthor; she may be identical to the author of a short study "*Über Dornfortsatzbrüche*" ("About Breaks of the Spinal Process") (1936). Athetoses (clinical pictures with incessant, slow, involuntary movements of the members) were the focus of interest in the cooperation between the Histopathological Department of the KWI for Brain Research unter Julius Hallervorden and the T4 "research center" in the State Institution at Brandenburg-Görden under Hans Heinze. Cf. Schmuhl, Hirnforschung. – Under the umbrella of the Department for Race Hygiene, in 1942 Werner Wolfslast published a genealogical study about Pelizaeus-Merzbacher syndrome, a condition

Yet the withdrawal from the areas of psychiatry and neurology was only in part the result of a conscious demarcation of the fields of work. It was much more a result of the fact that the large-scale project on the genetic biological inventory of 18 peasant villages in the Schwalm region of Hesse, which was begun in Frankfurt and was supposed to be continued in Dahlem, was not making any headway. Originally, the project was one of those taken on by Walter Scheidt as part of the "German Race Science" campaign. At Scheidt's request church records had been catalogued, scholars had begun to compile family tables from the around 65,000 excerpts from church records – the declared objective was to establish the genealogy of the peasant population of Schwalm from 1575 to the present. In 1935 Scheidt had turned the project over to the Frankfurt Institute for Genetic Biology and Race hygiene, which used funds from the Reich Committee for National Health Service to hire an assistant to complete the family tables. Further, Verschuer's institute set about recording the living population in these villages, whereby not only the usual anthropometric examinations took place, "but rather beyond these also comprehensive clinical-physiological and pathological findings [were to be] recorded" – this was the reason for handing the project's direction over to the physician Heinrich Schade. As the counterpart of the long-time resident population of the Schwalm region, a parallel genetic biological inventory of the city of Frankfurt south of the Main was to be carried out using the same methodology. The objective of the genetic biological inventories was to link together the fields of race anthropology, genetic pathology, and race hygiene:

> These studies are suitable for elucidating even precisely this difficult question of the meaning of race, miscegenation and constitution for pathological events in the human body. Moreover it will be possible, on the basis of such comprehensive genetic biological material on a population, to study theoretical issues of the genetic biology of humans, whose study was not previously possible because the necessary data were lacking. Genetic biological inventory goes beyond this in performing a quite essential service to the practical tasks of the care of genes and race, by supplying data for the further expansion of sterilization, marriage consultation and other measures.[203]

Schade and his staff had been working in the Schwalm region since winter 1935/36 – with the active support of the district administration, the mayors, the health offices, the schools, and the party offices. In February 1936 they had concluded their studies in two villages. In addition they evaluated the patient files of the relevant institutions of the treatment and care, hospitals, welfare offices, and practical physicians. All findings were recorded in the family tables and files, which were made accessible by a personal card index.

---

named for the neurologists Friedrich Pelizaeus (1850–1917) and Ludwig Merzbacher (1875–1942) with nystagmus, progressive psychomotoric retardation and other neurological symptoms (Wolfslast, Sippe). The paper was based on examinations Wolfslast had performed back in 1936 under the direction of Horst Geyer. In 1942 Grebe had published a case of "Dysplasia of the Right Half of the Body in One of Two Identical Twin Sisters" (Grebe, Dysplasie).

[203] Verschuer to Deutsche Forschungsgemeinschaft, 20/2/1936, BArch. Koblenz, R 73/15.341, pp. 31–34, quotes: p. 32 f.

In late 1937 Schade – with financial support from the DFG – had over 10,000 excerpts from the case histories of the University Clinics in Marburg and the files of the Ziegenhain Health Office, the State Insurance Institute in Kassel and from army physicals. Schade and his colleague Günter Burkert had personally examined 1,124 patients.[204] By March 1939 over 15,000 excerpts from patient files and Health Office certificates had been produced.[205] In the framework of the project, Schade's interest was directed primarily to the distribution of "feeble-mindedness" in the "inbreeding area" of the Schwalm region, with a strong practical orientation to race hygiene.[206] Burkert dealt with "acts of selection" through immigration to and emigration from the Schwalm region.[207] A study about the "character and aptitude of the Schwalm population" apparently was never concluded. The project staff member Heinz Koslowski performed anthropological studies in one of the region's communities, which had been founded as a Hugenot settlement, establishing there "demonstrable differences with regard to the population of purely German descent."[208]

Schade submitted his postdoctoral thesis about the genetic biological inventory of the Schwalm region in 1939 – it appeared in print in 1950.[209] At the beginning of the war the evaluation of the daunting mountain of material was far from concluded, however. Schade was drafted into military service. In December 1942 – on paper – he followed his mentor Verschuer to Dahlem, but continued to serve as a surgeon major on the front and was not able to work at the institute himself. A new "auxiliary statistical assistant" continued to evaluate the genealogical, anthropological and medical data.[210] Not until the turn of the year 1943/44 could Schade, as mentioned above, come to Berlin in the course of a military command and resume his work at the institute. In his activity report for the fiscal year 1943/44 Verschuer reported that Schade had

> [T]he essential task of processing the great amount of material on the genetic biological inventory for an old-established peasant population (from the pre-war period) continues to be sponsored. The population movements over 340 years have been established, the average burden with numerous illnesses determined and the question as to the importance of heredity for early invalidity investigated.[211]

---

[204] Verschuer, Bericht über die im Jahre 1937 durchgeführten und für das Jahr 1938 geplanten Forschungen, 21/12/1937, BArch. Koblenz, R 73/15.342, pp. 162–165, here: p. 164. Cf. Schade, Erbbiologische Bestandsaufnahme.

[205] Verschuer, Bericht über die im Jahre 1938 durchgeführten und für das Jahr 1939 geplanten Forschungen, 9/3/1939, BArch. Koblenz, R 73/15.342, pp. 125–127, here: p. 126.

[206] Schade, Häufigkeit; idem./Küper, Schwachsinn. Cf. also idem., Beitrag; idem., Befunde.

[207] Burkert, Auslesevorgänge.

[208] Verschuer, Bericht über die im Jahre 1938 durchgeführten und für das Jahr 1939 geplanten Forschungen, 9/3/1939, BArch. Koblenz, R 73/15.342, pp. 125–127, here: p. 127. Cf. Koslowski, Einfügung. The genetic psychology part of the project was to be carried out by a "Miss Dorer."

[209] Schade, Ergebnisse. Cf. also idem., Untersuchung.

[210] Verschuer, Bericht über das Projekt "Erbbiologische Bestandsaufnahme," n.d. [September 1943], BArch. Koblenz, R 73/15.342, p. 59.

[211] Verschuer, Tätigkeitsbericht 1943/44, MPG Archive, Dept. I, Rep. 3, No. 22.

Two publications that had been announced never came to be, however.[212] Consequently there was a great deal of material available at the KWI-A that could have been evaluated with regard to aspects of genetic pathology, in particular with regard to mental disabilities, had there not been a dearth of personnel.

In other areas, too, such as internal medicine, for practical reasons it was hardly possible to perform genetic pathology research during the war. Back in 1938 Grebe had begun a large-scale study in Frankfurt on the question of a "constitutional conditionality" of pneumoconiosis ("black lung" disease) on behalf of the Reich Labor Ministry. Through his service at the front this study was interrupted for 3 years and was supposed to be brought to its conclusion in Dahlem in 1943. But the air war made it impossible "to perform systematic examinations in the Ruhr area at this time, whence the majority of the cases originated,"[213] such that completion of the study was delayed even further. By then Grebe had recorded over 20,000 cases of black lung, which had been treated in social miner's hospitals or discovered during the series of x-ray examinations performed by the SS. The twins had been determined by means of inquiries at the offices of vital statistics. Grebe had contacted over 100 twins and requested file data and photographs. The clinical examination of the twins was interrupted by the start of the war, however, and the study had run aground in the second half of the war.

In the area of infectious diseases, genetic pathological research in the last 2 years of the war concentrated exclusively on tuberculosis research, after another long-term project had remained without any concrete results. Around 1937 Verschuer, together with the biologist Richard Prigge (1896–1967)[214] of the State Institute for Experimental Therapy in Frankfurt, had begun a "heredity experiment" on the "natural resistance of the guinea pig to diphteria toxin."[215] In their final report published in 1943, Prigge and Verschuer reached the conclusion that "the question of hereditary differences in resistance to diptheria toxin in the guinea pig clearly must be answered in the negative."[216] Of the 769 guinea pigs tested, only two survived, which had been "taken into breeding."[217] The production of a diptheria-resistant

[212] *Erkrankungsstatistik einer Wohnbevölkerung* ("Illness Statistics of a Residential Population") and *Bevölkerungsbewegung in drei Jahrhunderten in acht Dörfern* ("Population Movement over Three Centuries in Eight Villages"). Cf. Verschuer, Bericht über das Projekt "Erbbiologische Bestandsaufnahme," n.d. [October 1944], BArch. Koblenz, R 73/15.342, p. 40.

[213] Verschuer, Bericht über das Projekt "Staublungenkranke," n.d. [September 1943], ibid., p. 58. Through Grebe's move to Rostock the "Pneumoconiosis" and "Stillbirths" projects were delayed further. Cf. Verschuer to Präsident des Reichsforschungsrates, 4/10/1944, ibid., p. 37.

[214] For a biography: Klee, Personenlexikon, p. 473. On the inception of the project cf. Verschuer, Bericht über die im Jahre 1938 durchgeführten und für das Jahr 1939 geplanten Forschungen, 9/3/1939, BArch. Koblenz, R 73/15.342, pp. 125–127, here: p. 125 v (by this time ten pairs of twins with pneumoconiosis had been discovered).

[215] Verschuer, Tätigkeitsbericht 1942/43, MPG Archive, Dept. I, Rep. 3, No. 20.

[216] Prigge/Verschuer, Resistenzunterschiede, p. 162.

[217] Verschuer, Tätigkeitsbericht 1943/44, MPG Archive, Dept. I, Rep. 3, No. 22. Cf. Prigge/Verschuer, Resistenzunterschiede, p. 162 (note 1).

guinea pig through pure breeding was not successful, however. Instead, it seemed that the breeding of a tuberculosis-resistant rabbit was within reach.

4.3.2 Tuberculosis Research

Since tuberculosis research using the twin method had hit a dead end in the 1930s,[218] Karl and Anne Diehl – in close collaboration with Verschuer and Fischer – had been experimenting with rabbits in the "Waldhaus Charlottenburg" since 1934.[219] Diehl infected his experimental animals with a constant strain of the bovine tuberculosis bacillum by means of intravenous injection, but not until sufficient progeny were available for further breeding. The infected animals were held in a secluded stall and dissected after their death. Clear differences, interpreted as conditioned by heredity, became apparent as regarded the time of survival after injection and the infestation of the individual organs. Interest was directed primarily to two breeding lines – the one, "central" type developed a serious tuberculosis of the lung, while the other organs were hardly affected at all; in the other, "peripheral" type, by contrast, the sources of infection emerged primarily in the peripheral tissues, like in the kidneys or the nerve tissue. "A heritability of this organ specificity in the reaction to tuberculous infection," Fischer announced in his 1940/41 annual report, was "thus proven experimentally for the first time."[220] In the next annual report Fischer added that the results of the rabbit experiments could be "conferred without further ado […] to humans. This also yields important prospects for combatting tuberculosis in humans." According to Fischer, at the tuberculosis congress in Baden-Baden Diehl "held a lecture that aroused great attention and was received with much applause."[221] From 1940 on Fischer and Diehl presented the results of the rabbit experiments in Sommerfeld to the experts.[222]

The two different manifestations of tuberculosis could be bred constantly and in a pure form through eight generations of rabbits. The hereditary character of the clinical picture remained completely preserved, "even when animals were pre-treated with human tuberculosis bacilla, which are avirulent for rabbits."[223] Now Diehl attempted to get to the bottom of the riddle of organ resistance in animal experiments:

> It was attempted to modify the type of tuberculosis manifestation by inbreeding specimens. For this ink blocking, re-infections and organ transplants were performed. The persistence

[218] Twin research was not abandoned altogether, however. In his report to the Reich Research Council of September 1943, Verschuer wrote: "The research on the tuberculous twins are being continued through the collection of futher material and catamnestic analysis of the pairs of twins studied so far." Verschuer, Bericht über das Projekt "Tuberkulose," n.d. [September 1943], BArch. Koblenz, R 73/15.342, p. 54.

[219] Verschuer to Diehl, 29/6/1935, MPG Archive, Dept. III, Rep. 86 A (Münster), No. 7.

[220] Fischer, Tätigkeitsbericht 1940/41, MPG Archive, Dept. I, Rep. 3, No. 18.

[221] Fischer, Tätigkeitsbericht 1941/42, MPG Archive, Dept. I, Rep. 3, No. 19.

[222] Fischer/Diehl, Experimente; Diehl, Tierexperimentelle Erbforschung; idem., Erbe.

[223] Verschuer, Wirkung von Genen, p. 385.

> of the way in which tuberculosis is manifested appears to be very great in the bred specimens. These experiments will be expanded further, since it is possible that their result can be of fundamental importance for the medical therapy of humans. The endeavors to obtain clarity about the status of individual organs in the process of infection in the bred specimens aimed in the same direction. What was particularly interesting here was the status of the liver.[224]

In addition to these experiments, Diehl began crossing the two pure breeds with each other. When in the period from April to July 1943[225] he set about infecting the animals proceeding from the crossbreeding experiments with tuberculosis, he believed that his research was entering a decisive phase:

> Crossing the two pure breeds has now yielded a large F 1. Seventeen animals from the pure breeding experiments were taken as the point of departure. The F 1 amounts to around 50 animals. From the F 2, which we generated from animals born the previous year, we unfortunately lost quite a few because of the wet weather and the consequently wet feed. Now we have only about 40 animals. In the coming year the F2 will then appear in full force. I am glad that these animals were "vaccinated away"without having been able to reproduce. Only the desired "immune" animals will reproduce. I believe that if I aim for an F2 of about 300 animals that should be sufficient.[226]

The approach was clear: Through crossing the two pure breeds Diehl hoped to be able to cultivate "tuberculosis-resistant" rabbits. Thus, he continued working as if obsessed, although he felt miserable and exhausted at the time, since the late consequences of a lung tuberculosis contracted in his youth became noticeable. Diehl and Verschuer were feeling time pressure, too, not least because they had heard about rabbit experiments by the American tuberculosis researcher Max Bernhard Lurie (* 1893). In January 1943 – the catastrophe of Stalingrad was imminent – Diehl was still filled with hope by the sight of the dying rabbits:

> I go into the stalls in Sonnenberg often. Biological events are taking place there with a cruel consequence. It seems obvious that I hold the key in my hand. The decision will be made this summer![227]

The hope for tuberculosis-resistant rabbits was not fulfilled, however. Nevertheless, Diehl continued working doggedly on crossing the two pure breeds until the end of the Third Reich. In October 1944 he had dissected nearly 700 rabbits originating from

[224] Verschuer, Tätigkeitsbericht 1943/44, MPG Archive, Dept. I, Rep. 3, No. 22. In Diehl's draft the final point was explained in more detail: "In order to obtain clear results, pieces of the liver were surgically removed from members of the last two inbred generations, these pieces tested in terms of their antibacterial power and the animals, after recovering from surgery, tested as usual. The antibacterial power of the liver proved to be very different among the animals. The lung is currently being examined in the same direction." Diehl's draft for the Tätigkeitsbericht 1943/44, ibid. Similarly: Bericht über das Projekt "Tuberkulose," n.d. [March 1944], BArch. Koblenz, R 73/15.342, pp. 65–65 v.

[225] Diehl's draft for the Tätigkeitsbericht 1943/44, MPG Archive, Dept. I, Rep. 3, No. 22; Bericht über das Projekt "Tuberkulose," n.d. [March 1944], BArch. Koblenz, R 73/15.342, pp. 65–65 v.

[226] Diehl to Fischer, 10/8/1942, MPG Archive, Dept. III, Rep. 86 A (Münster), No. 7. Cf. also Fischer, Tätigkeitsbericht 1942/43, MPG Archive, Dept. I, Rep. 3, No. 20.

[227] Diehl to Verschuer, 23/1/1943, MPG Archive, Dept. III, Rep. 86 A (Münster), No. 7.

these crosses and performed statistical analyses of the results – but without attaining any final certainty. Nonetheless, Diehl – and Verschuer as well – still believed that this was the way to achieving a breakthrough in tuberculosis research.[228]

Biochemists were also interested in Diehl's experiments. The specifically genetic resistance of the organs had to be effected through the production of a substance in the organism, which prevented or hindered the tuberculosis bacillus from settling in certain organs. Thus the search was on for a biochemical compound. If they succeeded in isolating and identifying it, this would yield far-reaching consequences for tuberculosis therapy as well. This is the background of the lateral contacts between the external office of the KWI-A in Beetz and the KWI for Biochemistry under Adolf Butenandt. The connections extended all the way back to 1942. To keep from distorting the results of his study, Diehl required for his injections an emulsion that was completely dispersed, i.e. the tuberculosis bacilla had to be distributed as regularly as possible without any clumping. To solve this problem, Diehl had arranged with Gerhard Schramm (* 1910) of the KWI for Biochemistry to use the colloid mill located there, which could liquidify the tissue by rotating it at high speeds, in late August or early September 1942.[229] Verschuer made Adolf Butenandt himself aware of Diehl's experiments in July 1944, who proved to be "extraordinarily" interested. "Unfortunately," Verschuer reported to his friend Diehl, "we were interrupted, so that our conversation did not come to a conclusion. Thus I cannot give you any result yet today. Yet I will come back to the matter upon the next opportunity."[230] In a telephone conversation just a few days later, Butenandt expressed his wish to meet Diehl personally as soon as possible and learn about his experiments. Verschuer asked Diehl to bring "some of his family tables, tables, pictures or specimens."[231] The meeting was supposed to take place in July or August, but apparently was delayed until October.[232] This is in keeping with the comment in a report by Verschuer to the DFG of September 1944, that contact had been established with Butenandt in connection with the tuberculosis project in order to accomplish the biochemical analysis.[233] The collaboration between Diehl and Butenandt survived the Third Reich, but Diehl's research ultimately fizzled out.[234]

---

[228] In October 1944 they expressed themselves optimistically to the Reich Research Council – and not entirely truthfully – that "the heredity can already be surveyed" in the hybrids. Bericht über das Projekt "Tuberkulose," n.d. [October 1944], BArch. Koblenz, R 73/15.342, p. 48.

[229] Diehl to Fischer, 10/8/1942, MPG Archive, Dept. III, Rep. 86 A (Münster), Nr. 7.

[230] Verschuer to Diehl, 17/7/1944, ibid.

[231] Verschuer to Diehl, 20/7/1944, ibid.

[232] Verschuer to Butenandt, 30/9/1944, MPG Archive, Dept. III, Rep. 84/2, wiss. Korrespondenz. Verschuer's report to the Reich Research Council of October 1944 reads: "For the continuation of these studies using biochemical methods contact with Professor Butenandt has been established." Bericht über das Projekt "Tuberkulose," n.d. [October 1944], BArch. Koblenz, R 73/15.342, p. 48.

[233] Verschuer to DFG, Kennwort: Tuberculosis (Bericht für den Zeitraum vom 1. April bis zum 30. September 1944), 4/10/1944, ibid., p. 48.

[234] On this: Trunk, Zweihundert Blutproben, p. 47.

Through Diehl's rabbit experiments, Butenandt became aware of the entire range of genetic pathological research performed under the banner of phenogenetics at the KWI-A. On November 16, 1944 Verschuer held a lecture to the Prussian Academy of Sciences about "Heredity in Infectious Diseases,"[235] in which he attempted to link the results of the phenogenetically oriented research – especially Diehl's rabbit experiments – with the work by Alfred Kühn and Adolf Butenandt on gene action chains. It was no longer possible to publish the text of the lecture before the imminent collapse of the Third Reich, but it appeared – unaltered, as far as we can judge – in 1948 under the title *Die Wirkung von Genen und Parasiten im Körper des Menschen* ("The Effect of Genes and Parasites in the Human Body"). Verschuer's argumentation is still entirely fixed upon the problem of the interaction between infection and hereditary disposition, but also picked up on some thoughts of Butenandt's to touch on issues that are highly relevant today. For instance, Verschuer emphasized the importance of infectious diseases that jump from animals to humans, which is nothing short of prophetic in the age of BSE and SARS. He further indicated the similarity between viruses and genes – today we know that a significant part of the human genome consists of incorporated viral material. In Butenandt Verschuer found an attentive listener; the two even had "an especially pleasant (post-)meeting over a cup of tea at home."[236]

4.3.3 Experimental Genetic Pathology

On January 1, 1941 the Department for Experimental Genetic Pathology under the direction of Hans Nachtsheim began its work. In the meeting of the Board of Trustees on January 8, Fischer explained – presumably above all for Leonardo Conti – the key role of the new department. In the science of genetics, Fischer claimed, "the animal experiment [had] always been in the lead"; "*human* genetic research" had "always [received] directions and stimulation from the former." Nachtsheim's great service had been "to have recognized the fundamental importance of this wonderful research material and to have set about its evaluation [...]. This new ground he has broken must become ours."[237] This ground was not entirely new, as since the institute's founding in 1927 Fischer had occasionally provided for experiments to be performed on rats, rabbits or guinea pigs in order to get to the bottom of the genetics of normal attributes, especially *race* attributes, through the

[235] Verschuer to de Rudder, 4/10/1944, MPG Archive, Dept. III, Rep. 86 A (Münster), No. 8.

[236] Verschuer to de Rudder, 20/11/1944, ibid. Similarly, Verschuer to Fischer, 20/11/1944, MPG Archive, Dept. III, Rep. 86 A (Münster), No. 9: "I am pleased to have at least him [Butenandt] here from time to time. Otherwise things have become very quiet scientifically."

[237] Anlage 2 zur Niederschrift über die Sitzung des Kuratoriums des KWI-A am 9/1/1941: Bericht über die Neueinrichtung einer Abteilung für experimentelle Erbpathologie, erstattet vom Direktor, MPG Archive, Dept. I, Rep. 1 A, No. 2400, pp. 187–193, quotes: fol. 190 (original emphasis).

study of embryos.[238] But now Nachtsheim's preliminary works opened up the possibility of continuing this research on a grand scale and, what was even more important, expanding it to the area of genetic pathology. Although Nachtsheim's papers since 1941 fit into the leading research paradigm of phenogenetics and were not designed for rapid and direct application to race hygiene, Fischer never tired of emphasizing the practical importance of experimental genetic pathology before the board and also to the DFG, to whom he applied for 40,000 RM in research funds for Nachtsheim's department on March 13, 1941:

> What is most important is the corresponding examination of those genetic diseases that are important for humans. These are then model experiments for human genetic pathology. They will teach us why the same genetic condition often occurs in such different intensities; it will give us tips as to whether the development can be steered by external influences.

At the same time, Fischer stressed again that there was no alternative to the animal model:

> Such studies are practically impossible on humans, because it is never possible to know with any certainty what would have become of a dead embryo.

As Fischer elaborated to the German Research Association, Nachtsheim and his staff were to perform three parallel series of experiments on the diseases and anomalies to be studied: The first were breeding experiments. Nachtsheim's group of scientists was to detect rabbits with pathological genes, to propagate these "in pathologically pure culture" to the extent that pregnant females could be killed at all stages of embryonic development, and finally the heredity of the pathological genes be elucidated in crossbreeding experiments. Second, the dead embryos – in close collaboration with the Department for Embryology, still to be founded – were to be examined pathologically and histologically, in order to be able to study the inception of pathological processes during ontogenesis by comparing findings from various embryonic stages. Third and finally, it was to be attempted to influence the outbreak of disease by environmental stimuli (poisons, chemicals, feeding), not least in order to be able to differentiate between a "general" and a "genetically increased" susceptibility, which, according to Fischer, was "of particular importance in view of the most modern methods of treating diseased humans." At the close of his application Fischer stated his conviction "that these theoretical and experimental studies will be of benefit to suffering humanity and serve the preservation of the genetic health of our *Volk*." Mentioning discreetly that the Reichsgesundheitsführer shared his views, Fischer guaranteed that he could "carry the full responsibility" for the importance of Nachtsheim's research "even now at a time of war."

[238] Fischer referred to this in his application to the DFG of 13/3/1941, BArch. Koblenz, R 73/11.004. Among the works mentioned at this juncture Fischer were Kim's paper on embryonic pig skulls and Hauschild's on embryonic "negro skulls." All subsequent quotes also come from this application.

When Nachtsheim started in Dahlem, he had at his disposition, as Fischer informed the DFG, a series of rabbit strains that exhibited genetic diseases or disabilities: These were "genetic epilepsy [...], shaking palsies and other nervous diseases; glaucoma and other eye diseases; deformation of the limbs, the external sex organs (similar to those of humans), harelip and cleft palate and many others."[239]

From Nachtsheim's report to the German Research Association of January 14, 1941 – the first he submitted from his new position in Dahlem – proceed the work emphases of the group of scientists around Nachtsheim in 1940, that is, still at the Institute for Genetics and Breeding Research. At the very foreground was epilepsy research. Nachtsheim had bred from Vienna White rabbits a pure "strain of epileptics" and shown "through crossing with strains free of epilepsy [...] that *one* recessive gene [was] responsible for the increased convulsion-readiness." However, this gene was subject to certain fluctuations in manifestation: "In the pure-bred epileptic strain the condition becomes manifest in about 70% of individuals."[240] Nachtsheim had also made some progress in the search for a genetic marker, although no real breakthrough had been achieved. According to Nachtsheim's observations, the gene responsible for the increased convulsion-readiness must also grant leucism (the white color of the coat), although Nachtsheim was forced to admit that not every form of leucism could be traced back to this gene. Moreover, the influence of other genes, such as those for albinoism or "sooty coloring," could suppress the occurrence of leucism. The breeding experiments were complemented by a large-scale series of experiments on nearly 600 rabbits of different races, both from the "epileptic" and the "non-epilectic" lines, in which an injection of cardiazol induced convulsions to test their convulsion readiness.

In addition to epilepsy research, pathogenetic research on eye diseases, especially on the progressive heredity of certain forms of cataracts, constituted a second working emphasis, in collaboration with Hellmuth Gürich of the Charité Ophthamalogical Clinic. A third and final emphasis emerged from the work of the two doctoral students Christian Schnecke (* 1917) and Harry Suchalla (1912–1985), who concerned themselves with growth anomalies. Schnecke's studies on the "lethal dwarfism in rabbits led to the result, also important for the assessment of corresponding conditions in humans, that while the recessive dwarf factor in general may lead to a pathological form only in the homozygotes, but that there are genes that are harmless in and of themselves, which, when linked with the dwarf

[239] In his report to the Board (MPG Archive, Dept. I, Rep. 1 A, Nr. 2400, p. 190) Fischer further mentioned rabbits "with a kind of St. Vitus' dance." "Genetic St. Vitus' Dance," one of the indications in the GzVeN, was the contemporary term for Huntington Chorea. It is unclear whether this might have indicated the rabbits with "shaking palsy," which is a general lay term for Morbus Parkinson. In his activity report for the 1940/41 fiscal year Fischer also mentioned rabbits with "skin diseases" (Fischer, Tätigkeitsbericht 1940/41, MPG Archive, Dept. I, Rep. 3, No. 18).

[240] Nachtsheim, Bericht über die im Jahre 1940 mit Unterstützung des Reichsforschungsrates durchgeführten Untersuchungen zur vergleichenden Erbpathologie, 14/1/1941, p. 1, BArch. Koblenz, R 73/13.328 (original emphasis). For the basics on Nachtsheim's epilepsy research: Schwerin, Experimentalisierung, pp. 282–288.

factor in the homozygous form, yield a combination with lethal effects, even if the dwarf factor is only present in a single dose." Suchalla crossed giant and dwarf varieties of rabbits. This paper, Nachtsheim emphasized, represented "the first attempt to achieve an analysis of skull genetics by performing experiments with modern methods on large amounts of material."[241] Here the research projects of the Department for Experimental Genetic Pathology overlapped with Hans Grebe's studies on chondrodysplasia – Nachtsheim and Grebe did, in fact, work together closely, for instance, on a genetic biology dictionary.[242]

In 1941, the first year for the Department for Experimental Genetic Pathology, the group around Nachtsheim was able to continue its research only on a very restricted scale. Because it was increasingly difficult to obtain feed for the experimental animals, it was necessary to reduce their number and restrict the "consumptive research." The apparatus applied for arrived only after major delays – as, for instance, the "convulsator" for the generation of electric spasms – or were not delivered at all, as was the case for a slit lamp, a Zeiss microscope and a binocular eyepiece. The greatest problem was that all of the staff was called up for military service, such that the experiments could only be continued by Nachtsheim on his own. The fact that epilepsy research remained at the focus, although it had come to a preliminary conclusion in 1940, was grounded first of all in pragmatic reasons: The research on eye diseases could not be continued because Hellmuth Gürich, the partner in this collaboration at the ophthamalogical clinic, was drafted to the Wehrmacht. The same was true for research projects on growth anomalies. By this time Christian Schnecke and Harry Suchalla had also been drafted. The planned genetic pathological studies on a syndrome observed in dachhunds (characterized by hypodactyly or hyperdactyly, respectively, and hereditary blindness) never really got in gear. The resumption of research on the "Pelger anomaly" (today: Pelger-Huët nuclear anomaly), an autosomal-dominant hereditary anomaly of the leucocytes that occurs in both humans and rabbits, was just getting started – the first task that kept Nachtsheim busy was breeding a "pure Pelger-Huët strain,"[243] on the basis of which the characteristics of the gene it was based on could be studied.

When an epidemic broke out among the laboratory rabbits in summer of 1942, which necessitated halting the epilepsy experiments temporarily, the research on the Pelger anomaly shifted far into the foreground. Moreover, in this area Nachtsheim was able to present a sensational finding. While up until that time it had been assumed that "this deviation of the blood count from the normal [was] to be observed in both humans and animals as a harmless variety of blood without any further clinical manifestations," Nachtsheim produced evidence that in a rabbit

[241] Ibid., p. 5. Cf. Schnecke, Zwergwuchs; Suchalla, Variabilität.

[242] Schwerin, Experimentalisierung, pp. 278 f.

[243] Nachtsheim, Bericht über die im Jahre 1941 mit Unterstützung des Reichsforschungsrates durchgeführten Untersuchungen zur vergleichenden und experimentellen Erbpathologie, 27/3/1942, here: pp. 1, 5–7, BArch. Koblenz, R 73/13.328. Cf. Schwerin, Experimentalisierung, pp. 263–268.

which inherited the Pelger gene from the maternal *and* paternal side, and was thus homozygous with reference to the Pelger gene, "most serious impairments"[244] were to be expected. According to Nachtsheim's observations, most of the homozygous Pelger rabbits died in the womb. The few survivors – the "über-Pelger," as Nachtsheim called them – showed not only a changed blood count, but also a whole bundle of other clinical symptoms: "meager growth, serious deformation of the limbs, especially the forelegs, with shortening and twisting of the long, hollow bones and synostoses [fusion of bones], rashes of scurf around the muzzle and nose, salivation, anorexia."[245]

These findings, Nachtsheim explained, were of extraordinary importance for humans. True, no human "homozygous Pelgers" had been encountered as yet. Yet, trusting in the soundness of the animal model, Nachtsheim predicted that in humans, too, the homozygous carriers of the Pelger gene, if they were able to survive at all, would "thus certainly be greatly weakened in their vitality and deformed." In any case it is clear that the Pelger anomaly did not constitute a "harmless 'play of nature'" in humans either, but was an "erroneous mutation [...], whose propagation, in terms of race hygiene, [was] altogether undesirable." Here Nachtsheim opened up a new race hygiene perspective. The only way to be able to follow this perspective was to link the research on animal models with the genetic pathology of humans. The mission of the research would be, in Nachtsheim's words, "to carry out exhaustive surveys about the propagation of the Pelger gene in human populations." At the same time it would have to be investigated "whether among the stillbirths or behind an already familiar clinical picture, especially among cases with certain deformations of the limbs, homozygous Pelgers are to be found."[246] This suggested building a bridge from experimental genetic pathology to the genetic pathological research by Heinrich Schade and Hans Grebe, namely to Grebe's series of studies on stillbirths. In his report about the 1943 fiscal year Nachtsheim remarked that work in this direction had been "initiated," but had "not yet led to positive results." He further announced embryological studies in order to clarify in "which embryonic stage the homozygous Pelgers die and what the cause of this death" and "what, on the other hand, [is] the cause of the survival of individual homozygous Pelgers of certain parents."[247] The clinical and histopathological diagnostic

[244] Nachtsheim, Bericht über die im Jahre 1942 mit Unterstützung des Reichsforschungsrates durchgeführten Untersuchungen zur vergleichenden und experimentellen Erbpathologie, 22/3/1943, BArch. Koblenz, R 73/15.342, pp. 99–112 (no continuous pagination), quote: p. 100.

[245] Nachtsheim, Bericht über die im Jahre 1943 im Auftrage des Reichsforschungsrates durchgeführten Untersuchungen zur vergleichenden und experimentellen Erbpathologie, 15/3/1944, ibid., pp. 79–84, quote: p. 82.

[246] Nachtsheim, Bericht über die im Jahre 1942 mit Unterstützung des Reichsforschungsrates durchgeführten Untersuchungen zur vergleichenden und experimentellen Erbpathologie, 22/3/1943, ibid., pp. 99–112 (no continuous pagination), quotes: p. 100.

[247] Nachtsheim, Bericht über die im Jahre 1943 im Auftrage des Reichsforschungsrates durchgeführten Untersuchungen zur vergleichenden und experimentellen Erbpathologie, 15/3/1944, ibid., pp. 79–84, quote: p. 82. Cf. Nachtsheim, Pelger-Anomalie I und II.

picture of the homozygous Pelgers also demanded closer study. Since these formulations were repeated word for word in Nachtsheim's final report, which was dated March 15, 1944, it must be assumed that the studies never picked up speed.

In 1942 Nachtsheim turned his attention to another hereditary blood anomaly of the rabbit, which had its parallel in humans: Erythroblastosis, which occurs in rabbits as hereditary, general dropsy (*Hydrops universalis congenitus*). Today we know that this form of newborn jaundice in cases of incompatible rhesus factors in the blood of mother and child is caused by the formation of antibodies in the mother and their transition into the circulatory system of the fetus, where they destroy red blood cells. In 1942, however, Nachtsheim traced erythroblastosis in rabbits back to a gene that was "inherited recessively." But, Nachtsheim continued, this was "not a case of simple heredity" – indeed, it appears "that a wide variety of factors – besides the remaining genotype, also those of a peristatic nature – had an influence on the manifestation of the condition."[248] In 1943 Nachtsheim presumed that "several genes" were involved. Perhaps there must also be "a certain conditional factor present [...] so that the actual dropsy gene [could] become effective." Possibly, however – and Nachtsheim was on the right track here – "in addition still other factors located in the mother but outside the embryo [played] a role." As far as erythroblastosis in humans was concerned, Nachtsheim's judgement in 1943 was more cautious, stating that it was "still quite contested," whether a hereditary condition was involved or not – much spoke against, some for heredity.[249] In his penultimate report of March 15, 1944 Nachtsheim suggested that "certain observations on humans [made] probable a connection between the fetal blood diseases and certain serum characteristics of the blood." Thus it appeared desirable "to test experimentally for existing connections of this kind in animals, too." Again, he states that experiments in this direction had been "initiated."[250] By the way, the serological studies on *Hydrops universalis congenitus* in rabbits were conducted in collaboration with the Serological Department of the Reich Health Office, and the histopathological studies by Hans Klein (1912–1984). Klein was Senior Physician in the Pathological Department of the Rudolf Virchow Hospital in Berlin under Berthold Ostertag (1895–1975), who, however, was dispatched along with part of the Pathological Institute to the SS Sanatorium Hohenlychen at the time.[251]

[248] Nachtsheim, Bericht über die im Jahre 1942 mit Unterstützung des Reichsforschungsrates durchgeführten Untersuchungen zur vergleichenden und experimentellen Erbpathologie, 22/371943, ibid., pp. 99–112 (no continuous pagination), quotes: p. 111.

[249] Nachtsheim, Bericht über die im Jahre 1943 im Auftrage des Reichsforschungsrates durchgeführten Untersuchungen zur vergleichenden und experimentellen Erbpathologie, 15/3/1944, ibid., pp. 79–84, quote: p. 83.

[250] Nachtsheim, Bericht über die im Halbjahr 1943/44 im Auftrage des Reichsforschungsrates durchgeführten Untersuchungen zur vergleichenden und experimentellen Erbpathologie, 15/3/1944, ibid., pp. 61–63, quotes: p. 63.

[251] Verschuer, Tätigkeitsbericht 1943/44, MPG Archive, Dept. I, Rep. 3, No. 22; Nachtsheim, Bericht über die im Halbjahr 1943/44 im Auftrage des Reichsforschungsrates durchgeführten Untersuchungen zur vergleichenden und experimentellen Erbpathologie, 15/3/1944, BArch. Koblenz, R 73/15.342, pp. 61–63, here:p. 63; idem., Bericht über die im Halbjahr 1944 im Auftrage des

While the research on fetal blood diseases of the rabbit was an emphasis of the work in the Department for Experimental Genetic Pathology in 1942, in 1943 the research on the growth anomalies of the rabbit swung into full gear, when Wouter Ströer, the designated director of the planned Department for Embryology, took on the histological study of the rabbits with "lethal dwarfism" during his residency in Dahlem.[252]

In addition to these working areas, in the final 2 years of the war, epilepsy research moved back up to the top of the agenda of the Department for Experimental Genetic Pathology. In further breeding experiments Nachtsheim investigated the heredity of genuine epilepsy. Here it had become apparent, he reported in 1944, that the "epilepsy gene," although its behavior was "generally recessive," and thus had to be inherited homozygously in order to take effect, was also able "to let the diagnostic picture of epilepsy develop" even in cases of heterozygous heredity, "in combination with certain genes." As such, "the carrier of two albino genes and one epilepsy gene can become an epileptic." The situation was similar for the allele closest to the albino gene, the "black factor."[253] In addition to his breeding experiments, from 1943 Nachtsheim performed a great number of experiments on producing spasms through oxygen deprivation. Since these experiments led him directly into the research accompanying the Nazi "euthanasia" program, they are described in detail in another section.

4.3.4 *The Genetics of Normal Attributes*

This field of research now lagged behind, also and primarily because of problems acquiring material. Much of genetic pathology research was based on clinical material, which Verschuer and his staff had brought with them from Frankfurt, and which could be supplemented continuously through individual cases brought to the institute for evaluation. The genetic pathology research by Diehl and Nachtsheim also used the animal model – and the rabbit stocks were safeguarded by the elevation

---

Reichsforschungsrates durchgeführten Untersuchungen zur vergleichenden und experimentellen Erbpathologie, 15/3/1944, ibid., pp. 42–43, here: p. 63. The publication of the results took place in 1947 (Nachtsheim/Klein, Hydrops congenitus universalis). Nachtsheim also collaborated with Hans Klein in his research on the Pelger gene (cf. Klein, Pelger-Anomalie). – Ostertag and Klein performed autopsies of children from the Wittenau Sanatoriums murdered as part of the Nazi "euthanasia" program. In early 1945, Klein also studied the lymph nodes from the armpits of twenty Jewish children upon whom Kurt Heißmeyer (1905–1967), senior physician at the SS Sanatorium Hohenlychen, performed criminal human experiments for tuberculosis research in the period from December 1944 to April 1945 at the Neuengamme concentration camp near Hamburg. For details on these indirect connections: Weindling, Genetik und Menschenversuche.

[252] Verschuer, Tätigkeitsbericht 1943/44, MPG Archive, Dept. I, Rep. 3, No. 22.

[253] Nachtsheim, Bericht über die im Halbjahr 1943/44 im Auftrage des Reichsforschungsrates durchgeführten Untersuchungen zur vergleichenden und experimentellen Erbpathologie, 15/3/1944, BArch. Koblenz, R 73/15.342, pp. 61–63, quotes: p. 61.

of the KWI-A's status into that of a military economic enterprise. General human genetic research, in contrast, was based essentially on the combination of twin and family research. Yet these methodological approaches were nearly completely obstructed in the second phase of the war. As Fritz Lenz lamented in his annual report for the "Institute for Race Hygiene" in 1943/44:

> The work of research has been quite impeded by the circumstances of the war, especially since summer 1943. It is very difficult and frequently impossible to acquire sufficient observation material for certain essential scientific and practical problems. As a consequence of the evacuation of women and children, family research and twin studies are practically impossible. Not even surveys can be conducted any more.[254]

Conventional twin research in this area apparently did not come to complete standstill,[255] but the difficulty in acquiring subjects for both twin research and family research soon became an obstacle that could hardly be surmounted. Thus it is no coincidence that the research in the area of the genetics of normal attributes in the year 1943 was restricted to two projects – "Specific Proteins" and "Eye Color" – which made use of the unfettered access to subjects in the Auschwitz concentration and extermination camp – more on this later.

4.3.5 Genetic Psychology

In the Department for Genetic Psychology, Gottschaldt, who was called up to the Wehrmacht for a time, and his staff continued even after the start of World War II with the evaluation of the enormous amount of material they had compiled in the twin camps in 1936/37, and working through it "in a new methdological way." Preliminary results were published in 1942 in the first issue of *Erbpsychologie*, a new series of publications edited by Eugen Fischer and Kurt Gottschaldt. Through this work genetic psychology received a new foundation."[256] Analysis continued in the final war years, accelerated after the Department for Genetic Psychology was

[254] Lenz, Tätigkeitsbericht 1943/44, 28/3/1944, MPG Archive, Dept. I, Rep. 3, No. 21. Quite similarly, Verschuer: "The twin and family research projects in progress continue to be extremely restricted by the war conditions." Verschuer, Tätigkeitsbericht 1943/44, MPG Archive, Dept. I, Rep. 3, No. 22. Cf. Verschuer, Bericht über das Forschungsprojekt "Rassenhygiene," n.d. [September 1943], BArch. Koblenz, R 73/15.342, p. 56.

[255] Thus the Portuguese guest scholar José Ayres de Azevedo investigated "the quantitative course of blood-group reactions for a large amount of material from identical and fraternal twins," establishing "that in this regard, too, a clear hereditary conditionality is demonstrated" (Verschuer, Tätigkeitsbericht 1942/43, MPG Archive, Dept. I, Rep. 3, No. 20). Cf. on this Müller-Hill, Blut, p. 196. – "A paper initiated and advised by Lenz about the body length and weight of identical twins in comparison with fraternal twins, which yielded that identical twins are somewhat shorter and lighter on average than fraternal twins, was published by its Hungarian author Dr. L.[adislaus] Apor, unfortunately only in the Magyar language." (Lenz, Tätigkeitsbericht 1943/44, 28/3/1944, MPG Archive, Dept. I, Rep. 3, No. 21).

[256] Verschuer, Tätigkeitsbericht 1942/43, MPG Archive, Dept. I, Rep. 3, No. 20.

removed to Stavenhagen castle in Mecklenburg in September 1943.[257] These tasks of evaluation were extremely elaborate: In the 6 months from October 1943 to March 1944, Gottschaldt reported, "around 25 psychological analyses [were dictated], each of which was 200 pages long."[258] By October 1944 Gottschaldt and his staff had put to paper around 65 psychological analyses, "which cover extraordinarily comprehensive material, prepared for statistical evaluation, of more than 120,000 individual findings."[259] And for the coming 6 months Gottschaldt requested another 10,000 sheets of writing paper.

During the war period, Gottschaldt could not simply retire to his ivory tower. More and more he worked together with state and party offices, and endeavored to make the methods of genetic psychology useful for genetic health, race and colonial policy – be it voluntarily or under the pressure of the conditions must remain an open question. Even today, almost nothing is known about most of these projects. In the 1941/42 fiscal year, Gottschaldt's department, in collaboration with the Department for the Protection of Children and Youth (*Kinder- und Jugendschutz*) of the NSV, began with "catamnestic surveys of children formerly under the care of state welfare."[260] The 1942/43 business report also stated that the "Polyclinic for Nervous and Difficult Children," whose resources "increasingly [were] claimed for the scientific evaluation of the very extensive material on families that accumulates there,"[261] and that this would continue. From 1941/42 Gottschaldt held lectures and training courses, connected with the German Labor Front (*Deutsche Arbeitsfront*), the Department for Professional Training and the Improvement of Efficiency (*Abteilung für Berufsausbildung und Leistungsertüchtigung*) in the Reich Chamber of Commerce and the Colonial Policy Office of the NSDAP.[262] A deeper collaboration arose from the contact with the Colonial Policy Office – more on this later.

With the excursion into colonial science, Gottschaldt set out on the field of *race* psychology, which he had only skirted before World War II. Thus it was fitting that he prepared an article about "Race Psychology" for the fifth edition of "Baur-Fischer-Lenz."[263] Also to be viewed in this context are the *Untersuchungen über*

[257] Verschuer to Fischer, 20/9/1943, MPG Archive, Dept. III, Rep. 86 A (Münster), No. 9. The castle belonged to an acquaintance of Gottschaldt's. Fischer managed to wrangle an "allocation" with the assistance of the Gauleiter (District Leader) of Mecklenburg, Friedrich Hildebrandt (1898–1948). Cf. Gottschaldt, Bericht über das Forschungsprojekt "Zwillingslager," 25/9/1943, BArch. Koblenz, R 73/15.342, p. 70.

[258] Gottschaldt, Bericht über das Forschungsprojekt "Zwillingslager," 14/3/1944, p. 66.

[259] Gottschaldt, Bericht über das Forschungsprojekt "Zwillingslager," 4/10/1944, p. 49.

[260] Fischer, Tätigkeitsbericht 1941/42, MPG Archive, Dept. I, Rep. 3, No. 19. These studies were continued well into the final years of the war. Cf. Verschuer, Tätigkeitsbericht 1942/43 and 1943/44, respectively, MPG Archive, Dept. I, Rep. 3, No. 20 and No. 22, respectively.

[261] Verschuer, Tätigkeitsbericht 1942/43, MPG Archive, Dept. I, Rep. 3, No. 20.

[262] Fischer, Tätigkeitsbericht 1941/42, MPG Archive, Dept. I, Rep. 3, No. 19; Verschuer, Tätigkeitsbericht 1942/43 and 1943/44, respectively, MPG Archive, Dept. I, Rep. 3, No. 20 and No. 22, respectively.

[263] Gottschaldt's catchwords for the 1942/43 Jahresbericht, MPG Archive, Dept. I, Rep. 3, No. 20. An article about "genetic psychology" was also planned. Cf. Fangerau, Etablierung, pp. 60–62.

den Rassenruf mongolider Völker im Rassenbewußtsein von Japanern ("Studies about the Race Reputation of Mongoloid Nations in the Race Consciousness of Japanese"), which were carried out in the Department for Genetic Psychology in collaboration with the Cultural Department of the Japanese Embassy – presumably by the two Japanese guest scholars, the doctoral student Masataka Takagi and Professor Masaji Kamitake. The fact that Gottschaldt participated in a "German-Japanese Science Camp" together with the two guest scholars in summer 1942 suggests that he, too, was actively involved in these obscure studies. Finally, it must be added that a doctoral student of Gottschaldt's, Inez de Beauclair (1897–1981), carried out "Examinations of Physical Constitution on Southern and Northern Chinese" in Japanese-occupied China during World War II.[264]

4.4 The Kaiser Wilhelm Institute for Anthropology, Human Heredity and Eugenics and the State Crimes of the Nazi Regime, 1939–1945

4.4.1 Fischer, Verschuer, and the NSDAP

Since mid-1938 deliberations had been in progress about admitting Eugen Fischer and Fritz Lenz[265] to the NSDAP. It can be presumed that Fischer's political allies, Arthur Gütt and Walter Groß, exerted pressure on Fischer and his institute to this end once the NSDAP lifted its ban on admitting new members in 1937[266] – in any case at least ten members of the KWI-A staff joined the party on May 1, 1937.[267] That Fischer yielded to the pressure of his political patron and made active efforts to join the party from mid-1938 on was probably also a matter of calculation, and

[264] Fischer, Tätigkeitsbericht 1941/42, MPG Archive, Dept. I, Rep. 3, No. 19; Gottschaldt's catchwords for the 1942/43 Jahresbericht, MPG Archive, Dept. I, Rep. 3, No. 20. After 1945 de Beauclair lived in Taiwan. She published a series of ethnological works on China and the South Pacific.

[265] It is not entirely clear when Lenz joined the NSDAP. Klee, Personenlexikon, p. 367, names May 1, 1937 as his date of admission and adds that Lenz was also a member of the NSDÄB and the NS-Dozentenbund. Rissom, Fritz Lenz, p. 24, in contrast, cites a postwar deposition by Lenz, which states that he had become a party member in 1938 upon Gütt's urging. Kröner, Von der Rassenhygiene zur Humangenetik, p. 37, dates Lenz's admission to 1937. – Lenz's denazification document of 6/6/1949 states that Lenz had been a member of the NSV since 1935, and joined the NSDÄB, supposedly under compulsion, on May 1, 1937, and the NSDAP, also in the year 1937. MPG Archive, Dept. II, Rep. 1 A, PA Lenz. No indication of Lenz is included in the BDC Inventory 3100 (NSDAP central records) and 3200 (NSDAP local group file) in the Federal Archive in Berlin.

[266] According to Kröner, Von der Rassenhygiene zur Humangenetik, p. 28, who sees – besides Walter Groß – Leonardo Conti as the driving force in the background. However, Conti was not yet in such a key position in 1937/38 and did not participate actively in the fortunes of the KWI-A.

[267] Ibid., p. 39. Of the assistants at the KWI-A, it appears that Peter Emil Becker was the only non-member of the NSDAP after 1933. Verschuer's Frankfurt institute, in contrast, was "a real reservoir of old combatants" (ibid.).

closely connected with his plans for reorganizing the institute, which were not to be realized without strong political cover. An evaluation by the Race Policy Office of the NSDAP, no longer preserved in the archives, apparently reached a positive assessment. Reichsführer SS Heinrich Himmler, when asked for an opinion by the staff of the office of the Führer's deputy, offered support for Fischer and Lenz in 1938, arguing

> [T]hat through their scientific work in recent years both have made significant contributions to the fortification and scientific acknowledgement of the racial elements of the National Socialist world view. I am convinced that both Fischer and Lenz, despite a few remaining misgivings, can be admitted to the party. I even believe that the admission is a political necessity of sorts, for we cannot use the power of these two men for the scientific fortification of the party on the one hand, and reject them as party comrades on the other.[268]

Meanwhile, the admission proceedings dragged on exceedingly long. On December 12, 1939 the staff of the Führer's Deputy informed the Reich Treasurer of the NSDAP that Reichsgesundheitsführer Conti approved Fischer's application for admission and that Staff Chief Martin Bormann (1900–1945) had also given his go ahead. Rapid processing was requested.[269] According to this letter, Fischer had submitted his official application for admission on November 17, 1939. According to the files of the local NSDAP group, admission was not applied for until December 28, 1939. Whatever the date of the application: From January 1, 1940 on Eugen Fischer was a member of the NSDAP.[270] At the same time Fischer apparently induced his designated successor Verschuer to join the NSDAP. The latter became a party member on July 1, 1940, while still in Frankfurt.[271] Fischer congratulated Verschuer upon his admission to the party, commenting "I believe that the affiliation is correct and necessary, apart from the associated internal attitude [...]."[272] After taking over the institute in Dahlem, Verschuer went even further and joined the NSDÄB.[273]

At this point in time, Verschuer – despite his enduring links to the *Bekennende Kirche* ("Confessing Church") – had long since come to terms politically with the National Socialist regime. He had even made himself indispensable as a human geneticist, race hygienist, and "genetic physician" and cooperated with the regime. The lecture about "The Genetic Image of Humans," which Verschuer held for the

---

[268] Himmler to Stab des Stellvertreters des Führers, 17/8/1938, BArch. Berlin, BDC, DS G 117. Lösch, Rasse, p. 276, also cites this letter, but does not mention any date. And indeed, the date would not fit in well with his hypothesis that Fischer changed his opinion about joining the party only under the impression of the outbreak of the war, as he then had the need "to want to and have to prove his patriotism." (ibid., pp. 275 f.).

[269] Ibid., p. 276 (note 103).

[270] BArch. Berlin, BDC, 3200, E 0051.

[271] Ibid., S 0085 and X 0063.

[272] Fischer to Verschuer, 30/9/1941, MPG Archive, Dept. III, Rep. 86 A (Münster), No. 9. The membership card had been issued to Verschuer on April 30, 1941.

[273] Kröner, Von der Rassenhygiene zur Humangenetik, p. 33.

Main Assembly of the Kaiser Wilhelm Society in Breslau on May 24, 1939, culminated in a clear avowal to scientific policy consulting:

> The parallel progression of political and scientific thought is no coincidence, but an internal necessity. [...] We genetic biologists and race hygienists [...] remain in the peace of our scientific research activity from the interior conviction that on this field, too, battles of major importance are being fought for the continuity of our *Volk*.[274]

However, even as director of the KWI-A he was sometimes subjected to political pressures. With his very first lecture as a newly appointed member of the Prussian Academy of Sciences on November 10, 1943 he offended party circles. Under the title *Erbanlage als Schicksal und Aufgabe* ("Genetic Disposition as Fate and Function") Verschuer took his audience on a *tour d'horizon* through the regions of higher Mendelism and phenogenetics. Certainly: Verschuer criticized the naive dogma of heredity predominant in higher Mendelism. "For in many cases genotype and race were regarded far too simplistically in terms of materialistic determinism – as the *sole* source of all life performance, even of intellectual power, especially for culture and history."[275] But at the same time Verschuer made it perfectly clear that it was hardly his intention to explode the structure of genetic determinism:

> After these results of genetic and race research, is it justified to assert that genetic disposition is fate? Yes and no! *Through genetic disposition, certain fateful limits are determined for the development of each individual.* A Negro cannot produce any white children, the genetically feeble-minded have predominantly feeble-minded children, certain defects are passed down according to familiar rules, etc. These are limits that are becoming ever more clearly and definitely demarcated through our research. They cannot be transcended.[276]

What could appear offensive, however, were the social and moral conclusions Verschuer drew from the insights of higher Mendelism and phenogenetics: In terms of their genetic dispositions, Verschuer grouped people into a three-level model, arrayed between the two poles of "fate" and "function." Verschuer located the majority of people on the third and highest level: In their genetic dispositions lay "a *fateful predetermination* only very weakly [...] concealed," they had a "great breadth of possibilities for development."[277] The shaping of the phenotype on this third level was the task of the individual and of society. The people on the second, intermediate level may carry the disposition for serious diseases and disabilities with them, but these appear either not at all or only weakly in the phenotype due to the oscillation of manifestation, and in any case can be compensated for by measures of prevention or rehabilitation. On this second level the molding of the phenotype lay between fate and function. As examples Verschuer named club foot and congential hip luxation.[278] From his comments clearly proceeded that he believed that all possibilities for orthopedic rehabilitation must be exhausted – he was well

[274] Verschuer, Erbbild vom Menschen, p. 12.

[275] Verschuer, Erbanlage als Schicksal und Aufgabe, p. 24 (original emphasis).

[276] Ibid., p. 16 (original emphasis).

[277] Ibid., p. 19 (original emphases).

[278] Ibid., pp. 10f., 18.

advised to factor out the question of race hygiene sterilization at this juncture, since both conditions were officially considered to be indications for sterilization, a position that Verschuer and his colleagues, as we will show later, did not share. On the first and lowest level, finally, Verschuer placed people with serious genetic defects, whose manifestation was not mediated by other factors – "association with other genes," "course of development," "external influences."[279] On this level the phenotype was "to be accepted as *determined by fate*." Nevertheless, with a view to these humans as well, Verschuer argued in terms of the dualism of fate and function. Although their genetic dispositions had to be "accepted as given by fate," the affected confronted a double function:

> *First, even with a serious genetic defect, it is possible to give one's own life higher value and deeper meaning*. Just think of the extraordinary achievements of the blind and deaf-mute. Yes, even a mentally retarded person can still carry out useful work and distinguish himself through loyalty, love and the spirit of sacrifice. Second – and this demands a selfless readiness to make sacrifices – for the welfare of the *Volk*, the serious genetic defect must be eliminated by forgoing propogation.[280]

This passage could be understood as a criticism of the "euthanasia" under way since 1940, which had already claimed the lives of over 100,000 mentally ill and mentally disabled by this time – and it appears that party circles understood it as such. Even more important: It was probably so intended. At this juncture Verschuer made clear that he would continue to actively support eugenic sterilization that could be legitimated with the moral philosophy and theology of the idea of sacrifice – a position which Verschuer had advocated since the final years of the Weimar Republic –, but rejected for ethical reasons the murder of the mentally ill and mentally disabled. What's more, he openly repudiated the "breeding of the *Übermensch*" in Friedrich Nietzsche's terms as the basic motif for race hygiene – he wanted to restrict race hygiene to its function as "custodian of the genotype of the race."[281]

As demonstrated, Verschuer's lecture included some critical tones that could not have pleased the makers of National Socialist genetic health policy. But an entry in the diary of Ulrich von Hassell (1881–1944) shows that Verschuer's lecture could be interpreted differently as well:

> For me, a lecture on race policy for the Berlin circle of the German Academy was indicative of the level of some sectors of German science. The speaker was Prof. von Verschuer, the man whom E. Fischer dared to propose as his successor in the Mittwochs-Gesellschaft. Superficial prattle tailored to the purposes of party politics, truly a disgrace.[282]

Nevertheless: what appeared as pseudoscientific party propaganda to a member of the resistance provided for unrest in sectors of the party. On April 25, 1944,

[279] Ibid., pp. 7 f.

[280] Ibid., p. 19 (original emphases).

[281] Ibid., p. 22.

[282] Diary entry of 5/12/1943, Hassell-Tagebücher, p. 409.

that is, a considerable time after the lecture,[283] Verschuer related in a letter to Fischer:

> Yesterday afternoon I visited [Walter] Groß in his new office in Babelsberg, a country house in a beautiful setting. We conversed for 2 ½ hours in a very friendly and mutually obliging tone. He confirmed that he found nothing objectionable in the content of my lecture to the academy, and that my depiction and my standpoint were irreproachable. Incorrect reporting and the misleading interpretation of individual passages have caused political turbulence. However, I got the impression that he will put an end to the matter. I made an agreement with him to submit to him any publications that encroach upon the area of race policy for fine tuning. So I hope that our friendly terms are restored, and that in future he will not be so easily disquieted by such yapping and put the over-zealous curs back on their chains.[284]

The incident is further exemplary evidence of the fact that, within the alliance between science and politics, it was ultimately the political decision makers who made the rules. Hans-Peter Kröner's interpretation must be endorsed, that Verschuer's account of his meeting with Groß promoted his own self-deception: With the arrogant gesture of the academic, he required every effort to conceal from himself and his mentor that he – the director of a Kaiser Wilhelm Institute – had been "muzzled"[285] by one of the National Socialist satraps. After the war, Verschuer, together with his "whitewashers," greatly exaggerated the danger that threatened him from this direction.[286] Politically, he deviated from the line of state and party only in part. Aside from the issue of "euthanasia," broad consensus predominated in genetic health and race policy. As will be shown in the next section, Verschuer and his staff legitimated and propagated this policy, tended to the scientific substrate, provided practical support and did not hesitate to use the National Socialist politics of genocide in order to acquire scientific "material."

4.4.2 Lecturing Activity

Until well into the year 1944, the staff of the KWI-A, hardly hampered by the circumstances of the war, undertook lecture trips all over Germany and Europe. The lectures at universities and to scientific societies were attended by "science camps"

---

[283] Thus it is hardly possible to say that Verschuer was "invited to report" to Groß. According to Kröner, Von der Rassenhygiene zur Humangenetik, p. 34.

[284] Verschuer to Fischer, 25/4/1944, MPG Archive, Dept. III, Rep. 86 A (Münster), No. 9.

[285] Kröner, Von der Rassenhygiene zur Humangenetik, p. 34. In his memoirs, Verschuer stated that, because of his membership in the Prussian Academy of Sciences, Groß understood that he "could not muzzle" Fischer in a public lecture. Verschuer, Erbe – Umwelt – Führung, "Direktor des Kaiser-Wilhelm-Institutes für Anthropologie, menschliche Erblehre und Eugenik (seit 1942)" section, p. 11, MPG Archive, Dept. III, Rep. 86 A, Nos. 3–1.

[286] Cf. on this Adolf Butenandt/Max Hartmann/Wolfgang Heubner/Boris Rajewski, Denkschrift betr. Herrn Prof. Dr. med. Otmar Frhr. v. Verschuer, September 1949, Archive of the University of Frankfurt/Main, Dept. 13, No. 347, pp. 473–485, here: p. 477; Fricke, Kirchliches Urteil über die Persönlichkeit und die wissenschaftliche Arbeit von Herrn Professor Dr. Freiherr v. Verschuer, 26/10/1945, ibid., pp. 427–428, here: p. 428.

(*Wissenschaftslager*) – thus in April 1942 the entire staff of the institute participated in a "Science Camp of the Race Biologists" (*Wissenschaftslager der Rassenbiologen*) in Bad Nauheim, held by the *Reichsdozentenführung* ("Reich Leadership of Lecturers"), at which Fischer and Verschuer held the main lectures.[287] This scientific lecturing activity in the wider sense was supplemented with appearances at events organized by the Reichsgesundheitsführer, the Reich Youth Leadership and the Inspector of the Army Medical Corps.[288] In January 1942, for instance, Karin Magnussen gave a lecture about "Population Policy Problems in the War" to youth group leaders of the Nazi Women's Association (*NS-Frauenschaft*),[289] and in February 1944 Hans Nachtsheim spoke to the Greifswald branch of the German Society for Race Hygiene on the topic "What Can the Study of Genetically Diseased Mammals Contribute to the Investigation of Human Genetic Diseases?".[290]

Of particular importance were the lectures by the director and departmental chiefs of the KWI-A in occupied and allied Europe. In the 1941/42 fiscal year Eugen Fischer held lectures in Bucharest, Brasov, Sibiu, Temesvar, Cluj, Budapest, Paris, Paris, and Zagreb, Kurt Gottschaldt in Helsinki and Jyväskylä. "All of these lectures," which, as Fischer emphasized in his activity report, were held at the request of the Foreign Office and the Reich Ministry for Education, "served to promote cultural solidarity with these countries and spread propaganda for *Deutschtum* ('Germanness')."[291] Otmar von Verschuer continued the foreign activities of his predecessor. In December 1942 he held lectures in Brussels "on behalf of the military commander for Belgium" (to members of the military administration) and at the University of Ghent "to establish contact with Flemish cultural circles."[292] In

[287] Verschuer, Tätigkeitsbericht 1942/43, MPG Archive, Dept. I, Rep. 3, No. 20.

[288] Verschuer, Tätigkeitsbericht 1943/44, MPG Archive, Dept. I, Rep. 3, No. 22.

[289] According to her own statements, in addition to this talk Magnussen held only two other "purely scientific" lectures in 1944 (about "twin research and modern genetic research"). After leaving the KWI-A, from September 1944 to January 1945 she held "scientific lectures (as lecturer of the Race Policy Seminar) about genetic theory (cytology, chromosome theory, modification, and mutation, selection, the laws of heredity and speciation, the bastardization problem, heredity in humans, twin research, breeding research in animals and plants, etc.) for the staff of the Race Policy Office. Voluntarily!" Karin Magnussen, Verzeichnis meiner Veröffentlichungen, MPG Archive, Dept. I, Rep. 3, No. 26.

[290] Nachtsheim's notes for Tätigkeitsbericht 1943/44, MPG Archive, Dept. I, Rep. 3, No. 22.

[291] Fischer added: "Indirectly, in the same sense it must be designated as effective that a relatively large number of foreigners work at the institute as guest scholars." Fischer, Tätigkeitsbericht 1941/42, MPG Archive, Dept. I, Rep. 3, No. 19.

[292] Verschuer to Fischer, 22/10/1942, MPG Archive, Dept. III, Rep. 86 A (Münster), No. 9. Cf. Verschuer to Fischer, 7/1/1943, ibid.; Verschuer, Tätigkeitsbericht 1942/43, MPG Archive, Dept. I, Rep. 3, No. 20; Verschuer, Erbe – Umwelt – Führung, "Direktor des Kaiser-Wilhelm-Institutes für Anthropologie, menschliche Erblehre und Eugenik (seit 1942)" section, p. 3, MPG Archive, Dept. III, Rep. 86 A, No. 3–1. At this opportunity Verschuer made the acquaintance of the Belgian fascist leader Léon Degrelle (1906–1994). Cf. Verschuer, Stellungnahme zu den Angaben, die sich auf meine Person beziehen und in der "Neuen Zeitung" No. 35 of 3/5/1946 unter der Rubrik "Kunst und Kultur in Kürze" in der Notiz "Vertriebene Wissenschaft" erschienen sind, Archive of the University of Frankfurt/Main, Dept. 13, No. 347, p. 178.

1943/44 he spoke in Prague, Danzig, Zagreb, Graz, Vienna, and Mähritz.[293] Gottschaldt, too, continued to travel through occupied Europe. In 1942/43 lectures in Vienna, Budapest, Strasbourg and Innsbruck were on his itinerary; he further participated in the above-mentioned German-Japanese Science Camp in summer 1942.[294] In May 1943 and January 1944 Hans Nachtsheim appeared in Vienna.[295] In addition to their foreign travels, in February 1941 Fischer and Verschuer held lectures at the *Führerschule der deutschen Ärzteschaft* ("Leadership School of the German Medical Fraternity") on the occasion of a "joint camp" for physicians from Alsace, Luxembourg and the Netherlands in Alt-Rehse.[296]

Beyond their general foreign policy function, many of the lectures abroad apparently had the additional task of bringing functional elites from the field of medicine in the occupied and allied states "on course" with National Socialist genetic health and race policy. Thus in a dual sense they were a "service" the institute performed for the political rulers. However, they were also in the institute's own interest, as they can be regarded as part of a strategy to shape a continental European research alliance under German leadership after the collapse of the international scientific community. In this view, utilizing a large number of foreign guest scholars at the KWI-A, too, made a virtue of necessity. Beyond this Fischer and Verschuer endeavored to cultivate good relations with scientists from the allied and neutral countries, e.g. George Montandon, professor of ethnology at the Ècole d'Anthropologie in Paris, and Guido Landra of Italy.[297] When the Foreign Office of the Dozentenschaft in Frankfurt offered in November 1940 to cover the costs of sending scientific publications abroad as a gift to foreign scholars, Verschuer's list of recipients included Ernst Hanhart (Zurich), Elis Essen-Möller (* 1870, Lund), Torsten Sjögren

[293] Verschuer, Tätigkeitsbericht 1943/44, handwritten marginal, MPG Archive, Dept. I, Rep. 3, No. 22, Verschuer, Erbe – Umwelt – Führung, "Direktor des Kaiser-Wilhelm-Institutes für Anthropologie, menschliche Erblehre und Eugenik (seit 1942)" section, pp. 9f., MPG Archive, Dept. III, Rep. 86 A, No. 3–1. In Zagreb Verschuer spoke "to the Croatian medical fraternity" about "twin research," in Graz to the "local association of physicians," in Vienna to the "Freundeskreis der Deutschen Akademie". – In April 1943 Verschuer was apparently also on a lecture tour through Holland. Verschuer to Fischer, 31/3/1943, MPG Archive, Dept. III, Rep. 86 A (Münster), No. 9. In a further letter to Fischer Verschuer reports that he held two lectures at a congress of German physicians near Bratislava, Slovakia. Verschuer to Fischer, 20/9/1943, ibid.

[294] Gottschaldt, notes for Tätigkeitsbericht 1942/43; Verschuer, Tätigkeitsbericht 1942/43, MPG Archive, Dept. I, Rep. 3, No. 20.

[295] *Erbkrankheiten beim Tier in ihrer Bedeutung für die menschliche Erbpathologie* ("Genetic Diseases in Animals and their Importance for Human Genetic Pathology"), joint session of the Vienna Medical Society and the Vienna Veterinary Society on May 12, 1943, "Erbkrankheiten des Blutes in vergleichender Betrachtung" ("Genetic Diseases of the Blood in the Comparative Perspective"), lecture to the *Wiener Kulturvereinigung* (Vienna Federation of Culture) on January 27, 1944. These data are found in Nachtsheim's draft for Tätigkeitsbericht 1943/44, MPG Archive, Dept. I, Rep. 3, No. 22.

[296] Fischer to Verschuer, 20/2/1941, MPG Archive, Dept. III, Rep. 86 A (Münster), No. 9.

[297] Cf. the expert opion about Landa issued by Eugen Fischer, ibid. – Montandon was the translator of the French edition of Verschuer's *Leitfaden zur Rassenhygiene*.

(Göteborg), Tage Kemp (Copenhagen), Thordar Quelprud (Oslo),[298] Wouter Ströer (Groningen), Petrus J. Waardenburg (Arnhem) and Mihali Malán (Budapest).[299]

Contacts to scientists from the "antagonistic foreign countries" were disrupted as a natural course of the war. However, Otmar von Verschuer regarded science, too, as part of the military campaign. While he was at pains to keep himself up-to-date on the scientific production of the "enemy states," he did his best to conceal this from foreigners. This became particularly clear in March 1944, when he rejected Fischer's proposal to publish the Swiss guest scholar Erik Hug's summary of the last volumes of the most important anthropological and eugenic journals from the Great Britain and the USA in one of the journals he edited:

> I believe [...] that we [should] take cognizance of the journals from enemy countries for internal use only and evaluate the most important results of these in our own scientific work. In contrast, I have reservations about giving a complete summary of the journals of the enemy foreign states in a German journal. I hardly believe that the Americans or English, who have taken so little notice of German scientific production, now during the war present summaries about German scientific journals. Therefore my standpoint for the duration of the war is: To the extent it is possible, monitor the enemy foreign literature secretly, but give notice of this to the outside world only when citing especially important works in scientific journals.[300]

---

[298] In 1942 the National Socialists appointed Quelprud director of a new Genetic Biology Institute (*Arvebiologisk Institutt*) at the University of Oslo. Fischer pulled the strings in the background, whereby he did not shrink from denouncing Norwegian colleagues: "About him [Quelprud] I may inform you in confidence that he is repressed by the Oslo zoologist and geneticist Professor Mohr, who is a Socialist married to a Communist, and also by the professor for genetics in Oslo, Mrs. Kristine Bonnevie [1872–1949], a half-Jew, and by others of that circle because he is sympathetic to Germany [...]." Fischer to de Crinis, 2/2/1942, quoted in Kröner, Von der Rassenhygiene zur Humangenetik, p. 58. After belonging to the National Socialist Party of Norway (Nasjonal Samling) from 1933/34 to 1936, Quelprud left the party because of the biological ideas in the party newspaper, rejoining in 1941 only to leave again in 1944. Cf. Roll-Hansen, Nowegian Eugenics, pp. 179f., 193 (note 88).

[299] Verschuer to Fischer, 19/11/1940, MPG Archive, Dept. III, Rep. 86 A (Münster), No. 9.

[300] Verschuer to Fischer, 2/3/1944, ibid. It was, Verschuer continued, "extremely desirable" that knowledge about the research in the allied states made its way to Germany, and he supported all endeavors in this direction, e.g. "setting up an international, neutral, bibliographic service by Staehelin or Brugger in Switzerland." On the initiative of the director of the Department for Genetic Research at the Psychiatric Clinic of the University of Basel, John E. Staehelin, Verschuer remarked on March 27, 1943, that it meant "a gap for our research work that the literature of enemy foreign states and overseas is practically inaccessible. The collection of all relevant off-prints and monographs proposed by Professor Staehelin and the publication of a periodical, complete bibliographic directory therefore finds my support. In contrast, I reject the publication of an annual information bulletin about the works in progress of the individual scientists for the duration of the war. This would give insight into inter-German circumstances that could not be reconciled with the conditions of war" (Verschuer to the Reichsministerium für Wissenschaft, Erziehung und Volksbildung, 27/3/1943, BArch. Berlin, R 4901/3199, p. 5).

In his letter of March 2, 1944 Verschuer further informed Fischer that he had registered with the "Procurement Office of German Libraries for several journals and from there recently received 5 issues of the Annals of Eugenic[s], which were loaned out to me for 10 days. The time was just sufficient for me to have the works most important for us photocopied." After all, he was a subscriber to the journal *Fortschritte der Medizin* ("Progresses in Medicine"), which had developed into a "reference work of the foreign medical press." "However, this journal is only dispensed for

4.4.3 Race Hygiene, Sterilization, and "Euthanasia"

On May 10, 1940, Fritz Lenz addressed an extensive letter to the editorship of *Das Schwarze Korps*. He took reference to an article of April 11, 1940 entitled *Eine Frau hat das Wort* ("A Woman Has Her Say"). The anonymous author had demanded, in view of the surplus of women after World War I, that the state should create incentives for women of 30 who were still single to become unmarried mothers: Such single mothers should receive a higher income than childless married women of the same age; their dual role as mother and career woman should be accommodated by flexible working hours; they should receive a one-off benefit similar to the marriage loan; "for the less well off," further, "current state supports [were] to be guaranteed." Prerequisite for this benefit was the "genetic health" of the mother and the father – sperm donations were to come from single young men who had not yet started their own families. Finally, the anonymous author, "for the protection of the honor" of the single mothers, had demanded that "anyone who reproached the morals of a single mother" be sentenced to prison on principle.

Lenz responded to these proposals with sharp critique. Higher income for single mothers which, as Lenz emphasized, would have to be financed by state subsidies, was not only economically intolerable, especially since, for reasons of equal treatment, the demand would "as a consequence [...] would have to amount to ongoing support for all mothers." Under aspects of race hygiene, too, it would always make more sense to support *married* mothers. The particular displeasure of the Nestor of the race hygiene movement was evoked by the proposal to guarantee ongoing support for single mothers who were less well off: "A similar demand was raised in the Reichstag of the Weimar system by the Communists." Lenz warned that measures of this kind would "encourage the propagation of elements that [could not] demonstrate any sufficient performance as a result of mental or physical weakness and thus also [could not] exhibit any sufficient income." There was the danger of "adverse selection," which, as Lenz argued with reference to possible concealed genetic dispositions, could not be avoided by making the "genetic health" of the men and women involved a prerequisite, either: "By no means can the danger of a preferential propagation of inferior race elements be averted in this manner."

Under quasi educational aspects Lenz pointed out that the proposals would make necessary "special legislation" to expressly exempt the fathers of the children from support payments, "while in recent years the tendency has been to increase the responsibility of the father, in economic terms as well." The whole scheme boiled down to "state-approved temporary marriages of uncertain duration, which moreover would even be privileged by the fact that the state would take over the costs of

official use and under the obligation that it be kept under lock and key, to certain subscribers who must pledge their confidentiality with a signature on the back cover."

At another juncture Verschuer reported that he received, "from the exchange service," the *Eugenics Review*, of which he had photocopies made. Verschuer to Fischer, 11/8/1944, MPG Archive, Dept. III, Rep. 86 A (Münster), No. 9.

bringing up the children." This would have to weaken marriage as such. Here Lenz argumented quite conservatively and fundamentally, but not consistently in terms of race hygiene. He did see the childlessness of many women from the generation of the World War I as a race hygiene problem, but to him illegitimate motherhood did not seem a suitable solution under moral aspects – when, then polygamy instead: "Purely objectively" he would hold the "permission of a [...] limited number of second wives to be the relatively best solution; but the moral tradition of the German Occident appears to virtually exclude such a solution." "Breaking the moral tradition of a nation" was, however, "always perilous."[301]

Lenz did not insist that his reply be printed in *Das Schwarze Korps*. But he did call upon both the editorship and the author to enter into a critical dialog – to no avail, as the editorial board of *Das Schwarze Korps* did not react at all. Lenz addressed copies of his letter to the RuSHA, the Race Policy Office and the German Family League of the Reich (*Reichsbund Deutsche Familie*).

At the close of his letter Lenz illustrated his conception of state measures to increase the birthrate among married couples. A "state obligation to bring up children" should be introduced, in keeping with the principle "Every member of the nation capable of living has the duty to bring up at least four children." Anyone who did not fulfill this obligation should ante up "substitute payments in percentages of his income," which corresponded to "approximately the cost of bringing up children."

Lenz elaborated on this basic idea in an exposé about "Ways to Further Advance in Population Policy," which he wrote at the same time as the protest letter to *Das Schwarze Korps*. The French campaign had not yet entered its decisive phase, but Lenz appeared optimistic that the end of the war was immediately imminent and would offer "a unique opportunity for generous population policy." Because the birth cohorts since 1915, which were already not terribly strong as a consequence of World War I, had been weakened further by the losses in the war from 1939 to 1940, the idea was to induce the birth cohorts before 1914 to bring up as many children as possible. "Against birth premiums and child subsidies," Lenz announced yet again apodictically, there were "serious race hygiene objections." The experiences with marriage loans were, under quantitative aspects alone, "by no means encouraging." Each marriage supported by a marriage loan accounted for "slightly less than one child." What is more: "From the perspective of race hygiene there is hardly a reason to regret that only meager funds are available for such benefits at this time."[302] In contrast, Lenz expressly advocated burden sharing for families through higher taxes for families with no or only few children. The tax increases dictated by the war seemed to offer a convenient opportunity to engineer such a

[301] Lenz to the editorship of *Das Schwarze Korps*, 10/5/1940, MPG Archive, Dept. III, Rep. 86 B, No. 10, pp. 8–14, quotes: pp. 9–14.

[302] This view was backed up by a later study from the KWI-A: "Gründler performed follow-up examinations on pairs who were denied the marriageability certificate, and in doing so found important new aspects for marriage consulting and the execution of the marriage health law." Verschuer, Tätigkeitsbericht 1942/43, MPG Archive, Dept. I, Rep. 3, No. 20. Gründler was listed as a doctoral student of the KWI-A in this year.

burden-sharing scheme; the idea was "to make a demographic policy virtue out of the financial necessity of the war."[303] The core of the tax policy concept worked out to the last detail by Lenz was the proposal that the war surtax on income be eliminated after the end of the war only for families with four or more children. Tax advantages for childless married couples were to be omitted, just as the temporary tax relief for young couples and the enduring tax break for couples with grown children. Both parents of illegitimate children were to be allowed to deduct a child from their taxes, but only by half. Only families with many children were to enjoy full deductions from property and inheritance taxes. Finally, Lenz developed a mandatory savings system for peasant families, to finance the compensation paid out to the daughters and sons who did not inherit property – if a peasant family had fewer than four children, part of the money saved would fall to the state. In fact, Lenz's proposals – measured against the tax level before the beginning of World War II – amounted to a constant tax burden for families with many children and enormous tax increases for everyone else.

What is interesting is that Verschuer, to whom Lenz sent both of his documents for his perusal, responded immediately and declared himself in complete agreement with the contents. He expressly subscribed to Lenz's thesis that the propaganda for illegitimate children evoked the "race hygiene danger" of "adverse selection." He proposed publishing Lenz's tax policy exposé in the *Erbarzt*.[304] Yet Lenz had misgivings, since his proposals collided with the tax policy of the relevant state secretary in the Reich Finance Ministry, Fritz Reinhardt (1895–1969), whom he did not want to provoke by publishing the exposé. However, Lenz reported, it had been forwarded to Reinhard via the Race Policy Office.[305]

In 1943 Lenz decided to publish his ideas after all, in the *Archiv für Rassen- und Gesellschaftsbiologie* under the title *Gedanken zur Rassenhygiene (Eugenik)* ("Thoughts on Race Hygiene (Eugenics"). With this he spurred into action the press department of the Reich government in the Reich Ministry for Propaganda and Enlightenment of the Nation, which ordered that the passages about relieving families of the tax burden be struck.[306] New in this version of 1943 was that Lenz demanded compulsory employment for childless and "child-poor" women.[307]

While Lenz continued his efforts on the path of scientific policy consulting to secure recognition in population policy for a program of positive eugenics, Verschuer and his staff continued to be in demand as experts and evaluators whenever questionable cases arose in the application of the GzVeN. As has been shown above, after the changing of the guard at the head of the institute, genetic pathology research in Dahlem was consistently oriented toward this practical application, in

[303] Lenz, Wege weiteren Vormarsches der Bevölkerungspolitik, n.d. [June 1940], MPG Archive, Dept. III, Rep. 86 B, No. 10, pp. 15–21, quotes: p. 15.

[304] Verschuer to Lenz, 12/6/1940, ibid., p. 22f.

[305] Lenz to Verschuer, 17/6/1940, ibid., p. 24f.

[306] Cf. the documents printed as facsimiles in: Koch, Humangenetik, pp. 108–110.

[307] Lenz, Gedanken zur Rassenhygiene, p. 97.

order to "clear the complex jungle of the activity of producing expert opinions."[308] Due to the incomplete knowledge about the heredity of the diseases and disabilities listed in the catalog of indications of the GzVeN, since 1934 a jumble of contradictory decisions had resulted, and the initial enthusiasm of the race hygienists had given way to a kind of "hangover."[309] In accordance with the insights of higher Mendelism, since 1938 Verschuer had urged that the Hereditary Health Court proceedings be based not on the *clinical* diagnosis, but solely on the *genetic* diagnosis, to be reached through intensive genealogical studies.[310] This had the consequence that in some cases which had fallen under the GzVeN as a matter of course up to that point, Verschuer and his staff advised *against* sterilization. In one case this consistently pursued line went too far for even Verschuer's friend and teacher Fischer. When in 1940 Verschuer's employee Heinrich Schade advocated the interpretation that certain defects of the limbs were not genetic and thus, must be excepted from sterilization,[311] Fischer lodged vehement protest with Verschuer:

> With a degree of shock I read the paper by Schade about the defects of the limbs. I do not hold to be correct the conclusions that heredity was not present in general, presented here in apodictic form. Here Goethe is wrong!! All fun aside. Schade is, of course, right, that heredity is not proved in these cases. But for schizophrenia we also do not know what kind of and how many genetic factors are the basis. For the limbs there could hardly be any single, separate gene for each form and each location of defects, but rather different kinds, such as those which govern the development processes chemically. [...] There is not only this theoretical side to the matter, however. In terms of praxis, Schade arrives at the conclusion that one may not sterilize these cases unless another identical case was found to have occurred in the family. And this is extremely rare for today's small families. Then we would have the situation, which seems intolerable to me, that conditions like cleft palate, club foot and hip luxation are sterilized as serious physical defects, but when an entire extremity is missing, or when both hands are completely crippled, sterilization does not take place. The public knows both groups as congenital. Now the one defect, in fact the lesser, is regarded as congenitally inherited and thus to be sterilized; the other, more serious one, is regarded as congenital but not inherited, and thus not to be sterilized. The *Volk* does not understand this. And for us, too, it goes against every feeling of justice. Perhaps we were really somewhat hasty with the presumption of heredity in the case of defects. If one believes that, then the consequences must also be drawn for luxation, harelip, etc. I do not believe it personally. I am of the opinion that Schade went too far and allow only that while we do not know the individual genetic process, genes are the cause.[312]

Here Fischer pursued the dual logic of an applied science, which must always attempt to combine the logic of science with the logic of politics. Its recommendations to politics are always the product of several factors: scientific knowledge, the consideration of the practical utility of a measure, the expectation of its political feasability and its cultural acceptance, and finally its ethical admissibility. Thus,

[308] Weß, Humangenetik, p. 174.

[309] Ibid.

[310] Cf. e.g. Verschuer, Unfruchtbarmachung.

[311] Schade, Untersuchungen.

[312] Fischer to Verschuer, 2/7/1941, MPG Archive, Dept. III, Rep. 86 A (Münster), No. 9. Cf. also Weß, Humangenetik, p. 176.

many scholars of the human sciences in the Third Reich advocated large-scale eugenic sterilizations, although it was quite clear to them that they were on shaky ground scientifically. The heritability of one or the other clinical picture constituted a plausible assumption, but one that in many cases required further empirical confirmation. That this assumption was sometimes presented to the outside world as a fact – to the public, but also to the state – is by no means unusual, but rather is part of everyday science even today. They believed that the empirical proof for the seemingly evident facts of the case could be presented afterward, sooner, or later.[313] Fischer's stated opinion is practically paradigmatic for this position.

Yet Verschuer did *not* follow his mentor on this path. On the contrary: he countered Fischer's argument that if one did not want to sterilize people with deformed limbs, then "cleft palate, club foot and hip luxation" would also have to be deleted from the GzVeN's catalog of indications, by aspiring to that very end. As regarded cleft lip, cleft jaw, and cleft palate, since the late 1930s Verschuer had proceeded from the assumption of a high degree of heterogeny – based on Josef Mengele's dissertation, by the way. But the consequence of this was that heredity had to be checked for in each individual case. And according to the testimony of Gerhard Koch mentioned above, in July or August 1943 the Genetic Pathology Working Group under Verschuer's direction argued that club foot and congenital hip luxation should no longer be recognized as "genetic conditions" in the sense of the GzVeN. In the first, still-documented sessions of the Genetic Pathology Working Group, the participating scientists had also been extraordinarily reluctant to recommend sterilization in individual cases[314] – in principle they were in agreement that the heritability of a condition had to be proved with certainty on the basis of family anamnesis in the individual case, whereas the logic of the GzVeN had saddled the subject to be sterilized with the burden of proof: He or she had to prove that in his or her case the general assumption did not hold that the condition was hereditary.

In the judgement of concrete cases, Verschuer consistently held fast to the genetic diagnosis. As a rule, he was extremely reluctant to acknowledge nonscientific considerations, even when they spoke *for* the subject. This became apparent, for instance, in the different judgements of the very first case dealt with in the Genetic Pathology Working Group, by Verschuer and Lenz. It involved a 35-year-old man, who had gone completely blind at the age of 18 due to *Hydrophthalmus congenitus* (congenital glaucoma), but had nevertheless graduated from secondary school, worked as a music teacher and piano tuner and led the Association of the Blind of his district, and finally studied law. In 1939 he wanted to marry a healthy teacher, but he was denied the marriageability certificate and an exemption from the regulations of the marriage health law. An application for sterilization in accordance with the GzVeN was rejected, however. The case had been submitted to Verschuer for his expert opinion. In the ensuing discussion about the race hygiene consequences, Lenz argued that the risk of rare, recessive genetic conditions

[313] Cf. Roelcke, Programm, p. 57. In general: Schlich, Wissenschaft.

[314] Cf. Grebe, Erbpathologische Arbeitsgemeinschaft.

occuring in the couple's progeny was so low that the couple, which consisted of "personalities of above-average talent and prowess," should not be deprived of the possibility of having children. Verschuer argued against sterilization as well, but believed that according to the GzVeN the man "would have to be sterilized because of genetic blindness, if the hydrophthalmia was definitely genetic in his case, which he [Verschuer] personally did not hold to be proved." By contrast, Lenz was steadfast in his judgement "that even if the heritability of the condition was presumed, the man was not be sterilized, because there was no high probability that genetically ill individuals were to be expected among the progeny. The probability of this was less than 1%, and one may not do without 99 healthy children because of one that might be genetically ill." In the end Lenz and Verschuer came to an agreement, "albeit for slightly different reasons,"[315] that in the given case neither sterilization nor a marriage ban was justified. The salient point is that in his argumentation Lenz adhered closely to the letter of the law, to his official interpretation, and above all to the jurisdiction of the Hereditary Health Courts, which had long since begun to grant broader latitude to the criterion of "preservation of life" in their judgements. Verschuer's standpoint that upon proof of genetic blindness the subject was to be sterilized *in any case*, regardless of all other aspects, amounted in fact to an intensification of the existing legislation and administration of justice. That he decided *against* sterilization in this concrete case was due solely to the fact that he did not consider the proof of heritability to have been adduced. Here it becomes apparent how misleading it is to use the individual cases in which Verschuer and his staff advised against sterilization as an indication for the fact that the scientists of the KWI-A attempted, as a rule, to exert a moderating influence on the praxis of sterilization.[316] They wanted to place the sterilization program on a new scientific basis that did justice to the insights of higher Mendelism: Some of the subjects to be sterilized, who up until that time had been sterilized without any hesitation, were thus spared from sterilization. In other cases Verschuer and his staff judged even more harshly than the Hereditary Health Courts.

What is more: as a consequence of Verschuer's position, sterilization legislation had to be extended to the heterozygotic bearers of recessive genes, who had no

[315] Grebe, Hydrophthalmus, p. 93.

[316] So Koch, Humangenetik, p. 95. In a case of *Paramyotonia congenita* (Eulenburg syndrome), a dominant autosomal genetic disorder with occasional muscular rigidity, primarily induced by physical excercise and cold, Gerhard Koch indicated that sterilization was not appropriate. In his memoirs Koch emphasizes that this work "despite my critical opinions on the 'Gesetz zur Verhütung erbkranken Nachwuchses,' [was] released for publication in the Erbarzt by the military censors." (ibid., p. 104). Of course, from the perspective of Nazi genetic health policy, there could hardly have been anything objectionable about this article. The passage in question reads: "[Heinz] Boeters holds an application of the G.z.V.e.N. to be unnecessary due to the rarity of myotonical clinical pictures (1 of every 30,000–50,000 affected). In the case before us now, too, sterilization does not appear suitable because of the intellectual abilities of the subject. But since it is not to be expected that the disorder will become extinct through self-selection, as a rule marriage and having children should be urgently advised against. [...] Patients with myotonic apraxia are, of course, of no use for military service." Koch, Paramyotonia, p. 173.

clinical symptoms. Verschuer – in contrast to Lenz[317] – was too cautious to make such a demand publicly, for after all – to the disappointment of many a eugenicist and race hygienist – Hitler and the National Socialists had excepted the healthy bearers of genetic dispositions in their formulation of the GzVeN (which the Prussian draft of 1932 had not, by the way). Under conditions of war it was not to be expected that they would consent to such an explosive expansion of the sterilization program, which had been geared down in 1939 anyway. Nonetheless, Verschuer and his staff attempted to lay the scientific foundations for such an expansion, as shown by research like Grebe's search for stigmata that would allow healthy bearers of the genes for chondrodysplasia to be identified. In principle and potentially, Verschuer's position amounted to a further intensification of the sterilization legislation.

While the KWI-A continued to fulfill an important consulting function as regarded the sterilization program, as mentioned above, Verschuer distanced himself publicly from the Nazi "euthanasia." In contrast, Fritz Lenz cooperated with the "euthanasia" planning staff to create a legal foundation for the mass murder of the mentally ill and mentally disabled. It has been asserted on several occasions that Lenz had changed his opinion on the question of the "annihilation of life unworthy of life" at the beginning of World War II[318] – but this is only partially true. In the third edition of his work *Menschliche Auslese und Rassenhygiene (Eugenik)* – that is, the second volume of "Baur-Fischer-Lenz," which appeared in 1931 – Lenz had expressed his views on the issue of "euthanasia" at great length, after having made the impression on the public that he unreservedly advocated "euthanasia." On the contrary, Lenz emphasized "that so-called euthanasia [is out of the question] as an essential means of race hygiene." Taking reference to intensified postwar discussion about medically assisted suicide, killing on request and the "annihilation of life unworthy of life," Lenz pled for the view that "euthanasia" was "preeminently a question of humanity. Even the ancient Spartan abandonment of deformed children is still incomparably more humane than today's practice of rearing even the most unfortunate creatures in the name of 'compassion' [...]."[319] With reference to race hygiene, however, "euthanasia" had no great importance to the extent that the circle of those affected would hardly have the opportunity to propagate – if this danger existed, then it could be prevented by sterilization. What did speak for the painless killing of disabled children from the standpoint of race hygiene is that it would enable the parents to bring another, healthy child into the world. This would also mean that "the question of the marriageability of encumbered persons [could be] judged much more generously than it is today."[320] For instance, there would be

[317] Lenz, Gedanken zur Rassenhygiene, p. 100. At this juncture Lenz also openly demands the inclusion of "asocial" subjects in the sterilization legislation. Ibid., pp. 100 f.

[318] So runs the presumption of Rissom, Fritz Lenz, p. 69.

[319] Lenz, Auslese, 3rd edn, p. 306. Here Lenz took express reference to the publication by Karl Binding and Alfred Hoche about the "deregulation of the annihilation of life unworthy of life."

[320] Lenz, Auslese, 3rd edn, p. 307. Lenz argued very similarly in 1938. Cf. idem., Häufigkeit.

fewer misgivings about permitting a marriage between partners who were healthy themselves but known to be bearers of a recessive gene for deaf-muteness, because the deaf-mute children from such a marriage could be killed and thus the parents given the opportunity to have as many healthy children as their economic situation allowed. Although Lenz thus indirectly attributed a eugenic function to "early euthanasia," he persisted in his opinion that "euthanasia" was "hardly [...] so effective a means" under race hygiene aspects "that race hygiene must advocate it." Decisive for Lenz was that through "euthansia" the "respect for individual life, which is an essential foundation of our social order, would experience a critical loss." Although more than a few infanticides occurred for the purpose of family planning even in the Western cultures, "the moral consciousness in the Occident [excluded] a legal license for infanticide."[321]

In other words: Lenz did hold the killing of disabled newborns to be justifiable in principle as an act of "humanity," but in the early 1930s he still believed that deregulating "early euthanasia" would shock the "moral consciousness" and the "social order." It is presumably not incorrect to presume that Lenz feared "early euthanasia" could become the gateway to the deregulation of abortion for social indications. In principle Lenz adhered to his position. After including the passage about "euthanasia" unchanged in the fourth edition of his work in 1938, in the preface to the publication by Wolfgang Stroothenke about *Erbpflege und Christentum* ("Care of Genes and Christianity") published in 1940, in which "early euthanasia" was demanded on as a measure of caring for the genetic pool, Lenz reaffirmed his standpoint that "euthanasia" was primarily a question of "humanity."[322] What Lenz did not write openly, however: apparently he believed the time had come to set about resolving this question.

In any case Lenz – presumably in his capacity as a member of the Expert Council for Population and Race Policy – proved ready to participate in a commission, which probably convened in October 1940, to debate and finalize a draft law to legalize the "euthanasia" program under way since the change of years 1939/40.[323] A number of the physicians represented in this commission belonged to the medical staff of the "euthanasia" program – among them Georg Renno, who had passed through the KWI-A's first annual course for SS trainees – as well as several representatives of the medical administrations of the states, and, finally, the Chief of the Security Police and the SD, Reinhard Heydrich (1904–1942), who was interested in the procedure because at this time he was concerned with the planning for a "Community Alien Law" (*Gemeinschaftsfremdengesetz*).[324] The draft law

---

[321] Lenz, Auslese, 3rd edn, p. 307.

[322] Cf. the preface in Stroothenke, Erbpflege.

[323] On the previous history: Roth/Aly, "Gesetz über die Sterbehilfe"; Gruchmann, Euthanasie; Klee, "Euthanasie," pp. 241 f.; Schmuhl, Rassenhygiene, Nationalsozialismus, Euthanasie, pp. 291–297.

[324] Cf. on this: Ayaß, "Asoziale," pp. 202–209. Otmar von Verschuer also got involved in this discussion. He approved the establishment of a register of "community aliens" in order to attain a "differentiation between those to be eliminated and those to be supported" (Erbarzt 8, 1940, pp. 234 f.).

ultimately worked out by this body then presumably bore the title proposed by Fritz Lenz, "Law About Euthanasia for the Incurably Ill."[325] The final version of the draft was not preserved, but the contents of its six articles can be reconstructed on the basis of some of the surviving commission protocols. Apparently the preamble states that people "who because of an incurable disease long for [an end to their suffering]" or "as a consequence of an incurable chronic condition are incapable of productive life,"[326] were to be afforded assisted suicide. The first two articles represented slightly modified versions of a draft law debated by the official Criminal Law Commission on August 11, 1939:

> § 1: Anyone who suffers from an incurable disease [that presents a great burden to himself or others or is certain to lead to death] can receive euthanasia upon his express request with the approval of a specially authorized physician.[327]
> § 2: The life of a patient who otherwise would require lifelong custody as the consequence of incurable mental disease can be ended through medical measures unnoticeable for the patient.[328]

The following four articles regulated the process. The patients were to be registered – unless the patient himself submitted the petition – by the public health officers and institutional physicians. Thereupon an evaluation was to take place by "expert committees," each of which was to include a "specially authorized" public health officer and two medical assessors (psychiatrists). The committees were to be assembled by a "Reich Deputy" to be appointed for the execution of the law. This special agent, vested with far-reaching powers, was also to make the final decision about the petitions for "medical assistance" on the basis of the expert opinions submitted by the expert committees, and to appoint the physicians to perform the procedure. If an executing physician stated that he did not agree with the vote of an expert committee, he could submit a detailed written explanation of his reasons and apply for a new expert opinion by another expert committee.

What role did Lenz play on this commission? According to the protocol he was one of the most eager discussants, and many of the essential formulations of its content were based on his proposals. This began with the title of the draft law. Upon Lenz's suggestion, the word "deliverance" was struck, "which, originating from the world of Christian ideas, would evoke negative feelings against the law."[329] The wording of article 2 also came from Lenz. He had rejected the original term of "abnormal disposition" as "too indefinite and vague." In some cases it was

[325] Cf. Meinungsäußerungen zum Gesetz, BArch. Berlin, R 96 I/2, pp. 126.659–126.690, here: pp. 126.662 (as a facsimile, also in: Roth/Aly, "Gesetz über Sterbehilfe", pp. 140–172, here: p. 143).

[326] Ibid., p. 126.663. Formulation proposal by Ernst Wentzler, in square brackets: formulation proposal by Lenz. Wentzler's version read: "who long for deliverance because of an incurable disease."

[327] Ibid., p. 126.666. Formulation proposal by Lenz, in square brackets: formulation proposal by Kurt Pohlisch, in Lenz's version there is an omission here.

[328] Ibid., p. 126.668. Formulation proposal by Lenz.

[329] Ibid., p. 126.663.

"not at all clear whether abnormal genes or external damage was the basis"; "idiotic or seriously deformed children"[330] would thus not be included by the concept of abnormal disposition – for these, moreover, "a special legal determination was required."[331] Therefore Lenz pled for restricting the law only to the mentally ill for the time being. Presumably with a view to Heydrich and his interests, Lenz added that one would also have to define expressly "that criminal psychopathy is a mental illness in the sense of the law."[332] Lenz did not want to make exceptions, but he held the inclusion of "senility of the mind"[333] to be unsuitable. All in all the impression arises that Lenz was one of the driving forces in the discussion and that he left his mark on the draft law. Repeatedly he pushed for precise specifications – in the interest of legal certainty. It was also due to this interest that Lenz wanted to make sure that the law would be applied initially only to cases of serious mental disease, whereby he urged a regulation of "early euthanasia" at a later point in time. That it was by no means his concern to check the "euthanasia" program in progress is apparent in the fact that he was ready to include the "criminal psychopaths" in the sense of the "Community Alien Law" planned by Heydrich.

In the end the draft law discussed remained nothing more than paper. Hitler rejected a legal enclosure for the "annihilation of life unworthy of life." "Euthanasia" continued to proceed in the unlawful cavity of the National Socialist "prerogative state," flanked by an ambitious program of genetic psychiatry and genetic pathology research. In one case the KWI-A, too, profited from the unfettered access to human subjects in sanatoriums and hospitals.

4.4.4 Nachtsheim and the Low-pressure Experiments on Epileptic Children

Back in 1940 Hans Nachtsheim, as mentioned above, had performed a large-scale series of experiments on almost 600 rabbits of different races, both from the "epilectic" and the "non-epileptic" breeds, in which convulsions were induced through cardiazol injection to check their compulsion-readiness. As explained, the point was first to theoretically illuminate "the connections between convulsion-readiness and genotype." The series of experiments also pursued a second, entirely practical purpose, however: they were supposed "to provide a contribution to the question so debated in psychiatry, as to whether a genuine epileptic responds to a lower dose of cardiazol with convulsions than does a symptomatic epileptic or a non-epileptic, and thus whether inducing convulsions by cardiazol is of value for differential diagnostics." In this respect the experiments ended in failure. It became apparent that the convulsion-readiness of the rabbits was dependent on their age: Young rabbits

[330] Ibid., p. 126.668.

[331] Ibid., p. 126.673.

[332] Ibid., p. 126.668.

[333] Ibid., p. 126.688.

convulsed at a lower dose of cardiazol than older animals; convulsion-readiness in response to cardiazol thus, appeared to diminish with increasing age. This alone would not have debased cardiazol convulsions as a differential diagnostic instrument, as the rabbits of the purely bred "epilectic strain" reacted more sensitively to cardiozol at all ages than did other rabbits. However, at the same time it turned out that the convulsion-readiness of the "epileptic" rabbits was subject to frequently occurring, strong individual oscillations, so that the animals sometimes did not respond to a high dosage – at which even a high percentage of "non-epileptic" animals convulsed, and at other times convulsed even at low doses that would never have triggered convulsions in "non-epileptic animals." In view of these findings Nachtsheim had to admit that cardiazol convulsions possessed "only limited differential diagnostic value."[334]

Although the hope for a direct practical use of the convulsion experiments had not been fulfilled, Nachtsheim continued to grant high priority to his research on the "epileptic" Vienna Whites even after starting at the KWI-A on January 1, 1941. From this point on the research projects on epilepsy pursued modified research questions: The direct perspective on differential diagnostics was abandoned, and epilepsy research oriented instead entirely toward the paradigm of phenogenetics. What appeared as a mere disruptive factor in the experimental arrangement under the aspect of differential diagnostics – the modifying influence of age, time of year and season on the convulsion-readiness of the rabbits–, became the actual object of research when embedded in the paradigm of phenogenetics, for apparently the convulsion-readiness of the experimental animals was the result of an interplay among genetic and peristatic factors, which intertwined to cause, enable and induce. If research succeeded in exposing the complex reciprocal actions of genetic disposition, maturation, and environment, science would be much closer to illuminating the process of the pathogenesis of epilepsy.

In his activity report about the 1941/42 fiscal year, Eugen Fischer quite skillfully referred to Nachtsheim's "investigations about the epilepsy of rabbits, which corresponds completely to that of humans." The experiments "to use cardiazol to induce epileptic convulsions like those in humans" were "in full swing," and promised "a more precise analysis of the genetic and non-genetic conditions of the convulsion-readiness of vessels in the brain."[335] Here Fischer did imply that the studies in progress could contribute to the demarcation between genuine and symptomatic epilepsy, yet he painstakingly avoided the term "differential diagnostics," selecting a more open formulation. On the other hand, he left no doubt as to the applicability of the animal model.

However, Nachtsheim had to struggle with some exceptions voiced from the ranks of his critics on precisely this point. Concerns, such as the fact that the structures

[334] Nachtsheim, Bericht über die im Jahre 1940 mit Unterstützung des Reichsforschungsrates durchgeführten Untersuchungen zur vergleichenden Erbpathologie, 14/1/1941, pp. 2f., BArch. Koblenz, R 73/13.328. Cf. Nachtsheim, Krampfbereitschaft und Genotypus I-III.

[335] Fischer, Tätigkeitsbericht über das Geschäftsjahr 1941/42, MPG Archive, Dept. I, Rep. 3, No. 19.

of the human and rabbit nervous systems were too different for the findings obtained with rabbits to be applied to humans without further ado, were forestalled by Nachtsheim himself, who raised the conjecture that the pathological processes which also occurred in humans might possibly be better studied on rabbits, because in the rabbit the convulsion takes place in a much more primitive form, "without all of the accessory parts that have accrued in humans,"[336] – an argument based on an abridged mechanistic concept of the organism, which was not terribly solid. After brashly asserting the applicability of the animal model at the outset, over the course of time he sought refuge in more careful formulations:

> In this we are certainly aware that a result in an animal experiment can be translated to humans only with caution, especially when the substrates in question are as different as the rabbit brain and the human brain. However, a result for rabbits can be regarded as at least pointing the way to the conditions in humans.[337]

Nachtsheim also had to struggle with the clinicians' critical pointer to the polymorphy of the various clinical pictures subsumed under the concept of epilepsy. He thus toned down his pretense of using the animal model to explain *the* epilepsy of humans and hence aspired only to relate "rabbit epilepsy to a certain 'variety' of human epilepsy."[338]

The most serious was the objection of leading psychiatrists that cardiazol convulsions in rabbits were not comparable to spontaneous convulsions in humans – this critique was aimed straight at the experimental arrangement, which made the animal model organism available for human genetic research. Nachtsheim and his staff therefore, also tested other possibilities such as insulin and acetylcholine shock as well as electric spasms. In the course of these tests they concluded that the convulsions in Vienna White rabbits artificially triggered by cardiazol most closely approximated the spontaneous convulsions of the human epilepsy victim.

According to the concept of convulsion-readiness, every human could suffer convulsions – in the case of epilepsy Nachtsheim imagined the boundaries between health and illness to be fluid. Epileptics were different from other humans, he presumed, in that their convulsion threshold was significantly lower. Now, under the banner of higher Mendelism, this was no longer simply regarded as genetic, but rather – in terms of phenogenetics – as the result of a causal chain, the first cause of which is to be sought in the genotype, but which was also influenced by factors in the internal and external milieu. When an organism in a condition of heightened convulsion-readiness was subjected to an adequate environmental stimulus, this would trigger a convulsion – the epilepsy became manifested clinically. The point of Nachtsheim's experiments was thus to manipulate the convulsion-readiness of

[336] Nachtsheim, Krampfbereitschaft und Genotypus III, p. 60.

[337] Ruhenstroth-Bauer/Nachtsheim, Bedeutung des Sauerstoffmangels, p. 18. No qualification in Nachtsheim, Modelle menschlicher Erbleiden.

[338] Nachtsheim, Krampfbereitschaft und Genotypus II, p. 242. Cf. also Nachtsheim to Koch, 7/8/1944, MPG Archive, Dept. I, Rep. 3, No. 1: "What you write about the different types of fits in humans corresponds entirely with my opinion. 'Epilepsy' is a collective term, and it would be urgently required that a systematic order be brought into this area."

his rabbits by changing the intervening peristatic variables that modified the effect of the gene. The comparison of young and mature animals was an obvious choice for the first series of tests, as Nachtsheim believed he had determined in his cardiazol experiments, performed to resolve the differential diagnostic problem, that the convulsion-readiness was dependent on age. Therefore, in a paper for the *Zeitschrift für Altersforschung* in 1941 Nachtsheim reevaluated his previous experimental results under this aspect, and in so doing also introduced them into the still young discipline of geriatrics.[339] In the field of psychiatric research it was controversial at what point in time and in what form the various types of epilepsies became manifest, whether the clinical picture changed over the course of life, and whether such age differences occurred only in symptomatic epilepsy or in genuine epilepsy as well. Nachtsheim wanted to attempt to pursue these questions using a comparative experimental system on the animal model.

In summer 1942 an epidemic raged among Nachtsheim's experimental animals and forced him to temporarily suspend the experiments on epilepsy,[340] because the stocks had to recover before this form of "consumptive research" could be continued – in the artificially provoked convulsions, especially when cardiazol was used, it was not seldom for the experimental animals to suffer broken bones or collapse.[341] When Nachtsheim resumed the experiments in early 1943, he changed the method. He no longer resorted to cardiazol, which had the disadvantage in Nachtsheim's view that its toxic effect was superposed upon the convulsion events and made their observation more difficult.[342] Therefore, it must have been easy for him to give up this method, especially since cardiazol was difficult to obtain during the war as it was urgently needed for therapeutic purposes. The electric spasm attempts conceived as an alternative to the cardiazol experiments in 1942 had "not proceeded beyond certain preliminary tests,"[343] not least because the "convulsator" by the Siemens-Reiniger plant procured in 1941 proved unsuitable for animal testing.[344] A

[339] Nachtsheim, Krampfbereitschaft und Lebensalter. Cf. in general: Hahn, Altersforschung.

[340] Nachtsheim, Bericht über die im Jahre 1942 mit Unterstützung des Reichsforschungsrates durchgeführten Untersuchungen zur vergleichenden und experimentellen Erbpathologie, 22/3/1943, BArch. Koblenz, R 73/15.342, pp. 99–112 (incorrect pagination), here: p. 99.

[341] Nachtsheim, Bericht über die im Jahre 1941 mit Unterstützung des Reichsforschungsrates durchgeführten Untersuchungen zur vergleichenden und experimentellen Erbpathologie, 27/3/1942, BArch. Koblenz, R 73/13.328.

[342] A further practical disadvantage was that repeated injection of cardiazol destroyed the veins. Nachtsheim, Bericht über die im Jahre 1940 mit Unterstützung des Reichsforschungsrates durchgeführten Untersuchungen zur vergleichenden und experimentellen Erbpathologie, 14/1/1941, p. 4, ibid.

[343] Nachtsheim, Bericht über die im Jahre 1942 mit Unterstützung des Reichsforschungsrates durchgeführten Untersuchungen zur vergleichenden und experimentellen Erbpathologie, 22/3/1943, BArch. Koblenz, R 73/15.342, pp. 99–112, here: p. 112.

[344] In October/November 1942 this convulsator was loaned out to the "Waldhaus Berlin-Nikolassee" Clinic for Psychiatric and Emotionally Disturbed Patients, an institution that belonged to the Inner Mission of the Protestant church. In return, Nachtsheim received from the clinic a convulsator of Swiss manufacture, which, despite expectations, proved to be just as unsuitable. The trade was then reversed. Cf. on this the correspondence in BArch. Koblenz, R 73/13.328.

change in the experimental arrangement was thus essential, and this led Nachtsheim to high altitude medicine.

Crucial for the further development was the incipient collaboration between Hans Nachtsheim and Gerhard Ruhenstroth-Bauer (1913–2004).[345] After completing his studies of physics, in September 1939 Ruhenstroth-Bauer had come to Adolf Butenandt at the KWI for Biochemistry to write a dissertation in the area of hormone chemistry. When this dissertation project hit a snag due to the war, Ruhenstroth-Bauer turned to research on the regeneration of red blood cells (hemopoiesis). He had been forced to interrupt this research when he was drafted into the Luftwaffe as a military physician and sent to the Eastern Front. Butenandt lobbied Erich Hippke (1888–1969),[346] head of the Luftwaffe Medical Corps, to have Ruhenstroth-Bauer reassigned to Berlin, and Hippke – it is not clear whether upon Butenandt's urging or on his own initiative – ordered the young military physician to Berlin in June 1942, in order to perform special research on hemopoiesis. Ruhenstroth-Bauer was searching for a substance that was capable of effecting a prolonged propagation of the red blood cells (erythrocytes) – he assumed that it would be a hormone, which he intended to name hemopoietin. The potential military importance of the project for air warfare was obvious: pilots who were injected with the blood-enriching substance before takeoff would be able to fly at higher altitudes in air with less oxygen, without any decrease in performance. In a series of preliminary tests, Ruhenstroth-Bauer endeavored to research the process of generating blood cells in various experimental animals in different test arrangements, and in so doing also experimented with oxygen deficiencies and low air pressure. At this point the research interests of Ruhenstroth-Bauer and Nachtsheim overlapped. Convulsions are a characteristic symptom of altitude sickness – Ruhenstroth-Bauer's research practice was oriented around raising the threshold for altitude convulsions by increasing the number of red blood cells. For its part, epilepsy research was close to altitude research because oxygen deprivation had long been discussed as a possible trigger for epileptic convulsions. The possibilities of high altitude medicine to generate oxygen deprivation experimentally in vacuum chambers thus also opened up new ways for Nachtsheim to move his experiments with the "epilectic" Vienna Whites forward. Nachtsheim was interested in collaborating with Ruhenstroth-Bauer because the latter experimented with rabbits deprived of oxygen, was well familiar with the physiology of blood and respiration and brought along biochemical expertise. For Ruhenstroth-Bauer's part, Nachtsheim's research on the phenogenetics of convulsion-readiness must have been of fundamental importance – and added to this was the fact that Nachtsheim developed an interest in blood and hemopoiesis starting around mid-1942.

The initiative for collaboration probably came from Hans Nachtsheim, although it can be presumed that the two men were already acquainted due to the tight net of

[345] For a biography: Schwerin, Experimentalisierung, pp. 389–392; Klee, Personenlexikon, p. 514. As a physician of the Luftwaffe Ruhenstroth-Bauer had performed clinical experiments on "48 Russian edema patients" (Soviet prisoners of war). Schwerin, Experimentalisierung, pp. 302f. (note 91).

[346] For a biography: Klee, Personenlexikon, p. 258.

contacts between the Kaiser Wilhelm Institutes. Ruhenstroth-Bauer had contacted Hubertus Strughold (1898–1986),[347] the head of the Reich Ministry of Aviation's Research Institute for Aeronautical Medicine (*Luftfahrtmedizinisches Forschungsinstitut des Reichsluftfahrtministeriums*), for assistance in conducting the low-pressure experiments involved in his hemopoietis project. Nachtsheim, too, through his former doctoral student Harry Suchalla, who had found a position on the "top floor" of the institute,[348] had contacts to Strughold, whose institute, which was housed in the Military Physicians' Academy (*Militärärztliche Akademie*) along the bank of the Spandauer Schiffahrtskanal on Scharnhorststrasse in Berlin, had several vacuum chambers at its disposal. Around June 1943 Nachtsheim and Ruhenstroth-Bauer began with their rabbit experiments in the vaccum chamber of the Research Institute for Aeronautical Medicine.

Yet these low-pressure experiments constituted only a small portion of the around 150 experiments that Nachtsheim and Ruhenstroth-Bauer performed in 1943, with support from the Reich Research Council and the third-highest priority rating of "*S*," for the purpose of depriving their test subjects of oxygen in various ways. The two scientists advanced a concise justification for their experimental program:

> The results of the cardiazol experiments on epileptic and non-epileptic rabbits in previous years made it seem desirable to investigate the importance of oxygen deprivation for the inducement of the epileptic attack in special experiments on young and mature animals.[349]

Through the experiments in the vacuum chamber Nachtsheim saw his view confirmed that convulsion-readiness depended on age. "Normal mature animals" subjected to oxygen deprivation in the vacuum chamber, which corresponded to a height of 4,000–7,000 m, showed no reaction at all, and this was also the case for "normal young animals" and "mature epileptic animals." In contrast, "young epileptic animals aged 2–3 months" nearly always suffered at least a rudimentary epileptic attack under these conditions, and in cases of a "generalized attack with all phases of spontaneous convulsions" the frequent result was "the sudden death of the animals."[350]

Further series of tests "proved" to Nachtsheim that it was the oxygen deprivation ensuing from the low pressure that induced the convulsions: for one, the same result could also be attained when the test subjects were subjected not to low pressure, but to a mixed nitrogen-oxygen atmosphere that corresponded to a height of around 7,000 m. Second, it turned out that an epileptic attack could also be induced in the rabbits by interrupting the flow of blood to the brain, again, particularly "promptly and impressively" in the "young epileptic animals."

---

[347] For a biography: ibid., p. 610.

[348] Schwerin, Experimentalisierung, pp. 300, 312.

[349] Nachtsheim, Bericht über die im Jahre 1943 mit Unterstützung des Reichsforschungsrates durchgeführten Untersuchungen zur vergleichenden und experimentellen Erbpathologie, 21/9/1943, BArch. Koblenz, R 73/15.342, pp. 71–75, here: p. 71.

[350] Ibid., p. 71 f.

For a more precise analysis of the effect of oxygen deprivation, in further tests the rabbits were "set in part into a proconvulsant, in part into an anticonvulsant condition." So some animals were tested in the condition of alkalosis or acidosis (shift in the acid-base balance in the blood toward the alkaloid or acidic side, respectively). Others were placed in a mixture of air and carbonic acid, or treated with bromural, luminal, or caffeine before the oxygen deprivation test. It proceeded from all tests, Nachtsheim proclaimed, "that in the epileptic rabbit oxygen deprivation is the root cause for the inducement of the epileptic attack." The term "myoclonic threshold" is largely identical to the term "sensitivity to oxygen deprivation of the brain cells inducing the attack."[351]

This very assertion was disputed from an influential quarter. A group of scientists around Alois Kornmüller (1905–1968),[352] Director of the Department for the Experimental Physiology of the Brain at the KWI for Brain Research, had been studying epilepsy for a long time – also in collaboration with Strughold – and was already looking into the connections between epilepsy and altitude sickness. The junior physician J. Gremmler, who belonged to the "Brain Research Office of the Air Force" (*Gehirnforschungsstelle der Luftwaffe*) under Hugo Spatz, performed a series of experiments in which (adult) epileptic patients from sanatoriums and hospitals were experimentally put into a condition of hypoxemia and then their brain waves measured. This experiment brought Gremmler to the conclusion that oxygen deprivation must be excluded as the trigger for epileptiform convulsion fits.[353] This result constituted a double challenge to Nachtsheim: not only were Gremmler's findings on the importance of oxygen deprivation diametrically opposed to his own, but Gremmler also cast doubt as to whether the convulsions in altitude sickness could be equated with the epileptic attack at all. In so doing, he also questioned the very foundations of the animal model developed by Nachtsheim and Ruhenstroth-Bauer, for if the convulsions generated in rabbits by low pressure were not epileptiform, then the results on varying convulsion thresholds in young and mature animals could not be translated to human epileptics. Unless Nachtsheim and Ruhenstroth-Bauer wanted to call Gremmler's findings into question in principle – and they did not, because they saw nothing objectionable in the experiments – there was only one way for them "to salvage" their own research findings:

> Elsewhere it has been proved for humans that adult epileptics do not respond to oxygen deprivation with an attack. Since a significant difference in the behavior of mature and young epileptics was yielded in our animal experiments, we tested epileptic children at low pressure in a similar manner.[354]

---

[351] Ibid., p. 72.

[352] For a biography: Klee, Personenlexikon, p. 332.

[353] Gremmler, Beziehungen.

[354] Nachtsheim, Bericht über die im Jahre 1943 im Auftrage des Reichsforschungsrates durchgeführten Untersuchungen zur vergleichenden und experimentellen Erbpathologie, 21/9/1943, BArch. Koblenz, R 73/15.342, pp. 71–75, here: p. 73. Cf. also Ruhenstroth-Bauer/Nachtsheim, Bedeutung des Sauerstoffmangels, p. 20: "The characteristic difference in the behavior of young

If they were successful in inducing epileptic attacks in epileptic children through low pressure, Gremmler's negative findings would be relativized – in Gremmler's experimental arrangement, it could be argued, the oxygen deprivation was simply not great enough to induce a convulsive attack in the adult test subjects – and the hypothesis of oxygen deprivation as the trigger of the epileptic attack would be saved. Beyond this, if the epileptic children reacted to low pressure in the same way as the young epileptic rabbits, this would furnish impressive evidence of the animal model's applicability. Paradoxically, in this case the human experiment was to function as the confirmation for the animal experiment, which was originally conceived of as a substitute for human experiments.

There are only two written sources on the further course of events, both of them quite meager – a report by Nachtsheim to the Reich Research Council of September 21, 1943 or March 15, 1944, respectively, and a short letter from Nachtsheim to Gerhard Koch of September 20, 1943 – as well as several testimonials put down in writing by Gerhard Ruhenstroth-Bauer at great intervals of time. These sources document without a doubt that at least one such human experiment took place. However, we know hardly anything about how this experiment came about and how it proceeded in detail. For the present we also remain in the dark about what happened to the human "guinea pigs" later and whether further experiments of this kind followed.

Apparently Nachtsheim, in his search for epileptic children for the planned tests, turned to Gerhard Koch, who was convalescing in Berlin from June to August 1943 and worked as a guest scholar at the KWI-A during this period. At the time Koch's research included work on "residual epilepsy." As he wrote in his memoirs, he and Nachtsheim "repeatedly [conducted] instructive and useful conversations about the etiology and heritability of the various epileptic convulsive conditions in humans and animals and about the convulsion-readiness behind these conditions which is so different for each individual."[355] It was presumably Koch who drew Nachtsheim's attention to the Berlin-Wuhlgarten Sanatorium and Hospital (*Heil- und Pflegeanstalt Berlin-Wuhlgarten*), in which a large number of epileptics were housed. Koch had worked there from 1937 to 1939 on "family studies" in the context of his dissertation about Sturge-Weber disease (today: Sturge-Weber-Krabbe syndrome). He had maintained contact afterward – as late as 1943 Julius Hallervorden sent Koch the pathological report of a test subject who died after the family study was concluded.[356] While still working in Dahlem in summer 1943, Koch, assisted by Hans Grebe and

---

and mature epileptic animals in response to oxygen deprivation made it appear desirable to investigate on humans a comparison of young and adult epileptics. Gremmler investigated only adults, and was not successful in inducing an epileptic attack in them through hypoxemia. After conclusion of our own studies of young epileptics, which are also interesting to the clinic, we intend to report about the detailed results." – On the human experiment described in the followiing, cf. Müller-Hill, Genetics after Auschwitz; Deichmann, Biologen, pp. 308–314; idem., Hans Nachtsheim, pp. 146–148; Koch, Humangenetik, pp.120–148; Knaape, Medizinische Forschung, p. 227; Klee, Auschwitz, pp. 228–230; Proctor, Adolf Butenandt, pp. 18–20; on the basics, see the recent work: Schwerin, Experimentalisierung, pp. 302–319.

[355] Koch, Humangenetik, p. 103.

[356] Schwerin, Experimentalisierung, p. 308 (note 115).

the photographer of the KWI-A, Ingeborg Hellhoff, produced photographs of the patient with Sturge-Weber disease who had been the point of departure for his genealogical studies over the course of his dissertation.[357] On September 20, 1943 Nachtsheim informed Koch, who had since moved on to Neubrandenburg:

> Dr. *Ruhenstroth* had, in the meantime, already established contact with Wuhlgarten and learned that only adult epileptics are there. He was referred from Wuhlgarten to Görden, and from there, through the kindness of Senior Medical Councilor Dr. [Karl] *Brockhausen* [* 1890], received 6 epileptic children (4 genuine, 2 symptomatic epileptics), with whom we did experiments last Friday [September 17, 1943] in Prof. Strugholt's [sic] vacuum chamber. Yet the tests came out just as negative as those *Gremmler* performed on adult epileptics. But at the moment it is not possible to say that rabbits and humans respond differently to low pressure, for the children we tested were aged 11-13, which corresponds to a rabbit aged 5-6 months. However, epileptic rabbits of 5-6 months do not show the reaction-readiness of 2-3-month-old animals, which nearly always had attacks. We would have to be able to test epileptic children of 5-6 years of age, but this is not possible at the moment because this age group is not present at Görden.[358]

So from Wuhlgarten, Ruhenstroth-Bauer and Nachtsheim had been referred to the State Institute in Brandenburg-Görden directed by Hans Heinze, which played an important role in the Nazi "euthanasia" program. It remains unclear who ultimately established contact with Görden. In the 1990s Ruhenstroth-Bauer claimed that Nachtsheim had enjoyed good contacts to Görden and was involved in the treatment of epileptic children there, so that he addressed the children by their first names and elucidated their anamneses,[359] while he – Ruhenstroth-Bauer – met the children for the first time in the vacuum chamber on September 17, 1943, had never seen them before and did not even know where they came from. Considering the letter from Nachtsheim to Koch, there is certainly reason to regard this testimony as an attempt at self-justification, but it is indeed conceivable that Nachtsheim had been in contact with Görden for some time previously. This could have come about via the KWI for Brain Research, which was, for its part, linked closely with Görden through Julius Hallervorden, who was both Director of the Department for Histopathology at the KWI for Brain Research and Prosector of the Brandenburg State Psychiatric Institutes from 1938 onward – in fact the Department of Pathology located in Görden from 1938 had been officially transferred to the KWI for Brain Research in Berlin-Buch, and the laboratory in Görden was run as an outpost of the KWI. Through the department of pathology in Görden and other channels, over 700 brains of "euthanasia" victims made their way to the KWI for Brain Research, where they were subjected to pathological examination by Julius Hallervorden and Hugo Spatz. Nachtsheim had good contacts to the KWI for Brain Research – for years he had sent his rabbits from the epilepsy experiments to Gerd Peters (1906–1987) for postmortem examination. As mentioned above, Nachtsheim also had

[357] Koch, Humangenetik, pp. 101, 104.

[358] Nachtsheim to Koch, 20/9/1943, printed as a facsimile in: Koch, Humangenetik, pp. 125 f., quote: p. 125 (original emphases).

[359] Schwerin, Experimentalisierung, p. 307 (note 111).

close contact with the pathologist Hans Klein, who performed postmortem examinations on rabbits with dropsy for him starting in 1943, but also participated in the autopsies of the victims of the "Special Children's Department" at Wiesengrund. What should not be forgotten is that Fritz Lenz was familiar with a number of physicians from the staff of the "euthanasia" program from his consulting activities on the draft law for euthanasia, including Hans Heinze. Pointing out these entanglements is important to the extent that it can be presumed with a high degree of security that Nachtsheim was aware of the "euthanasia" program still in progress.

As the available sources testify unanimously, the experiment did not produce any tangible result – it did not succeed in inducing an epileptic fit in the children through low pressure. Consequently it did not cause them any suffering – but Ruhenstroth-Bauer and Nachtsheim could not have foreseen this. According to Nachtsheim's account, the children were subjected to a low-pressure situation that corresponded to an altitude of 6,000 m (not to mention the mental strain of being locked into the vacuum chamber). According to the knowledge available to altitude medicine at the time, at this altitude the onset of threatening conditions had to be expected even for adults – all the more so for children. Moreover, there was no possibility of resorting to any previous experience with epileptic humans in low-pressure situations. Furthermore, Ruhenstroth-Bauer and Nachtsheim knew from the animal experiments that young epileptic rabbits reacted to low pressure with violent, often fatal convulsions – and they expected (and hoped!) that the children would react like the rabbits. In other words: the scientists knowingly accepted the risk that the children could be placed in fatal danger. Ruhenstroth-Bauer's reassuring statement that he himself, Nachtsheim and an additional physician of the Luftwaffe had been in the vacuum chamber with the children and had been able to abort the experiment at any time – as could the children themselves – thus fails to get at the root of the matter.

There is no doubt that Ruhenstroth-Bauer and Nachtsheim planned further tests with younger children after the failed first experiment. Whether these came about cannot be determined with any certainty. However, it is probable that it was no longer possible to realize these tests. Of the six children in the first experiment, there is proof for only one having survived the Third Reich;[360] the fate of the other children must remain an open question. Perhaps they fell into the gears of the "euthanasia" program – in contrast to the clinical examinations and tests in the two "research departments" of the "euthanasia" apparatus in Brandenburg-Görden and Wiesloch/Heidelberg, however, it went against the logic of the experiment to kill the children and examine them pathologically as long as they had not suffered an epileptic attack.

Nonetheless: the low-pressure experiments by Nachtsheim and Ruhenstroth-Bauer ignored the Reich Health Council's regulations on human experiments from the year 1931 as a matter of course. For the most part, these regulations, as adduced elsewhere, had already been ignored by research back in the 1930s. Yet this experiment

[360] Knaape, Medizinische Forschung.

marked a further boundary crossing, as the experimenters unscrupulously subjected the children to an incalculable health risk, even accepting a potentially fatal outcome of the test – and all of this *needlessly*, for the utilization of the vacuum chamber was by no means imperative. Oxygen deprivation could have been effected in other ways, especially as Gremmler, upon whose work the experiment was based, had not worked with low pressure. Apparently the standards of scientific ethics had shifted further. A comment with which Nachtsheim and Ruhenstroth-Bauer preceded their short report about the low-pressure experiments on rabbits implied as much:

> For the clinician working on patients experimentally, the possibilities are always restricted, for he has to take the welfare of his patients into consideration. Only in exceptional cases will a researcher dare to perform an experiment on a patient in the interest of future patients, the outcome of which cannot be predicted with any certainty. Here a method assists the field of medicine, which allows these difficulties to be circumvented at least for a few genetic illnesses, the *model experiment on animals*.[361]

Alexander von Schwerin is correct to emphasize that this opens up a new moral dimension. While up to this point Nachtsheim had designated the human experiment as morally inadmissable without restriction, and recommended the animal experiment as a morally unobjectionable alternative, he now no longer categorically excluded the possibility of research on humans for the benefit of others, even if the outcome was uncertain. In this case human and animal experiments no longer appear as mutually exclusive alternatives; on the contrary, it suggests a complementary relationship. Schwerin lists a number of factors that contributed to the erosion of the ethical standards of science: the objectifying linguistic usage, which not only blurred the boundaries between humans and animals (Nachtsheim, for instance, referred to both as simply "epileptics") and transformed both into "material," but also elevated the "genotype epilepsy," detached from the human patient, to the actual scientific object; and also the "militarization" of altitude research.[362] Two other aspects deserve special emphasis: First it must be kept in mind that the newly developed coma and shock therapies (insulin coma treatment, cardiazol convulsion treatment, and electric shock therapy in the first years of World War II) had been widely adopted in German institutional psychiatry since the mid-1930s, although these "heroic therapies" put the patients in horrible states of anxiety, often inflicted serious injury to their health, and in some cases even resulted in their deaths. Therapeutic ambition was willing to accept high risks[363] – thus, it is no wonder that artificially inducing convulsive fits in epilepsy research was not questioned. Second it must be considered that by 1943, somewhere around 100,000 mentally ill, epileptic, or mentally disabled patients from the sanatoriums and hospitals of the German Reich already had been murdered in the course of "euthanasia" – and thousands of infants, children, and teenagers had also been killed in the course of the children's "euthanasia," the "Aktion T4" and "decentral euthanasia"

[361] Ruhenstroth-Bauer/Nachtsheim, Bedeutung des Sauerstoffmangels, p. 18 (original emphasis).

[362] Schwerin, Experimentalisierung, pp. 312f., 318f.

[363] Cf. Kersting/Schmuhl, Einleitung, pp. 37f.

since August 1941. This undermined the moral status of the children from Görden. Now they were little more than readily available, not terribly valuable "material" for "consumptive research."

4.4.5 *"The Problems of White Africa": Colonial Science Ambitions*

With the entry of the German Reich into the circle of colonial powers, German anthropology and ethnology – like the other sciences – felt challenged to make their knowledge useful for the justification and legitimation, execution, and consolidation of colonial rule. A relationship of mutual engagement emerged: The sciences aligned themelves with colonial interests in their selection of subjects and objects, their theoretical and methodological approaches, and made the knowledge thus obtained available to the colonial administration. In return, the colonial state furnished the colonially oriented scientific disciplines and subdisciplines with financial resources, granted them privileged status in the institutional structures and raised their value in the public. Colonial interest groups mediated between state and science. "In this system of mutual obligations between state, political parties, interest associations, and sciences after 1885, a spectrum of new areas of knowledge developed in the German science landscape, which was known as the 'colonial sciences' in the contemporary diction [...]."[364]

Fischer's study on the "Bastards of Rehoboth" of the year 1913 was conceived and intended as a contribution to colonial science, apparent in the fact that the author drew practical consequences for colonial policy from his research findings in its concluding chapter, *Die politische Bedeutung der Bastards* ("The Political Importance of the Bastards"). Despite his heterosis thesis, according to which a "population of bastards" is located between the "source races" as regarded their physical, mental, and intellectual characteristics, he took a clear position on the ban on mixed marriages in the colonies so hotly debated at the time:

> Every European nation without exception [...] that has assimilated the blood of inferior races – and that Negroes, Hottentots and many others are inferior can be denied only by dreamers – has paid for this assimilation of inferior elements with intellectual, cultural decline.

At the end of his colonial policy conclusions, Fischer designed a system of apartheid for German Southwest Africa, long before such a system was introduced in South Africa: The Ovambo and Herero were to be deployed as agricultural laborers, the Hottentots as herders. The "bastards of Rehoboth," in contrast, were assigned an important function as a privileged intermediate class, "as native craftsmen and manual laborers [...], as policemen, i.e. minor officers, foremen, and leaders of the entire supply lines and vehicle pool of the government, troops and

[364] Grosse, Kolonialismus, p. 35.

private persons, in part as small farmers in their bastard country, to which everyone returns after serving their time." Despite his paternalistic attitude toward the "little nation of bastards," Fischer regarded the Rehoboths from the perspective of the colonial masters:

> So they will be granted just that degree of protection which they *need* as a race inferior to us, in order to endure, no *more* and *only* as long as they are useful to us – otherwise free competition, i.e. in my opinion, here downfall![365]

This last comment by Fischer reads like a retrospective justification of the war of extermination the German colonial troops had led against the rebellious Herero and Nama from 1904 to 1908.[366] Fischer had profited from this genocide directly, for he apparently brought skulls and skeletons of "Hottentots" with him from Southwest Africa,[367] which may have come from the internment camps on Shark Island, where people died like flies.[368] The skeleton of the Nama leader Cornelius Frederiks († 1907) also supposedly came into Fischer's collection in this way.[369]

As mentioned above, Fischer continued his studies of the "bastards of Rehoboth" until 1942.[370] Yet the "bastard studies" by Fischer and his pupils were no longer embedded in a colonial science and colonial policy context, but rather in the concept of anthropobiology: With its particular methodology, which combined anthropometry, genealogy, genetics, and ethnology, they were supposed to bring together anthropology and human genetics. However, "bastard research" had not lost its practical application, as the role of the institute in Dahlem in the sterilization of the "Rhineland bastards" quite impressively evinced.

After World War I the colonial sciences became part of the "colonialism without colonies," which blossomed so lushly in Germany between 1918 and 1943. Colonial research did not simply cease after the loss of the colonies. On the contrary, in view of a future German colonial empire, it was even intensified. Until the German defeat in Stalingrad, when colonial planning was officially discontinued, a perfect colonial empire had been designed on the drawing board. "One can only ask with astonishment," in the words of Wolfe W. Schmokel, "whether at any point in history a non-existent empire had ever been so well administered [...]."[371] Increasingly, colonial planning was based on a scientific foundation. Tropical medicine,

---

[365] Fischer, Rehobother Bastards, pp. 302, 305, 302 (original emphases). Cf. also Gessler, Eugen Fischer, pp. 73 f.; Lösch, Rasse, pp. 72–77.

[366] On this, still the best: Drechsler, Südwestafrika.

[367] Cf. Uhlebach, Messungen, p. 449, where the author thanks Eugen Fischer for "material" from his collection. Apparently this collection was destroyed by the bombing of the Freiburg Anatomical Insitute in 1917. In 1921 Fischer addressed the readers of the *Deutsche Kolonialzeitung* asking them to provide him with material. Cf. Eckart, Medizin, p. 257.

[368] Cf. ibid., pp. 283–290.

[369] Dierks, Chronologie, pp. 98, 101. It is possible that this is mistake. Cf. Fetzer, Untersuchungen.

[370] On the history of the "bastard studies of Rehoboth" up to Fischer's death, cf. also Schmuhl, "Neue Rehobother Bastardstudien."

[371] Schmokel, Traum, p. 159. On this, now also: Kundrus (ed.), Phantasiereiche.

tropical technology, geography, regional development, demography, anthropology, social hygiene, and eugenics dealt intensively with colonial policy issues.

At this time there is no indication that the KWI-A was included in colonial policy planning in the late 1930s – be it by the Race Policy Office of the NSDAP, which was closely linked with the institute in Dahlem through its director Walter Groß, and which had presented the main features of a future National Socialist race policy in the colonies in 1938 with a tract entitled *Kolonialfrage und Rassegedanke* ("The Colonial Question and Race Theory"):[372] The plan was for strict race segregation, a ban on mixed marriages, the restriction of contact between blacks and whites to a minimum, and so on. In an article for the periodical *Rassenpolitische Auslandskorrespondenz* (Race Policy Foreign Correspondence), Eugen Fischer legitimated such forms of apartheid with reference to "indisputable and provable facts, to the fact that mental attributes are based on genetic dispositions, that race differences are genetic differences, that mental attributes are different for each races, and that there are thus mental differences between races."[373] Fischer's admonition to investigate such "mental race differences" scientifically and to lay a scientific foundation for race policy fell on deaf ears, however. That the "Law for the Protection of Colonial Blood" drafted by the Colonial Policy Office in 1940 equated the "half-breeds with an admixture of native blood" with the population of color as regarded the ban on mixed marriages[374] was certainly in accordance with Fischer's wishes, but the notion upon which this passage was based, that the "half-breed" was under *both* "source races" in terms of his or her mental and psychic attributes,[375] stood in blatant opposition to Fischer's theory of heterosis.

In the course of World War II, however, as mentioned above, collaboration developed between the Colonial Policy Office and Gottschaldt's Department for Genetic Psychology. In September 1940 Gottschaldt took on the article about "Psychological Problems and Methods in Colonial Science" for the *Afrika Handbuch der angewandten (kolonialen) Völkerkunde* ("Africa Manual of Applied (Colonial) Ethnology")[376] contracted by the Colonial Policy Office of Hugo Adolf Bernatzik (1897–1953). The manuscripts were ready for printing in fall 1943, but were destroyed as a result of a bombing, so that the *Handbuch der angewandten Völkerkunde* could not be published until 1947 – including the article by Gottschaldt along with a "Questionnaire for the Psychological Evaluation of Native Workers" he had developed.[377]

---

[372] Hecht, Kolonialfrage.

[373] Fischer, Geistige Rassenunterschiede, p. 4. In 1936 Fischer felt compelled to protest vehemently in *Volk und Rasse*, the organ of the National Committee for the National Health Service, against an article by the Catholic theologian Theodor Gentrup (Berlin), who had advocated "racially mixed marriage" in the colonies on the authority of Fischer's work on the "bastards of Rehoboth." Fischer, Frage "Rassenmischehe."

[374] Reprinted in: Kum'a N' Dumbé III, Pläne, p. 179. On the political background: Hildebrand, Reich.

[375] Hecht, Bedeutung, here: No. 11, p. 8.

[376] Byer, Fall Bernatzik, p. 295.

[377] Gottschaldt, Psychologische Probleme in der Kolonialforschung (the questionnaire is on pp. 180–186).

Another, entirely unexpected possibility to reestablish himself in the field of colonial science emerged from one of Fischer's other research interests: his search for the "Cro-Magnon race," whose traces he believed to have discovered back in 1908 upon his return journey from Southwest Africa, and then on a further research trip in 1925 in the population of the Canary Islands, and finally also in the contemporary European "Phalian type."[378] On a research journey to Spanish Morocco, planned for the 1934/35 fiscal year, Fischer apparently had intended to track down the Cro-Magnon type in Northern Africa as well, yet this research plan was delayed indefinitely because of Fischer's heavy workload at the rectorate.

With the formation of the German Africa Corps in January 1941 and the conquest of Cyrenaica in March/April 1941, when the plans for founding a German colonial empire in Northern Africa took on more concrete shape, Cro-Magnon research, little more than a hobbyhorse of Fischer's for so many years, quite surprisingly took on political importance. The virtuosic research strategist Eugen Fischer immediately recognized the emerging possibilities. On May 8, 1941 he lectured to the Prussian Academy of Sciences about "The Problems of White Africa." Proceeding from the term "White Africa," coined by Dominik Josef Wölfel (1888–1963), Fischer claimed that the part of Africa located north of the Sahara, in terms of climate, geology, zoology, and botany, but above all "according to human races and cultures, clearly and fundamentally departs and stands out from the remainder of Africa, from the Africa of the Negroes, from Black Africa."[379] Fischer presumed that the entire Mediterranean region was settled by a "Mediterranean race," in which shares of other races had been incorporated in the historical era – Arab, Nordic, Alpine, Negroid. Fischer saw one of the roots of the "Mediterranean race" in the prehistoric Cro-Magnon race, which was characterized by "blondness and blue eyes."[380]

The line of attack is clear: Through the anthropological-ethnological differentiation between Black and White Africa, Fischer supplied the scientific basis to legitimate pushing forward the borders of the emerging Greater European Empire under the hegemony of National Socialist Germany to the northern edge of the Sahara, without any race policy scruples. It can come as no surprise that Fischer, in return, demanded funds to accelerate the advancement of the scientific exploration of Northern Africa.

Following his lecture, Fischer – along with the Africanist Dietrich Westermann (1875–1956) and the Egyptologist Hermann Grapow (1885–1967) – thus proposed to the Prussian Academy of Sciences the establishment of an interdisciplinary research commission on White Africa. In their proposal the three scholars urged for haste, for "after the war the development of the Sahara areas with automotive and aeronautic routes, and through the construction that has just commenced of a [...] trans-Saharan railway, will certainly restart in full strength, and thus an increasing

[378] Cf. also Ritter, Cro-Magnon-Merkmale.

[379] Quoted in Lösch, Rasse, p. 376. Reworked version of the lecture: Fischer, Weißafrika.

[380] Ibid., p. 132.

destruction of the remaining witnesses of the White African past set in."[381] The academy approved the proposal immediately. In 1942/43 Fischer held "soliciting lectures"[382] on the topics surrounding White Africa.

The commission instigated by Fischer had no opportunity to develop any activities of note – Germany's colonial dreams were over too soon. When the commission's three subject groups convened for the first time at the invitation of the Colonial Science Department of the Reich Research Council and the German Research Association on January 27, 1943 – a few days before the defeat in Stalingrad – for a 3-day conference about "Colonial Ethnology, Colonial Linguistic Research and Colonial Race Research" in Leipzig, all of the colonial science plans were already scrap.

The speakers in the "Colonial Race Research" section – besides Otto Reche, Director of the Institute for Race Science and Ethnology at the University of Leipzig, and Egon von Eickstedt (1892–1965),[383] Director of the Institute for Anthropology at the University of Breslau – were Eugen Fischer and Wolfgang Abel. Based on a reworked version of his lecture for the academy, Fischer outlined the anthropological concept of White Africa once again. Abel dealt with "Race Problems in Sudan and Its Borderlands." This harmless title concealed extremely explosive subject matter. Abel presented numerous photographs of anthropological types from the Sahel zone, most of which depicted French prisoners of war. As Abel mentioned in passing, he had been detailed to the "Inspection of the Personnel Controlling of the Army (Army Psychology)" (*Inspektion des Personalprüfwesens des Heeres (Heerespsychologie)*, to perform series of anthropological examinations of French colonial soldiers in a number of war prison camps.[384] "Hereby the residents of different areas or different tribes of Sudan were put together in large groups and the number of the persons best rendering the type were always photographed." Thus, "good illustrative material"[385] was created, comprising the photographs of around 350 persons. According to statements made in the 1980s, in the context of this activity Abel was also at a "leper station in Bordeaux"[386] – what was probably meant was the Special Military Hospital for Colonial Medicine in St. Médard near Bordeaux – in order to examine the changes in the pattern of fingerprints caused by the disease. The footprints of "Guinea Negroes" from the Special Military Hospital for Colonial Medicine in Georg Geipel's estate were quite probably taken by Abel.[387]

[381] Quoted in Lösch, Rasse, p. 376.

[382] Fischer, Tätigkeitsbericht 1942/43, MPG Archive, Dept. I, Rep. 3, No. 20.

[383] Eickstedt himself was not present in Leipzig, his lecture must have been read by a deputy. Cf. Verschuer to Fischer, 9/2/1943, MPG Archive, Dept. III, Rep. 86 A (Münster), No. 9.

[384] On Hitler's orders, a large portion of the African prisoners of war – around 80,000 men – had been deported to Southern France (Bordeaux). Cf. Klee, Auschwitz, p. 257.

[385] Abel, Rassenprobleme im Sudan, here: p. 144.

[386] In an interview with Benno Müller-Hill, Abel stated that he was "in a leper station in Bordeaux visiting Dr. Weddingen in the tropical medicine military hospital." Müller-Hill, Tödliche Wissenschaft, p. 146.

[387] MPG Archive, Dept. III, Rep. 48, Box 4, "Pictures for the Work on Pygmy Soles of the Foot".

Abel was not alone: Otto Baader, too, combed through the war prison camps in France in his search for Cro-Magnon types.[388] Both scientists presented their findings to the Berlin Anthropological Society.[389] Those involved apparently had no grasp of the fact that such examinations in prison war camps signified a subtle, but nevertheless fundamental boundary crossing – for the first time, scientists of the KWI-A researched on people who were capable of giving consent, but whose possibilities for refusing the examination were at least restricted because they were imprisoned. Even though the examinations as such were harmless, and the probands had to suffer neither pain and fear nor abasement and were not subjected to any health risks, abandoning the principle of informed consent signified a deep rupture.

The war prison camps of the French campaign amounted to a sort of laboratory for race anthropology research. The special conditions of such research resulted in a process of radicalization, which is to be illustrated with a further example: Robert Stigler (1876–1975),[390] Director of the Institute for the Anatomy and Physiology of Domestic Mammals at the University of Vienna, and his five assistants, performed a series of race anatomy and race physiology tests in a war prison camp near Vienna in July 1940.

> In the camp, besides around 15,000 white French, Alsatians, Flemish and Walloons there were also around 2,400 Moroccans, Tunesians and Algerians, 63 Tonkinese, 2 Annamese, 12 Negroes, among them 9 from West Africa, 3 from tropical America, and 16 European Jews, among them several diamond merchants from Antwerp.

As in the examinations by Abel and Baader, here to the question as to the consent of the probands was not posed at all. Even so:

> Our examinations met with no resistance at all from the prisoners. The colored were intially very shy, but soon began to trust us and many cheerful scenes ensued. I had the Negroes perform their dances and sing their songs for us. The Moroccans, Tunisians and Algerians were much more negative, the little yellow Tonkinese were the shyest of all [...].[391]

Recorded in the examinations was the clotting time of the blood, the sinking speed of the blood, the viscosity of the blood, blood pressure, pulse rate, respiration, the upper hearing limit, the threshold of the sense of touch, reaction speed, right-handedness and left-handedness, hair growth on the genitals, and sexual characteristics – the last of these substantiated by numerous photographs. In comparison to the examinations by Abel and Baader, further boundary trangressions can be determined: Not only would the measurements of the naked body and the photographing of the genitals have been perceived by the probands as humiliating and a violation of modesty. In taking blood samples the scientists had gone a step further – this was

---

[388] Baader, Cro-magnide Typen. Baader summoned colonial soldiers from Morocco, Algiers and Tunis. Cf. also Fischer, Tätigkeitsbericht 1942/43, MPG Archive, Dept. I, Rep. 3, No. 20: "Examinations were made of Northern African prisoners (Dr. Baader), a possibility for examining the bones of Guanche limbs as compared to Nordic ones is being worked out in detail."

[389] Fischer, Weißafrika, p. 133.

[390] For a biography: Klee, Personenlexikon, pp. 603 f.

[391] Stigler, Untersuchungen, quotes: pp. 26, 27.

a first, albeit minimal, *invasive* approach. The examinations of the group of researchers around Stigler were thus positioned between those of Abel and Baader and the examinations and experiments of Karl Horneck, which will be depicted at a later juncture.

4.4.6 Fischer, Verschuer, and the "Final Solution" to the Jewish Question

Even after the start of World War II, Eugen Fischer and Otmar von Verschuer brought their national and international reputations to bear in order to provide a scientific foundation to legitimate the "total solution to the Jewish question" tackled by the National Socialists, which by late 1941 early 1942 had taken on the character of the "Final Solution" once and for all. For Fischer and Verschuer there could hardly have been a doubt as to what the measures aimed to achieve. They were guests of honor to a working congress at the inauguration of the "Frankfurt Institute for the Investigation of the Jewish Question" (*Frankfurter Institut zur Erforschung der Judenfrage*) on March 27/28, 1941. The aspired goal of the "total solution" to the "Jewish question," as was bluntly stated here, was the *Volkstod* ("death of the nation"). The economist Peter-Heinz Seraphim (1902–1979)[392] pointed out for consideration that the deportation for forced labor in camps in Poland or an overseas colony could also have the consequence of "social pauperization and upheaval," but "by no means the physical self-disintegration of Jewry, for the death of a nation is never a fast death."[393] The logical conclusion from these comments was, as Benno Müller-Hill emphasizes correctly, that the "physical self-disintegration" would require some assistance. When the deportation of the German Jews began in October 1941, nobody who had participated in the congress in March could have been in doubt as to what was in store for the Jews deported to the East.

This did not prevent Fischer from making an appearance in late 1941/early 1942 as part of a lecture series organized by the German Institute in Paris. In his lecture about "Race and German Legislation," Fischer certified that the "Bolshevist Jews" were of "monstrous mentality" and assigned them to a "different species." Fischer himself emphasized in a report about his trip to Paris that he had found much acknowledgement among the attendant French scientists for his discussion held of the "Negro problem" and the "Jewish problem" in a "very candid, but in purely scientific form"[394] – and this right before the deportation of 165,000 Jews from France was discussed at the Wannsee Conference.

[392] For a biography: Klee, Personenlexikon, pp. 579 f.

[393] Quoted in Müller-Hill, Tödliche Wissenschaft, p. 48.

[394] Quoted in Weiss, Sword. A German translation of the French text of the lecture is also in Müller-Hill, Tödliche Wissenschaft, p. 49.

As mentioned previously, in 1944 Fischer and the theologian Gerhard Kittel published a book about the "world Jewry of antiquity," essentially a selection compiled by Kittel of ancient sources with a decidedly anti-Semitic perspective. Kittel supplemented the written sources with illustrations of Egyptian mummy tablets, which supposedly constituted further evidence for the worldwide propagation of Jewry.[395] At Kittel's request, Fischer undertook to determine the "race type" of the persons illustrated. This was not the first time Fischer had done something like this (for instance, he had studied the illustrations on Etruscan tombs and the masks found during excavations in Mycenae), yet in this publication it was practically tangible that Fischer's interpretations of the pictures completely abandoned the basis of precise anthropometry and relied only on intuition – and that his intuition was distorted by anti-Semitism:

> Granted, the expert sees for all races, and also for the basic races of the Jews, a number of physiognomic details which we cannot name and fit into the usual model: shape of nose, shape of face, shape of skull, etc. Often a Jew is recognized as a Jew with complete certainty even though he does not have [...] a so-called "Jewish nose." There is something [...] in the Jewish physiognomy that cannot be measured, and can hardly be described in detail such that the reader or listener can visualize it clearly. But no one will doubt that very many Jews can be picked out from groups of non-Jews with complete certainty. [...] It is not permissible to disqualify as unscientific the statement of a general "impression" of "Jewish" in the evaluation of the pictures.[396]

The attempt by Niels C. Lösch to play down Fischer's participation in this anti-Semitic pamphlet as an expression of senility[397] deserves vehement contradiction – Fischer, 69 years old at the time, was of remarkably fresh intellect, and his scientific publishing activities extended well into the 1960s. It must also be kept in mind that Fischer's studies on "Jewish physiognomy" were by no means the concern of an individual scholar in retirement, but rather were based on preliminary work performed at the KWI-A at the beginning of World War II.

In late 1939/early 1940 – probably in the first 3 months of 1940 – one assistant and three students made several trips to Łódž ("Litzmannstadt") on Fischer's behalf,[398] where the group – in a cauldron of executions, pogroms, and synagogue

[395] Cf. also Heiber, Walter Frank, p. 463. Upon Fischer's request, Verschuer had copies of the "Jew pictures of Egypt" slides made for the collection of the KWI-A. Cf. Verschuer to Fischer, 31/3/1943, MPG Archive, Dept. III, Rep. 86 A (Münster), No. 9.

[396] Fischer/Kittel, Weltjudentum, p. 113.

[397] Lösch, Rasse, pp. 291 f.

[398] The travel costs are accounted for in the cost report for the closing of accounts on March 31, 1940 (for the 1939 fiscal year). The controlling report states: "For a trip undertaken to Lodsch [sic!], now Litzmannstadt, 1853.96 RM were spent. There are discrepancies in the accounting, which could no longer be clarified, as the participants are no longer in the institute's service." On this Telschow commented in the margins: "Seems very expensive to me. Duration how long?" (Bericht über die Prüfung des Rechnungsabschlusses des KWI-A zum 31. März 1940, MPG Archive, Dept. I, Rep. 1 A, Nr. 2409, pp. 74–76, quotes: p. 76). Fischer explained the costs, stating that "all four gentlemen occasionally traveled and worked independently, so that I was not able to give the entire sum of the expenses to the assistant, so that he could pay out the individual amounts, but occasionally had to pay all gentlemen individually. [...] And in so doing the account between their own funds and those of the institute occasionally were somewhat mixed up." Fischer to Generalverwaltung, 11/10/1940,

desecrations, while at the same time ten thousands of Jews were deported from the city and carried off to concentration camps – performed series of anthropological examinations on more than 250 Jews. Among the students were also Harry Suchalla and Christian Schnecke, who were still working as doctoral students of Nachtsheim's at the Institute for Genetic and Breeding Research and were "loaned out" by Fischer, presumably for want of manpower of his own.[399] This circumstance indicates that the opportunity was favorable and time was pressing. It can be presumed that Herbert Grohmann, a graduate of the first annual SS course at the KWI-A and an assistant to Fischer in the years 1938/39, made the "field research" in Łódž possible, having held the position of senior medical councilor at the newly founded Health Office of "Litzmannstadt" since September 1939.

In a letter of the year 1950 Suchalla willingly provided information about the kind of material collected on these visits and what became of it:

> Prof. Dr. Eugen Fischer had authorized me to evaluate the fingerprint and handprint material in 1940 and thus it initially remained in my possession. When I was called up for military service the further analysis remained unfinished and the documents remained in my apartment in Berlin until I found time to hand over the complete materials to the Anthropological Institute in fall 1944. Among the material was a directory of persons with complete data about gender, age, place of birth, descent and social status, ordered sequentially. The material also included the standard anthropometric data recorded in series of anthropological examinations, including index calculations. At the time there was talk of moving the material with the institute archives to somewhere near Neuruppin.[400]

Not only photographs taken by Suchalla and his comrades on their trips to Łódž were included in Fischer's and Kittel's book about "Ancient world Jewry." After the war the fingerprints and handprints from the Łódž ghetto were recovered in Hans Nachtsheim's Institute for Comparative Genetic Biology and Genetic Pathology of the German Research Academy in Dahlem, where they were discovered by Georg Geipel, who, as elaborated elsewhere, had worked at the KWI-A as an expert for dermatoglyphics (and who had introduced Suchalla to the technique of dactyloscopy in 1940). As late as the end of 1950, the publication of this material was discussed in all earnestness in the context of dermatoglyphic race research, but was stopped, presumably upon Fischer's advice. The fingerprints and handprints themselves are untraceable today. However, the anonymized fingerprint formulas of 169 "Litzmannstadt Jews" are included in Geipel's scientific estate.[401]

The incident shows that Fischer was willing to use the "total solution of the Jewish question" at short notice in order to obtain research material, and in return

---

ibid., pp. 80a–80a v, quote: p. 80a. – The trips must have taken place between September 8, 1939, the day on which Łódž was occupied by the Wehrmacht, and the close of accounts on March 31, 1940. In retrospect, Harry Suchalla dated his stay in Łódž to the year 1940.

[399] The incident shows that a close connection between Fischer and Nachtsheim must have existed before October 1940.

[400] Suchalla to Geipel, 29/12/1950, MPG Archive, Dept. III, Rep. 48, No. 6.

[401] Portfolio "Material Suchalla. Polnische Juden. Fingerleistenmuster, 20/11/50," MPG Archive, Dept. III, Rep. 48, box 3.

to place the results of his research, as dubious as they might have been, unquestioningly at the service of "Jewish policy." That Fischer regarded his anthropological studies as a contribution to the "total solution of the Jewish question" was demonstrated quite clearly in June 1944, when Alfred Rosenberg (1893–1946), Minister for the Occupied Eastern Territories, invited Fischer to act as one of the presidents of an international "Anti-Jewish Congress" to be convened in Kraków. Fischer accepted the invitation, explaining:

> I hold [...] your intention to found a scientific front to defend against the influence of Jewry on European culture and to call together the scientists of all of the nations in conflict with Jewry to be very good and altogether necessary. Yes, it is high time for such an action, for Jewry has been battling us for decades not only politically, but quite certainly in terms of pure intellectural history as well.[402]

The congress never took place. Nevertheless the incident shows how loyally Fischer supported the "Final Solution" even at a point in time when the collapse of the National Socialist state was already clearly imminent. This was also true for his successor, friend and pupil Otmar von Verschuer. In late 1941/early 1942 – the deportation of German Jews had begun a few months previously – he wrote in the *Erbarzt*:

> Never before in history has the political importance of the Jewish question emerged so clearly as today: The whole of Europe in alliance with Japan-led East Asia is battling against the English-American-Russian world power jointly led by Jewry. The nations unified with us recognize more and more that the Jewish question is a question of race, and that they therefore must find a solution like the one we initially introduced for Germany.[403]

This was open approval for the deportation of Jews from the Third Reich, and pled for its expansion to German-dominated Europe. In evaluating this statement it must be kept in mind that the mass murder of mentally ill and mentally disabled people in the gas chambers of the "Aktion T4" in 1940/41 was known to large sectors of the German population, and that information had leaked quickly about the massacres committed by the task groups of the security police and the SD in the occupied territories of Poland and the Soviet Union. When the systematic deportation of German Jews began in October 1941, a significant degree of self-deception was needed to accept the official version of "resettlement" and "work assignment in the East."

As late as 1944 Verschuer, as mentioned above, demanded a "new total solution of the Jewish problem," now that the "historical attempts at solution" – "absorption of the Jews," "seclusion of the Jews through the ghetto" and "the emancipation of Jewry" – had failed.[404] As to what the "total solution" looked like in the ghettos and extermination camps, there was hardly a scientist in the German Reich who had such profound information as Otmar von Verschuer. Yet, with his research on the development of a serological race test all the way into the final months of the war he made his contribution to this "total solution."

---

[402] Fischer to Rosenberg, 10/6/1944, quoted in Müller-Hill, Tödliche Wissenschaft, p. 80. At Fischer's suggestion Lothar Loeffler also took part in the organization of the congress.

[403] Verschuer, Erbarzt an der Jahreswende, p. 3.

[404] Verschuer, Leitfaden, 2nd edn., pp. 138 f., 137.

In addition, under Verschuer's directorship the KWI-A continued, albeit to a diminished degree, to contribute practical legwork on "Jewish policy" in the form of certificates of race and descent. In 1942/43, for instance, Verschuer and his staff members Schade, Grebe, Mengele, Fromme, Baader, and Liebau produced evaluations bringing in a total of 2,340 RM.[405] "As a special war service" the institute also provided "Certificates of Wehrmacht Members (racial descent, marriage permits)."[406]

That Verschuer used his activities as an evaluator to help those suffering racial persecution *methodically and systematically*, as a war legend claimed,[407] must be challenged on the basis of today's state of knowledge. It is indisputable that he delivered a judgement advantageous for the individual involved *in individual cases*. It is also obvious that the test subjects enjoyed his sympathy in these cases. Yet for the assertion that one of his closest friends, the Frankfurt pastor Otto Fricke, made in his denazification testimony of October 1945, that Verschuer had gone "to the limit of scientific credibility [...] in order to prevent people from fall victim to the methods of the National Socialist state,"[408] there is no believable source evidence.

Important in this context is a letter by Verschuer to Karl Diehl of February 12, 1942. The subject was the case of the "half-Jewish" physician Werner Wund (1911–1990), to whom approbation was denied in National Socialist Germany and who had found employment in May 1941 as an intern in the remote Eckardtsheim Branch Institute, one of the Von Bodelschwingh Bethel Institutes. His situation had become precarious in September 1941, when the Reich Ministry of the Interior had rescinded the employment permit it had initially granted for Wund. Thereupon Bethel endeavored to procure a certificate of exemption from the Reich Chamber of Physicians.[409] In this context a file on the "Wund case" must have made its way via the channels of the deaconry to the practicing Catholic Karl Diehl. Diehl had submitted the case to his friend Verschuer, requesting his assistance. However, in a letter of February 12, 1942 Verschuer expressed his regret that he could not undertake anything, "as the question of the racial descent is undisputed." "For such applications a race biology certification plays no role. For it is of no consequence whether or not the individual involved looks Jewish." Verschuer recommended a

---

[405] Lösch, Rasse, p. 407.

[406] Verschuer, Tätigkeitsbericht 1942/43, MPG Archive, Dept. I, Rep. 3, No. 20. The same formulation is included in Verschuer's Tätigkeitsbericht for 1943/44 (MPG Archive, Dept. I, Rep. 3, No. 22), only here he elaborates that this certificate concerned "racial descent, paternity evaluations, marriage permit."

[407] Cf. e.g. Adolf Butenandt/Max Hartmann/Wolfgang Heubner/Boris Rajewski, Denkschrift betr. Herrn Professor Dr. med. Otmar Frhr. v. Verschuer, September 1949, Archives of the University of Frankfurt/Main, Dept. 13, No. 347, pp. 473–485, here: p. 476; Dekan der Medizinischen Fakultät Frankfurt to Hessisches Staatsministerium für Kultus und Unterricht, 9/9/1949, ibid., pp. 414–422, here: p. 418.

[408] Fricke, Kirchliches Urteil über die Persönlichkeit und die wissenschaftliche Arbeit von Herrn Professor Dr. Freiherr v. Verschuer, 26/10/1945, ibid., pp. 427 f., here: p. 428.

[409] On this in detail: Schmuhl, Ärzte in der Anstalt Bethel, pp. 63–66.

"clemency plea to the Reich Chancellory," whereby he was skeptical about the success of such a petition from the outset. Interesting in this context is a passage of his letter in which Verschuer went into his role as an assessor:

> Only in those cases in which doubts exist as to the correctness of the blood descent am I consulted as an expert, and in many such cases I have been able to help the people involved decisively. Just recently, for instance, a physician from Stuttgart came to me, whose wife was hitherto supposed to be a full-blooded Jew. From her appearance alone doubts as to this descent were justified. The couple also had four children, who now were supposed to be taken out of school and who would be banned from all higher professions as "1st degree mixed-race." In this case I could supply evidence that the woman was not descended from her Jewish father, but had a German physician, since deceased, as her biological father. This just as an example of the cases in which my involvement can be successful.[410]

Here Verschuer was probably alluding to the case of the professor's wife Luise S., in which he had been consulted as an assessor. Verschuer's expert opinion had in fact been successful in declaring Mrs. S., who had been considered a "full Jew" until that time, to be a "half-Jew," by abnegating the biological paternity of her legal father. By no means did this close the case, however, for the husband of Mrs. S. fought for the recognition of his wife as "German-blooded" by questioning the biological maternity of her legal mother as well. The Race Policy Office, to which he addressed his petition, called in the race biologist Wolfgang Lehman from Strasbourg, who, as already mentioned, was a member of the "Dahlem circle." Lehman was to examine photographs to ascertain whether they yielded "indications for a Jewish descent" of Mrs. S.[411] Since he gathered from the files that Verschuer had already submitted an expert opinion, Lehmann turned to Verschuer first before delivering an opinion himself. The characteristic style of Lehmann's letter to his former colleague makes apparent that he was disposed to agree with the standpoint of Professor S., and that he proceeded from the assumption that Verschuer would agree as well.

The response was different than expected, however. While Verschuer allowed that Mrs. S. belonged to the cases "in which nobody would suspect a Jewish influence. As such one can concede to her husband that she appears to be a 'pure German woman' […]." But in his expert opinion at the time he had not been able to "lend support for the assumption that she was not the child of her mother. She received notification from the Reich Heritage Office that she was […][412] degree mixed race, and I believe," Verschuer added with slight irritation, "Professor S.[413] should be content with this."[414] He had already delivered a second expert opinion to this end in February 1943. Keep in mind: In June 1943 it would have been easy for Verschuer – based on her external appearance and seconded by a further full

[410] Verschuer to Diehl, 12/2/1942, MPG Archive, Dept. III, Rep. 86 A (Münster), No. 7.

[411] Lehmann to Verschuer, 23/6/1943, MPG Archive, Dept. III, Rep. 86 A (Münster), No. 5.

[412] Omission in the original. From a handwritten marginal it can be induced that Verschuer submitted this information later along with a copy of his expert opinion on July 2, 1943. It should read: "1st degree mixed race."

[413] Verschuer accidentally used Mrs. S.'s maiden name here.

[414] Verschuer to Lehmann, 26/6/1943, ibid.

professor for race biology and the Race Policy Office of the NSDAP – to relieve Mrs. S. of the stigma of being a "half-Jew," and her children "quarter-Jews" – had he only been willing, despite his scientific conviction, to depart from the result of his first expert opinion in the interest of this human being. This case confirms the judgement Hans-Peter Kröner made about Verschuer as an evaluator on the basis of a case of "race treason" from 1937:[415] Verschuer was neither one of those scientists who provided incorrect opinions knowingly and deliberately in order to save people, nor one of those who interpreted the race laws extensively to the disadvantage of their subjects. Verschuer was the type of the "correct, law-abiding but merciless evaluators."[416] Of course, Verschuer abetted the emergence of his legend by portraying to his friend Diehl his first expert opinion as emergency assistance for a subject suffering racial persecution, although he did *not* claim at this juncture to have *falsified* the findings of the paternity examination. Through this it was possible for the impression to emerge in Verschuer's circle of friends – and also among the affected[417] – that he used his position as an evaluator to help the persecuted. However, all cases documented by sources prove that Verschuer followed the exact letter of the law and that "scientificity" was the only criterion for his expert opinions.

4.4.7 "Generalplan Ost" and Wolfgang Abel's Research on Soviet Prisoners of War

With his appointment to "Reich Commissioner for the Fortification of German *Volkstum*" Himmler was entrusted with the "ethnic cleansing" of the occupied Eastern territories. Consequently he set a mighty population transfer in motion, whereby the settlement of German nationals in the conquered areas and the deportation of Poles and Jews from these spaces drove each other like cogs in a powerful machine. Yet the forced migration, which was effected starting in winter 1939/40, constituted only a fraction of the planned resettlement program, which was worked out by Himmler's accomplices between 1940 and 1942 and entitled the *Generalplan Ost* ("General Plan for the East").[418]

The original Generalplan Ost, which was reworked several times, has not survived. But through two written position papers from the pen of the head of the Race Division in Alfred Rosenberg's Reich Ministry for the Occupied Eastern Territories, Erhard Wetzel (1903–1975),[419] we know its contents down to the details. Within 20 years at least 10 million Germans were to be resettled in the East. The territories

[415] This case was mentioned for the first time by Müller-Hill, Tödliche Wissenschaft, p. 39. A detailed depiction in: Kröner, Von der Rassenhygiene zur Humangenetik, pp. 42–44. Cf. also Meyer, "Jüdische Mischlinge," p. 130.

[416] Kröner, Von der Rassenhygiene zur Humangenetik, p. 42.

[417] Cf. the "Denazification Certificate" printed as a facsimile in Koch, Humangenetik, p. 94.

[418] Cf. in general Heiber, Generalplan Ost; Rössler/Schleiermacher (eds.), Generalplan Ost; Aly/Heim, Vordenker, pp. 394–440; Madajczyk (ed.), Generalplan Ost.

[419] For a biography: Klee, Personenlexikon, p. 673.

slated for settlement were the occupied areas of Poland, the Baltic countries, Belarus, parts of Russia, Ukraine and Crimea. The population in these areas was estimated at about 45 million, including 5–6 million Jews, whose extermination Wetzel presupposed as a matter of course in his position paper of April 27, 1942. In total, 31 of the 45 million people who lived in the territory destined for German settlement were categorized as "racially undesirable." They were to starve or be expelled to Siberia. According to Generalplan Ost, 80–85% of the population of Poland, 64% of the population of western Ukraine and 75% of the population of Belarus was to disappear. The remainder was to be "Germanized" or to serve the German "master race" as "helot folk." As such, Generalplan Ost was the blueprint of a gigantic program of extermination, expulsion and enslavement. Three scientists from the KWI-A were involved directly or indirectly in elaborating the plan.

Eugen Fischer took part in a meeting in the Ministry of the East protocolled by Wetzel on February 4, 1942 "about the questions of Germanization, especially in the Baltic countries," in which a draft of Generalplan Ost, presumably worked out by Group III B of the RuSHA in late 1941 was discussed – by the way, side by side with his old nemesis Bruno K. Schultz, by now head of the Race Office in the RuSHA. According to the protocol, in this meeting Fischer gave one of the introductory position papers and spoke once during the discussion, when Wetzel asked the group to consider "whether through the industrialization of the Baltic region it might not be possible to scrap the racially undesirable sectors of the population," rather than forcibly deporting them to Siberia. With the formulation "scrapping through industrialization" Wetzel did *not* mean "extermination through labor." By way of explanation he added, namely: "If they [the 'racially undesirable' sectors of the population] were given suitable pay, in particular, if their cultural condition were to be raised, a drop in the birthrate would be expected." Wetzel thus set his hopes in the regularly observed drop in birthrates in the industrialized states as a means of making the sectors of the Baltic population that were not to be "Germanized" vanish in subsequent generations. In opposition Fischer expressed misgivings: The "better standard of living" could, contrary to Wetzel's expectation, "easily lead to a rise in birth rates." Wetzel admitted that Fischer's view was "correct to the extent that [...] those concerned are unmistakably anti-social."[420] In short the brief exchange between Fischer and Wetzel amounted to the scholar coming out *against* the representatives of the Ministry of the East and *with* the representatives of the SS for large-scale deportations from the Baltics to Siberia, and thus giving preference to a more radical variant of "ethnic cleansing." In spite of this, Fischer's consulting activity continued to enjoy high estimation in the Ministry for the East. He played a key role in Rosenberg's plans for the founding of a "Reich Headquarters for Research on the East." In a file note for Hitler of March 23, Rosenberg informed the Führer that he had "thought of" Fischer to fill the position at the Reich Headquarters, "as a representative personality for biological research and a leading member of the Kaiser Wilhelm Society."[421]

[420] Quoted in Heiber, Generalplan Ost, p. 295.

[421] Quoted in Müller-Hill, Tödliche Wissenschaft, p. 52.

In a detailed position paper on Generalplan Ost of April 27, 1942, Wetzel took reference to Fritz Lenz and Eugen Fischer in the section about "German Settlement Issues." This section concerned the question as to whether southern Ukraine and Crimea would come into question for German settlement because of the climate there. Lenz, as Wetzel reported, had "taken the standpoint that the climatic conditions in these regions were detrimental for the settlement of the Nordic-Phalian appointed race." In this Wetzel must have referred to Lenz's exposé submitted to the RuSHA in January 1940, *Bemerkungen zur Umsiedlung unter dem Gesichtspunkt der Rassenpflege* ("Remarks on Resettlement under the Aspect of the Care for the Race"). Lenz continued to concern himself intensively with the issues involved in "East settlement," and advised the SS physician Hellmuth Thieme (* 1917), who had been involved with the processing of marriage applications at the RuSHA since 1942, on his dissertation on the topic of "The Selection of New Peasants and Their Importance for a Race Hygenic Population Policy."[422] In December 1941, Eugen Fischer, too, expressed his opinion on the question of German settlements in southern Ukraine and on the Crimean peninsula. Wetzel cited him with the words "that a settlement of German people in these regions could only be considered if there was a conscious effort to create rich wooded regions all over and thus bring about a change in climate."[423]

In his exposé of April 27, 1942 Wetzel finally cited a third scholar from the institute in Dahlem: Wolfgang Abel. After being called up for military service, Abel had first seen action in the Luftwaffe, but after he was wounded he was transferred to the Department for Army Personnel Controlling as a consulting anthropologist. As mentioned above, in this capacity he had examined colonial soldiers held in war prison camps in occupied France in 1940. In winter 1941/42, accompanied by two army psychologists, he then visited various war prison camps in which soldiers of the Red Army were crowded together in close quarters.[424] On behalf of the Superior Command of the Wehrmacht, he subjected the Russian prisoners of war to crude anthropological evaluation. His findings, which he presented to a larger public in a lecture at the "East Conference of German Science" (*Osttagung*

---

[422] Thieme, Neubauernauslese. For a biography: Lösch, Rasse, p. 575. However, Lenz was disappointed with his dissertation, completed in 1943. Cf. ibid., pp. 378 f.

[423] Quoted in Heiber, Generalplan Ost, p. 322. Wetzel's comments on the "streaming in of alien, non-European blood into our *Volkskörper* (national body)" could also be traced back to research performed at the KWI-A: "The investigations performed by the Race Policy Office of the NSDAP have yielded the fact that interbreeding with Germans takes place continuously as a consequence of the foreigners residing in the territory of the German Reich. The German woman plays a role in this that is anything but honorable. At issue here are Chinese, Near Easterners, Indians and other kinds of foreigners, who engage themselves with German women today in a manner that has provoked great strife in the *Volk*. Hundreds of births of half-breed children have already been registered. Very numerous are the half-breed children of Chinese and Near Easterners. Here one can almost establish the rule of thumb that every foreigner leaves at least one illegitimate child here "(quoted in ibid., p. 323). This passage apparently incorporates the findings of the dissertation by Yun-kuei Tao about "Bastards of Chinese Men and European Women."

[424] Müller-Hill, Tödliche Wissenschaft, p. 141.

der deutschen Wissenschaft) flowed directly into Wetzel's exposé. According to Wetzel, Abel had reached the conclusion:

> [T]hat in the Russians much stronger Nordic race elements are present than had been presumed up to this time. In addition to these truly Nordic race elements, the great mass of which probably have been located in this region for some time now, especially in the northwestern areas of Russia, and which cannot be traced back solely to Germanic, especially Varangian immigrants, there are a predominant number of light-skinned, primitive Caucasian, more or less long-headed race types, who by no means fall under the 6 races of Günther's system,[425] and cannot be explained as Caucasian-Mongoloid hybrid forms either, but rather constitute the undoubtedly ancient Caucasian race forms that have yet to be described in detail. Also present among the Russians, primarily in the western regions, are eastern Baltic influences. However, these eastern Baltic race characteristics are by no means as strong as was previously presumed.

At the congress, Wetzel continued to relate, Abel had proposed two different "solutions" to this delicate race question, which showed how Abel's advantageous judgement about the racial composition of the Russian nation ultimately cut both ways:

> Either the eradication of the Russian nation, or alternatively, the Germanization of the portion of the Russian nation with Nordic characteristics. [...] This concerns not only the annihilation of the culture of the Muscovites [...]. Rather, it concerns the annihilation of Russian *Volkstum* itself, splitting it up. Only if the problems here are viewed consistently from the biological, especially race biological standpoint, and if, accordingly, German policy is established in the East, will we have the possibility of meeting the danger threatening us from the Russian *Volk*.

The "very serious comments by Abel," according to Wetzel, deserve "the very greatest attention." The "path of liquidating Russian *Volkstum*" suggested by Abel, however, aside from the fact that its "execution [was] hardly possible," was "out of the question for political and economic reasons."[426] However, the strategy that Wetzel himself developed in the following – fragmentation of the Russian population, "racial lixiviation of Russian *Volkstum*," the "singling out the Nordic clans present in the Russian nation and gradual Germanization,"[427] sinking of the Russian birthrates – was largely oriented to Abel's biologistic perspective. What is more: Because Abel in all seriousness posited the idea of physically exterminating many millions of people, he set a negative precedent against which all other proposed solutions, no matter how radical they were, seemed moderate.

Abel continued his anthropological examinations of Russian prisoners of war, intensifying his connection to the SS *Ahnenerbe* ("Ancestral Heritage Society") at the beginning of 1943 – presumably not least with a view to his own uncertain future prospects, as the call to Fischer's vacated professorial chair met with

---

[425] Hans F. K. Günther differentiated between the Nordic, Phalian, Mediterranean, Dinaric, Alpine and eastern Baltic races.

[426] Wetzel, Stellungnahme und Gedanken zum Generalplan Ost des Reichsführers SS, 27/4/1942, reprinted in: Heiber, Generalplan Ost, pp. 297–324, quotes: p. 313.

[427] Ibid., p. 315.

unexpected resistance.[428] Abel sought cover with the ornithologist and SS Sturmbahnführer (Major) Ernst Schäfer (1910–1992).[429] After three expeditions to Tibet – he had just returned from the last in August 1939, Schäfer had taken over the "Department for the Central Asian Research and Expeditions" of the Ahnenerbe society in 1940, which developed to become a "Reich Institute" of its own under his direction. The "Sven Hedin Institute for Central Asia and Expeditions," soon the largest department of the Ahnenerbe, with its own domicile in the medieval castle of Mittersill in Pinzgau, had been opened on January 16, 1943 on the occasion of the 470th anniversary of the University of Munich and the awarding of the honorary doctorate to Sven Hedin (1865–1952). One week later, on February 23, 1943, Abel, who had performed anthropological examinations of around 7,000 Soviet prisoners of war by this time, turned to Schäfer with a request for support – a clever move, as Schäfer had since encroached on the entire area of the natural sciences within the Ahnenerbe organization. Abel's concrete request was that the anthropologist and SS Hauptsturmführer (Captain) Bruno Beger (* 1911)[430] be assigned to him.[431] Beger had originally belonged to the RuSHA, then was transferred to Himmler's personal staff, took part in Schäfer's expedition to Tibet in 1938/39 as an anthropology student, entered Schäfer's Department for Central Asian Research and Expeditions in the Ahnenerbe in 1940, and took his doctorate in anthropology with Ludwig Ferdinand Clauss.

On March 8, 1943 Schäfer forwarded Abel's remarks to Himmler's personal assistant Rudolf Brandt (1909–1948),[432] with the request that he report them to the Reichsführer SS.[433] After intial skepticism, Himmler's staff received Abel's research plans quite positively. Not wanting to make a decision without consulting the directors of Ahnenerbe, however, the managing director of this organization, Wolfram Sievers (1905–1948) was called in.[434] Sievers first consulted with a number of staff members at the "Institute for Military Science Application Research" (*Institut für wehrwissenschaftliche Zweckforschung*) under his direction, which, founded in 1942, functioned like a "state within the state of the 'Ahnenerbe',"[435] which also approved Abel's research plans. On May 3, 1943 he wrote to Brandt that he held

---

[428] Lösch, Rasse, pp. 401 f. Even though Abel had been drafted into military service, he was still a departmental director at the KWI-A and maintained constant contact with the institute. As such, his examinations of Soviet prisoners of war and the demands and recommendations derived from these studies cannot be factored out of the KWI-A's responsibility, especially since Abel's anthropological studies in war prison camps were a direct continuation of his prewar work in the context of the institute.

[429] For a biography: Kater, "Ahnenerbe," pp. 79 f., 211–218; Klee, Personenlexikon, p. 523.

[430] For a biography: Kater, "Ahnenerbe," pp. 208–211; Klee, Personenlexikon, p. 36.

[431] Kater, "Ahnenerbe," p. 208.

[432] For a biography: Ebbinghaus/Dörner (eds.), Vernichten, p. 626; Klee, Personenlexikon, p. 71.

[433] Schäfer to Brandt, 8/3/1943, Cf. Lösch, Rasse, p. 402.

[434] For a biography: Ebbinghaus/Dörner (eds.), Vernichten, pp. 643 f.; Klee, Personenlexikon, p. 583.

[435] Kater, "Ahnenerbe," pp. 255–257, quote: p. 257.

> [T]he evaluation of the examination material to be very imporant in order to obtain reliable documents and then affect the execution of the measures. Therefore we should make the anthropologists slated for the Caucasus project, Dr. [Heinrich] Rübel, Dr. [Hans] Endres[436] and Dr. [Hans] Fleischhacker,[437] available until the analysis has been completed. If the Reichsführer-SS approves the application, however, then we must come to a precise agreement with Prof. Abel as to how long the designated anthropologists will have to be available for the evaluation.[438]

What was the "Caucaus project" mentioned here? On August 10, 1942, 2 days after the Wehrmacht had captured the oil fields of the Caucasus, Heinrich Himmler ordered the "Ahnenerbe" to prepare a scientific expedition led by Schäfer in order to explore the Caucasus under the aspects of botany, zoology, entomology, geophysics, and also anthropology. The planning for this *Unternehmen K* as Michael H. Kater establishes, "exceeded in scale everything that came before it."[439] With the defeat of Stalingrad the plan for an SS expedition to the Caucasus may have lost any basis in reality, but Unternehmen K was not abandoned for good until January 1944. Thus it was from the pool of scientists involved in this planned Caucasus expedition that three anthropologists were detached temporarily for Abel's project of an anthropological study of Russian prisoners of war.

In a further letter to Brandt of May 22, 1943 Sievers stated more precisely that the evaluation of the material from the study was "extraordinarily important, because labor is to be assigned, and also for demographic, economic and cultural reasons. […] However, Prof. Dr. Abel should be disposed to concentrate his work abovc all on the question of the individual groups' treatment and utilizability for labor in the war and to orient his work toward the solution of these questions."[440] As such, Abel's examinations were embedded in a new context. After the defeat at Stalingrad, the labor administration under the direction of the "General Deputy for the Employment of Labor," Fritz Sauckel (1894–1946), made every effort to effect the deployment of foreign forced laborers under the banner of "European Workers against Bolshevism."[441] Anthropological expertise was welcome in the attempt to

---

[436] The psychologist Hans Endres (* 1911) was employed by the RuSHA in 1942. Cf. Klee, Personenlexikon, p. 135.

[437] Hans Fleischhacker (* 1912) worked at the RuSHA in 1941 and served as an aptitude tester for the Germanization of Poles, especially in Łódž. Cf. Klee, Personenlexikon, p. 155.

[438] Sievers to Brandt, 3/5/1943, quoted in Lösch, Rasse, pp. 402f.

[439] Kater, "Ahnenerbe," p. 214. – Independent of this large-scale project, the Institute for Political Geography of the NS Ordensburg (official SS training center) "Falkenburg am Krössinsee" addressed the KWI-A in June 1942, requesting that it make available "for preparatory works for deployment in the future Reich Commissariat in the Caucausus […] data and material about population density, races, nations and religions in the Caucasian and central Russian areas, respectively." Institut für politische Erdkunde to KWI-A, 9/6/1942, MPG Archive, Dept. I, Rep. 1 A, No. 2400, p. 230.

[440] Sievers to Brandt, 22/5/1943, quoted in Lösch, Rasse, p. 403. Sievers had expressed himself quite similarly in a letter to Richard Korherr (* 1903), the Inspector for Statistics at the Reichsführer-SS Office, on 29/4/1943. Cf. Kater, "Ahnenerbe," p. 209.

[441] For background: Schmuhl, Arbeitsmarktpolitik, pp. 307–317.

differentiate from the giant army of "Eastern workers" individual "racially more valuable" groups, who were supposed to be spurred on to higher performance by offering them better living and working conditions, or so the apparent calculation of the Ahnenerbe. By way of precaution, Sievers had Rübel, Endres, and Fleischhacker exempted from the staff of Unternehmen K on the very same day.

Sievers further proposed in his letter to Brandt of May 22, 1943 that the three anthropologists to be detached to Abel could take care of an additional mission on this opportunity:

> Once access to the Auschwitz camp is possible again, these anthropologists could also perform the examination there for that collection of 150 persons of which you are familiar. Since at this time, as SS Obersturmbannführer (1st Lieutenant) [Adolf] Eichmann informed me, there is especially suitable material available, the time would be particularly opportune for this examination.[442]

In this Abel's project was linked with another one that had been pursued for quite some time by the Ahnenerbe: the erection of a "Jewish skeleton collection."[443] The first impetus for this project proceeded from Bruno Beger in December 1941. On the search for a scientist who was to take control over the setting up of the collection, the organization quickly hit on August Hirt (1898–1945),[444] who held the chair for anatomy at the newly founded "Reich University" in Strasbourg. From late 1941/early 1942 he was courted by Sievers, Brandt, and Himmler, so that he took over a primary role in the framework of the natural science research empire that was to emerge under the protectorate of the SS. On December 29, 1941 Brandt passed on to Sievers a generally formulated directive of Himmler's that Hirt should be "given the possibility to engage in experiments of every kind that could support his research work, using prisoners, professional criminals who will never be released anyway, and persons awaiting execution."[445] Initially the Jewish skeleton collection was an issue. Rather, the hope was to win over Hirt for the directorship of a planned Institute for Entomology. The anatomist had made a name for himself with his work in the fields of the sympathetic nervous system and intravital microscopy, and especially this latter area was to be used in the research of new possibilities for pest control. In the end, Hirt's criminal experiments with poison gas (mustard gas) on prisoners at the Natzweiler concentration camp emerged from these plans.[446] Yet back in January 1942 there was also talk of "anthropological studies"[447] Hirt was supposed to perform. Probably the Jewish skull collection was meant with this. In any case, a report by Hirt about his research fields, which

[442] Sievers to Brandt, 22/5/1943, quoted in Lösch, Rasse, p. 403.

[443] On the following: Mitscherlich/Mielke (eds.), Medizin, pp. 174–182; Kater, "Ahnenerbe," pp. 245–255; Klee, Auschwitz, pp. 356–391; Wojak, Das "irrende Gewissen"; Lang, Grab.

[444] For a biography also: Klee, Personenlexikon, p. 259.

[445] Brandt to Sievers, 29/12/1941, BArch. Berlin, BDC, Wi A-0494. Cf. Sievers to Hirt, 3/1/1942, BArch. Berlin, BDC, B-254.

[446] Cf. on this: Klee, Auschwitz, pp. 361–366; Schmaltz, Kampfstoff-Forschung, pp. 349–584.

[447] Quoted in Kater, "Ahnenerbe," p. 247.

Sievers forwarded to the Reichsführer SS on February 9, 1942, was appended by an exposé in which the plan for establishing the Jewish skull collection was explained in greater detail:

> Comprehensive skull collections exist for nearly all races and nations. Only of the Jews are there so few skulls available to science that their processing does not permit any certain results. The war in the East now offers us the opportunity to remedy this lack. In the Jewish-Bolshevist commissars, who embody a disgusting, but characteristic class of sub-humans, we have the possibility to acquire a concrete scientific document by securing their skulls.[448]

The plan was frustrated by the reality of the war. Hirt, Sievers, and Beger thus agreed to procure the material not from the front, but from a concentration camp – and then not just skulls, but entire skeletons. On November 2, 1942, in a secret letter to Brandt, Sievers wrote, "for certain anthropological examinations [...] 150 skeletons of prisoners (Jews) [were] required, which are supposed to be provided from the Auschwitz concentration camp."[449] The Head Office for Reich Security was instructed to issue a corresponding directive. Brandt forwarded this request to the SS Obersturmbannführer Adolf Eichmann (1906–1962), head of the Department for Jews (*Judenreferat*) IV B 4 in the Head Office for Reich Security.

As the letter from Sievers to Brandt of May 22, 1943 indicates, Eichmann had just sent word that "at present especially suitable material" was available in Auschwitz for the Jewish skeleton collection. On June 6, 1943 Bruno Beger arrived in Auschwitz, surveying technician Willi Gabel having been sent ahead. On June 10 Fleischhacker followed, temporarily detached from the RuSHA to the Ahnenerbe. By June 15 Beger had selected and, assisted by Fleischhacker, Gabel, and several prisoners, measured the victims. In total Beger had selected, as Sievers wrote Eichmann on June 21, "115 persons, of which 79 were Jewish men, 2 Polish men, 4 central Asian men, and 30 Jewish women."[450] The unfortunate were deported to the Natzweiler concentration camp in August 1943 and murdered there in a specially furnished gas chamber under Hirt's direction, and some parts of their bodies conserved, others preserved.

So how was Abel's project of anthropological examinations of Soviet prisoners interlocked with this complex of crimes? And how did it continue? At present these questions can be answered only in part due to the fragmentary sources available. What is clear is that Abel, armed with a research contract from the Reich Research Council,[451] continued working on his "race biological studies of Eastern nations." In September 1943, with Sievers' help, he managed to extend his "indispensable"

---

[448] Quoted in Mitscherlich/Mielke (eds.), Medizin, p. 174.

[449] Quoted in Kater, "Ahnenerbe," p. 249.

[450] Quoted in Mitscherlich/Mielke (eds.), Medizin, p. 175. Beger's own interest focused on the four "central Asian" prisoners. "Two Usbeks, 1 Usbekian-Tadjik mixed-race and 1 Chuvash from the Kasan region [were] measured and cast," Beger reported to his superior Schäfer on June 24, 1943. "In addition, just for our institute" (quoted in Kater, "Ahnenerbe," p. 251).

[451] Abel to Breuer, 13/12/1944, BArch. Koblenz, R 73/10.005. Cf. Müller-Hill, Tödliche Wissenschaft, p. 109 (note 74).

status. This was justified with the fact that it was absolutely necessary that the examinations of war prisoners be concluded, "since it is imperative that the race-biological selection and evaluation of the Great Russians be clarified for later deployment, for up to this point we knew almost nothing about them and were misguided by incorrect conceptions."[452] Sievers' intercession can be interpreted as an indication that Abel's research on Russian prisoners of war continued to be performed in cooperation with the Ahnenerbe. This fits in with the fact that in October 1943, Beger suggested continuing the anthropological studies begun in Auschwitz on the "Mongoloid" types among the Soviet prisoners of war "by taking advantage of the material handed to us by this war in the form of prisoners."[453] With the help of Schäfer and Sievers, in spring 1944 Beger succeeded in deploying the wounded anthropologist Rudolf Trojan (* 1917)[454] to various camps in order to measure "central Asian" prisoners of war.

Another question is whether and to what extent the SS was involved *directly* in Abel's anthropological examinations. In response to his letter of May 3, 1943, in which he suggested providing Abel with three assistants, the anthropologists Rübel, Endres, and Fleischhacker, Sievers received an answer on June 23, 1943. Brandt had presented the plan to Himmler and now imparted the decision of the Reichführer SS:

> One of the 3 anthropologists can be detached for the short term, for 3, 4 or 5 weeks, while instead of the other two suitable inmates of the Sachsenhausen concentration camp should assist. For this it would be necessary that Prof. Dr. Abel and one of the three SS Führer move out to Sachsenhausen for this period to take care of their work there [...].[455]

Whether Fleischhacker – for only he came in question under the circumstances, as Endres and Rübel were no longer available[456] – was actually dispatched to Abel's project after his assignment in Auschwitz, and whether Abel actually set up a base in the Sachsenhausen concentration camp, must remain an open question at the current state of knowledge. The assumption that Abel's examinations of Soviet prisoners of war took place in the Sachsenhausen camp is highly plausible. For one, Sachsenhausen was very conveniently located, not only near Berlin, but more importantly, not far from the KWI-A's External Department for Tuberculosis Research in Sommerfeld/Beetz. Secondly, a large number of Soviet prisoners of

---

[452] File note by Sievers about a meeting with Abel on 18/9/1943, 30/9/1943, BArch. Berlin, R 26 III/122. I thank Helmut Maier for the reference to this document. – In a letter of June 23, 1943 Sievers mentioned that Abel "occasionally is mobilized for collaboration by the 'Ahnenerbe'" (Sievers to Persönlicher Stab des Reichsführers-SS, 23/6/1943, IfZ, MA 287, p. 9443). Sievers was referring to a devastating expert opinion by Abel about abstruse proposals by the self-proclaimed breeding researcher Kurt F. König of the "Internal Office for Speciation Research" (*Eigenstelle für Artungsforschung*) (cf. ibid., pp. 9444–9448).

[453] Beger to Sievers, 25/10/1943, quoted in Kater, "Ahnenerbe," p. 211.

[454] For a biography: Klee, Personenlexikon, pp. 630f.

[455] Quoted in Lösch, Rasse, p. 403.

[456] Cf. ibid., p. 404.

war were held in Sachsenhausen, such that this would open up a further field of activity for Abel's ambitions. Since 1942 Abel had pursued the idea "of an instructive collection for the race history of Europe and the world, the development of race, population movements domestic and international, etc."[457] Should this idea have taken on shape over the course of the war, a portion of the material could have come from Sachsenhausen. From the testimony of witnesses we know that skulls, skeletons, and other body parts were sent from the Sachsenhausen concentration camp to universities and other anatomical institutes.[458] But specimens could also have come from Auschwitz – at least the former prison physician Miklos Nyiszli (1901–1956) mentions in his memoirs that Jewish skeletons were sent from Auschwitz to Berlin for a "race exhibition."[459]

4.4.8 The Kaiser Wilhelm Institute for Anthropology, Human Heredity and Eugenics and the Research Accompanying the Genocide of the Roma and Sinti

In the first years of the Third Reich, "Gypsy policy" for the most part remained in the trails blazed for it in the Weimar Republic.[460] The outlines of a new "Gypsy policy" began to emerge as individual Sinti and Roma were subjected to compulsory sterilization in accordance with the GzVeN. From fall 1935 on they also fell under the "blood protection" law, which enacted bans on marriage between "Germans" and "members of alien races" – besides Jews, as the commentators of the Nuremberg Race Laws emphasized expressly, this meant above all the Sinti and Roma – and also under the "marriage health law," which prohibited marriage for the "inferior," regardless of their ethnic heritage. This complex of laws signaled a shift in "Gypsy policy." Had the "Gypsy question" been conceived as a problem of regulatory policy up to that time, now it was reinterpreted, like the "Jewish question," as a "race problem." As such the Sinti and Roma found themselves doubly suppressed: Like the Jews they were stigmatized as an "alien race" in terms of race anthropology; as mentally ill and mentally disabled they were also considered to be "genetically inferior aliens to the community" in terms of race hygiene. The supposed

[457] Abel to Sievers, 6/6/1942, BArch. Berlin, BDC, Research Wolfgang Abel.

[458] Hrdlicka, Alltag, p. 107. – Prisoners with unusual physical attributes like bone deformations or abnormalities of the limbs were also murdered in Sachsenhausen in order to make their corpses available to science. Cf. e.g. Naujoks, Leben, p. 81. In these cases the KWI-A was one of the potential recipients. In this the specialist was Hans Grebe, who worked at the reception center of the KWI-A in the "Haus am See" in Beetz starting in September 1943, in the direct vicinity of the External Department for Tuberculosis Research and not far from the Sachsenhausen camp. Cf. Verschuer to Fischer, 20/9/1943, MPG Archive, Dept. III, Rep. 86 A (Münster), No. 9.

[459] Nyiszli, Jenseits, pp. 125–128. Cf. also Sachse/Massin, Forschung, pp. 27 f.

[460] On the preliminary history: Hehemann, "Bekämpfung des Zigeunerunwesens." For the basics of the following: Zimmermann, Rassenutopie, esp. pp. 125–152.

"anti-sociality" of the Sinti and Roma was interpreted to be the consequence of a genetic defect, which, in term, was traced back to the interbreeding of the "genuine Gypsy lineage" with the "German-blooded" lower classes.

The racist conception of National Socialist "Gypsy policy" necessitated the collaboration of scientific functionary elites. The scientific center to "combat the Gypsy nuisance" was the "Race Hygiene and Population Biology Research Office," which was founded in spring 1936 at the Reich Health Office in Berlin-Dahlem. It was headed by Robert Ritter (1901–1951),[461] who was chief physician in the Youth Department of the Tübingen University Psychiatric Clinic before turning to the research of "vagabond stock" and "Gypsy half-breeds" full time in 1934/35. From spring 1937 on, the research office dispatched "mobile working groups," which sought out Sinti and Roma at gathering places, in camps, prisons, and institutions, subjected them to anthropometric examination and interrogated them – even under the application of threats and violence – to ascertain their family backgrounds. This information was supplemented by genealogical material from church and civic registries, private and state archives, as well as communications from the police, the courts, community authorities, welfare institutions, prisons, and penitentiaries. The information was compiled into family tables at the "Gypsy Clan Archive" of the Research Office, which, in turn, served as the data source for the expert opinions produced by the Research Office. By March 1944 the Research Office produced almost 24,000 such expert opinions, in which the subjects were classified according to a sophisticated system as "Gypsies" or "Gypsy half-breeds" of various degrees. The staff of the Research Office was aware of the deportation of the German Sinti and Roma to the Auschwitz concentration and extermination camp in March 1943. In spite of this they continued to write their certified expert opinions, which constituted a decisive foundation for internment in Auschwitz.

The Research Office also took on consulting duties. It advised the offices of the criminal police on the application of the "Gypsy legislation," the Wehrmacht and the Reich Labor Service on physical inspections, the groupings of the NSDAP on the admission and expulsion of members, the school boards on school admissions, factory managers on hiring and labor offices on the provision of labor, rural authorities on the issuing of peddling licenses, mayors, National Socialist Welfare Offices in welfare questions, and tax offices regarding the granting of child subsidies. Above all, however, Ritter and his staff instructed medical officers and registry officials about how to behave when Sinti and Roma applied for banns and marriage loans. By the second half of the war, the Research Office also delivered recommendations for sterilizations and abortions among Sinti and Roma. Finally, by 1938 at the latest, Ritter intervened in the discussion about a "Reich Gypsy Law," but this legislation was never introduced.

Shortly after the founding of the Race Hygiene and Population Biology Research Office, close connections developed with the nearby KWI-A. Wolfgang Abel, who, as mentioned above, had undertaken a "study trip" to Romania in 1935/36 in order

---

[461] For a biography: Hohmann, Robert Ritter, esp. pp. 133–184; Zimmermann, Rassenutopie, pp. 127–130; Klee, Personenlexikon, pp. 499 f.

to examine the Roma living there with regard to "the question crossbreeding," established contact with Ritter in March 1937 in order to draw his attention to the supposed importance of fingerprints in differentiating between "purebred Gypsies" and "Gypsy mixed-breeds". At Abel's instigation, from this time forth the anthropological files of the Sinti and Roma collected by the "mobile working groups" of the Research Office also included fingerprints, which were registered by the police as a matter of routine.[462] At the beginning of World War II, two doctoral students and scientific staff members of the KWI-A moved to Ritter's Research Office: Adolf Würth,[463] as mentioned above, had earned his doctorate under Eugen Fischer in 1937 with a dissertation on the emergence of flexion creases on the human palm. Immediately thereafter, Würth, who had also been interested in the "Gypsy question" since 1931/32, started at Ritter's Research Office. Working independently on Ritter's behalf from the Criminal Police Office in Karlsruhe, in 1937/38 he examined of the Sinti and Roma in southern Germany. On September 17/18, 1937 he represented his boss at the annual conference of the German Society for Race Research in Tübingen. At the conclusion of his lecture, Würth expressed his conviction that the National Socialist state, just as it had "solved the Jewish question," would "also settle the Gypsy question in principle."[464] In 1939 Ritter's group of scientists was joined by Brigitte Richter (married: Hercher), who, as also mentioned above, had earned her Ph.D. in 1936 in the context of the the large-scale "German Race Science" project with a dissertation about the Upper Hessian villages of Burkhard and Kaulstoß. Until her departure in 1943 she was responsible for registering the descent of Sinti and Roma.[465] Gerhart Stein (1910–1971) also had connections to Ritter.[466] In 1936 he had approached Verschuer with the request that he serve as advisor for a dissertation on the "Gypsy question." In summer 1936 Stein performed the first anthropological examinations for his dissertation in Berlin, where around 600 Sinti and Roma had been arrested and sent to the newly established internment camp in Marzahn in preparation for the Olympic Games.[467] In late 1938 Stein submitted his dissertation, which was based on the anthropometric measurement of 247 Sinti and Roma that he had performed in summer 1937, predominantly in Berlin,[468] but in part in Frankfurt. Stein determined the blood groups of samples from 244 Sinti and Roma in the second half of 1937 in the laboratory of the Institute for Genetic Biology and Race Hygiene. In 1939, even before the dissertation was printed in 1941, Verschuer used Stein's study to extol the pioneering

---

[462] Hohmann, Robert Ritter, p. 221.

[463] On the following: ibid., pp. 275–280.

[464] Würth, Bemerkungen, p. 98.

[465] Lösch, Rasse, p. 572; Hohmann, Robert Ritter, p. 314.

[466] On the following, ibid., pp. 291–296; Sandner, Frankfurt. Auschwitz, pp. 184–196.

[467] Cf. Milton, Vorstufe; Brucker-Boroujerdi/Wippermann, "Zigeunerlager" Berlin-Marzahn.

[468] There is no concrete evidence that Stein was in contact with the KWI-A while performing research in Berlin in 1936/37, yet in view of the close collaboration between Fischer and Verschuer this may be presumed.

role of his own institute as opposed to Ritter's Research Office.[469] What Verschuer neglected to mention at this juncture: After he left the institute in Frankfurt, Stein had participated in one of Ritter's "mobile working groups" from January to April 1938. The entanglements of the various personnel show that there were close contacts between the KWI-A and the Reich Health Office's Research Office, even though Fischer and Verschuer had strong reservations about Ritter.

Under shady circumstances, Eugen Fischer helped Eva Justin (1909–1966),[470] Ritter's "right hand," obtain her Ph.D. in 1943. After training as a nurse, Justin had begun work as an intern in Ritter's Genetic Biology Laboratory at the University of Tübingen Clinic in 1934. In the Race Hygiene and Population Biology Research Office she effectively acted as Ritter's deputy. In 1937 she registered as a student at the University of Berlin, where she supposedly studied anthropology, genetic psychology, race hygiene, criminal biology, and ethnology – although she could not provide evidence of a methodical program of study when she registered for her doctorate in 1943. She had accepted a dissertation topic proposed by Kurt Gottschaldt, but then changed it without consulting Gottschaldt, and then on her own, so to speak, written her dissertation about "The Fates of Gypsy Children Raised as Aliens and their Progeny." Upon Fischer's recommendation, on the basis of this dissertation she was permitted to register for the doctorate with a major in anthropology and minors in ethnology and criminal biology. Fischer, Ritter, and the ethnologist Richard Thurnwald (1869–1954) passed the dissertation, which quite obviously did not meet basic scientific standards. The oral examination by Fischer, Abel, Thurnwald and Ritter took place on March 24, 1943 in Ritter's private residence. Why Fischer, Abel, and Thurnwald were willing to issue positive evaluations as an obvious favor to Ritter becomes clearer upon perusal of Justin's references: The documents include letters of recommendation from Hans Reiter, President of the Reich Health Office; Herbert Linden, Hitler's Reich Deputy for Sanatoriums and Hospitals, one of the key figures in the NS "euthanasia" program; as well as the ministry official Paul Werner (1900–1970),[471] Deputy Director of the Reich Criminal Police Department, responsible for "preventative crime-fighting," "Gypsies," "juvenile delinquency" as well as concentration camps for juveniles, and who had assisted the "euthanasia" planning staff by procuring medications for the murder of patients. Justin's dissertation picked up directly on the debates about the "limits of educability" underway since the 1920s in the area of corrective training. She subjected Sinti children, who were accommodated in the Catholic St. Josefspflege home in Mulfingen/Württemberg, because their parents were interned (most of them in the concentration camps at Buchenwald and Ravensbrück), to "psychological" tests. On May 9, 1944, 2 months after Justin's dissertation appeard in print, the 39 children were deported to Auschwitz-Birkenau – only four survived.[472]

---

[469] Verschuer, Vier Jahre Frankfurter Universitätsinstitut.

[470] On the following: Gilsenbach, Lolitschai; Hohmann, Robert Ritter, pp. 238–271.

[471] For a biography: Klee, Personenlexikon, p. 670.

[472] On this also: Meister, "Zigeunerkinder."

Also in 1943, Georg Wagner (* 1898)[473] submitted his dissertation about "Race Biology Observations on Gypsies and Gypsy Twins." The trained farmer had spent the years from 1923 to 1939 abroad, where he apparently worked as a correspondent for German newspapers and as a "nationalist political writer." In 1940 he began studying natural sciences at the University of Berlin. He must have joined up with Ritter's Research Office shortly thereafter, for the material upon which Wagner's dissertation was based had been collected in the framework of the total inventory of the Sinti and Roma in Germany and the occupied territories initiated by Ritter. As such, Wagner – like Eva Justin – was at the same time an employee of Ritter's Research Office and a doctoral student at the KWI-A. His doctoral research was advised by Verschuer, although Verschuer requested that Fischer step in as the official doctoral advisor for Wagner, whom he described as "a somewhat peculiar fellow."[474]

In the introduction to his dissertation Wagner proudly remarked "that for the examinations of the probands around 14,000 km had to be traveled, and over 100 locations of the old Reich and the protectorate had to be visited."[475] He had examined 209 persons and categorized them according to Ritter's classification system. He characterized the "pure Gypsies" as the descendants of the Aryans. Thus, Wagner was the right man for the SS Ahnenerbe, which was searching for a scientist to "research the Gypsy attributes derived from Aryans" in November 1943. On behalf of the Ahnenerbe and with the consent of Arthur Nebe (1894–1945), Head of the Reich Criminal Police Department until 1944, Wagner settled in Königsberg, intending to survey the "Gypsies" in Latvia, Estonia, Lithuania, and Finnland first, and to visit the "Gypsy settlements" in the Białystock district. Hence, Wagner, despite an unmistakable fondness for the "pure Gypsies," was party to creating the scientific foundations for the extermination of the Sinti and Roma. Wagner was still working on his research project in March/April 1945. On this Joachim S. Hohmann fittingly remarks, "Apparently Wagner was to merely record the evidence of life of an ethnic minority sentenced to extinction, before its genocide was completed. That he obstinately continued working on this just days before the end of the war is presumably one of the many paradoxes of the racist 'Third Reich'."[476]

Wagner drew the attention of twin researchers to the Sinti and Roma, as he had examined 74 "Gypsy twins" himself as part of his dissertation, and had reached the conclusion that twin births occur nearly twice as often among Sinti and Roma than in the remaining population.[477] Wagner reported to his colleague Karin Magnussen about "Gypsy twins" among whom he had noticed certain eye anomalies – Wagner's scientific curiosity brought these people directly to their death.

---

[473] On the following: Hohmann, Robert Ritter, pp. 281–286; Lösch, Rasse, pp. 383–387.

[474] Verschuer to Fischer, 9/3/1943, MPG Archive, Dept. III, Rep. 86 A (Münster), No. 9.

[475] Wagner, Beobachtungen, p. 1.

[476] Hohmann, Robert Ritter, p. 286.

[477] Wagner, Beobachtungen, pp. 56, 58, 61. Cf. Massin, Mengele, pp. 234f.

4.4.9 *Verschuer, Mengele, and the Interconnections Between Dahlem and Auschwitz*

Josef Mengele,[478] born in Günzburg, Bavaria in 1911 as son of an agricultural machinery manufacturer, studied medicine at the Universities of Munich and Bonn from 1929 to 1932. He passed the intermediate examinations to become a doctor in Bonn in 1932, before continuing his study of medicine, and now of anthropology as well, in Vienna and Munich. In 1935 he received his Ph.D. in Munich with a dissertation in the field of classical physical anthropology – on the morphology of the anterior section of the mandible in four racial groups.[479] Mengele's doctoral advisor Theodor Mollison, Director of the Anthropological Institute of the University of Munich, one of the joint editors of the *Archiv für Rassen- und Gesellschaftsbiologie* and Chairman of the Munich Society for Race Hygiene, was one of the leading race hygienists of the Third Reich. He advocated an interlocking of race hygiene and race anthropology. Accordingly, at the course for psychiatrists organized by the German Research Institute for Psychiatry in January 1934, he warned against the "invasion of races of Asian origin" (with this he meant both "the yellow race" and "the Jews")[480] to Europe. Mengele must have been just as well acquainted with his doctoral advisor's race policy positions as he was with Mollison's attempt to develop a serological race test using the "precipitine reaction" – more on this later.

In 1936 Mengele passed his state medical examinations and received his approbation as a physician in 1937. After a four-month internship at the University Clinic in Leipzig, in 1937 Mengele became an assistant at Verschuer's Frankfurt Institute for Genetic Biology and Race Hygiene. Here he took his second, medical Ph.D. with a dissertation about "family examinations in cases of cleft lip, cleft jaw, and cleft palate."[481] Mengele proceeded from a group of 17 children with cleft lip, cleft jaw and cleft palate, who had undergone surgery at the Surgical Clinic of the University of Frankfurt/Main between 1925 and 1935. For these 17 children Mengele produced "family tables" covering a total of 1,222 "clan members," 583 of whom Mengele visited personally. He had the remainder examined by their local health offices. The genetic evaluation of the genealogical material, as Mengele summarized his results, made it possible "to recognize an irregular, singly dominant heredity of the disposition, whereby the manifestation depends on other developmental disorders" – among others, Mengele mentioned serious defects of the limbs, the lack of a closed spine and closed cranial bones, "as well as feeble-mindedness and mental disorders."[482] In addition, Mengele established the frequent occurrence

---

[478] For a biography: Lifton, Ärzte, pp. 392–449; Zofka, Josef Mengele; Posner/Ware, Mengele; Roth, Normalität; idem., Josef Mengele; Kubica, Mengele; Völklein, Josef Mengele. On the "Auschwitz-Dahlem connection": Massin, Mengele; Trunk, Zweihundert Blutproben.

[479] Mengele, Untersuchungen des vorderen Unterkieferabschnittes.

[480] Mollison, Rassenkunde und Rassenhygiene, p. 44.

[481] Mengele, Sippenuntersuchungen.

[482] Ibid., pp. 23–25, 41 f., quotes: p. 42. Cf. also Massin, Mengele, pp. 209 f.

in the families he examined of rudimentary forms of clefts in the area of the lips, the jaw and the palate, which suggested strong variations in the manifestation of the gene. Mengele's work made an important contribution to the elucidation of the disputed question as to the heredity of cleft lip, cleft jaw, and cleft palate,[483] whereby the evidence of variations in manifestation fit in well with the recent findings of higher Mendelism. Moreover, the work was located in the area of arrested development malformations, in which a certain embryonic state of development remains intact, even when development ceases prematurely. Such arrested development malformations were of central interest under the aspect of phenogenetics, however – Mengele's later attachment to the KWI-A was due not only to his personal relationship with Verschuer, but also predisposed by his research emphasis.

The "cum laude"[484] dissertation met the scientific standards of the time and was published in 1939 in the renowned *Zeitschrift für menschliche Vererbungs- und Konstitutionslehre*. It immediately attracted considerable attention, not only on the national, but also on the international level, after Verschuer referred to Mengele's findings in his paper at the International Congress for Genetic Science in Edinburgh.[485] Well into the 1960s Mengele's dissertation was well-received internationally and considered to be the standard work on its topic.[486]

Until 1941 Mengele published several short articles and reviews in Verschuer's journal *Der Erbarzt*.[487] Interesting to note is that he worked not only in the field of hereditary defects, but also undertook an excursion into neurology: after the death of the assistant Ottwil Reichert in 1939, Mengele completed Reichert's genealogical study "On the Heritability of Thrombangitis obliterans"(today: thrombangiitis obliterans, Winiwarter-Buerger disease), which was oriented toward the question of the heritability of rheumatism.[488] Mengele also produced expert opinions in Frankfurt. Verschuer even entrusted him with the scientific evaluation of the comprehensive material that accrued in the "certificates of race and descent."[489] In one case this yielded a short genealogical study on the heredity of *Fistula auris congenita* (branchiogenic syrinx, a special form of the cervical syrinx).[490]

---

[483] Mengele Sippenuntersuchungen, pp. 18–20.

[484] Massin, Mengele, p. 219.

[485] Verschuer, Bemerkungen zur Genanalyse, p. 67.

[486] Seidelman, Mengele Medicus, p. 604.

[487] For example, Mengele, Tagung der Gesellschaft für physische Anthropologie; idem., Review: Lothar Stengel von Rutkowski, Grundzüge der Erbkunde und Rassenpflege; idem., Review: Gerhard Venzmer, Erbmasse und Krankheit; idem., Review: Gottfried Pressler, Untersuchungen über den Einfluß der Großstadt; idem., Review: Georg von Knorre, Vererbung angeborener Herzfehler.

[488] Reichert, Erbbedingtheit, p. 53 (note 1): "Completed and published by Dr. J. Mengele after the death of Ottwil Reichert." On the discussions about the naming of this disease in the Third Reich, cf. Schmuhl, Ärzte in der Westfälischen Diakonissenanstalt Sarepta, pp. 72f.

[489] By January 1941 Verschuer and his staff had produced 448 paternity opinions. Verschuer, Vaterschaftsgutachten, pp. 25f. Cf. also idem., Vaterschaftsbestimmung.

[490] Mengele, Vererbung der Ohrfisteln.

Whether Mengele was actually Verschuer's "pet pupil,"[491] as Hans Grebe asserted in the 1980s, remains to be seen. Certainly Verschuer saw in Mengele great promise for the future. It was Verschuer's suggestion that Mengele attend the International Congress for Anthropology and Ethnology in Copenhagen in 1938, and the International Congress for Genetics in Edinburgh in 1939[492] – and that in both cases Mengele was not able to participate was due to foreign exchange difficulties, but changed nothing about Verschuer's special esteem, which left no doubt that the young scientists included in his list of proposals were the only ones who came into question for him as future university instructors.[493] The judgement of Benoît Massin, that Mengele, had there been no war or had Germany not lost the war, in all probability would have made the leap to a professorial chair – like his associates in Frankfurt, Ferdinand Claußen, Heinrich Schade, and Hans Grebe – must be confirmed wholeheartedly.[494]

It was probably Mengele's tremendous ambition that led him into the temptation to take a shortcut against the background of World War II to drive his career forward more quickly and further than his associates, by unscrupulously taking advantage of the unfettered access opened up to him by the world of the National Socialist camps. His close connections to the SS constituted free admission to this world. From 1931 to 1934 Mengele was a member of the *Stahlhelm*; after this organization was subsumed by the SA he remained a member until October 1934. In 1937 he joined the NSDAP, in 1938, the SS. Called up to the Wehrmacht in 1940, Mengele volunteered for the Waffen-SS, where he was assigned to the Medical Corps Inspection Office. In November 1940 he was transferred to the RuSHA, where he worked in Department II of the Family Office, responsible for "care of genetic health" and "genetic health tests."[495] What his job was and where he was deployed has yet to be clarified conclusively. Presumably, for a time at least, he wrote expert opinions about the "Germanizationability" of "German national" resettlers at an office of the Reich Commissioner for the Fortification of German *Volkstum* in Posen.[496]

In late 1941/early 1942[497] Mengele was sent to the eastern front with the SS Division "Viking," after he had been promoted from SS-Untersturmführer (lieutenant)

---

[491] Quoted in Müller-Hill, Tödliche Wissenschaft, p. 158.

[492] Massin, Mengele, p. 221. Extensive documents in: BArch. Berlin, R 4901/3016.

[493] Massin, Mengele, p. 221.

[494] Ibid., pp. 221 f.

[495] Sanitätsinspektion der Waffen-SS to RuSHA, 5/11/1940, BArch. Berlin, BDC, SSO Mengele, p. 403.

[496] Such the portrayal by Zofka, Josef Mengele, p. 254, and Völklein, Josef Mengele, pp. 90 f., which was not documented, however. Heinemann, "Rasse," p. 626, states – here, too, without any proof – that Mengele's job at the RuSHA predominantly concerned the "genetic health examination" of candidates for the SS, but that he was detached to Posen for a short period to evaluate the racial status of "German nationals." Cf. Trunk, Zweihundert Blutproben, pp. 12 f., which, however, also believes that Mengele was temporarily employed in this function in Posen.

[497] The central file card in Mengele's personnel file as an SS officer lists January 1942. In the recommendation for promotion of 24/11/1941, however Mengele, was already designated as "physician SS Viking Division." BArch. Berlin, BDC, SSO Mengele, p. 395, 405.

to Obersturmführer (1st Lieutenant). During his military deployment he received the Iron Cross, 2nd and 1st class, the Eastern Campaign 41/42 Medal and the *Kriegsverwundetenkreuz* (Germany's Purple Heart), 2nd class with swords. In July 1942 – probably due to a wound – Mengele was transferred to the office of the "Reich Physician SS and Police" Ernst Grawitz in Berlin,[498] i.e. to the office responsible for oversight of the concentration camps and the human experiments performed there. It is questionable whether he actually reported for service there, however, for he apparently remained with the "Viking" division – perhaps he was posted to the "Viking" Division by the Reich Physician SS and Police. In any case he was still on the rolls of this unit as physician for the troops in October 1942 and recommended for a further promotion.[499] He also participated in the Battle of Stalingrad.[500] As proceeds from a letter by Verschuer to Fischer of January 25, 1943, Mengele did not return to Berlin until early 1943:

> A few days ago my assistant Mengele flew 2 days long from Salsk [a city east of Rostow on the Don] to Germany. He took part in all of the battles with the SS division Viking, was decorated with the Iron Cross and has been transferred to an office here in Berlin for the time being, so that he can also be active at the institute in addition to his duties there.[501]

In February 1943 Mengele was assigned to the SS infantry substitute batallion "East," which was stationed in Berlin. He used his time in Berlin – from late January to late May 1943 – to consolidate his relationship with his doctoral advisor Verschuer. As mentioned above, Verschuer already had the intention of bringing Mengele to Dahlem once he had established himself there. At the institute Mengele was apparently regarded as a guest scholar, although he did not sign a regular employment contract with the KWG. His name appeared on an internal list of birthdays, apparently as a matter of course.[502] As also mentioned above, Verschuer again entrusted him with expert opinions.[503] However, the official version was that Mengele was on combat leave from the University of Frankfurt/Main until the end of the war – the position as a regular assistant under Verschuer's successor Heinrich

---

[498] Personnel command of 17/7/1942 (signed by Siegfried Liebau), ibid., p. 406.

[499] Recommendation for promotion of 13/10/1942 by Battallion Commander Schäfer of SS Pioneer Dept. 5, a sub-division of the SS Division "Viking," ibid., p. 412.

[500] Mengele was not deployed *in* Stalingrad (as stated in Kröner, Von der Rassenhygiene zur Humangenetik, p. 53), but took part in the battles *around* Stalingrad.

[501] Verschuer to Fischer, 25/1/1943, MPG Archive, Dept. III, Rep. 86 A (Münster), No. 9. Verschuer continued: "He [Mengele] accounted very interestingly that the entire disaster came about to the right and left of Stalingrad through a collapse of the Romanian Army. The officers ran away and left their soldiers in the lurch. Thus in the end everything degenerated into chaotic flight. Several German tank divisions then stuck into this confusion, but they were forced to retreat in the face of immense Russian superiority." – In his response of February 2, 1943 Fischer remarked, "I am very happy for you that you have Mr. Mengele, at least part time." Fischer to Verschuer, 2/2/1943, ibid.

[502] Reprinted as a facsimile in Müller-Hill, Blut, p. 197.

[503] Verschuer to the KWG, 29/6/1943, MPG Archive, Dept. I, Rep. 1 A, No. 577.

Wilhelm Kranz was reserved for him.[504] Nevertheless one may presume that Mengele saw his future in Berlin. Verschuer certainly regarded him as a candidate for professorship, and it probably can be assumed that the two discussed possible topics for his postdoctoral dissertation during Mengele's stay in Berlin.

On May 24, 1943 Mengele, who had been promoted to captain of the reserves of the Waffen-SS shortly before, was transferred effective May 30, 1943 to the SS Main Economic and Administration Office, Group D III (Medical Care and Camp Hygiene for Concentration Camps) and sent to the Auschwitz concentration and extermination camp,[505] where he worked as Executive Camp Physician in the "Gypsy camp" (Section B II e Auschwitz-Birkenau).[506] Whether Mengele was assigned to Auschwitz through no fault of his own, or, as Verschuer claimed after the war, against his will,[507] or whether, on the contrary, he took steps himself to effect a transfer to Auschwitz, and whether Verschuer pulled some strings[508] – these questions cannot be answered conclusively based on today's state of knowledge. Ulrich Völklein argues that Mengele ended up in Auschwitz more or less by coincidence: The SS physician initially assigned to the "Gypsy camp" at Auschwitz-Birkenau, Benno Adolph (1912–1967), had fallen ill with scarlet fever in April 1943 and was unable to work until November 1943 – thus a short-term replacement was sought, and Mengele was available at the time. Völklein can support his argumentation with the fact that Mengele's transfer orders expressly noted "Reference: None."[509] This can be assessed as a certain indication that no written transactions existed. In other words: A voluntary enlistment by Mengele *in written form* was in all probability *not* submitted. But this was not absolutely necessary. It cannot be excluded that Mengele – possibly with Verschuer's support – contrived behind the scenes for a transfer to Auschwitz. This is the gist of Benoît Massin's argument, whereby he assigns a key role to Siegfried Liebau and even alleges that there was an "alliance between Verschuer and the SS" arranged by Liebau.[510] It is documented

---

[504] Cf. excerpts from the course catalog of the University of Frankfurt in Koch, Humangenetik, pp. 130f. – Koch believes it is possible that Mengele wanted to keep open his options for qualifying as a professor in Frankfurt (ibid., p. 133). There is no evidence of collaboration between Mengele and Kranz, however. It is much more probable that Mengele wanted to qualify under Verschuer.

[505] SS-Führungshauptamt to WVHA, 24/5/1943, BArch. Berlin, BDC, SSO Mengele, p. 409.

[506] From August to November 1944 Mengele also held the position of 1st Camp Physician of the Auschwitz II-Birkenau concentration camp. After the re-integration of the Auschwitz II camp into the main camp in November 1944 he became Executive Camp Physician in the men's hospital area B II f. Kubica, pp. 377f.; Lasik, p. 314.

[507] Verschuer, Stellungnahme zu den Angaben, die sich auf meine Person beziehen und in der "Neuen Zeitung" Nr. 35 vom 3.5.1946 unter der Rubrik "Kunst und Kultur in Kürze" in der Notiz "Vertriebene Wissenschaft" erschienen sind, Archives of the University of Frankfurt/Main, Dept. 13, No. 347, p. 178f.

[508] Thus, e.g. Zofka, Josef Mengele, p. 255; Posner/Ware, Mengele, p. 37; and quite emphatically: Massin, Mengele, pp. 224–233.

[509] Völklein, Josef Mengele, p. 92.

[510] Massin, Mengele, p. 228.

that Liebau, in his capacity as head of the personnel division in the Office of the Waffen-SS Medical Corps, signed the order of July 17, 1942 which provided for Mengele's transfer from the "Viking" division to the "Reich Physician SS and Police." Also documented is the fact that Liebau, at Verschuer's request, was detached to the KWI-A for specialized training from December 1942 to October 1943 and thus present there in the period when Mengele was a regular guest at the institute. Finally, it is also documented that Liebau spent the first half of 1943 there – *before* Mengele's transfer to Auschwitz – and brought with him photographs of a "Gypsy clan" with heterochromous eyes for Karin Magnussen. Massin finds support for his theory in a statement by Hans Münch (1911–2001), who was the Director of the SS Hygiene Institute in Auschwitz from 1943 to 1945 and worked closely with Mengele in this capacity. Münch, as he stressed later in an interview with Robert Jay Lifton, had the impression that Mengele had "requested his transfer to Auschwitz, apparently because of the great research possibilities."[511]

Münch further testified that Mengele had worked on a postdoctoral project in Auschwitz – and this claim, regardless of how Mengele ended up at Auschwitz, can arrogate a high degree of probability. In retrospect Münch described Mengele's mentality with the words, "it would be a sin, it would be crime … that it was irresponsible not to take advantage of the opportunity presented by twin research in Auschwitz. If they were going to be gased anyway … This comes around only once, this chance."[512] Regardless of whether Mengele caught wind of this chance on the basis of information from the office of the "Reich Physician SS and Police" and thus actively instigated his transfer to Auschwitz, or whether he did not recognize this chance until he reported for duty – it is clear that Mengele unscrupulously exploited the opportunities presented to him there. Before long he built up his own research empire. From among the prisoners, he recruited a group of medical specialists for pathology, pediatrics, gynecology, ophthamology, ear, nose and throat medicine, and dentistry, along with technical assistants, nurses, kindergarden and nursery-school teachers, and secretaries. Mengele's laboratory barracks in the "Gypsy camp" – after its liquidation the laboratory was moved to Block 15 in Section B II f of the camp – was directed by the internationally respected pediatrician Berthold Epstein (1890–1962) from the University of Prague, supported by Charles Sigismund Bendel from the University of Paris. For the analysis of blood, urine, feces, saliva and tissue, Mengele had the SS Hygiene Institute in Rajsko at his disposal.[513] But above all, the selection of new arrivals on the platform gave him unlimited possibilities to access humans completely devoid of rights and protection. From the endless stream of deportation trains he could single out any human "guinea pigs" he pleased – Jews, "Gypsies" and other "alien nationals," people with

[511] So Lifton paraphrased Münch's testimony. Lifton, Ärzte, p. 398.

[512] Quoted in ibid., p. 418 (original omissions). This also explains Mengele's obsession, who – in contrast to his colleagues – often came to the platform even when he was not on duty. Ibid., pp. 399–401.

[513] Cf. Kieta, Hygiene-Institut.

physical anomalies, entire families[514] and, best of all, twins. Mengele created a "twin camp" in Auschwitz, the sheer population of which exploded all dimensions previously known. The exact number of twin pairs that fell into his hands in Auschwitz is unknown – Massin estimates that at least 900 children went through Mengele's "twin camp."[515] Moreover, twin research under the conditions of the Auschwitz concentration and extermination camp presented the unique opportunity to supplement the clinical and anthropological examination of twins with the pathological examination of their corpses, as Mengele could murder, or have murdered, *both* twins at any time. Miklós Nyiszli already pointed out this circumstance:

> An event never before experienced in the history of medicine worldwide is realized here: Twins die at the same time, and there is the possibility of subjecting their corpses to an autopsy. Where in normal life is there the case, bordering on a miracle, that twins die at the same place at the same time? [...] A comparative autopsy is thus absolutely impossible under normal conditions. But in the Auschwitz camp there are several hundred pairs of twins, and their deaths, in turn, present several hundred opportunities![516]

Massin characterizes Mengele's research empire at Auschwitz as the "Auschwitz Branch Office"[517] of the KWI-A. I find this analysis problematic for two reasons. First, it suggests a formal institutional connection, which certainly did not exist in this form – Mengele's position in Auschwitz did not at all correspond to Diehl's position in Sommerfeld. Secondly, Massin's interpretation constructs all too great a dependence of Mengele's on Verschuer. Certainly: Mengele's interests in twin research, in chondrodysplasia, in physical defects and in eye anomalies were oriented toward the model of the Frankfurt Institute for Genetic Biology and Race Hygiene, according to which the institute in Dahlem was also reorganized from 1943 on. Even his interest in *Noma facies* (gangrenous stomatitis, water cancer), a rare deficiency disease caused by extreme hunger, which raged among the children in the "Gypsy camp," presumably had a genetic pathology background. In this case Mengele probably continued with his mentor Verschuer's research strategy of scrutinizing all kinds of forms of disorders – from cancer to tuberculosis, to diabetes, to diptheria, and pneumoconiosis – to see if they were hereditary. This orientation is ultimately not surprising. Mengele shaped his own research empire in accordance with the institutes at which he had worked before, but in Auschwitz he was his own master.

On the other hand it is indisputable that Mengele, at his outpost in the no-man's-land of the world of National Socialist camps, was interested in being integrated into the scientific community and sought contact and collaboration with other scientists and research institutions – consider, for instance, his pharmacological investigations for

[514] Massin, Mengele, p. 234, points out that Auschwitz, in contrast to all other concentration camps, had sections of the camp in which entire families were imprisoned together: the "Gypsy camp" (from February 1943 until late July 1944) with around 23,000 inmates and the "family camp" for the Jews from the Theresienstadt ghetto (from September 1943 to July 1944) with more than 18,000 inmates. This was an essential aspect for a scientist interested in "family research."

[515] Ibid., pp. 235 f.

[516] Nyiszli, Jenseits, p. 42. Cf. Massin, Mengele, pp. 210–217.

[517] Massin, Mengele, p. 236.

I.G. Farben.[518] His most important cooperation partners by far, however, were and remained Verschuer and his group of researchers in Dahlem (all the more so if the assumption is correct that Mengele intended to write his postdoctorate dissertation under Verschuer). In any case Mengele upheld contact with Verschuer from Auschwitz, and paid at least one visit to the institute in Dahlem during this time – in his memoirs based on his diary of the time, Gerhard Koch reports meeting Mengele sometime around July 1943 in the KWI-A library.[519] In at least two cases this contact resulted in concrete collaboration: In the first case, in 1943/44, Mengele delivered the heterochromous pairs of eyes belonging to several members of a Sinti family to Karin Magnussen, on the other, between October 1943 and März 1944, he joined in Verschuer's "Specific Proteins" project, providing his mentor with around 200 blood samples from persons of various races. As the prisoners' physician Miklós Nyiszli reported, Mengele was also interested in inmates with growth anomalies ("dwarfism" or "gigantism") or physical defects. According to Nyiszli, Mengele picked out such persons during the selections on the platform, and then had his assistants examine, kill and dissect them. Mengele ordered that some of the specimens obtained from these autopsies be sent to Dahlem:

> The scientifically interesting parts of the corpse are preserved, so that Dr. Mengele can take a look at them. I have to keep anything that could be of interest to the institute in Dahlem. These specimens then come into a package for the journey, and a special stamp sees to it that it is dispatched more quickly: 'Urgent, contents of strategic importance for the war.' During my stay at the crematorium I expedited countless packets of this kind to Berlin-Dahlem, in response to which extensive replies with scientific opinions or instructions came in. I put together a special dossier for the purpose of storing this correspondence. For the rare materials we sent, the institute almost always expressed its deepest thanks to Dr. Mengele.[520]

Elsewhere Nyiszli depicts the case of two Jews, father and son, who were deported to Auschwitz on a train from the Łódž ghetto and had piqued Mengele's scientific interest. The father suffered from scoliosis as a long-term consequence of rachitis, the son from hypomelia (a disorder that affects the development of the limbs). After a clinical examination, Mengele had them murdered and commanded that "the skeletons [must be] dissected and sent to the Anthropological Museum in Berlin'."[521] On the basis of this testimony, the authenticity of which is estimated to be very high, there is the suspicion that Mengele's deliveries to Dahlem took place on a large scale, and that not only eyes and blood, but also internal organs and skeletons found their way to the KWI-A. The most likely recipients are Hans Grebe and Wolfgang Abel: Grebe as a specialist for chondrodysplasia and physical defects,[522] Abel with his plan for an "instructive collection for the race history of Europe."

---

[518] On this, the letter by Wilhelm R. Mann, the director of I.G. Farben, to Verschuer, in: Koch, Menschen-Versuche, p. 179 (without annotation of its source).

[519] Koch, Humangenetik, p. 130.

[520] Nyiszli, Jenseits, pp. 45 f.

[521] Ibid., pp. 123–128, quote: p. 126.

[522] Müller-Hill, Blut, p. 205, writes, one can "presume" that the specimens ended up on Grebe's desk. It must be emphasized that this is merely a justified suspicion. Also along these lines: Klee, Auschwitz, p. 473.

In addition to Mengele, two other scientists connected with Verschuer's institute worked at Auschwitz. One of them was Siegfried Liebau, who performed research on "Gypsy" twins in Auschwitz during the period when he was detached to Dahlem for further training – as mentioned above, he may have been the one who initiated the contact to Auschwitz and arranged for Mengele's transfer there. The other was the SS-Obersturmführer and physician at the SS military hospital in Berlin-Lichterfelde, Erwin von Helmersen (1914–1949), who took his doctorate with Fritz Lenz in August 1943, with a dissertation on "The Descendants of an Armenian Family in a Village of the Bukovina Settled by Germans."[523] Helmersen had been listed as a doctoral student since 1942, and thus was connected with the KWI-A at the same time Mengele was spending time as guest scholar there. After receiving his Ph.D., followed by a short assignment in the Oranienburg concentration camp, Helmersen went to Auschwitz, where he worked as camp physician in the "Gypsy camp" in section B II e and in the prisoners' hospital B II f. Helmersen, who was also involved in medical experiments at Auschwitz, was thus one of Mengele's subordinates for a time.[524]

Consequently, a network of lines emerge connecting Dahlem and Auschwitz, which cannot yet be tracked down into its finest nooks and crannies on the basis of the contemporary state of research. Clear is that Magnussen received from Mengele a series of pairs of eyes for her "eye color" project, and Verschuer around 200 blood samples for his "specific proteins" project. It is also clear that the two "not only knew of the provenance of these specimens, but that, in this knowledge, they used their contact to Mengele in order to obtain these specimens."[525] What is not clear, however, is the question as to whether they knew under what circumstances and in what manner the specimens were extracted in Auschwitz. During interrogation by the American military authorities on May 13, 1947, Verschuer denied that he had known about the events in Auschwitz, but did admit to having heard the rumors abounding at the time. During a visit by Mengele, Verschuer testified further, he had asked Mengele "just to explain if there was actually anything true about these rumors." In response Mengele had reported "about factories located there, his camp hospital, the excellent harmony that existed between him and his patients." "He knew absolutely nothing about inhuman treatment or any other kinds of atrocities."[526] Whether Mengele completely denied the industrially mass murder perpetrated in Auschwitz, or simply let the matter rest with these sinister intimidations,[527] it is highly improbable that he confronted his collaborative partners at the KWI-A with the whole truth.

[523] Helmersen, Nachkommenschaft. Cf. also Lösch, Rasse, p. 379.

[524] Lasik, Personalbesetzung, pp. 307 f.

[525] Sachse/Massin, Forschung, pp. 24.

[526] Quoted in Kröner, Von der Rassenhygiene zur Humangenetik, p. 129.

[527] In an interview with Benno Müller-Hill, Verschuer's son Helmut recalls that his mother had told him about dining with Mengele in Berlin. In response to her inquiry, Mengele is supposed to have replied that his job in Auschwitz was "horrible," but that he could not talk about it. Müller-Hill, Tödliche Wissenschaft, p. 129.

However, it can also be assumed that they did not harry him to do so. They certainly knew enough to not *want to know* more. In general it can be said – with all due reservations – that only a few Germans knew *everything* about the "Final Solution," but likewise, only very few knew *nothing at all*. Those working at the KWI-A certainly did not know everything, but quite a bit indeed: "In hardly any other scientific institution in Germany," in the judgement of Carola Sachse and Benoît Massin fittingly, "could access to knowledge about the crimes in Auschwitz have been so easy [...]."[528] As discussed extensively elsewhere, the genocidal character of the "total solution of the Jewish question" must have been altogether clear to Fischer and Verschuer. Further, Fischer was familiar with the details of the Generalplan Ost, for which the extermination of the eastern European Jews was a prerequisite. Beyond this, the KWI-A enjoyed excellent connections to the RuSHA through Herbert Grohmann, Günther Brandt and Helmut Thieme. At least two scientists who worked at the KWI-A, Harry Suchalla and Christian Schnecke, knew about the crimes against Jews in occupied Łódź. Siegfried Liebau, Josef Mengele, and Erich von Helmersen amounted to three scientists with contact to the KWI-A who actually worked at Auschwitz. Wolfgang Abel, too, had indirect contacts at Auschwitz, and perhaps at the Sachsenhausen concentration camp as well, but in any case he had insight into the stalags for Soviet prisoners of war.

4.4.10 Karin Magnussen and the Project on Heterochromia

Karin Magnussen was born in Bremen in 1908.[529] Her mother was a sculptor, her father a teacher at the Bremen School for Applied Arts. After graduating from secondary school, in 1928 she began studying biology, chemistry, geology, and physics – still quite unusual for a woman at the time – at the University of Göttingen. In 1929 she spent two semesters at the University of Freiburg, where she was influenced above all by the lectures of the developmental physiologist Hans Spemann. In 1932, back in Göttingen, she took her doctoral examinations in the subjects zoology, botany, and geology. Her dissertation, entitled *Untersuchungen zur Entwicklungsphysiologie des Schmetterlingsflügels* ("Studies on the Developmental Physiology of the Butterfly Wing") was advised by Alfred Kühn and published in 1933. This work concerned artificially induced defects in the rudimentary origins of the organs in caterpillars and their effects on the patterns and pigmentation of the fully developed butterfly wing – the influence of Spemann and Kühn is obvious.[530]

Göttingen had been a stronghold of National Socialism of the first hour, and the student body in Göttingen was particularly involved in the earliest successes of the NSDAP in the university town.[531] Walter Groß, who had been involved in the very

528 Sachse/Massin, Forschung, p. 26.

529 For the basics on the following: Hesse, Augen, pp. 32–46; Klee, Medizin, pp. 357–371.

530 Magnussen, Untersuchungen zur Entwicklungsphysiologie.

531 Tollmien, Nationalsozialismus in Göttingen; Dahms, Universität Göttingen, p. 408.

founding of the Göttingen branch of the NSDAP back in 1922, one of the first in northern Germany, established a university group of the National Socialist League of German Students (*Nationalsozialistischer Deutscher Studentenbund*) in Göttingen in 1926/27. The students in Göttingen who were active in the party included Rudolf Mentzel – as District Leader of the NSDAP[532] –, who became president of the German Research Association in 1937; and Achim Gercke, the organizer of the "Archive for Race Science Statistics on Professions" (*Archiv für rassenkundliche Berufsstatistik*) who later became "Race Science Expert" (*Sachverständiger für Rasseforschung*) to the Reich Minister of the Interior before advancing to the top of the Reich Genealogical Office. Magnussen found admittance to this network during her days as a student in Göttingen. In 1931 she became a member of the NSDAP – this, too, highly unusal for a woman, as only very few women joined the party at this time[533] – and was active in the National Socialist League of German Students. Her associates later remember that Magnussen had attracted notice in Göttingen as a "fanatic National Socialist."[534] After the Nazis took power she resolutely pursued her party career. In 1934 she became a member of the BDM, for which she held training lectures about race and demographic issues in the district of Bremen. From 1935 on she was also an employee of the Race Policy Office of the NSDAP in the district of Hanover.

Magnussen apparently also had contact with the Bremen branch of the German Society for Race Hygiene under its chairman, the lecturer Hans Duncker (1881–1961) since the Weimar Republic.[535] In this framework, Magnussen may even have experienced the greats of Weimar Eugenics first hand – Fritz Lenz, Hermann Muckermann, Eugen Fischer, and Otmar von Verschuer – all of whom held lectures to the Bremen branch.[536]

There is no doubt that Karin Magnussen was a fervent National Socialist, race hygienist and anti-Semite. In 1936 – barely 27 years of age – she published her 150-page book *Rassen- und bevölkerungspolitisches Rüstzeug* ("The Munition of Race Policy and Population Policy"). The third edition appeared in 1943, now swollen to 230 pages. The tract, which propagated "population statistical and race statistical material" and discussed relevant "legislative measures," was conceived – as stated in the preface to the first edition – as an overview for "biology teachers and trainers for instruction in the graduating classes and for race biology training."[537] In 1943 the author designated the solution of the "Jewish question" as the "core racial problem in Europe":

> From the European standpoint the Jewish question is not solved by the circumstance that Jews emigrate from the racially thinking states to the other states. We see that these

---

[532] Popplow, Machtergreifung, p. 180.

[533] Only 5–8% of the new members of the NSDAP before 1933 were women. Cf. Falter, Hitlers Wähler, pp. 146 f.

[534] So Georg Melchers (1906–1997) in an interview with Müller-Hill, Tödliche Wissenschaft, p. 164.

[535] Nitschke, "Erbpolizei," esp. pp. 63 ff., 79, 89 ff.

[536] Also Walter Groß and the later "Gypsy researcher" Robert Ritter held lectures there.

[537] Quoted here in: Magnussen, Rüstzeug, 3rd edn. p. 5. Hesse, Augen, pp. 45 f., presumes that this was a "contracted work," arranged by the director of the Race Policy Office, Walter Groß.

> emigrants merely breed unrest and incite the *Völker* ("nations") against each other. [...] The race policy goal of this struggle of nations thus must be: The spatial separation of the European races and nations from all aliens (Jews, Gypsies, Negroes) [...].[538]

With her book Magnussen supplied an apparent basis of scientific legitimation to the National Socialists' gigantic deportation program, which reckoned with the decimation of the deported from the very outset.

In December 1933 Magnussen passed the state examinations for secondary-school education in the subjects biology, chemistry, and geology and began teaching. Her last position – from 1939 to 1941 – was as a secondary school teacher in Hanover, in line for a civil service post. On September 15, 1941 she began her work at the Department for Experimental Genetic Pathology at the KWI-A – initially as a scholarship student. None of the details of the circumstances of her hiring are known. Hans Hesse conjectures that she was hired "because of her old connections and early party membership." Walter Groß, whom she knew well from her days in Göttingen, had played an important role on the Board of Trustees of the KWI-A since 1935, as discussed previously. Hesse further assumes that the drafting of male employees to the Wehrmacht had created gaps in the scientific staff that were difficult to close, so that Eugen Fischer had been forced to appoint a "woman and not particularly established scientist."[539] This thesis can be concurred with only in part. That political protection played a role in Magnussen's appointment may well be true. That a woman was hired and entrusted with a research project central to the conception of the institute, however, was not as rare an exception as Hesse apparently assumes, even before 1939 – remember Rita Hauschild. What is least true of all is the assumption that Karin Magnussen was not sufficiently qualified for her post at the KWI-A. True: She had worked 8 years as a teacher, far removed from research. But for Eugen Fischer she was nevertheless a very interesting candidate – especially against the background of the paradigm shift to phenogenetics: The complex interplay of genes and environmental factors in the pigmentation of the iris constituted one of the central research fields in Fischer's conception of phenogenetics, and he oriented his focus above all on Alfred Kühn's and Adolf Butenandt's research on the flour moth *Ephestia.* A pupil of Kühn, Magnussen had worked on the influence of genes and pharmacologically effective agents on eye color, and after earning her Ph.D. in July 1932 with Butenandt as her advisor, worked on pituitary hormones.[540] According to later testimony by Magnussen, back in 1938 she was already researching the pigmentation of the eyes, and the phenomenon of heterochromia (the occurrence of two irises with different colors) in particular. In other words: Magnussen was well-versed in a research field that was of the greatest interest to Fischer in the course of restructuring his institute, and she had collaborated with the researchers to whom he had sought closer contact since 1938.

---

[538] Magnussen, Rüstzeug, 3rd edn., pp. 202f.

[539] Hesse, Augen, p. 46.

[540] Moreover, as a student in Freiburg she had also learned from Hans Spemann the techniques of producing, dyeing and embedding microtome cuts.

The new institute director Otmar von Verschuer also thought the world of Magnussen, and gave her an assistantship in November 1943, when she was unable to extend her leave from the school system.[541] From that point on, Magnussen was listed in Nachtsheim's official reports as a staff member of his Department for Experimental Genetic Pathology, but Nachtsheim neither went into her research in his annual report of 1943/44, nor did he include her works in the list of publications from his department[542] – Verschuer did this instead in his own report. It seems that Nachtsheim and Magnussen followed different paths in their research,[543] although Magnussen took up work in one of Nachtsheim's main areas of research – in fact, she started at the KWI-A shortly after Nachtsheim's eye research had begun stagnating as a consequence of his collaboration partner Hellmuth Gürich's being drafted. There are also numerous potential points of contact: both used rabbits as their animal model, both were interested in the pigmentation of the rabbit eye (Nachtsheim's "epileptic" Vienna Whites had blue eyes as a consequence of a pigment deficiency), both directed their attention to the effects of the aging process on genetic attributes. Yet their research projects, as far as we can tell, proceeded in parallel and without any connection: Nachtsheim worked on genetic pathology in the strictest sense, while Magnussen understood her work as a contribution to the phenogenetics of normal attributes. Nachtsheim stated after the war that he had not wanted to work with Magnussen because of her political views. He also claimed to know about her connections to Auschwitz, which was the "greatest shock"[544] he experienced during the Third Reich.

In three of Magnussen's progress reports of September 1943, March and October 1944, the contours of the research project "On the Investigation of the Heritability of the Development of Eye Color as the Basis for Examinations of Race and Descent"[545] became clear, which covered six different, clearly demarcated subareas:

First, in preparation for all other subareas, as it were, Magnussen dealt with methodological questions on the determination of the structure, color, and pigment distribution of the human iris. She published the results of this work in 1943 in *Der Erbarzt*.[546] This clarification of preliminary methodological questions pertained directly to application: "In the paternity expert opinions, new experiences are being collected constantly and already obtained experience evaluated."[547] In the very

[541] Verschuer to Fischer, 13/11/1943, MPG Archive, Dept. III, Rep. 86 A (Münster), No. 9.

[542] Cf. Unterlagen der Abteilung für experimentelle Erbpathologie für den Jahresbericht [1943/44], MPG Archive, Dept. I, Rep. 3, No. 22.

[543] As such: Schwerin, Experimentalisierung, pp. 270–273.

[544] Nachtsheim to L. Dunn, 14/2/1961, MPG Archive, Dept. I, Rep. 20 A, No. 22.

[545] Magnussen, Bericht über die Durchführung der Arbeiten zur Erforschung der Erbbedingtheit der Entwicklung der Augenfarbe als Grundlage für Rassen- und Abstammungsgutachten, 24/9/1943, BArch. Koblenz, R 73/15.342, p. 69.

[546] Magnussen, Bestimmung der Irisstruktur; idem., Bestimmung der Farbe.

[547] Magnussen, Bericht über die Durchführung der Arbeiten zur Erforschung der Erbbedingtheit der Entwicklung der Augenfarbe als Grundlage für Rassen- und Abstammungsgutachten, 13/3/1944, BArch. Koblenz, R 73/15.342, p. 68.

heading of her first research report Magnussen had emphasized that her research, as applied genetics, was of importance for the praxis of National Socialist race policy. It can be presumed that in Magnussen's case this was more than the rhetoric necessary to get a research application approved, and that she placed her research at the service of National Socialist race policy out of true conviction. But this was only one side of the coin: her research on eye pigmentation was also, and above all, conceived of as theoretical research, as an important building block of phenogenetics.

Second, Magnussen bred strains of rabbits with certain eye colors "for the purpose of determining the influence of certain hereditary dispositions on eye pigmentation." The animals were under constant observation so that the development of pigmentation could be followed over time. From the report of October 1944 it is clear that Magnussen was also busy with "breeding for the investigation of heterochromia."[548]

Third, Magnussen performed experiments on rabbits "to physiologically influence pigment development." This project was initially delayed by the war. "The series of experiments planned," Magnussen reported in September 1943, "could not be performed, since the compound required, which is manufactured in Hamburg, was lost in the terrible attack [...]."[549] In October 1944 Magnussen was able to report on the first results of these experiments:

> During the main breeding period in the summer months, several series of examinations on the physiology of pigment development were performed, in which the action of several hormones and pharmacologically effective substances on the development of pigment in the eyes of different races was studied. Here certain races whose pigment development showed certain similarities with that of humans, the influence of such substances was determined and thus the foundation laid for larger series of experiments in the coming year.[550]

It is no longer possible to reconstruct which substances were used in these series of experiments – possibly a conversation which Magnussen conducted with Adolf Butenandt on December 2, 1942 concerned the question as to which hormones should be applied in the animal experiments.[551] It proceeds from an essay fragment – which presumably originated in 1948 – that Magnussen dropped adrenaline into the eyes of several chinchilla rabbits in experiments performed privately from 1946 to 1948, as well as the extraneous substances physostigmine, atropine, and doryl.[552]

---

[548] Magnussen, Bericht über die Durchführung der Arbeiten zur Erforschung der Erbbedingtheit der Entwicklung der Augenfarbe als Grundlage für Rassen- und Abstammungsgutachten, 2/10/1944, ibid., p. 38.

[549] Magnussen, Bericht über die Durchführung der Arbeiten zur Erforschung der Erbbedingtheit der Entwicklung der Augenfarbe als Grundlage für Rassen- und Abstammungsgutachten, 24/9/1943, ibid., p. 69.

[550] Magnussen, Bericht über die Durchführung der Arbeiten zur Erforschung der Erbbedingtheit der Entwicklung der Augenfarbe als Grundlage für Rassen- und Abstammungsgutachten, 2/10/1944, ibid., p. 38.

[551] Massin, Mengele, p. 242 (note 142); Klee, Medizin, p. 370.

[552] Hesse, Augen, p. 96.

Fourth, Magnussen conducted series of experiments on the development of pigment in childhood at schools, combined with genealogical studies. Due to the war these serial studies rapidly became more difficult. In September 1943 Magnussen reported on this:

> In summer 1943 the studies about the pigment development in childhood and the required genetic inquiry among the families could be performed only on a smaller scale than previously, due to the drafting of fathers for military service and the evacuation of children. Yet the observations are still in progress and, as soon as the situation in the air permits, will be continued on a larger scale. [...] As the start of a larger series of observation series, serial examinations about iris structure and eye pigmentation were performed on over 1,300 children in Holstein.[553]

Half a year later she sounded less optimistic: "The remaining observations of humans had to be discontinued for a time for reasons concerned with the war, but are to be resumed in summer, to the extent possible."[554] In addition to the serial examinations in Holstein (Eutin), two further were performed in Baden (Wolfach) and Upper Bavaria (Mittenwald), and moreover "in Eutin and Mittenwald family studies to investigate the heredity of certain structural attributes [...] (especially important for opinions on descent)."[555] Further studies of schoolchildren planned "in several cities of the Reich"[556] had to be discontinued in September 1944, since they were not categorized as of strategic importance, and because "by no means [were] they to interfere with lessons."[557]

Magnussen's research report of October 1944 shows that, once the serial examinations of the German population had been disrupted, Magnussen intended to start a parallel study in the world of the National Socialist camps: "The first series of observations of alien nations in a prisoner of war camp, planned as a comparison with the German population, was prevented by enemy operations. Further series of

---

[553] Magnussen, Bericht über die Durchführung der Arbeiten zur Erforschung der Erbbedingtheit der Entwicklung der Augenfarbe als Grundlage für Rassen- und Abstammungsgutachten, 24/9/1943, BArch. Koblenz, R 73/15.342, p. 69.

[554] Magnussen, Bericht über die Durchführung der Arbeiten zur Erforschung der Erbbedingtheit der Entwicklung der Augenfarbe als Grundlage für Rassen- und Abstammungsgutachten, 13/3/1944, ibid., p. 68.

[555] Magnussen, Bericht über die Durchführung der Arbeiten zur Erforschung der Erbbedingtheit der Entwicklung der Augenfarbe als Grundlage für Rassen- und Abstammungsgutachten, 2/10/1944, ibid., p. 38. On this also, Verschuer to Fischer, 12/7/1943, MPG Archive, Dept. III, Rep. 86 A (Münster), No. 9: "Miss Magnussen was just here. She performed iris examinations of schoolchildren in Eutin (Schleswig-Holstein) and collected very interesting findings. It is necessary for her to examine populations in Germany of different racial composition in the same manner. Presumably she will best find the Alpine groups in the Black Forest. I am writing to my sister in Wolfach, which seems to me a suitable location. Do you perhaps have any other suggestions and connections to arrange Miss Magnussen's study? She could undertake the journey in either late August or October."

[556] Reichsministerium für Wissenschaft, Erziehung und Volksbildung to Reichsforschungsrat, 8/9/1944, BArch. Koblenz, R 73/15.342, p. 51.

[557] Graue to Reichsministerium für Wissenschaft, Erziehung und Volksbildung, 21/9/1944, ibid., p. 50.

studies of this kind are planned for the coming months."[558] At the current state of knowledge we have no more details about this first serial study in a war prison camp, for which there were already concrete plans, but which never took place – which camp was involved, who the prisoners were whose eyes were to be measured, in what manner and with whose help Magnussen intended to gain access to the camp. Neither do we know whether such studies ultimately did take place in other camps and, if so, what happened.

The fifth subarea of Magnussen's project was histological, in which she dissected the eyes of rabbits, and of humans as well, in series of microtome cuts. As proceeded from the progress report of September 1943, this area had high priority:

> At the moment, primarily the histological works are being performed, so that the irreplaceable material is processed and thus not subjected to the risk of loss due to long storage.[559]

From this emerged a paper "About the Relationships between Iris Color, Histological Pigment Distribution and the Pigmentation of the Bulbus in the Human Eye," which was completed in 1944, but not published until 1949.[560] For this study Magnussen used the eyes of "31 subjects from central Europe" and compared them with "specimens from the institute's collection, of the dissected eyes of colored races and of a Papuan eye."[561] Magnussen failed to make any mention of who those 31 people from central Europe were. Hans Hesse suspects that they could also have been concentration camp inmates. In procuring her material, Magnussen had few scruples: according to Benoît Massin's account, Magnussen also worked on the eyes of people who were murdered by the Gestapo and made available to the anatomist Hermann Stieve (1886–1952) in Berlin.[562] By the way, it is equally unclear where the dissected specimens of "races of color" contained in the institute's collections came from, which Magnussen studied comparatively.[563]

Sixth and finally, Magnussen was interested in anomalies of the eye, such as corneal conjunctivalization,[564] but above all heterochromia. In October 1944 she

---

[558] Magnussen, Bericht über die Durchführung der Arbeiten zur Erforschung der Erbbedingtheit der Entwicklung der Augenfarbe als Grundlage für Rassen- und Abstammungsgutachten, 2/10/1944, ibid., p. 39.

[559] Magnussen, Bericht über die Durchführung der Arbeiten zur Erforschung der Erbbedingtheit der Entwicklung der Augenfarbe als Grundlage für Rassen- und Abstammungsgutachten, ibid., p. 69.

[560] Magnussen, Beziehungen. Cf. also idem., Beitrag.

[561] Magnussen, Beziehung, p. 296.

[562] Massin, Mengele, p. 246.

[563] The "Papua eye" had been provided to Magnussen by Eugen Fischer.

[564] Magnussen had established corneal conjunctivization in several animals of one strain of rabbits from Nachtsheim's breeding experiments. At the same time, Georg Wagner, in his examinations of "Gypsy twins" apparently discovered in East Prussia two "clans" of "Gypsy half-breeds of predominantly Gypsy descent" in which this anomaly occurred with unusual frequency (Wagner, Partielle Irisfärbung, pp. 62, 64). Thereupon Magnussen systematically investigated a series of human eyes provided by Hermann Stieve for the occurrence of this anomaly and found several cases. She also found "during the systematic investigation of the eye specimens of colored races at the institute [...] a corresponding tissue fold in a Negro." She concluded that the conjunctivization "is propagated more widely

announced: "A paper about the histology of total heterochromia in humans is about to be concluded."[565] This work had become possible through one of the most monstrous medical crimes by Josef Mengele in the Auschwitz concentration and extermination camp.

In 1942 "a member of the institute's staff who worked on criminal biology issues"[566] – from another source it is clear that this meant Georg Wagner –, in one of the "mixed-breed Gypsy clans" he had examined,[567] which also included several pairs of twins, established an increased frequency of heterochromia, "in addition to other supplementary findings interesting in terms of genetic biology."[568] It is highly probable, as Hans Hesse convincingly portrayed, that the family in question was that of the Sinto Otto Mechau of Oldenburg.

Verschuer commissioned Magnussen with the task of taking on the "genetic biological analysis"[569] of this case of heterochromia. To the institute director, this Sinti clan must have seemed a rare godsend in several respects at once: first, as mentioned above, heterochromia represented an extremely interesting anomaly within the phenogenetic concept, which promised information about the way gene action chains worked in phenogenesis. Second, in this case – as a consequence of the complete inventory of Sinti and Roma aspired to by the "Reich Central Office of Gypsies" and the "Race Hygiene Research Office" – it was possible to create a complete family table, clarify the genealogical relationships of the "clan" and thus

---

and occurs in various races" (Magnussen, Hornhautüberwachsung, p. 62). On these two papers, cf. also Verschuer to Fischer, 19/11/1943, MPG Archive, Dept. III, Rep. 86 A (Münster), No. 9.

565 Magnussen, Bericht über die Durchführung der Arbeiten zur Erforschung der Erbbedingtheit der Entwicklung der Augenfarbe als Grundlage für Rassen- und Abstammungsgutachten, BArch. Koblenz, R 73/15.342, p. 38 f.

566 Protokoll der Vernehmung Karin Magnussens durch die Spruchkammer Bremen am 25.5.1949, MPG Archive, Dept. I, Rep. 3, No. 26.

567 Ibid.

568 Magnussen to Schwartz, 2/3/1949, quoted in Klee, Medizin, pp. 363. Cf. also Verschuer's assertion that Wagner had "left [him] 2 interesting twin cases from his material for utilization elsewhere." Verschuer to Fischer, 31/3/1943, MPG Archive, Dept. III, Rep. 86 A (Münster), No. 9.

569 That she researched independently, but on Verschuer's orders, that she kept him constantly informed about the progress of her work and could not publish her research results without Verschuer's approval, Magnussen left no doubt in her defense testimony after the end of the war: "Special works at scientific institutes are in the hands of specialists, who perform this work alone. – The directors of the institute have knowledge of the execution of these projects, which are published under the name of the institute and the director, who must give his approval for publication. – The directors have a direct stake in the performance of the research only when they sign as authors themselves." Protokoll der Vernehmung Karin Magnussens durch die Spruchkammer Bremen am 25/5/1949, MPG Archive, Dept. I, Rep. 3, No. 26. A letter by Magnussen to Viktor Schwartz, an associate of Alfred Kühn, of March 2, 1949 includes the passage: "The genetic biological analysis of the matter was thus [because she had worked on the phenomenon of heterchromia "for years" already] conferred to me by Prof. von Verschuer." Quoted in Klee, Medizin, p. 363. In her written statement of November 22, 1966 she again wrote: "Since he [Wagner] had neither time nor interest in this problem himself, the institute director delegated the scientific analysis of this matter to me, for it fell to my area of specialization (time: winter semester 1942/43?)."Quoted in Hesse, Augen, p. 92.

possibly obtain conclusions about the heredity of heterochromia. Third, because the Sinti and Roma were "locked up" in "Gypsy collection points," the test subjects were together at close quarters and – even more important – in the completely lawless area in which Sinti and Roma were now compelled to live, there were practically unlimited possibilities for access. The scientists could thus perform all examinations and collect all data they liked – even against the will of the affected. Fourth, the circumstance that this family included several pairs of twins with heterochromous eyes presented a truly unique opportunity to combine family and twin research in order to investigate the developmental physiological events in the emergence of heterochromia. For this, however, an imperative prerequisite was the histological examination of the heterochromous eyes of the twins – and that meant: the death of these children. Fifth, it must have seemed an advantage that the affected family was also the object of criminal biology research – as such, supposedly genetic physical stigmata like heterochromia or deaf-muteness potentially could be linked with supposedly genetic social deviance. Sixth, the circumstance that the family was classified in the group of "Gypsy half-breeds" built a bridge between phenogenetics and race research, and race diagnostics in particular. Thus it is no surprise that Verschuer strived to obtain additional financing from the German Research Association for this central research project.

Clear is that Magnussen performed eye examinations of members of the Mechau family before their deportation to Auschwitz in March 1943.[570] After the war she gave contradictory accounts about the exact course of events. In her interrogation by the Bremen Denazification Commission on May 25, 1949 she testified: "In spring 1943 I took my own photographs of the eyes of such twins at the institute in Dahlem, before the twins ended up at Auschwitz."[571] A short time before, on March 2, 1949, in contrast, Magnussen had written in a letter to Viktor Schwartz, an associate of Alfred Kühn's:

> The only thing I got to see of the entire clan was two young boys without an eye anomaly, for all of the clans were interned around that time, namely in Auschwitz. Since civilians were not admitted, any inspection of the people was made impossible during the period of their prophylactical internment. I had a "criminal" twin researcher, who was allowed to visit the camp in his capacity as a officer of medical corps, bring me back color photographs of a number of the people at the time, so that I had a small foundation, but it was very imprecise."[572]

The officer of the medical corps described here ironically as a "'criminal' twin researcher" was, as mentioned above, Siegfried Liebau.[573] A written statement by Magnussen of November 22, 1966, submitted in the course of the preliminary

---

[570] According to the *Gedenkbuch. Die Sinti und Roma im Konzentrationslager Auschwitz-Birkenau, München 1993*, the Mechau family was admitted to the Auschwitz extermination camp on March 14, 1943. Cf. Hesse, Augen, p. 22.

[571] Protokoll der Vernehmung Karin Magnussens durch die Spruchkammer Bremen am 25/5/1949, MPG Archive, Dept. I, Rep. 3, No. 26.

[572] Quoted in Klee, Medizin, pp. 363 f.

[573] Liebau confirmed this himself in later testimony. Cf. Klee, Auschwitz, p. 483.

proceedings of the Münster Public Prosecutor versus Otmar von Verschuer, states in contrast:

> At the time I had seen only one pair of twins from this clan, which had come to the institute in Dahlem for the doctoral student [Wagner]. I was able to photograph the eyes of both children on this occasion. At the time it was said that these twins (like a major portion of this clan already) were supposed to be sent to an internment camp. At the time I was told that antisocial clans were to be interned in this camp for the duration of the war as a preventative measure.[574]

Thus, it is probable that Magnussen herself photographed the (homochromous) eyes of one pair of twins from the Mechau family, which Georg Wagner had brought to the institute in Dahlem for examination before their deportation to Auschwitz in March 1943, and that she received additional photographs of the (heterochromous) eyes of members of the family in Auschwitz from Liebau.

Through the deportation, Magnussen had lost direct access to the Mechau family, since she could not be admitted to the "Gypsy camp" in Block B II e of the concentration and extermination camp at Auschwitz-Birkenau. Initially Liebau continued to help her. Then, effective May 30, 1943, Josef Mengele was assigned to Auschwitz. Magnussen related how the collaboration with Mengele came about in her denazification testimony of 1949:

> I met Dr. Mengele, who had been drafted as a Medical Corps officer, in Dahlem during the war, when he visited the institute while on leave. – I spoke with him a few times during such visits to the institute about scientific papers and scientific problems. [...] I performed my research tasks, despite the fact that any access to members of the heterochromia clan was barred after its internment in Auschwitz, and this was only possible with the help of Dr. Mengele, who was coincidentally detached to the camp as a physician. – At the time he made it possible for me to elucidate heredity by determining the eye colors and relationships between the members of the clan. – Through him I also learned that one of the most important families of the clan was contaminated with pulmonary tuberculosis. – Thereupon I requested that, if possible, he should send me the autopsy and the pathological eye material if anyone in this family died.[575]

At the same time, she expressed to Viktor Schwartz:

> Had I not heard from Prof. von Verschuer that a previous associate of his [Mengele] had been ordered to the camp as a physician, I would have been able to base the further genetic biology of the attribute only upon police files. Through this colleague, who recorded the attributes precisely and compiled the family tables, I learned that a part of the clan was contaminated with pulmonary tuberculosis, above all a family with 12 children.[576]

In her written testimony from 1966 Magnussen went into this point in greater detail:

> My demand to be able to visit the remaining members of the clan in the camp was rejected as impossible. At that time I was forced to conclude that women were strictly forbidden

[574] Quoted in Hesse, Augen, p. 92.

[575] Protokoll der Vernehmung Karin Magnussens durch die Spruchkammer Bremen am 25/5/1949, MPG Archive, Dept. I, Rep. 3, No. 26.

[576] Quoted in Klee, Medizin, p. 364.

from setting foot in the camp. Prof. von Verschuer thus referred me to Dr. Mengele, who had been his assistant in Frankfurt, and who had been ordered to the camp as a physician (officer of the Medical Corps). I did not know Dr. M. at the time; I only knew from the literature that he worked in the field of genetics. During a visit to the institute in Dahlem (summer 1943?) I made his acquaintance and discussed with him the possibility of performing the research task. I asked him first of all to make sure that this clan not be lost track of after the war, so that the research on this very rare mutation could then be continued intensively. The work of the geneticist is entirely dependent on the life of the carriers of the gene, who hand their genes down to the next generation. At the time Dr. M. told me that a particularly important family was subject to tuberculous contamination (with 12 children). Everyone knew what that meant in those days, when there was still no remedy for pulmonary tuberculosis, especially for young people under 21 years of age. Thus I asked Dr. M., whenever any of the carriers of heterochromia should die, to send me protocols of the autopsy and the pathological eye material for examination if possible, – just as I would have in any other case.[577]

Magnussen's account cannot be confirmed, supplemented, or contradicted by other sources. As far as the framework of facts is concerned, it appears to be coherent and fits in logically with the reconstruction of the project on heterochromia: By this account Magnussen, when the project slowed down as a consequence of the deportation of the Mechau family, was alerted to Mengele by Verschuer. Had Mengele made efforts of his own to be ordered to Auschwitz, and had Verschuer known of these or even actively encouraged Mengele's assignment to Auschwitz, the tip to Magnussen might have been issued *before* the posting was ordered on May 24, 1943; on the other hand, had Verschuer been surprised by Mengele's transfer he might have informed Magnussen *immediately after* the command was issued. In either case it is conceivable that Magnussen, even *before* Menegele's departure from Berlin to Auschwitz, reached an agreement with him that he would compile the "family table" of the Mechau family on location in the "Gypsy camp" and – presumably using the eye-color table developed by Magnussen – determine the eye color of the family members. Otherwise such a deal must have been made during one of Mengele's visits to Dahlem on leave. At this or a further meeting then, the arrangement must have been made concerning the family members supposedly suffering from pulmonary tuberculosis.

With some degree of certainty it can be presumed that Magnussen rendered the events by and large correctly in her postwar testimony. Her *interpretation* of what happened, however, can be scrutinized with a critical view to the sources. In her defense after World War II she made every effort to portray her arrangement with Mengele as entirely normal cooperation among colleagues. In 1949, she stated on record to the Bremen Denazification Commission:

I would have directed the same request as a matter of course to the hospital, the treating physician or the responsible pathologist, had I learned of a critical illness of a non-interned person from this clan. Naturally, I could not dispense with the evaluation of such valuable pathological material, which I would have obtained in any case under the circumstances.

---

[577] Quoted in Hesse, Augen, pp. 92f.

In her letter to Schwartz, Magnussen vested this argument in the form of a rhetorical question:

> In cases of death should I thus dispense with the histological analysis of the unique, abnormal material just because the people happened to die in the camp? Otherwise I could have arranged to get the material from the responsible pathological institute.[578]

In her statement from 1966 she added that in 1942 she had also received from the Charité Hospital the heterochromous eyes of a deceased patient.[579] The linchpin of this argumentation is Magnussen's assertion that she did not know that Auschwitz was an *extermination* camp. In her testimony to the Denazification Commission she claimed to have been unsuspecting:

> According to the impression I had of the case histories and of the extremely responsible and humanly decent attitude of Dr. Mengele toward his interned patients and staff (after the war, he hoped to win over for the institute a certain female Polish physician interned in Auschwitz, as he told me), the idea would never have entered my mind that anything could occur at the Auschwitz camp that might violate state, medical or human laws.[580]

To Schwartz, too, Magnussen asserted that "nothing awful [was] known" to her; "on the contrary – from the case histories, the colleague's stories and from his human attitude to the inmates I could only have the impression of proper and decent treatment."[581] Nevertheless, it seems more than improbable that Karin Magnussen, a scientist at an institute whose leading members had been involved in the discussions about the *Generalplan Ost* and who made no secret of their attitude toward the "Final Solution" of the "Jewish question" and the "Gypsy question," and, moreover, an active National Socialist with close contacts to the Race Policy Office of the NSDAP, had no knowledge at all of the genocide of Jews, Sinti and Roma and no conception at all of the conditions in a concentration camp; accordingly, this must be dismissed as nothing more than an attempt to rationalize her behavior. Rather, it can be assumed that it was altogether clear to Magnussen that the Sinti and Roma had hardly a chance of survival in Birkenau, and that this came in handy due to her interest in the eye specimens.[582] Mengele's indication that a family was "contaminated" by tuberculosis could even be interpreted as a discreet offer to assist in their demise, and Magnussen's request to send her the specimens as consent.

---

[578] Quoted in Klee, Medizin, p. 364.

[579] Hesse, Augen, p. 93.

[580] Protokoll der Vernehmung Karin Magnussens durch die Spruchkammer Bremen am 25/5/1949, MPG Archive, Dept. I, Rep. 3, No. 26. At this juncture Magnussen also claimed that she did not know that other categories of prisoners besides "Gypsies" and "mixed-race Gypsies" were held at Auschwitz.

[581] Quoted in Klee, Medizin, p. 365. Magnussen went so far as to assert that even former prisoners from Auschwitz had "nothing awful" to report.

[582] The following testimony in her defense from 1949 must be relativized in this respect: "My work about the genetic biology of heterochromia was not performed on this clan *because* it was in a concentration camp, but rather *even though* it was interned in a camp, which made my work extraordinarily difficult, and almost impossible, had not a scientist coincidentally been assigned as a physican there. (Protokoll der Vernehmung Karin Magnussens durch die Spruchkammer Bremen am 25/5/1949, MPG Archive, Dept. I, Rep. 3, No. 26, original emphases).

Magnussen herself admitted that such deliveries took place. In her Denazification Commission testimony of 1949 she stated:

> Of the twelve children from the one family, initially two died in the years 1943 and 1944 (one of erysipelas and one of a pulmonary TB on both sides). A child from a related family died of pulmonary TB as well. The case histories and autopsy reports I received were painstakingly recorded.[583]

At this juncture the testimony of two surviving prison physicians can be drawn upon for further information. In his memoirs, Miklós Nyiszli reports that on several occasions he was required to dissect the corpses of "Gypsy twins." Once the corpses were of four pairs of twins, that is, eight children who were under 10 years of age. Nyiszli established the cause of death to be injection of chloroform to the heart – therefore the children were murdered systematically. Nyizli related the further events as follows:

> Of the four pairs of twins, three have eyes of different colors. One is blue, the other brown [...]. I dissect the eyes out and lay them, each separately, in formalin, whereby I note precisely all information about them, so that they cannot be mixed up. [...] In two pairs I also find an active pulmonary tuberculosis. [...] In the afternoon Dr. Mengele makes his rounds of the ward. [...] He is very interested in the heterochromia of the eyes [...]. Right away he instructs me to prepare the entire material for dispatch, along with the protocols, but I should also list the causes of death. He leaves it up to me what I write, but the causes of death must be different. Almost by way of apology he says that these children, as I could see myself, suffered from syphilis or tuberculosis and would not have been able to survive anyway, so ... He does not say anything more.[584]

This account is confirmed in its entirety by testimony from the SS commander Erich Mussfeld.[585] Whether the children's corpses autopsied by Nyizli were members of the Mechau family must remain an open question. The children of the Mechau family may also have been autopsied by the Jewish Romanian prison physician Iancu Vexler, who worked in the "Gypsy family camp" from June 2, 1943. At least, this is what Hermann Langbein, himself a prisoner at Auschwitz reports:

---

[583] Ibid. Magnussen continued: "Before Prof. von Verschuer was relocated (early February 1945) I gave them [case histories and autopsy reports] back to his head nurse [Emmi Nierhaus], after I had completed the analysis of the histological material." By contrast, Magnussen wrote in her written testimony of 1966: "Extensive case histories and dissection protocols were sent to Dahlem with the eye specimens for inspection. After processing they had to be returned, as case histories are the property of the hospital or the treating physician as a matter of principle." Quoted in Hesse, Augen, p. 93.

[584] Nyiszli, Jenseits, pp. 46 f. (original omission). The original Hungarian edition appeared in 1946 – that makes this report so valuable, for it was created *before* Magnussen's and Verschuer's defense testimony. – If Nyiszli (pp. 44–46) gave the "Race Biology and Anthropological Institute in Berlin-Dahlem" as the address for dispatch, this could – if it is not simply a mistaken memory – have been because one of the possible addressees, Wolfgang Abel, was both Director of the Department for Race Science at the KWI-A and, since 1943, Director of the Institute for Race Biology at the University of Berlin. Abel used letterhead with the address: "Institut für Rassenbiologie der Universität Berlin, Direktor: Professor Dr. W. Abel, z. Zt. Bln.-Dahlem, Ihnestr. 22 (Kais. Wilh. Inst. f. Anthropologie)." Cf. Sachse/Massin, Forschung, p. 24, note 49.

[585] Reprinted in: Hesse, Augen, p. 21.

> Vexler also reports about Mengele's laboratory in the sauna block of B II e, in which anthropological examinations took place, especially on hair and eye color. [...] Once a Gypsy family by the name of Mechau with striking heterochromia of the eyes was brought in. Mengele drew Vexler's attention to the family and ordered him: 'Well, good, when it's time – when that happens, understand? – you will carefully take out the eyes and put them in bottles prepared for you. They will go to Berlin for the investigation of the iris pigment. You understand, genetic issue, heredity dominant, recessive, etc. highly interesting.[586]

The parallel account suggests that *several* prison physicians were occupied with the autopsies of (twin) children with heterochromous eyes from the "Gypsy family camp" at Auschwitz-Birkenau – and there are additional indications that point in this direction.[587] In any case the reports from the prison physicians confirm the suspicion that Mengele killed many more twin children from the "Gypsy family camp" because of their eyes, and delivered many more eyes to the KWI-A than Magnussen testified in her defense after World War II – in 1949 she spoke of three, in 1966 of five pairs of eyes that she received from Mengele.[588] Nyiszli's report further suggests that several heterochromous pairs of eyes from twins ended up in Dahlem *simultaneously*. By this time at the latest it must have been clear to Magnussen that the children had to have died a violent death. Benoît Massin is justified to establish:

> The case of the three heterochromous pairs of eyes from twins, which died more or less on the same day, must have been conspicuous and surprising. The very case that two twin siblings die "a natural death" on the same day and in the same place is a statistical rarity. Moreover, twins with heterochromous eyes are extremely seldom. But the death of six twin children with heterochromous eyes on the very same day or in the very same week is well outside the bounds of statistical probability and clearly points to a crime.[589]

The above-mentioned publication of her research results about the "Heredity and Histology of a Total Heterochromia of the Iris in Humans" failed in late 1944/early 1945, because from the article it was clear – at least, according to testimony by Georg Melcher (1906–1997) of the KWI for Biology and coeditor of the *Zeitschrift für induktive Abstammungs- und Vererbungslehre* at the time, in an interview with Benno Müller-Hill in the early 1980s – that all subjects died at the same time and thus it stood to reason to suspect that they had fallen victim to a crime.[590] In 1953 Karin Magnussen herself claimed that her essay did not appear because the printing plates, ready to go to press at the time, were destroyed in an air raid[591] – this was probably much closer to the truth. Then, after World War II, Magnussen's persistent attempts to place the manuscript after all – perhaps in the *Zeitschrift für*

---

[586] Quoted in Münzel/Streck (eds.), Kumpania, p. 123 (note 13). Cf. also Massin, Mengele, p. 240.

[587] According to Ernst Klee, the mother of the Hungarian composer György Ligeti also performed such autopsies. Klee, Medizin, p. 360.

[588] Benoît Massin must be confirmed (Mengele, p. 243), in that Magnussen, in view of possible fluctuations in manifestation, had to have been interested in the non-heterochromous eyes of the members of the Mechau family murdered at Auschwitz as well.

[589] Ibid., pp. 244f.

[590] Müller-Hill, Tödliche Wissenschaft, p. 164.

[591] Magnussen to Fischer, 17/5/1952, reprinted in part in Klee, Medizin, p. 367.

Morphologie und Anthropology, which was then edited by Hans Weinert – failed due to misgivings about the origin of the material investigated.

In Auschwitz, Mengele also performed experiments "about the possibility of a change in iris color," by dropping substances into the eyes of numerous children. The consequences of these experiments ranged from slight itchiness to swelling, inflammations, and suppurations of the eyes, in some cases the subjects lost their eyesight. Mengele even performed such experiments on newborn infants – with fatal results. According to the prison physician Ella Lingens (1908–2002), in 1944 a newborn died after Mengele injected a substance into its eye "because he was attempting to induce a change in eye color. Little Dagmar was to get blue eyes."[592] In 1944 Irmgard Ludwig had her newborn child taken away from her. When she saw it again, the eyes looked "like a crude clump." It is not known what substance Mengele dropped into the children's eyes. According to a report by the Polish prison physician Rudolf Diem, Mengele claimed that the drops he had administered to persons with heterochromous eyes contained adrenaline: "He believed that the application of these drops would cause the iris color to change."[593]

What at first glance seems to be a mad, pseudoscientific experiment to instantly "Aryanize" brown-eyed "Gypsy children" by injecting substances like methylene blue[594] takes on a new meaning against the background of Magnussen's project on heterochromia. These experiments investigated the pigmentation of the human eye under phenogenetic aspects. Mengele did not perform these experiments "single-handedly [...], but rather in 'teamwork' with Magnussen. Mengele was thus more than a passive supplier of dead 'human material,' and by no means did Magnussen research on dead objects only; she was actively involved in Mengele's human experiments."[595] After the war Magnussen confirmed that she had taken part in Mengele's eye experiments, whereby she boldly redefined heterochromia as a disease and passed off the human experiments performed by Mengele in coordination with her project as an attempt to cure the subjects:

> The histological investigation succeeded in obtaining an indication of the presumptive cause of the disturbing anomaly. – We decided to apply the results immediately in the interest of the same family as an attempt to cure the anomaly in one of the children. – Since animal experiments of this kind had already been performed with success by other scientists, and since we had received previous assurance from the University Ophthamological Clinic in Berlin that no unpleasant side effects of any kind were to be expected from the treatment planned (adding drops of a bodily substance for the purpose of restoring a disturbed function), the treatment was started. – Since the child, unfortunately, died after a few months, it was not possible to perform the treatment for a sufficient length of time to yield an externally visible success. – Shortly thereafter I received the eyes of this child for histological analysis, along with the eyes of another deceased child from this family contaminated with TB, – (i.e., 4 of 12 children) and performed the histological analysis of these eyes as well, although the histological work on the first three pairs of eyes had been

[592] Langbein, Menschen in Auschwitz, p. 383.

[593] Quoted in Hesse, Augen, p. 78.

[594] For instance, Lifton, Ärzte, p. 423. In opposition: Massin, Mengele, p. 247.

[595] Ibid.

concluded long ago. – The possibility that an advantageous effect of the treatment is present after two months is given in the histological examination, so that I would advise this treatment even today in a case with similar conditions.[596]

4.4.11 Verschuer's "Specific Proteins" Project: The Attempt to Develop a Serological Race Test

Also closely linked with the Auschwitz concentration and extermination camp was that secretive research project of Verschuer's, funded by the DFG and listed in the files of the Reich Research Council as "Experimental Research on the Determination of the Heredity of Specific Proteins as the Foundation of Genetic and Race Research."[597] This project was long regarded as a contribution to genetic pathology research under the aspects of race and implicated with Diehl's tuberculosis research in the external office of the KWI-A in Beetz. One of the connections existed on the *organizational* level, for Verschuer's project used rabbits from Diehl's breeding; their sensitivization with human blood sera took place in the KWI-A "reception center" in *Haus am See* in Beetz, that is, in the direct vicinity of the External Department for Tuberculosis Research, and technical problems were resolved in part thanks to the know-how of the KWI for Biochemistry, which also provided consulting for genetic tuberculosis research. Yet there was also a connection on the *conceptual* level, to the extent that both projects fit in to the paradigm of phenogenetics. There may have been a *practical* connection beyond this, as it cannot be excluded that blood samples of sick patients were also targeted for use in the "Specific Proteins" project – more on this later. In this case there would have been the prospect of genetic pathology findings becoming available as a kind of "byproduct" of the "Specific Proteins" project.

However, it must be emphasized that the two projects did *not* constitute a single unit and that it is by no means correct to conceive of the tuberculosis project as a preliminary phase of the "Specific Proteins" project. Benno Müller-Hill and many others after him advanced the theory that the "Specific Proteins" project concerned the investigation of race-specific susceptibility or resistance to tuberculosis, and this, in turn, was connected with the theory that Josef Mengele had purposely infected inmates of the Auschwitz camp with tuberculosis on behalf of the KWI-A.[598] In contrast to this, Bernd Gausemeier formulated the theory that the project pursued the goal of developing a serological

[596] Protokoll der Vernehmung Karin Magnussens durch die Spruchkammer Bremen am 25/5/1949, MPG Archive, Dept. I, Rep. 3, No. 26. The written testimony by Magnussen of 1966 reads: "After discussion with Dr. M. (1944), next a medical treatment was performed on a child-age member of the clan. According to previous consultation with an ophthamological clinic, there were no side effects of any kind to be feared from the drop-by-drop addition of a bodily substance. [...] After the death of the patient some time later there was a histological finding that could be interpreted as the first successful cure, but which bears no conclusiveness." Quoted in Hesse, Augen, p. 93.

[597] Thus the datum in the file of the Reich Research Council. BArch Berlin, R 26 III/6, p. 82.

[598] For the first time in: Müller-Hill, Tödliche Wissenschaft, pp. 71–75. Most recently in: idem., Blut, esp. pp. 204–212.

race test.[599] Achim Trunk recently subjected both positions to meticulous critique, reaching the conclusion that Gausemeier's theory can be reconciled with the scant source material available much better than that of Müller-Hill.[600] I concur with Gausemeier's position, whereby I can submit a document not yet taken into consideration that provides magnificent confirmation for Trunk's analysis.

The "Specific Proteins" project was presumably launched in spring 1943. It was acknowledged as strategically important and thus received special funding as regarded material procurement, but yet – in contrast to the tuberculosis project – it was rated as the lowest priority *S*. The first indications as to what the project was about appeared in Verschuer's interim report to the DFG of September/October 1943:

> Once all materials for the performance of this research had finally been delivered, the first preliminary examinations were begun and the methods tried out in consultation with Privy Councilor Abderhalden, Halle. An interruption in the work occurred when this branch of research was moved to the reception center of the institute in Beetz, but by now the laboratory there is completely equipped. Work can be continued.[601]

Two things can be taken from this report: The project had been temporarily moved to Haus am See in Beetz, and it used a method that was widespread in Germany at the time, although controversial, "Aberhalden's reaction."[602] This procedure, developed by the Swiss biochemist Emil Abderhalden starting in 1909, proceeded from the basic assumption that an animal organism can recognize and destroy a foreign protein that penetrates it – such as those of bacteria in the case of an infection – by manufacturing enzymes (at that time they were known as "ferments") that catalyze a defensive reaction against the foreign protein. The "protective ferments," the presence of which Abderhalden and his pupils believed they could demonstrate in blood, and since 1930 in urine, ultimately turned out to be chimerical. The entire edifice of teachings constructed by Abderhalden was founded on either fraud and/or – scholars are still arguing about this – on error and self-deception. In the early 1940s Abderhalden had faced increasing criticism, but his method had not yet been clearly disproved. Broad hopes were still linked with the method; it was believed that it might open up new possibilities for the diagnosis of infectious diseases, cancer, and even psychoses. What is more: in the second half of the 1930s, Abderhalden and his pupils attempted to use protective ferments for the determination of race in sheep and pigs. "This important research," Verschuer commented in his review, "finds the greatest interest of the genetic biologist [...]."[603] In 1940 Abderhalden

---

[599] Gausemeier, Radikalisierung.

[600] Trunk, Zweihundert Blutproben (edited and abridged version: idem., Rassenforschung).

[601] Verschuer, Bericht über das Projekt "Spezifische Eiweißkörper" an die DFG, n.d. [September/October 1943], BArch. Koblenz, R 73/15.342, p. 55. On September 20, 1943 Verschuer reports to Fischer that he now had the technical assistent Irmgard Eisenlohr and that "with her the research with Abderhalden's reaction, now finally picking up steam," was moved to Beetz. Verschuer to Fischer, 20/9/1943, MPG Archive, Dept. III, Rep. 86 A (Münster), No. 9.

[602] Müller-Hill/Deichmann, Fraud; Deichmann, Flüchten; Kaasch, Sensation; Lindemann, Abderhaldens Abwehrenzyme.

[603] Verschuer, Review Emil Abderhalden, Rasse und Vererbung, pp. 91 f.

and Verschuer exchanged several letters, in which it was Abderhalden's idea to investigate the reactions of protective ferments in twins. Verschuer rejected this for the time being, referring to the difficulties of obtaining blood samples.[604]

The correspondence between Verschuer and Abderhalden shows that in 1943 the latter was training one of Verschuer's medical-technical assistants to use his methods in Halle – Müller-Hill's investigations produced evidence that the assistant in question was Irmgard Eisenlohr (from 1944: married Haase).[605] The second interim report by Verschuer to the DFG of March 1944 confirms clearly that Abderhalden's method was applied in the "Specific Proteins" project:

> In the trials of the methods new difficulties have come to light, which were resolved in consultation with Privy Councilor Abderhalden, Halle. Series of rabbits are being subjected to thorough testing in order to find animals free of spontaneous ferments and thus suitable for the experiments. My assistant Dr. med. et Dr. phil. Mengele has come in as an associate in this branch of research. He is stationed in the Auschwitz concentration camp as Hauptsturmführer and camp physician. With permission of the Reichsführer SS, anthropological examinations are being performed on this camp's many different racial groups and the blood samples sent to my laboratory for analysis.[606]

This is the first evidence that the "Specific Protein" project used blood samples from people of different races, which came from Josef Mengele. In two letters by Verschuer to the pediatrician Bernhard de Rudder, a close friend of his and Diehl's, he goes into greater detail about the blood samples supplied by Mengele. In October 1944 Verschuer wrote:

> Plasma substrates were produced from over 200 people of various races, pairs of twins and a number of families.[607]

And a letter of January 1945 reads:

> I have the substrates from the blood sera of over 200 persons of various racial descent and also of pairs of twins and a few families ready, so that it is now possible to start the actual comparative studies.[608]

---

[604] Kaasch, Sensation, pp. 183 f.

[605] Ibid., p. 186. Müller-Hill, Tödliche Wissenschaft, pp. 162 f.

[606] Verschuer, Bericht über das Projekt "Spezifische Eiweißkörper" [for the period from October 1, 1943 to March 31, 1944], March 20, 1944, BArch. Koblenz, R 73/15.342, p. 64.

[607] Verschuer to de Rudder, 4/10/1944, MPG Archive, Dept. III, Rep. 86 A (Münster), No. 8.

[608] Verschuer to de Rudder, 6/1/1945, ibid. In a position paper of 1946 Verschuer claimed that he received these blood samples from various sources: "To perform Abderhalden's protective ferment reaction in order to study the individual specificity of the serum proteins, I received blood samples sent from several hospitals, like those taken for most clinical investigations (Wassermann's reaction, the erythrocyte sedimentation rate [ESR]), around 5–10 ccm, without harming the health of the patient in any way. Among these were also blood samples from the sick bay where that assistant from Frankfurt worked in Auschwitz [i.e. Mengele]. In this manner I received – over the course of time, not at regular intervals – several dispatches of 20–30 blood samples each, also amounting to around 5–10 ccm each." Verschuer, Stellungnahme zu den Angaben, die sich auf meine Person beziehen und in der "Neuen Zeitung" Nr. 35 vom 3/5/1946 unter der Rubrik "Kunst und Kultur in Kürze" in der Notiz "Vertriebene Wissenschaft" erschienen sind, Universitätsarchiv Frankfurt/Main, Dept. 13, No. 347, p. 179.

What these "actual comparative studies" involved and what purpose they pursued is a matter of great controversy in historical research. Achim Trunk reconstructed the course of events as follows: From the blood samples serum was extracted and sent to Berlin. From this, dry preparations were produced in the laboratory, which were then supposed to serve as the "substrate" converted by the protective ferment in the defense reaction. Meanwhile, the search was on for suitable test rabbits, i.e. ones that did not already have the capability to decompose the foreign protein before they were inoculated with this protein. This is what Verschuer was referring to with his comment that "series of rabbits [had been] subjected to thorough testing in order to find animals free of spontaneous ferments." When suitable rabbits were found, researchers injected them subcutaneously with a portion of the dried substrate and waited until they developed protective ferments against the race-specific human proteins. In the next step the protective ferments supposedly created had to be isolated. To do this, the urine of the rabbits was collected in special apparatus, as it was believed that the protective ferments were excreted with the urine. All substrates were then subjected to cross-reactions by adding the ferment solutions extracted from the rabbit urine in order to determine similarities and differences between the substrates. The final objective was to identify a panoply of protective ferments, each of which reacted with the proteins from the blood of a very specific human race. This would indeed have opened the way for a biochemical race test that would have eclipsed all anthropometric methods of race diagnostics attempted up to that time.

A letter from Verschuer to Karl Diehl of July 17, 1944 presents impressive confirmation for Trunk's version of events:

> The experiment about the serum proteins with Abderhalden's reaction has proceeded to the point where *I have copious material at hand in the form of substrates.* A conversation I conducted a few days ago with one of Butenandt's assistants gives me occasion to begin now *with the actual experiment, i.e. the reaction with the protective ferments generated in the rabbit.* To do this, first of all, as last fall, the rabbits must be subjected again to thorough tests for the presence of spontaneous ferments, so that we can finally arrive at an animal that tests negative for ferment. Therefore I request of you, just as you did last fall, to allow that a few animals from each of your *normal breeds* be placed into the special cages so that urine can be collected for the examination.[609]

This clearly proves that which Trunk held to be very probable: the "Specific Proteins" project quite clearly did *not* concern protective ferments against tuberculosis or any other infectious disease in the blood samples taken by Mengele in Auschwitz. Rather, these served without a doubt as *substrates*, which were to be converted by protective ferments extracted from rabbits.

Also worthy of emphasis in this letter by Verschuer is the term *normal breeds*. What must be kept in mind here is that Diehl held not only rabbits from the two pure breeds in his stalls – the ones he called "lung and belly rabbits" – and from the cross of these two breeding lines, but beyond these also a great number of other

---

[609] Verschuer to Diehl, 17/7/1944, MPG Archive, Dept. III, Rep. 86 A (Münster), No. 7 (my emphasis, HWS).

rabbits, of which many gave their lives for "preliminary experiments," for instance, by using glycerine to extract from their lungs a culture medium for tuberculosis bacilla.[610] The normal breeds were very valuable for Diehl – in December 1942 he answered the Verschuer family's inquiry as to whether he could spare a rabbit or two for the Christmas feast in the negative.[611] Presumably with a heavy heart, he declared himself willing to make rabbits from the "normal breeds" available for the "Specific Proteins" project, in return for Verschuer's submitting a DFG application to obtain funding for his "Tuberculosis" project.

The point here is that the "Specific Proteins" project had nothing to do with Diehl's pure breeds and the crosses between these pure breeds. Thus, we can exclude with a high degree of probability that the human tuberculosis bacilla, with which a portion of the pure breeds were pretreated according to Verschuer's statement, came from Auschwitz.

In Verschuer's letter to Diehl of July 17, 1944 the "Specific Proteins" project's connection with the KWI for Biochemistry becomes clear for the first time. The report by Verschuer to the DFG of October 1944 provides further explanation:

> The research has continued to enjoy intensive support. Blood samples of over 200 persons of various racial descent were processed and substrates of the blood plasma produced. The further research will be continued in collaboration with Dr. Hillmann [Günther Hillmann (1919–1976)], a staff member of the Kaiser Wilhelm Institute for Biochemistry. Dr. Hillmann is a biochemical specialist for protein research. With his help Abderhalden's original method has been perfected, so that now the actual experiments on the rabbits finally can be started.[612]

Much has been written about Hillmann's involvement, his position at the KWI for Biochemistry, his relationship to Butenandt and the responsibility of the latter for Hillmann's commitment to the "Specific Proteins" project.[613] Here just this much: from November 1944 to February 1945 Irmgard Haase, advised by Hillmann, continued working on the project. When it had to be disrupted due to the war, she had "just sensitivized the first rabbits with the dried sera."[614] As late as February 1945, Verschuer addressed Butenandt from his family estate in Solz, Hesse, to which a large portion of the KWI-A had been relocated by then, "because of his problem child, protein research,"[615] inquiring about possibilities to continue his research – ultimately in vain. In the end, Butenandt wrote to Verschuer on October 18, 1945:

---

[610] Diehl to Fischer, 10/8/1942, ibid. At this point in time Diehl had around 450 live rabbits at his disposal, although around 100 young animals had died in the previous weeks.

[611] Diehl to Verschuer, 20/12/1942, ibid.

[612] Verschuer, Bericht über das Projekt "Spezifische Eiweißkörper" [for the period from April 1 to September 30, 1944], 4/10/1944, BArch. Koblenz, R 73/15.342, p. 47.

[613] Cf. Trunk, Zweihundert Blutproben, pp. 16–23, 50–53, 67–73, and Schieder/Trunk (eds.), Adolf Butenandt. On the friendship between Verschuer and Butenandt cf. also Sachse, Adolf Butenandt.

[614] Müller-Hill, Tödliche Wissenschaft, p. 163.

[615] Verschuer to Butenandt, 28/2/1945, MPG Archive, Dept. III, Rep. 84/2, general correspondence.

"It is too bad that our shared work plans cannot be continued at the time being, but I hope it will be possible later."[616]

From what was said it should have become obvious that while the "Tuberculosis" and "Specific Proteins" projects were closely linked together on the practical level, they pursued different objectives and were located in different fields of research: Diehl's tuberculosis research fit into the long series of genetic pathology projects at the KWI-A, while the "Specific Proteins" project concerned *human races*. Nevertheless Verschuer, as proceeds clearly from his letters to de Rudder, saw a close connection between the two projects. This is evinced particularly by a letter of October 4, 1944, in which Verschuer named the two projects in one breath with regard to his impending lecture to the academy:

> Diehl obtained new, and as I believe, fundamentally very important research in his tuberculosis research. [...] I believe that my research about the question as to the heritability of specific proteins is also connected to the problem as a whole. [...] The goal of my different endeavors is now not only to establish *that* the influence of heredity is important in some infectious diseases, but in what manner it takes action and what kind of processes occur in these cases.[617]

At first glance this passage appears to speak for Müller-Hill's reconstruction of events, according to which the "Specific Proteins" project did have the object of race-specific dispositions or resistances to tuberculosis. It is clear that Verschuer was interested in such issues. In his lecture to the academy on November 16, 1944 he hit on the gradual differences in the susceptibility and frailty of various human races with regard to infectious diseases – here he also mentioned the supposedly greater resistance of Jews to tuberculosis – but he added, with reference to a publication by de Rudder, that it had yet to be elucidated "whether these differences are truly conditioned by the different genetic disposition of the races and not by other influences."[618] The "Specific Proteins" project actually promised to contribute to the clarification of this question *indirectly*, and thus there was an internal connection to tuberculosis research. The conceptual brackets around the two projects were constituted by the paradigm of phenogenetics. While each of the projects had a practical application in mind: Diehl's attempt to breed rabbits resistant to tuberculosis was borne by the hope of discovering a biochemical substance that could also give humans protection from tuberculosis – and this without inoculation. At the forefront of Verschuer's protein project was the development of a biochemical race test. However, it must not be overlooked that both projects were apparently understood as complementary contributions to theoretical research in the area of phenogenetics, as they both aimed at the level of the *proteome*, where the gene action chains proceeding from the genome are set into action and shape the phenome, where race attributes are developed and many of the dispositions for diseases were also established.

---

[616] Butenandt to Verschuer, 18/10/1945, MPG Archive, Dept. III, Rep. 84/1, No. 601.

[617] Verschuer to de Rudder, MPG Archive, Dept. III, Rep. 86 A (Münster), No. 8.

[618] Verschuer, Wirkung von Genen, p. 383.

The "Specific Proteins" project shows – as did Karin Magnussen's project on heterochromia, by the way – that research guided by the paradigm of phenogenetics, although it left behind the genetic determinism of the old race hygiene and race anthropology, was by no means invulnerable to drifting into the zone of crime. My theory is that one can conclude from this that in the area of the science of humans there can be *no* security against crossing scientific boundaries inherent in the paradigm guiding research – whatever shape it may take –. *Every* form of research on humans takes place in the tug of war between the researcher's interest in scientific knowledge and the human and civil rights of the person researched, regardless of their idea of man, that is, no matter whether they regard the individual as a pure product of his or her genetic information, as result of the interplay between heredity and environment, or as a *tabula rasa* that is marked by the environment. Drawing science-ethics boundaries takes its basis of legitimation from values held beyond the sphere of science.

4.4.12 Model and Competition: Karl Horneck and his Research Project About "The Serological Race Differentiation of Humans"

Otmar von Verschuer was neither the only one nor the first to work on a serological race test. The anthropologist Theodor James Mollison had long been concerned with the question as to whether serological race diagnostics was possible. Mollison attempted to reach his goal using the "precipitine reaction." This reaction involved the formation of precipitation that occurred when blood serum from another animal, for example, from a chimpanzee, was injected into a rabbit, and the antiserum, which was produced from the blood of a rabbit immunized in this fashion, was mixed with the original serum of the chimpanzee. If the same antiserum was allowed to react with sera of related species – like macaques, gibbons, orangutans, or humans – the precipitin reaction varied in strength. Mollison traced this back to proteins in the blood serum specific to each species. What was true for animal species, Mollison deduced, must also be applicable to human races. Therefore, since the 1920s he had been attempting to develop serological race diagnostics on the basis of the precipitin reaction.[619]

Other scientists in the Third Reich picked up on this approach. Werner Fischer (1895–1945)[620] from the Scientific Department of the Institute for Experimental Cancer Research in Heidelberg under Ernst Rodenwaldt, who collaborated with Benno Raquet in 1938 to submit a paper "On the Question of the Proof of a

[619] Cf. Trunk, Zweihundert Blutproben, p. 10. On Mollison also: Kröner, Von der Rassenhygiene zur Humangenetik, p. 3. From Mollison's work: Mollison, Serodiagnostik; idem., Verwandtschaftsforschung; idem., Anthropologisches Institut der Universität München.

[620] For a biography: Klee, Personenlexikon, p. 154.

Serological Differentiation of the Human Races," also employed the precipitin reaction. He believed that he had established quantitative differences in the abilities of "white serum" and "Negro serum" to react with a certain "white serum-antiserum," which could be demonstrated using precipitation. Fischer qualified his conclusion, however, adding that "before the potential perspective of a serological race diagnosis using such antisera can be considered, [...] *numerous control and supplementary experiments*[would be] *essential*."[621]

At this point Karl Horneck and his mentor Lothar Loeffler entered the stage. Horneck, an Austrian citizen, was born in Graz in 1894. In World War I he fought in the ranks of the Tyrol Kaiserjäger regiment. In 1919 he fought in the "defence of Kärnten," the guerilla war of Austrian snipers against the annexation of part of Kärnten by Slovenia. In the following year Horneck completed his dissertation in medicine. Between 1920 and 1924 he worked in various clinics as "operation disciple," intern and "secondary physician;" from 1924 to 1931 he was a general physician in Feldbach, and from 1927 to 1930 he also directed a small hospital during the construction of the Feldbach-Bad Gleichenberg rail line. At the same time he was also active in the "Protection of the Steyr Homeland" (*Steirischer Heimatschutz*) and joined the Austrian National Socialists. In 1931 he found a position at the Medical Clinic of the University of Graz, from early 1932, however, only as an unpaid assistant. In 1933 he applied for a position as railway physician, but his application was denied due to his membership in the NSDAP. His application as a panel doctor was not processed for the same reason. In 1934 Horneck applied for a position as chief physician at the Elisabethine Hospital in Klagenfurt, but here, too he was rejected – this time because of his membership in the Protestant Church, as Horneck speculated.[622] 1936 brought the shift in Horneck's career path. Lothar Loeffler brought him to the Race Biology Institute of the University of Königsberg as chief physician, "since he was in a position devoid of prospects in Austria." Horneck had not made any prominent contributions as a scientist at the time – his appointment to Königsberg was due solely to his political loyalty. But then, as Loeffler established in retrospect, he had "worked tremendously hard to become acquainted not only with the scientific questions of the care of genes and race, but also performed practical work from the outset." Although at this point in time he had produced only one scientific paper,[623] Horneck obtained his professorial qualification before the beginning of World War II, whereby, in Loeffler's words, "the faculty in Königsberg took into account his past and the necessity of practical training." In World War II Horneck was drafted again, held the rank of a staff surgeon, later chief staff surgeon, and earned "the EK [Iron Cross] II in Poland and the EK I at Dunkirk," before he was transferred to the Eastern front.[624]

---

[621] Fischer/Raquet, Beitrag, p. 121 (original emphasis).

[622] Personnel questionnaire on the request for a research stipend, BArch. Koblenz, R 73/11.807.

[623] Horneck received a research stipend from the German Research Association during his time in Königsberg, for "Microscopic Studies about the Structure of the Capillaries Using Infrared Photography." This also concerned changes in the vessels due to "genetic diseases." Cf. Arbeitsplan, BArch. Koblenz, R 73/11.807.

[624] Loeffler to Blome, 17/10/1942, BArch. Koblenz, R 73/12.756.

Horneck continued to work as a scientist while in military service. "In personal, official visits" with Werner Fischer, by this time Director of the Serological Department of the Robert Koch Institute for Infectious Diseases in Berlin, he had arranged to participate, under Fischer's "guidance," in the control experiments Fischer had declared necessary in his essay of 1938.[625] Possibilities for this of which he never could have dreamed presented themselves in occupied France. Horneck took blood samples from two "Moroccans," one "Annamese" and one "Senegalese Negro" from the ranks of the colonial troops held in war prison camps, and conducted serological investigations on these and other blood samples while on leave, assisted by a French laboratory technician in the Serological-Bacteriological Laboratory of the *Hospice Générale du Havre*.[626] From the blood samples taken in the war prison camp, Horneck produced "Moroccan, Annamese and Negro sera." These he compared to various "European sera." Over a period of two months, Horneck injected five to seven intravenous injections of these sera in five rabbits, in order to immunize each of them against a specific serum.[627] Then he killed the animals, let them bleed to death and in this manner obtained a "precipitating antiserum" for each serum injected. A precipitin reaction was induced for each antiserum by combining them with all sera – i.e. with "white, Annamese, Senegalese Negro, and Moroccan serum." Horneck arrived at the conclusion that the "white serum" reacted more weakly in the two cases portrayed in detail – both in the precipitation with "Moroccan serum-antiserum" and with "white serum-antiserum" – and thus possessed less "precipitating antigens" than the other sera. This, Horneck stated, could mean "that the differences present were *not actually of race*, i.e. based on the circumstance that whites, besides the antigen for the human species, also possess a white antigen, while the Moroccan, Negro, Annamese also possess a Moroccan, Negro, or Annamese antigen in addition to the antigen for the human species; rather, there may merely exist certain differences between whites and the other races in the amount of precipitatable antigens."[628] The "determination of the protein content" and the determination of the composition of the protein were thus an imperative prerequisite "for a serological race diagnosis."[629]

At Werner Fischer's urging, for this research in 1941 Horneck also began "to attempt the immunization of human to human in different races."[630] In a later research report Horneck mentions incidentally that these first immunization

---

[625] Horneck, Nachweis, p. 309.

[626] Ibid., p. 318 f.

[627] Two rabbits who were treated with "Senegalese Negro serum" perished of peritoneal tuberculoses during this procedure. Since there was not sufficent "Senegal Negro serum" available, Horneck dispensed with immunizing the third animal, so that no antiserum was available for this serum. Ibid., p. 310.

[628] Ibid., pp. 316 f. (original emphasis). It could be that other races possessed more easily precipitatable (lyophobic) serum protein (euglobulins), while Europeans had more strongly lyophilic serum protein (pseudoglobulins, albumins).

[629] Ibid., p. 318.

[630] Ibid., p. 309.

attempts, which had not produced any "usable results,"[631] were performed *on himself*. However, this account must be cast in doubt – Horneck, who had worked as a general physician for 7 years, after all, must have been aware of the great risks involved with such immunization experiments. It is highly improbable that he bore this risk himself.

"As a consequence of his [...] deployment on the Eastern front,"[632] in 1941 Horneck was forced to discontinue his experiments for the time being. He published his results in a paper, which he submitted to the editorial board of the *Zeitschrift für menschliche Vererbungs- und Konstitutionslehre* on April 20, 1943, and was published in October of that year.

Even before this paper appeared in print, Lothar Loeffler submitted an application for research funding to the German Research Association, in order to allow Horneck to continue his project. The objectives of future investigations were, according to Loeffler, "absolute exclusion of individual differences, especially diseases, within one and the same race," as well as "determination of the protein factions of the antigens." Independent of this, the "experiments about immunization from human to human" were to be continued. Research on "Negroes" was to be continued, "as initially only significantly different races come into question."[633] The medical faculty of the University of Königsberg, he stated further, soon will apply for a military exemption or "working leave" for Horneck, which had good prospects for success. Since the institute in Königsberg was not equipped for such extensive examinations, and the required apparatus could not be procured during the war either, and because the Race Biology Institute being set up in Vienna (Loeffler was just about to move from Königsberg to Vienna at the time) did not yet have a serological workplace, Werner Fischer expressed himself willing to grant Horneck a temporary workplace at the Robert Koch Institute.

Since 1933, Loeffler had first propelled ahead with his research in the field of radiation genetics. Called upon by Alfred Kühn to take part in a joint project for the investigation of genetic damage through x-rays, in December 1933 he had requested a considerable sum from the Emergency Committee of German Science for radiation genetics experiments on mice, which Loeffler wanted to perform in collaboration with Paula Hertwig (1889–1983) of the KWI for Biology and Nikolaj Timofféeff-Ressovsky of the KWI for Brain Research.[634] The Emergency Committee actually approved a credit of up to 7,000 RM for this project.[635] Around October 1935 Loeffler had his assistant Karl-Heinz Koch, who had been "initiated in drosophila genetics" by Timofféeff-Ressovsky at the KWI for Brain Research,

[631] Karl Horneck, Bericht über die von mir im Januar 1943 begonnenen Untersuchungen über die serologische Verschiedenheit der menschlichen Rassen, n.d. (April 1943), BArch. Koblenz, R 73/12.756.

[632] Grawitz to Himmler, 20/7/1942, quoted in Klee, Auschwitz, p. 166.

[633] Loeffler to DFG, 6/10/1942, BArch. Koblenz, R 73/12.756.

[634] Loeffler to Schmidt-Ott, 9/12/1933, ibid.

[635] Notgemeinschaft to Loeffler, 9/2/1934, ibid.

perform studies on the "question of generating mutations in drosophila through X-rays." "In these studies [work was performed] in very close coordination with Timoféeff-Ressovsky, in order to achieve as great a consistency as possible in work methods and work orientation."[636] In June 1936 Loeffler was able to report that the studies, which were part of the joint project sponsored by the German Research Association and the Reich Health Office, already covered 1,100 cultures with over 50,000 animals.[637] Yet Loeffler had strong competition in this field he had originated, for instance, from the Genetic Biology Department of the Reich Health Office.[638] Thus, it was altogether logical that he assigned his right hand Horneck to the explosive problems surrounding a serological race test, which – as Loeffler had recognized clearly – touched on not only "important fundamental issues of our science," but also was aimed "at purely practical matters."[639] Thus, Loeffler could open up a new field of research, which was not only of scientific interest, but – more importantly – also of immense importance for National Socialist race policy. In this he attached himself to Werner Fischer, who had been working on the complex of themes for some time and had both the required know-how and the necessary infrastructure at his disposal.

For his part, Fischer was happy to accept Loeffler's advances, and bound Horneck's project to his institute, as in this manner he could take advantage of Loeffler's far-reaching political connections in the National Socialist state. This was of particular interest to Fischer because, at around the time Horneck published his first results, he had begun with serological examinations of concentration camp inmates. On this, a report by the Reich Physician SS Ernst Grawitz to Reichsführer SS Heinrich Himmler of July 20, 1942 states:

> May I send word that by now Prof. W. Fischer has begun his experiments about serum differences in the human races in the Sachsenhausen concentration camp. The first examinations are being performed on 40 Gypsies. Afterward the experiments are to be expanded to Jews.[640]

Fischer – like Günther Just – was willing to offer a positive opinion about Horneck's research plans. The other personal reference listed besides Fischer was Maximinian

[636] Loeffler to DFG, 22/5/1936, ibid.

[637] Loeffler to DFG, 30/6/1936, ibid.

[638] Hans Reiter, President of the Reich Health Office, was noticeably reserved in his position paper for the German Research Association about an application by Loeffler for additional funding. He stated his wish to "emphasize in principle that the radiation genetic experiments on drosophila Professor Loeffler intends to perform also belong to the work area of the Genetic Biology Department of the Reich Health Office. The x-ray apparatus procured for radiation genetic experiments of all kinds, and the manpower available, guarantee a generous execution of radiation genetic work. At this time studies are in progress about the intensification of the genetic radiation effect through pretreatment with chemicals, especially with heavy metal salts, on drosophila." Reiter to DFG, 27/7/1936, ibid.

[639] Loeffler to Blome, 17/10/1942, ibid.

[640] Grawitz to Himmler, 20/7/1942, quoted in Klee, Auschwitz, p. 166. Cf. also idem., Medizin, pp. 163 f.

de Crinis (1889–1945),[641] a National Socialist emigrated from Austria whose curriculum vitae exhibited many a parallel to Horneck's, and who had succeeded Karl Bonhoeffer as full professor for psychiatry and neurology at the Charité hospital in Berlin.

Loeffler bestowed particular urgency upon his application by following it with a letter to Kurt Blome (1884–1969),[642] Deputy Director of the Main Office for National Health at the NSDAP, a liaison of the German Research Association for the subject area "Population Policy, Care of Genes and Race," who was certain to be one of the people evaluating Horneck's research plan. Loeffler supplied a short synopsis of Horneck's biography, summarizing that he had "proved his worth both politically and in the war." According to report, Loeffler continued, Horneck had been listed in third place for two pending appointments, and it was to be expected that he would move up to positions with more prospects in later appointments, so that, also in view of the "lack of truly good new blood," it was important to give Horneck the opportunity to perform scientific work in the future. Moreover, Horneck was "almost the only race biologist performing serological work at this time."[643] Blome actually did send Loeffler's letter immediately to the Reich Research Council with a request for review.[644] One month later – in November 1942 – the Reich Research Council approved the application for a grant of 2,600 RM.[645]

In January 1943 Horneck, who was working in the Special Colonial Medicine Military Hospital in St. Médard near Bordeaux at this time,[646] resumed his research. In his first preliminary report, Horneck once again described the point of departure of his study: The purpose was to establish whether the varying intensity of precipitin reactions to human sera was influenced by individual factors, especially by diseases, in such a way that the race differences were blurred. Therefore sera of both healthy and sick "Negroes" as well as of whites – for the purpose of comparison – were tested using the precipitin method (optimal precipitation), whereby the same blood groups were used in each test.[647] "With the enormous material" Horneck had "at his disposal an abundance of the most varied diseases, some of which hardly occur at all in our country (like leprosy)." Further diseases he named included

[641] For a biography: Jasper, Maximinian de Crinis.

[642] For a biography: Klee, Personenlexikon, p. 54.

[643] Loeffler to Blome, 17/10/1942, BArch. Koblenz, R 73/12.756.

[644] Blome to Breuer, 20/10/1942, ibid.

[645] Reichsforschungsrat to Loeffler, 24/11/1942, ibid.

[646] Horneck to Breuer, 12/1/1943, ibid.

[647] The blood group to which each of the rabbits belonged was also taken into consideration. There were "as we know, rabbits – known as 'A rabbits,' who possess an anti-A factor. Upon pretreatment with A serum, these rabbits give a much more strongly precipitating antiserum, and that is why this fact must be taken into consideration." Karl Horneck, Bericht über die von mir im Januar 1943 begonnenen Untersuchungen über die serologische Verschiedenheit der menschlichen Rassen, n.d. (April 1943), ibid.

tuberculosis, *typhus abdominalis* and the worm infection *Filaria Bancrofti*. Horneck summarized his preliminary results as follows:

> To the extent that anything at all can be said about them, these studies showed that individual differences do exist, but that they are expressed only in the time and intensity of flocculation. Thus in such a manner that the serum of a certain subject, e.g., a typhus patient, flocculates earlier and more intensively than the serum of a healthy subject of the same race. However, in all experiments it could be confirmed that with regard to the concentration at which the best (optimal) flocculation occurs, fundamental differences exist between white serum and Negro serum.[648]

In a later, brief interim report Horneck portrayed this preliminary result as already proven and declared categorically: "Differences in the optimal stage of flocculation may thus be based only on race differences."[649]

The studies to "determine the protein factions" had not been tackled yet in the first quarter of 1943. On the other hand, Horneck had resumed his experiments on immunization from human to human:

> This time I began the immunization experiments on several Negroes with various blood groups. Before the first injection, about 30-55 ccm blood was taken from each of the Negroes, in order to obtain a serum of the species before treatment. Then the Negroes received a total of 80-100 cmm white serum in four intravenous injections. Twenty-four hours after the final injection and one week after the last [sic, presumably must mean: first] injection, another 50-60 ccm blood was taken from the Negroes. The sera of the pre-treated and those of the non-pre-treated were evaluated for their optimal precipitation and interesting differences were established in this anaylsis.[650]

A portion of these sera was filled into sterile test tubes and sent by courier to Fischer in Berlin, where control tests were to be undertaken. Ernst Rodenwaldt showed animated interest in the immunization experiments in particular. He visited Horneck on location in the Special Colonial Medicine Military Hospital – as mentioned, Werner Fischer had worked as Rodenwaldt's assistant at the Institute for Experimental Cancer Research in Heidelberg from 1933 to 1938. In his later report Horneck noted with disappointment that "extensive attempts at a direct immunization from human to human [proceeded] completely in vain." "The proof of an immunization can only be furnished indirectly via rabbits […]."[651] With the immunization experiments on war prisoners of color, Horneck clearly transgressed the boundary to criminal human experiments, for hereby he not only disregarded his proband's right to self-determination – as in taking blood samples for the precipitin reaction, but he also subjected them to serious health risks. For with the injection of the foreign serum Horneck assented to hazard the potential occurrence of allergic

[648] Ibid.

[649] Karl Horneck, Bericht über die Arbeit "Serologische Differenzierung der menschlichen Rassen," n.d., ibid.

[650] Ibid.

[651] Karl Horneck, Bericht über die Arbeit "Serologische Differenzierung der menschlichen Rassen," n.d., ibid.

shock, of hemolysis (dissolution of the red blood cells), of intravascular clotting events and thromboembolism with consequent circulatory failure and death.

Apparently Horneck's research came to a standstill when he was transferred from France to Italy in 1943. In November 1943 Loeffler reported to the Reich Research Council that Horneck "has received a command from the Wehrmacht, which now puts him in the position to continue his scientific work despite his continued military service status."[652] Horneck intended to travel to France in the near future to resume the interrupted research, Loeffler continued. In February 1944 Horneck applied for a further grant of 1,500 RM, which was approved in March.[653] In October 1944 he informed the Reich Research Council that he had completed a paper "About the Possibility of a Serological Race Differentiation" and sent it to Fischer for appraisal – "with consideration of the fact that this paper contains many new aspects,"[654] Fischer expressed the wish to talk through it with Horneck personally before it went to print, a plan that was frustrated for the time being by the fact that Horneck was denied special leave. The account ends abruptly at this point; the project must have run aground.

Horneck's and Fischer's project is of fundamental importance with respect to Verschuer's project for several reasons:

First, it *temporally preceded* the "Specific Proteins" project. Verschuer had dealt with the proteins of human serum back in his dissertation in 1923 and showed his lively interest in the possibilities of serological race diagnostics in the late Weimar Republic. In a short paper about "Physiology and Pathology in Anthropology" of 1930, Verschuer had regretted that there was still no success in using the precipitin reaction to "establish with certainty protein differences between the human races."[655] One must presume that he observed the developments in this field of research attentively. When he succeeded Eugen Fischer in 1942, the race to develop a race test on a serological basis was in full swing – and the institute in Dahlem had not left the starting blocks. Engelhardt Bühler's project begun in 1935, on the heritability of the isoagglutinin content of human blood serum, which – as Eugen Fischer had implied to the German Research Association – also was to open up possibilities for a serological race test, had come to a complete standstill when Bühler was called up to the Wehrmacht at the beginning of World War II. Certainly it can be assumed that Verschuer had taken notice of Fischer's and Horneck's work, and it can also be assumed that he knew about the series of experiments in progress at the Sachsenhausen camp and in the Special Colonial Medicine Military Hospital, perhaps from Lothar Loeffler, who was, after all, a member of the "Dahlem circle," and – as portrayed elsewhere – probably remained in constant contact with the KWI-A because of the fingerprints and handprints from the

[652] Loeffler to Breuer, 21/11/1943, ibid.

[653] Horneck to Reichsforschungsrat, 11/2/1944; Reichsforschungsrat to Horneck, BArch. Koblenz, R 73/11.807

[654] Horneck to Breuer, 4/10/1944, ibid.

[655] Verschuer, Physiologie und Pathologie, p. 351.

Wittenau Sanatoriums,[656] but perhaps through Wolfgang Abel, who, we recall, also had spent time at the Special Colonial Medicine Military Hospital in St. Médard, possibly worked in the Sachsenhausen concentration camp as well, and could have stumbled over Horneck's and Fischer's tracks. Whatever the sources from which Verschuer obtained his knowledge about the competition's project: with the "Specific Proteins" project he attempted to regain the upper hand in this field of research.

Second, the experimental design of the "Specific Proteins" project, as reconstructed by Achim Trunk, corresponded to Fischer's and Horneck's approach down to the last detail – only that Verschuer pursued his goal using Abderhalden's protective ferments rather than the precipitin reaction. Verschuer was familiar with Mollison's attempts to make the precipitin reaction useful for race diagnostics, just as he was with Abderhalden's works, but, as his reviews show, since the late 1930s he granted the better chances to Abderhalden's method. Since he enjoyed a direct connection to Abderhalden, he believed his hand held a decisive trump over Fischer and Horneck.

Third it becomes apparent that the competition between the scientists corresponded to the rivalry between the politicians involved – Conti and Blome. It was all the easier for Verschuer to win over Conti for his plans because the competing undertaking was located in Blome's sphere of influence.

Fourth and finally, against the background of Horneck's project it cannot be excluded that the "Specific Proteins" project utilized the blood of subjects who were *ill*, perhaps even that of humans who were *made ill*. Since 1942 Horneck had researched on blood samples of diseased subjects of various races to investigate whether the serum of sick members of a race behaved differently in the precipitin test than did that of healthy members of the same race. This question was also posed in principle by Verschuer in his application of Abderhalden's reaction – in competition with Horneck and Fischer he could not really afford to leave this question unanswered. Thus, it is possible that the 200 blood samples Mengele sent to Dahlem include some originating from diseased inmates. Similarly, against the background of Horneck's research, the suspicion that Mengele purposely could

---

[656] In 1938 Fischer ceded a part of his skull collection to Loeffler, so that he would have demonstration material for the Race Biology Institute at the University of Königsberg. Cf. Fischer to Reichserziehungsministerium, 3/1/1938, MPG Archive, Dept. I, Rep. 1 A, No. 2399, p. 116 a. – Verschuer's stance on Loeffler proceeds from a letter to de Rudder of 10/7/1944: "You know that I actually have him [Loeffler] 'on my conscience' to the extent that I brought him with me from Tübingen to the institute in Dahlem in 1927. But he was an assistant to Dr. Fischer and me for only one year before going to Aichel at the Anthropology Institute in Kiel, where scientific leadership was lacking, and then he drifted off entirely into the political direction, scientifically sterile since 1932, but altogether successful in his career. Now a mammoth institute in Vienna has been approved for him. It remains to be seen whether he will succeed in establishing it. In any case he has set up a fine department for the proficient experimental geneticist Gottschewski, in which work is proceeding intensively. Of late Loeffler has treated me with striking courtesy, while for years he had believed that he could take no notice of me along his path. Even so, he has remained that kind of small mind which likes to make an appearance with arrogance, as a know-it-all claiming that "Goethe is wrong here'." Verschuer to de Rudder, 10/7/1944, MPG Archive, Dept. III, Rep. 86 A (Münster), No. 8.

have infected humans with infectious diseases, such as typhus, cannot be dismissed completely in the context of the "Specific Proteins" project. Moreover, Verschuer was interested in questions of "race pathology," as his paper from 1930 already demonstrated. In this paper he presented "as certain pathological facts" that "numerous genetic disorders like diabetes, deaf-muteness, endogenous psychoses [occurred] in Germany more frequently among Jews than non-Jews and "amaurotic idiocy [had] been observed only in Polish Jews.[657] As we have seen elsewhere, he was still concerned with "race pathology" in 1939 and 1944 – to that extent it is quite conceivable that Verschuer followed Horneck's experimental design and opened up the "Specific Proteins" project toward a genetic pathology orientation.

4.5 Relocating and Closing the Institute

As heavy air attacks on Berlin increased in mid-1943, Verschuer began to look around for possibilities to relocate part of the institute. He managed to rent a house in Sommerfeld, on Lake Beetz, in the direct vicinity of the External Office for Tuberculosis Research in Waldhaus Charlottenburg, which had been expanded and equipped as an auxiliary hospital, but had never been moved into. From July 1943 this *Haus am See* had been used as a "receiving office for the institute"[658] – it was staffed by Hans Grebe with his secretary, nurse Emmi Nierhaus, Karl Diehl's technical assistant Charlotte Gruetz and the technical assistant Irmgard Eisenlohr, who was involved with the "Specific Proteins" project.[659] Verschuer had thought about relocating the entire institute to Beetz, but for tactical considerations he dispensed with this idea. If the institute were vacated voluntarily, Verschuer wrote in a letter to Fischer, it might be lost, even if it were not destroyed by air raids. "I could not take responsibility for being at fault myself." In order "to prevent the destruction of our scientific body of thought" he had a wagon load brought to Beetz and Sommerfeld, and in Beetz a hutch for Nachtsheim's rabbits was to be built as well.

> With von Wettstein, Butenandt, Heisenberg [Werner Heisenberg (1901-1976), Director of the KWI for Physics] and Eitel [Wilhelm Eitel (1891-1979), Director of the KWI for Silicate Research] (I name only those with whom I actually spoke) we are in agreement that we must defend our institutes here, for they constitute the core of the KWG. If they were to be lost, the entire future of the KWG would but put into question.[660]

Back in September 1943 Gottschaldt, as mentioned above, sent the materials from the "twin camps" from Dahlem to Rottmannshagen Castle near Stavenhagen in Mecklenburg, at the same time further research materials were sent to *Haus am See* in Beetz. Part of the institute library was brought to Beetz, part to Rottmannshagen.

---

[657] Ibid., p. 351.

[658] Verschuer to Fischer, 30/6/1943, MPG Archive, Dept. III, Rep. 86 A (Münster), No. 9.

[659] Verschuer to Fischer, 20/9/1943, ibid.

[660] Verschuer to Fischer, 25/8/1943, ibid.

In the Dahlem institute, which officially bore the name "Eugen Fischer Institute" since its founding director's 70th birthday in June 1943,[661] of the valuable goods only the "photograph collection" remained, which was stowed in the air raid shelter. The only staff that continued working in Dahlem were the department heads, Karin Magnussen and a few auxiliary assistants. "This dispersion of items is not ideal," Verschuer wrote, "but provides great reassurance."[662] In order to preserve coherence, he set up a standing "courier service" between Dahlem and Beetz.[663]

In February 1944 the institute was damaged in a heavy bombing attack. Yet Verschuer still regarded the situation to be "by no means so grave that a relocation of the complete institute would come into question."[664] In the provisionally repaired building[665] he kept operations afloat for the time being. In September 1944, however, first signs of deterioration became apparent. Lenz, as Verschuer reported, after having brought his wife and children to relatives in Obernfelde near Lübbecke in Westphalia, fell deeper and deeper into depression and could hardly work any longer – shortly thereafter he took leave for reasons of poor health and followed his family to the West, such that the Institute for Race Hygiene ceased to exist in fact.[666] Abel had "left his people more or less to themselves and consumes aerated baths in Bad Ischl." According to Verschuer, Abel managed "excellently to put his personal affairs in order as advantageously as possible. Now he is shifting ever further away from the institute and has become a rare guest." Karl and Anne Diehl, despite health problems, continued their rabbit research, "albeit often by summoning their last strength;" the same was true of Hans Nachtsheim, who had been declared fit for combat in his army physical, so that it was only a matter of time before he was called up. Gottschaldt was the only one who exhibited "an active demeanor loaded with energy," and pushed ahead "the evaluation of his twin findings with extremely hard work and great energy." However, he was often in Rottmannshagen, where he had lodged his wife and children. The situation there, Verschuer warned, was "by no means harmless, for in the sparsely settled land the foreign workers constitute a majority, which could easily seize power for themselves if enemy pilots were to furnish them with ringleaders."[667]

---

[661] On this celebration in detail: Lösch, Rasse, pp. 417–422.

[662] Verschuer to Fischer, 20/9/1943, MPG Archive, Dept. III, Rep. 86 A (Münster), No. 9.

[663] Verschuer to Fischer, 2/3/1944, ibid.

[664] Verschuer to Fischer, 15/2/1944, ibid.

[665] The broken window panes could not be replaced by September 1943; the empty window frames were sealed with cardboard. Verschuer to Fischer, 20/9/1943, ibid. In January 1945 the coal deliveries stopped, so that only the basement and ground floors could be heated. Verschuer to Lenz, 26/1/1945, MPG Archive, Dept. III, Rep. 86 B, No. 12.

[666] Cf. Verschuer to Lenz, 15/1/1945, ibid. Kröner, Von der Rassenhygiene zur Humangenetik, p. 63, is presumably correct in viewing Lenz's " 'vacation' " as a "move to withdraw from Berlin." As early as September 20, 1943, though, Verschuer wrote to Fischer: "Lenz is not doing well at all. He suffers greatly from these times and the conditions and has lost even more weight." Verschuer to Fischer, 20/9/1943, MPG Archive, Dept. III, Rep. 86 A (Münster), No. 9.

[667] Verschuer to Fischer, 29/9/1944, ibid.

On February 3, 1945 a directive was issued by the Reich Minister for Armaments and War Production, Albert Speer (1905–1981), to the operations staff of the KWG, instructing that the institutes under its control be relocated from endangered areas. Ernst Telschow forwarded this directive to the KWI-A, where it arrived on February 5, delivered by a courier. Had it been, up to New Year's 1945, Verschuer's express goal to hold out in Dahlem as long as possible and await the further course of events in order to defend the institute building against competing claims, by February 1945 it must have been clear to him that the fall of Berlin was merely a matter of time. Relocating the institute appeared imperative, and in secret Verschuer already had begun the preparations for a move.[668] So Speer's directive came at just the right time, although initially appearances suggested that it was already too late, for an execution of the directive appeared "impossible."[669] In addition, between February 5 and 12, 1945 Telschow, as he claimed angrily after the fact, informed Verschuer orally that Speer "in retrospect [had] not desired" the "application of the relocation directive" to the KWI-A.[670] Although Verschuer later vehemently denied ever having received such a communication,[671] Telschow's account is confirmed by other sources.[672] Thus, it can be presumed that Verschuer was quite aware that he had received a green light to relocate his institute neither from the General Administration nor from the Armaments Ministry. However, when Engelhardt Bühler, who had been assigned to the institute a short time before,[673] managed to organize a trailer truck around February 9, 1945 – to everyone's surprise,[674] Verschuer acted without delay, supported by Speer's *written* command to relocate, abruptly overrode the *oral* counter-command communicated by Telschow and set the relocation in motion. On February 12, 1945, when part of the material sent to Beetz had already been loaded on the truck, he sent a circular to the department heads Abel, Diehl, Gottschaldt, Lenz, and Nachtsheim, officially informing them that the majority of the institute's inventory was to be relocated to his family estate in Solz near Bebra. The Department for Experimental Genetic Pathology remained in Dahlem, since the extensive animal breeds could not be taken with the institute. Nachtsheim was appointed Verschuer's deputy and entrusted with the oversight of

[668] Verschuer to Lenz, 9/2/1945, MPG Archive, Dept. III, Rep. 86 B, No. 12.

[669] Verschuer to Lehmann, 23/2/1945, MPG Archive, Dept. III, Rep. 86 A (Münster), No. 5.

[670] Telschow to Verschuer, 12/3/1945, MPG Archive, Dept. I, Rep. 1 A, Ni. 2400, pp. 272–272 v, quotes: p. 272.

[671] Verschuer to Generalverwaltung, 21/3/1945, ibid., pp. 273–273 v.

[672] Kröner, Von der Rassenhygiene zur Humangenetik, pp. 80–83.

[673] After spending time at a military hospital, in January 1945 Bühler was ordered to the KWI-A for 2 months. Verschuer wanted to have Karl Diehl give him a topic from tuberculosis research and station him at the "Haus am See." In 1944 Bühler had submitted a postdoctoral thesis, but this had been rejected by the anatomist Hermann Stieve after Fritz Lenz had refused to head the examination committee. Verschuer endeavored in vain to obtain a professorial qualification for Bühler on the basis of the works he had published so far. Cf. Verschuer to Lenz, 15/1/1945; Lenz to Verschuer, 18/1/1945; Verschuer to Lenz, 26/1/1945, MPG Archive, Dept. III, Rep. 86 B, No. 12.

[674] Verschuer to Lenz, 9/2/1945, ibid.

the institute building and the inventory remaining there. The External Department for Tuberculosis Research stayed in Sommerfeld, as Diehl was indispensable as the senior physician of the Waldhaus Charlottenburg Hospital, and he, too, did not want to leave his animal breeds behind. By contrast, the alternative location Rottmannshagen, under Gottschaldt's direction, was to be dissolved and also relocated to Solz as soon as possible. Some of the "followers" were supposed to remain in Dahlem, some were to move to Solz, and some sent home on leave.[675]

On February 13, 1945 the inventory of the Dahlem institute was loaded on the trailer truck provided. In a letter of February 12, 1945 Verschuer informed the General Administration in writing of the relocation already in progress.[676] Immediately before his departure, on the afternoon of February 13, 1945, Verschuer must have had another meeting with Telschow, in which the General Secretary vented his anger, but he was not able to stop the operation in progress.[677] Quite obviously, Verschuer used the chaos predominant in the final phase of the war, above all "the nearly complete collapse of the standard paths of communication," to remove himself from Berlin and in this manner present the General Administration of the KWG with a "fait accompli."[678] How hasty Verschuer's departure was is apparent in the fact that he did not even find the time to contact Günther Hillmann to discuss the continuation of the "Specific Proteins" project. He left the inventory of the laboratory with the "special rabbit cages for the collection of urine" in Butenandt's institute in Dahlem. "I brought with me only the especially valuable and irreplaceable protein substrates,"[679] Verschuer reported from Solz – thus it is possible that some of the sera that ended up in Solz came from the blood samples taken by Mengele in Auschwitz. On the other hand, the written documentation on the "Specific Proteins" project may have been left in Dahlem. On March 12, 1945 Nachtsheim wrote to Verschuer:

> From Miss Jarofki [Ruth Jarofki, one of the institute's secretaries] I learned that many files remained here, which should, or must, be destroyed before falling into enemy hands. While I have not yet taken a look to see what and how much is concerned, I presume that Miss Jarofki knows this exactly. You did not speak about this with me, otherwise I would have advised that the things be taken to Solz. In any case we may not choose too late a point in time for their destruction, and I thus consider myself authorized to make the decision on this matter.[680]

---

[675] Kröner, Von der Rassenhygiene zur Humangenetik, pp. 78 f.

[676] Verschuer to Generalverwaltung, 12/2/1945, MPG Archive, Dept. I, Rep. 1 A, No. 2400, pp. 265–265 v. On Verschuer's news that a trailer truck had been "made available" to the institute, a marginal comment reads: "without our knowledge and assistance – against the will of R.M. Speer (apparently procured unofficially by Dr. Bühler)."

[677] Verschuer to Geschäftsführender Vorstand der Generalverwaltung, 21/3/1945, ibid., pp. 273–273 v. Here Verschuer also claimed that he had been in contact with Speer by telephone before his departure.

[678] Kröner, Von der Rassenhygiene zur Humangenetik, pp. 83 f.

[679] Verschuer to Butenandt, 19/2/1945, MPG Archive, Dept. III, Rep. 86 A (Münster), No. 12.

[680] Nachtsheim to Verschuer, 12/3/1945, quoted in Kröner, Von der Rassenhygiene zur Humangenetik, p. 82.

Verschuer confirmed that some of the material involved was "secret files, which by no means may fall into enemy hands," asked Nachtsheim to attend to the matter and to give the caretaker the order to burn the material "in good time."[681]

On February 17, 1945 Verschuer laconically informed the General Administration that the relocation of the institute to Solz had been completed "without significant inconvenience."[682] On March 1, 1945 Gottschaldt arrived in Solz as well, with the material that had been stored in Rottmannshagen.[683] In a letter to his friend Karl Diehl of March 12, Verschuer appeared optimistic that the institute would be able to continue its scientific work in Solz:

> From here I have good news as far as it goes. It is certainly an enormous luxury not to have any sirens in the village and not to feel like a direct target of enemy pilots. As such I manage more positive work than was possible during the last phase in Berlin. The establishment of my small institute here is making progress, although all sorts of difficulties must be overcome. Someday I would like to give you a tour of my facilities here, my Director's study (also living room and bedroom for Erika and me); the library, in which all of the books brought from Beetz have been arranged, which is also the study for Miss Sesselberg (not to mention the group dining room); to the church hall in the manse, which I have furnished as a study for Miss Lüdicke and Nurse Emmi, in which thus the twin files are being analyzed and the institute's administration and treasury are located; and, finally, in a restaurant hall where the institute property is stacked (including that which Gottschaldt has since brought here from Rottmannshagen).[684]

Shortly before Christmas 1944 Eugen Fischer and his wife had fled from Freiburg before the approaching allied troops to their daughter Gertrud in Sontra, near Bebra, so that Fischer and Verschuer found themselves just a few kilometers distance from each other at the end of the war.[685]

In Berlin Nachtsheim had to struggle with increasing signs of dissolution. In fear of the approaching Red Army, many staff members refused to work. Most of Nachtsheim's rabbits had to be slaughtered once the plan to bring the animals to Switzerland had been discarded. Some of the institute's rooms had to be yielded to the Reich Office for Land Use Planning, the Reich Ministry for Church Matters and to a department of the University of Posen. Finally, on March 13, 1945 the institute building was requisitioned as a reserve military hospital. The General Administration of the KWG, angered by Verschuer's going it alone, undertook nothing to prevent the requisitioning.[686] The General Administration also took a passive stance in the conflict about the *Haus am See* that broke out in March 1945, when the responsible local group leader requisitioned the building to accommodate refugees. In the end, the KWI-A was left with two rooms of the *Haus am See*,[687] in which institute

---

[681] Verschuer to Nachtsheim, 24/3/1945, quoted in ibid., pp. 82f.

[682] Verschuer to Generalverwaltung, 17/2/1945, MPG Archive, Dept. I, Rep. 1 A, No. 2400, p. 267.

[683] Kröner, Von der Rassenhygiene zur Humangenetik, p. 84.

[684] Verschuer to Diehl, 12/3/1945, MPG Archive, Dept. III, Rep. 86 A (Münster), No. 7.

[685] Cf. Lösch, Rasse, pp. 426f., 432f.

[686] Kröner, Von der Rassenhygiene zur Humangenetik, pp. 79f.

[687] Diehl to Verschuer, 17/3/1945; Verschuer to Diehl, 29/3/1945, MPG Archive, Dept. III, Rep. 86 A (Münster), No. 7.

property – "numerous scientific apparatus, including special fabrications […], valuable optics, microtome, projection equipment, part of the scientific library, the twin archive and additional scientific materials"[688] – were stored. Karl Diehl managed to rescue some of these materials in September 1945 when the building was requisitioned for good by the Soviet military authorities.[689]

Along with Nachtsheim, Bühler, Baader, and Magnussen stayed in Berlin. In the end, Verschuer fell out with Magnussen in the course of a conflict within the institute[690] – she went to Bremen and ultimately returned to teaching after leaving the KWG for good.[691] Mengele went underground. At the end of the war Abel withdrew to his estate at Mondsee[692] and dropped out of sight. Heinrich Schade was still a prisoner of war in Yugoslavia. Lenz initially remained in eastern Westphalia, and – as the first of the "Dahlem circle" – was appointed associate professor for human genetic theory at the University of Göttingen in October 1946.[693] With this it appeared that the institute's "political baggage" had been swept under the carpet. Verschuer indulged himself in the hope that he would be able to reestablish the Kaiser Wilhelm Institute for Anthropology, Human Heredity and Eugenics, whereby of the former department chiefs he wanted to reappoint only his friend Karl Diehl with his tuberculosis research. His relationship to Gottschaldt, who became Director of the Institute for Psychology at the Humboldt University in East

---

688 Verschuer to an Bürgermeister der Gemeinde Beetz, 9/3/1945, MPG Archive, Dept. I, Rep. 1 A, No. 2400, pp. 270–270 v. Cf. also Verschuer to Generalverwaltung, 9/3/1945, ibid., pp. 269–269 v; Telschow to Verschuer, 13/3/1945, ibid., p. 271.

689 Diehl to Verschuer, 11/9/1945, MPG Archive, Dept. III, Rep. 86 A (Münster), No. 7.

690 On the background: Magnussen had become friends with Dorothea Michaelsen, Eugen Fischer's secretary of many years (cf. Verschuer to Forstmann, 6/1/1945, MPG Archive, Dept. I, Rep. 1 A, No. 2400, p. 261). In the course of the change in leadership at the institute, Michaelsen had been displaced from her position as executive secretary by nurse Emmi Nierhaus, had taken a long leave of absence for health reasons, and had found herself in fierce conflict with Verschuer in December 1944, who charged her publicly with kleptomania and forced her to resign from the institute (cf. Michaelsen to Generalverwaltung, 21/12/1945, MPG Archive, Dept. I, Rep. 1 A, 3026). Magnussen had sided with her friend (Verschuer to Fischer, 11/10/1951; Fischer to Verschuer, 13/10/1951, MPG Archive, Dept. III, Rep. 86 C, No. 9) with the result that Verschuer fired both women "in the final days in Berlin with the words […] they all had to leave Berlin; all should take care of themselves" (file note by Telschow of 2/4/1945, MPG Archive, Dept. I, Rep. 1 A, No. 2400, p. 274). In April 1945 Magnussen – together with Michaelsen – moved to her parents' home in Bremen (cf. file note by Telschow of 26/9/1945, MPG Archive, Dept. I, Rep. 1 A, No. 3026). That Verschuer, and later, Fischer as well, broke off contact with Magnussen was clearly because of this "Micha case" (cf., for instance, an undated postcard by Fischer to Verschuer [June 1952], MPG Archive, Dept. III, Rep. 86 C, No. 10), *not* because of Magnussen's entaglement in Mengele's medical experiments!

691 Until late 1945 Magnussen was officially listed as an assistant at the institute (cf. Notetat für das KWI-A für das Rechnungsjahr 1945/1946, MPG Archive, Dept. I, Rep. 1 A, No. 3026). Suttinger, "returned from captivity," was supposed to take Magnussen's place from January 1, 1946.

692 Verschuer to Lehmann, 23/2/1945, MPG Archive, Dept. III, Rep. 86 A (Münster), No. 5.

693 Extensively on this: Kröner, Von der Rassenhygiene zur Humangenetik, pp. 63–77.

Berlin in 1946, and also with Nachtsheim, who received a chair for genetics at the Humboldt University in 1946, deteriorated visibly.

Verschuer's hope for the reestablishment of the institute was to prove deceptive. His past caught up with him when the physicist Robert Havemann (1910–1982), who had spent the final years of the Third Reich as a political prisoner in the Brandenburg Penitentiary and been appointed by the City Council of East Berlin (*Magistrat*) as the provisional Director of the Kaiser Wilhem Institutes remaining in Berlin in 1945, exposed Verschuer's connections to National Socialism and his state crimes in 1946. Although at times he was in danger of criminal prosecution and temporarily banned from professional activity, Verschuer ultimately survived the critical situation undamaged[694] and did well in postwar Germany. However, the Kaiser Wilhelm Institute for Anthropology, Human Heredity and Eugenics, which the General Administration had counted among the core inventory of the Kaiser Wilhem Society immediately after 1945,[695] was implicated so heavily that, although it was never officially dissolved, it was never reopened.[696] Only Nachtsheim's Department for Experimental Heredity Pathology was recognized by the Max Planck Society, the legal successor to the Kaiser Wilhelm Society, as the Max Planck Institute for Comparative Genetic Biology and Genetic Pathology in 1953 (since 1964: Max Planck Institute for Molecular Genetics).[697] Hermann Muckermann, who had hibernated in interior emigration for most of the Third Reich, founded a "Kaiser Wilhelm Institute for Applied Anthropology" in Dahlem in 1947/49. Muckermann's application for admission to the Max Planck Society was dragged out by the General Administration. While the "Research Office for Applied Anthropoligy" received financial support from the Max Planck Society, it did not receive the title of a Max Planck Institute. The "Institute for Natural Science and Humanities Anthropology" (*Institut für natur- und geisteswissenschaftliche Anthropologie*), as it was known from 1952 on, never developed noteworthy activities; it was dissolved without further ado in 1961.[698] In the end Verschuer was appointed to the newly created chair for human genetics at the University of Münster in 1951, which long remained the only one of its kind.

Even though the Dahlem institute fell apart after the end of the war: the "Dahlem circle" of Verschuer, Lenz, Lehmann, Schade, Weinert, and Schaeuble constituted an "'invisible' institution" well into the 1960s, a "network of vertical (teacher/pupil) and horizontal (colleagues/associates) relationships." Hans-Peter Kröner is correct to warn against a sweeping thesis of continuity, but he draws an equally valid balance, that the "Dahlem circle" exerted "a decisive influence on human genetics and anthropology in the young Federal Republic."[699]

---

[694] Ibid., pp. 97–149.

[695] Ibid., p. 175.

[696] Ibid., p. 1.

[697] Ibid., pp. 209–221.

[698] Ibid., pp. 195–208.

[699] Ibid., p. 2.

Chapter 5
Boundary Transgressions

In its endeavor to merge together the biosciences and biopolicy, the group of researchers at the KWI-A did the legwork for the National Socialist regime knowingly and willingly. The institute's scientists provided the genetic health and race policy of the "New Germany" with a fixed basis of legitimation, defended it with the weight of their scientific authority at international congresses, and in this manner helped to reduce foreign political pressure on the National Socialist state. Their popular science writings and lectures for the German public were propaganda for the biopolicy of the Nazi rulers. In a multiplicity of courses and training sessions they made a vital contribution to the instruction and further education of professionals – above all health officers and judges – who played an important role in implementing biopolicy. Through scientific consulting as members of advisory boards and expert staffs, the institute's scientists collaborated directly in laying the foundation for this policy and keeping it up to date. As experts, they played a key role in the registration and selection of Jews, Sinti, and Roma, "Rhineland bastards," "foreign nationals," "the genetically ill," and people with physical or mental disabilities.

The scientists of the KWI-A thus took part in the state crimes of National Socialist Germany on various levels and in different functions: in the compulsory sterilization of several hundred thousand people imprinted with the stigma of "genetic inferiority," in the policy of apartheid in the National Socialist race state, in the mass murder of mentally ill people and those with mental disabilities, in the genocide of Europe's Jews, Sinti, and Roma, in the planning for the reestablishment of a German colonial empire (this time under a clearly racist banner), and in the Generalplan Ost. Convinced of the *legitimacy* of their actions, they had no scruples about transgressing the boundaries of *legality* and violating valid criminal law or generally recognized international law – as, for instance, in the participation of Fritz Lenz, Wolfgang Abel, Engelhard Bühler, Herbert Göllner, and Eugen Fischer in the sterilization of the "Rhineland bastards," Fritz Lenz's consulting activity in the discussions about legalizing "euthanasia," Wolfgang Abel's recommendation for the treatment of the "foreign nationals" in the occupied eastern territories, as well as Eugen Fischer's cooperation on the Generalplan Ost. All of these abetting services, it must be emphasized expressly at this juncture, took place *voluntarily*: no scientist of the institute was subjected to pressure, let alone coercion, to participate in the state crimes of the National Socialist regime, although some, keeping an eye on posts and sinecures, found it useful to place their expert knowledge at the disposal of the political decision makers. However, decisive for cooperation and complicity was not the mutual exploitation of science and policy in pursuing the goals and purposes of each side, but rather, the broadly based congruence of biopolicy ideas.

Conversely, the bioscientific research at the institute in Dahlem profited from the unfettered possibilities to access human subjects held prisoner in the world of National Socialist camps. The institute's scientists had no problem with carrying out anthropological examinations on humans whose right to dispose over their own bodies was severely curtailed or completely revoked: on Sinti and Roma in "Gypsy collection camps" (Adolf Würth, Brigitte Richter, Eva Justin, Georg Wagner, Karin Magnussen), on Jews in German-occupied Łódź (Harry Suchalla, Christian Schnecke), on "colored" colonial soldiers and soldiers of the Red Army in war prison camps (Wolfgang Abel, Otto Baader), or on Jews, Sinti, and Roma in the Auschwitz concentration and extermination camp (Siegfried Liebau, Karin Magnussen). Otmar von Verschuer and Karin Magnussen carried out research on human material Josef Mengele supplied from Auschwitz. In this Verschuer apparently preferred not to ask for any details about the circumstances under which Mengele had taken the blood samples he sent to Dahlem, while Magnussen practically incited Mengele to "assist" in the demise of the Sinti and Roma with heterochromous eyes in whom she was interested. Institute scientists also – deliberately disregarding the standards of science ethics valid at the time – performed research on children and youths *for the benefit of others*, whereby they subjected their probands to unpleasant and painful procedures (as in the case of Martin Werner's physiological examinations of teenage twin pairs in 1935), exposed them to an incalculable risk and thus potentially placed them in mortal danger (in the case of the low-pressure experiments on epileptic children from the Brandenburg-Görden State Sanatorium performed by Hans Nachtsheim and Gerhard Ruhenstroth-Bauer in 1943), or even willingly accepted the possibility of serious disabilities and even death (in the case of the eye experiments performed on children in the Auschwitz concentration and extermination camp by Josef Mengele in connection with Karin Magnussen's project). Again, it must be emphasized that all contacts to the world of National Socialist camps were *initiated* by the scientists.

The facts are clear: the transgression of even the last boundaries drawn by law and convention, morals, and ethics, had its source in the dual hubris of a form of bioscience, which on the one hand did not accept that its thirst for knowledge be fenced in by any means, be it through law, politics, culture, or religion; while on the other hand asserted its claim to reshape state and society, even the human condition, in the comprehensive sense – birth and death, sexuality and reproduction, body and germ line, variability and evolution – in the course of a rigorous biopolicy. The question here is not where this hubris comes from – it is *one* of the possibilities of the Modern Age. In this case the destructive potential of human beings materialized behind the mask of reason. Zygmunt Bauman masterfully elaborated the logic underlying this phenomenon by drawing an analogy to the gardener who *weeds out* the garden of humanity and *wipes out* vermin[1] – there is no need to repeat this yet again. Rather, here the question is why precisely *this* possibility of the Modern Age blazed the way, why the many safeguards with which modern society constrains the traits of violence and destruction always inherent in modern science failed in this very special case.

First it must be stated that "scientificity" in and of itself offers no guarantees against transgressing ethical boundaries.[2] It is a tenacious bias that science, or at least "good" science – in the sense of "methodologically appropriate" science – must always be "good" in a moral sense as well, because the strict methodological standards of "pure science" prevent any research from drifting into barbarism. Inversely, this implies that morally reprehensible and ethically inadmissible research, from its very disposition, can be *no* science, but merely pseudoscience or, at best, "bad" science – in the sense of "methodologically inadequate." The history of the KWI-A demonstrates once more that this conception of an internal interconnection between methodology and ethics is more wish than reality: in terms of their methodological standards, the group of researchers around Eugen Fischer and Otmar von Verschuer were doubtlessly at the pinnacle of their age – even on the international scale, as the positive reception of the Verschuer's speech read *in absentia* at the Seventh International Congress for Genetics in Edinburgh proved in August 1939. It could be objected that a considerable share of the research at the institute in Dahlem was based on the twin method, which was no longer universally accepted. Despite the case where Verschuer attempted to use his academic authority to silence the methodological criticism of an outsider, while avoiding any discussion about its content; the publicly aired methodological controversies about twin research *within* the institute – between Lenz and Verschuer, or between Lenz, Kurt Gottschaldt, and Kurt Wilde – demonstrated that the scientific praxis at the KWI-A was accompanied by continuous and altogether self-critical methodological reflection – this functioned as an effective corrective against "methodologically inadequate" research.[3]

---

[1] Bauman, Moderne, pp. 43–56. Cf. also idem., Dialektik.

[2] The following according to Potthast, "Rassenkreise," pp. 304–308.

[3] Abderhalden's protective ferment reaction, which was applied in Verschuer's "Specific Proteins" project, also had become a matter of controversy in the early 1940s. We know too little about the project to decide whether critical reflection about the methods took place. However, the connection to the KWI for Biochemistry suggests that Verschuer and his associates hardly dealt with this method uncritically. Against the "scientificity" of the research at the institute in Dahlem it could also be objected that Fischer and Verschuer made several decisions about research strategy that can be interpreted as wrong from the perspective of the decades that followed: for instance, they resolved not to push ahead with blood group research in Dahlem after 1933, to abandon paleoanthropology for the most part (and thus miss the boat on the synthetic theory of evolution) and to ignore important fields of "classical" genetics like mutation research and population genetics (Benoît Massin argues this case rigorously in *Rasse und Vererbung*, pp. 213, 215, 224–227). On the one hand it must be remarked that the basic research strategic decision for "phenogenetics" in the years 1938–1942, which could be conceived as setting the course for a dead end in view of the

What follows from the fact that research in Dahlem nevertheless started down the slippery slope despite all methodological diligence is thus: science is not invulnerable to drifting into the sphere of crime by mere virtue of the fact that it adheres to methodological standards. Inversely, morally reprehensible and ethically inadmissible research – including that which exposes humans to debasing procedures, scares them or inflicts pain upon them, which willingly accepts the risk of a human subject's death or even brings it about – is not methodologically inadequate per se. Painstaking scientific work alone, without ethical reflection on a metalevel, generates at most "scientific integrity," which may prevent fraud (fabricating and doctoring data, manipulating methods or suppressing, falsifying or inventing results) but still cannot answer any of the ethical questions that concern the *context* of scientific praxis.

One critical objection could be raised against what has been stated so far: while the *praxis* of the institute in Dahlem may have fulfilled the methodological standards of "serious" research, the very *presentation* of the research results transgressed the boundaries of "scientificity." Admittedly, from the outset Fischer, Verschuer, and their associates understood their research, no matter how removed from praxis its experimental design, as *applied* science. In this understanding science and policy were two areas of society intimately connected with each other and oriented toward each other. Political guidelines flowed into the axioms, premises, concepts, and questions guiding race research, and politics was eager to "focus [their findings immediately] in the sense of possible guidelines for political behavior."[4] In their public position papers on issues of population, health, social, and race policy, the scientists of the institute in Dahlem postponed many a differentiation, many a reservation, many an open question, to present as hard "scientific facts" findings that were not (yet) considered certain in the scientific discourse, and cloaked scientific insights that were not quite compatible with the guidelines of National Socialist genetic health and race policy in nebulous formulations. Did not scientists who argued thus leave the ground of science? The question misconstrues the character of an applied science, which always follows a dual logic: the logic of science *and* the logic of politics. Its recommendations to politics are always the product of multiple factors: scientific knowledge, weighing the practical benefit of a measure against whether it is politically feasible and whether it will find cultural acceptance, and finally, the assessment of its ethical admissibility. As scientists, Fischer, Verschuer, and their staff felt bound to "truth;" in their function as political consultants and "genetic physicians," however, the "health of the *Volk*" and "purity of the race" enjoyed the highest priority. In this double bind constellation many

triumph of *molecular* genetics that soon followed, appears interesting again from today's perspective, as questions of *developmental* genetics pose themselves more urgently than ever since the successful decoding of the human genome. On the other hand, I must persevere in the view that decisions about research strategy – regardless of whether they are right or wrong – say nothing about methodological standards.

[4] Roelcke, Programm, p. 44.

race scientists at the beginning of the Third Reich pled for large-scale eugenic sterilization, although they were quite aware of the fact that in this they were treading on shaky scientific ground. At this point in time the thesis of the heritability of mental illnesses, and mental and physical disabilities constituted a plausible assumption, but one that was in need of empirical confirmation in many cases. The fact that this assumption sometimes was presented to the outside world as conclusive – to the public, but also to the state – is hardly unusual, and remains part of everyday science even today. They believed that, sooner or later, they would be able to provide empirical proof for the supposedly evident facts of the case.[5]

In the tug of war between scientific truth, political calculation, and personal professional ambition, Eugen Fischer was willing to make greater compromises than Otmar von Verschuer. Fischer, under open attack from the National Socialists for his views on miscegenation, saved himself in an uneasy compromise of formulation, ultimately without sacrificing his scientific position. However, on the international stage, subjected to the attacks of American scientists and responding with a polemical fundamental critique of the studies by Franz Boas, he made an intellectual sacrifice, having acknowledged the validity of Boas' studies on race anthropology internally and directed the work of his institute along its lines. In the issue of sterilization he quite bluntly argued, as we saw above, for performing sterilizations on a large scale, although the scientific foundations for this had not yet been laid.

In this respect Verschuer was, on the whole more, scrupulous. However, it is interesting that he, too, despite having arrived at a more skeptical view of National Socialist genetic health policy from his watchtower of higher Mendelism, did not draw from this the consequence to withdraw from the function of advisory expert and assessor, but on the contrary attempted to close the gap between political practice and the scientific basis – a gap that widened increasingly as scientific knowledge progressed – by intensifying research in the field. The high priority he placed on field research, in connection with the unfettered barriers to accessing human subjects in hospitals and asylums, in the ghetto, in war prison, concentration and extermination camps, resulted in accelerating the rate at which it drifted into the zone of crime.

Here it becomes clear that not even a more differentiated conception could protect the biosciences from this slippery slope. With the paradigm shift to phenogenetics, Verschuer, and his staff had left the reductionism of classical genetics behind them. Verschuer never entirely abandoned the idea of the primacy of heredity, which granted to the multifarious environmental influences the function of mere modifying factors – to that extent his position can be described appropriately as "extended reductionism."[6] The paradigm of phenogenetics, which guided research projects at the KWI-A from 1938 to 1942, was still well removed from the currently discussed conceptions of *epigenetics*, and yet, the group of scientists around

[5] Ibid., p. 57.

[6] I thank Katrin Grüber for this critical reference. Cf. also idem., Plädoyer.

Verschuer stood out in clear relief from the doctrinaire genetic determinism that was well alive elsewhere in National Socialist Germany. From this follows the connection between bioscience and biopolicy, regardless of whether the bioscience proceeds from the primacy of heredity, the primacy of environment or an interdependence between the two, holds enormous destructive potential.

Ultimately it is decisive whether bioscientific research is constrained by a wreath of juridical, political, social, and cultural safeguards, or whether it simply follows its own logic, subject only to self-checks through a more or less reflected interior ethics. From the standpoint of an internal scientific ethic, nothing spoke against taking advantage of the freedoms hardly conceivable up to that time, which opened up for bioscientific research in the structure of the National Socialist "dual state." On the contrary, because science is bound only to the guiding concept of "the truth," as far as internal scientific ethics are concerned, it is practically imperative to exhaust the possible barriers to access subjects that had been removed because of the cognitive gains to be expected. Josef Mengele expressed this concisely, it would have been a "sin," a "crime," *not* to seize the "chance" Auschwitz offered to bioscientific research. With the same perfidious logic, it would have been practically morally reprehensible for Magnussen and Verschuer not to evaluate the eyes and blood samples supplied by Mengele, especially since pathologists had never been known for their scruples in procuring their "material" and bore no misgivings about researching on execution victims. Besides, dealing with human specimens in the field of pathology was not a matter that opened up questions of ethics in and of itself.

For the area of human experimentation, the "Guidelines for Novel Therapy and Execution of Scientific Experiments on Humans" (*Richtlinien für neuartige Heilbehandlung und Vornahme wissenschaftlicher Versuche am Menschen*), were worked out by the Reich Health Council in 1930 and published 1 year later in a circular by the Reich Ministry of the Interior and in the *Reichgesundheitsblatt* ("Reich Health Newsletter"), constituted an orientation aid for the human sciences, which outlined the boundaries of ethically admissible human experimentation with sufficient clarity.[7] However, these guidelines were barely acknowledged, which can be interpreted as an indication of the widespread skepticism as to whether fixed rules could be set up at all for the highly sensitive area of human experimentation, or whether it must simply be left to medical ethos. On the other hand, these guidelines, which had come about upon the initiative of the Social Democratic Reichstag deputy Julius Moses, were understood from the outset – and even more so after 1933 – to be an expression of the Weimar system.

In National Socialist Germany there were tentative steps toward renewed reflection on the professional ethics of physicians. "Each generation," as Joachim Mrugowsky (1905–1948), the highest-ranking hygienist in the Reich SS Medical Corps, wrote

[7] Winau, Versuche, pp. 173–175; Saretzki, Reichsgesundheitsrat, pp. 289–295; Frewer, Medizin und Moral, pp. 139–145.

in the introduction to a volume he edited in 1939 with texts by Christoph Wilhelm Hufeland (1762–1836), had "its own outlook on the problems of life [...] It would not be youth if it did not laughingly leap over the restrictive reservations of the older generation [...] Every other problem of philosophy and biology is subject to the same transformation as this question about one's outlook on life in general. There is nothing eternally constant, and even the facts that appear to be set in stone today may tomorrow be connected up in a different fashion and in new display [...]."[8] Seemingly effortlessly, Mrugowsky succeeded in placing Hufeland's ideas in the service of the National Socialist "health leadership." We do not know whether Eugen Fischer and Otmar von Verschuer read Mrugowsky's book, but it was included in the KWI-A library.[9] The guiding ideas of the "new ethics" are found once more in Verschuer especially: the natural order of what happens in society, the principle of selection and the degeneration thesis, entailing the absolute equation of the living organism with the social structure regarded to be the actual object of the "genetic scientist" and "genetic physician," the devaluation and objectification of the individual, the priority of collective ethics over individual ethics. Because the focus shifted to the "*Volkskörper* (body of the nation)" as the object of treatment, the consequence for the ethic of the "genetic physician" was that all values had to be reassessed. Against this background, participation in a biopolicy oriented toward large-scale social engineering, and the exploitation of the unfettered access to "human material" opened up by this policy, appeared not only ethically admissible, but, as it were, ethically required.

Verschuer's case poses the question of how this position could be reconciled with his Christian faith – Verschuer was a practicing Protestant and closely linked with the Confessing Church from its very beginnings – a circumstance that was suspect to the National Socialists. Since World War II, the apparent contradiction between Verschuer's support for the Confessing Church and his participation in National Socialist biopolicy has been pointed out repeatedly, whereby this contradiction was usually appraised as an indication that Verschuer got involved in the crimes of the National Socialist state against his inmost conviction, "in order to prevent the worst." This argumentation – usually implicit – presupposes as a matter of course that the Confessing Church stood in fundamental opposition to the genetic health and race policy of the National Socialists. However, recent research on current church history suggests that large sectors of German Protestantism were by no means negatively disposed to the genetic health and race policy of the

[8] Mrugowsky, Ärztliches Ethos, pp. 7 f.

[9] According to the Inventory List; cf. MPG Archive, Dept. I, Rep. 1 A, No. 3034. Mrugowski's book is located under the signature F II, where biographical references on scientists and physicians, autobiographical and biographical works, science history works, but also travel literature are collected. Among the books in this stock with a similar thematic emphasis are *Das Weltbild des Arztes und die moderne Physik* by the internist Gustav von Bergmann (1878–1955), *Der Arzt* by the internist Ludolf von Krehl (1871–1937), and *Der Arzt und seine Sendung* by the surgeon and writer Erwin Liek (1878–1935).

National Socialist regime, and this was true not only for the German Christians, but also well into the ranks of the Confessing Church.[10] Public criticism by the Confessing Church was oriented against the race ideology of Alfred Rosenberg and his adherents, who wanted to make racism into a substitute religion or a religious substitute, but not against scientifically grounded biopolicy. Large sectors of German Protestantism understood "race, *Volk* and state," as Walter Künneth formulated in his introductory essay to the omnibus volume *Die Nation vor Gott*, as "preservation orders of God," such that "the belief in God the creator [included] the affirmation of race, *Volk*, state as God's gifts."[11] On this basis Verschuer had no problem reconciling with each other bioscience, biopolicy, and Protestant Christianity – as mentioned above, he, too participated willingly in the omnibus conceived of as a "message of the Church in the Third Reich."

Starting in the 1920s, Verschuer developed his own rudimentary strategy of justification, which revolved around the concept of "sacrifice." In his inaugural speech as a newly appointed member of the Prussian Academy of Sciences on November 10, 1943, Verschuer had elaborated that even the people whose fate was restricted by their genetic disposition could give their lives value and meaning – through "selfless readiness to make sacrifices." On the one hand, this readiness to make sacrifices meant the willingness to volunteer for sterilization, but on the other hand it also meant the willingness to place one's own body at the disposal of science – both with the objective of "wiping out" in the next generations that genetic disposition which had staked such close boundaries for their own lives. In Verschuer's eyes the human beings who *fell victim* to the bioscientific research at his institute had *made a sacrifice* for science.[12] And since in Verschuer's world-view the researcher – in the sense of interior asceticism – *sacrificed* his life for science, the unfettered research on humans constituted a mutual act of sacrifice by the researcher and his research object: on this side the heroic figure of the scientist, who, in the service of science, overcame his own fears and sense of guilt to advance into ethical boundary regions like the priest and executioner, *dehumanized* himself or herself to a degree; and on the other side the people who are *genetically inferior*, whose existence receives value and meaning only through their being prepared as objects of science, that is, also through *dehumanization* (albeit in the sense of objectification), bound together in a ceremony characterized

[10] Research on the contemporary history of the church still has a broad field to plow on this subject. The first steps toward a new interpretation: Gailus, Antisemitismus; idem., Sozialpfarrer; Lindemann, "Typisch jüdisch"; idem., Antijudaismus; Nowak, "Euthanasie"; Kaiser, Innere Mission; idem., Protestantismus, pp. 316–390; Schleiermacher, Sozialethik, pp. 223–254.

[11] Künneth, Offenbarung, p. 41, 42.

[12] The German language does not distinguish between these two semantic aspects of the word *Opfer*, which can be discerned in English by using "victim" (from the Latin "victima," which is used in particular for an *animal* to be sacrificed) and "sacrifice" (from the Latin "sacrificium," which emphasizes the sanctification of the victim through the ceremony of sacrifice). The German language is thus particularly suitable for blurring the difference between these two levels of meaning.

by concealed purification rituals and gestures of reconciliation. This attempt to give the events a "sacred meaningfulness"[13] was a version of that connection, observed since the Early Modern Age, between execution and torture, anatomy and experimental medicine, created from elements of the cult of the dead and the cult of sacrifice, which continues to be (subliminally) effective even after the rationalist turn of science in the period saddling the outgoing eighteenth and incoming nineteenth centuries. This relationship of scientists like Verschuer to their human research objects is thus strangely ambivalent: on the one hand it is governed by the "methodological demand that scientific objectivity could only be guaranteed under circumstances of the absolute suspension of feelings."[14] This leads to a process of dehumanization, which makes material out of humans and subjugates those who exploit this material to moral anesthesia. On the other hand, however, this process of dehumanization is surmounted sacredly – it is the means to a higher end, and this end justifies the means.

What follows from all of this? Bioscientific research is safe from drifting into barbarism only when it maintains respect for life. Finding the location of the Archimedes' point upon which this respect for life is based – a universal ethical principle derived from the world religions, Western humanism, an Enlightenment aware of its own depths, a form of postmodern consensual ethics – that is the great challenge for bioethics in the coming decades. Only if we succeed in solidly anchoring the respect for life in the biosciences will they surmount the level of "false learnedness" – as Fyodor Mikhailovich Dostoevsky (1821–1881) once called it – and achieve "true learnedness":

> True learnedness is not antagonistic to life, but rather ultimately always accords with life, which it gives *new revelations* that it discovers in life itself. That is the essential and marvelous characteristic of true learnedness. False learnedness, on the other hand is, no matter how great it may be, somehow always antagonistic to life and, if anything, can lead to the negation of life.[15]

[13] Bergmann, Patient, p. 270. The following considerations are from this new, exciting study, although it often drifts into speculation.

[14] Ibid., p. 272.

[15] Dostoevsky, Tagebuch, pp. 614 f. (original emphasis).

Sources and Literature

Unprinted Sources

Archive on the History of the Max Planck Society, Berlin (MPG Archive)

Dept. I, Rep. 1 A:	General Administration of the Kaiser Wilhelm Society
Dept. I, Rep. 3:	Institute Files of the Kaiser Wilhelm Institute for Anthropology, Human Heredity and Eugenics
Dept. I, Rep. 20 A:	Estate of Hans Nachtsheim (office area)
Dept. II, Rep. 1 A:	General Administration of the Max Planck Society, personnel files
Dept. III, Rep. 48:	Estate of Georg Geipel
Dept. III, Rep. 84:	Estate of Adolf Butenandt
Dept. III, Rep. 86 A:	Partial Estate of Baron Otmar von Verschuer
Dept. III, Rep. 86 A (Münster):	Partial Estate of Baron Otmar von Verschuer, Formerly Archive of the University of Münster
Dept. III, Rep. 86 B:	Correspondence Fritz Lenz – Baron Otmar von Verschuer
Dept. III, Rep. 86 C:	Correspondence Eugen Fischer – Baron Otmar von Verschuer
Dept. V a, Rep. 16:	Collection of Baron Otmar von Verschuer
Dept. IX, Rep. 1:	Documentation on Baron Otmar von Verschuer
Dept. IX, Rep. 2:	Documentation on KWI for Anthropology, Human Heredity and Eugenics

Federal Archive Berlin-Lichterfelde (BArch. Berlin)

BDC: Former Berlin Document Centre
NS 11: Parteiamtliche Prüfungskommission zum Schutze des NS- Schrifttums (Official Party Board of Examiners for the Protection of NS Literature)
NS 15: Der Beauftragte des Führers für die Überwachung der gesamten geistigen und weltanschaulichen Schulung und Erziehung der NSDAP (Deputy of the Führer for the Monitoring of the Entire Intellectual and Ideological Training and Education of the NSDAP)
R 26 III: Reichsforschungsrat (Reich Research Council)
R 43 II: Reichskanzlei (Reich Chancellory)
R 49: Reichskommissar für die Festigung des deutschen Volkstums (Reich Commissioner for the Consolidation of German Nationhood)
R 96 I: Reichsarbeitsgemeinschaft für Heil- und Pflegeanstalten (Reich Working Committee for Institutions of Healing and Care)
R 1501: Reichsministerium des Innern (Reich Ministry of the Interior)
R 4901: Reichsministerium für Wissenschaft, Erziehung und Volksbildung (Reich Ministry for Science, Training and Education of the Nation)

Federal Archive Koblenz (BArch. Koblenz)

R 73: Deutsche Forschungsgemeinschaft/Reichsforschungsrat (German Research Association/Reich Research Council)

Archive of the Diakonisches Werk, Berlin (ADW)

CA, G 381

Archive of the Evangelischer Diakonieverein Berlin-Zehlendorf

W 3848 (pre-archive)

Archive of the University of Frankfurt/Main, Files of the President

Dept. 13, No. 347: Personnel file for Baron Otmar von Verschuer

State Archive of Greifswald

Rep. 76 G, No. 536

Institute for Contemporary History, Munich (IfZ)

MA 116/16
MA 287

Max Planck Institute for Psychiatry, Historical Archive, Munich (MPIP-HA)

Genealogisch-Demographische Abteilung (Department of Geneaology and Demography)

Estate of Leonardo Conti (private property, soon in the BArch.)

Printed Sources

Abel, Wolfgang, *Bastarde am Rhein*, in: Neues Volk 2, 1934, pp. 4–7.

Abel, Wolfgang, Die *Erbanlagen der Papillarmuster*, in: Just (ed.), Handbuch der Erbbiologie, vol. III, pp. 407–440.

Abel, Wolfgang, Die *Erbanlagen des normalen Stützgewebes*, in: Just (ed.), Handbuch der Erbbiologie, vol. III, pp. 1–45.

Abel, Wolfgang, Über *Europäer-Marokkaner-* und Europäer-Annamiten-*Kreuzungen*, in: ZMA 36, 1937, pp. 311–329.

Abel, Wolfgang, Ein *Fall von Vererbung* gestörter Papillarmuster, in: ZMA 41, 1944–1949, pp. 73–86.

Abel, Wolfgang, *Finger- und Handlinienmuster*, in: Wegener (ed.), Wissenschaftliche Ergebnisse der Deutschen Grönland-Expedition, vol. VI, pp. 1–23.

Abel, Wolfgang, Über die *Frage* der Symmetrie der menschlichen Fingerbeere und der Rassenunterschiede der Papillarmuster, in: Biologia Generalis 9, 1933, pp. 13–32.

Abel, Wolfgang, Das *Gebiß* der Feuerland-Indianer, in: ZMA 38, 1940, pp. 349–358.

Abel, Wolfgang, *Kritische Studien* über die Entwicklung der Papillarmuster auf den Fingerbeeren, in: ZMVKL 21, 1938, pp. 497–529.

Abel, Wolfgang, *Physiognomik* und Mimik, in: Just (ed.), Handbuch der Erbbiologie, vol. III, pp. 407–440.

Abel, Wolfgang, *Rassenprobleme im Sudan* und seinen Randgebieten, in: Koloniale Völkerkunde, pp. 140–151.

Abel, Wolfgang, Über *Störungen der Papillarmuster*. I: Gestörte Papillarmuster in Verbindung mit einigen körperlichen und geistigen Anomalien, in: ZMA 36, 1936, pp. 1–38.

Abel, Wolfgang, Die *Vererbung* von Antlitz und Kopfform des Menschen, in: ZMA 33, 1934, pp. 261–345.

Abel, Wolfgang, *Vererbung normaler morphologischer Eigenschaften* des Menschen (I. Teil), in: Fortschritte der Erbpathologie, Rassenhygiene und ihrer Grenzgebiete 4, 1940, pp. 211–238.

Abel, Wolfgang, Über die *Verteilung der Genotypen* der Hand- und Fingerbeerenmuster bei europäischen Rassen, in: ZIAVL 70, 1935, pp. 458–460.

Abel, Wolfgang, *Zähne* und Kiefer in ihren Wechselbeziehungen bei Buschmännern, Hottentotten, Negern und deren Bastarden, in: ZMA 31, 1933, pp. 314–361.

Abel, Wolfgang, *Zwillinge bei Mantelpavianen* und die Zwillingsanlage innerhalb der Primaten, in: ZMA 31, 1933, pp. 266–275.

Agthe, Margaret; Gertud v. Poehl, Das *Judentum*, das wahre Gesicht der Sowjets, Berlin 1941.

Ammon, Otto, Die *natürliche Auslese* beim Menschen, Jena 1893.

Aus der Gesellschaft für Rassenhygiene und Eugenik. Die Denkschrift Muckermanns an die Regierung, in: AfRGB 26, 1932, pp. 231–233.

Baader, Otto, *Cro-magnide Typen* aus Nordafrika, in: ZMA 41, 1944, pp. 155–159.

Baur, Erwin, Die *biologische Bedeutung* der Auswanderung für Deutschland, in: Archiv für Frauenkunde und Eugenetik, Sexualbiologie und Vererbungslehre 7, 1921, pp. 206–208.

Baur, Erwin, Die *Erhaltung* und Stärkung unserer Volkskraft nach dem Kriege, in: Der Tag, Berlin, 9/6/1916.

Baur, Erwin, *Rassenpolitik* und Rassenhygiene, in: Tägliche Rundschau, Unterhaltungsbeilage, Berlin, 23/3/1917.

Baur, Erwin, Der *Untergang* der Kulturvölker im Lichte der Biologie, in: Deutschlands Erneuerung 6, 1922, pp. 257–268.

Baur, Erwin; Eugen Fischer; Fritz Lenz, Menschliche *Erblichkeitslehre* [in later editions: Erblehre] und Rassenhygiene. vol. I: idem., Menschliche Erblichkeitslehre [in later editions: Erblehre], Munich 1920; vol. II: Fritz Lenz, Menschliche Auslese und Rassenhygiene (Eugenik), Munich 1921.

Becker, Ernst-Georg, *Pneumonien* bei Zwillingen, in: ZMVKL 22, 1938, pp. 77–95.

Becker, Peter Emil, Zur *Erbbiologie* der Speiseabneigungen (Ein Beitrag zur Zwillingsforschung), in: AfRGB 32, 1938, pp. 223–237.

Becker, Peter Emil, Zur *Erblichkeit der Ischias* (Zwillingsstudien über die Erbanlagen bei der Neuritis lumbosacralis), in: Zeitschrift für die gesamte Neurologie und Psychiatrie 161, 1938, pp. 183–201.

Becker, Peter Emil, Zur *Erblichkeit der Motorik*, in: Zeitschrift für die gesamte Neurologie und Psychiatrie 161, 1938, pp. 374–378.

Becker, Peter Emil, *Zwillingsstudien* zur Strichführung, in: ZIAVL 73, 1936, pp. 517 f.

Becker, Peter Emil; Fritz Lenz, Die *Arbeitskurve* Kraepelins und ein psychomotorischer Versuch in der Zwillingsforschung (Zugleich ein methodischer Beitrag zur Zwillingsforschung), in: Zeitschrift für die gesamte Neurologie und Psychiatrie 164, 1938, pp. 50–68.

Behr-Pinnow, Carl von, Der *Deutsche Bund für Volksaufartung* und Erbkunde, in: Kultur und Leben 2, 1925, pp. 410 f.

Bergmann, Gustav von, Das *Weltbild des Arztes* und die moderne Physik. Ein Ausgleich alter Widersprüche, Berlin 1943.

Bericht über die Verhandlungen psychiatrischer Vereine: Deutscher Verein für Psychiatrie, in: Allgemeine Zeitschrift für Psychiatrie 76, 1920/21, p. 598

Berliner, Max, Einige konstitutionelle und umweltbedingte *Besonderheiten des Kleinkindesalters* in ihren Beziehungen zu Krankheitszuständen, in: Gesundheitsfürsorge für das Kindesalter 3, 1928, pp. 305–319.

Berliner, Max, *Blutgruppenzugehörigkeit* und Rassenfragen, in: ZMA 27, 1929, pp. 161–170.

Berliner, Max, *Hochwuchs* und Breitenentwicklung, in: Zeitschrift für klinische Medizin 108, 1928, pp. 378–385.

Berliner, Max, Vergleichende *Untersuchungen über die Weichteilhärte* am Lebenden, in: Zeitschrift für Konstitutionslehre 15, 1929, pp. 114–126.

Berliner, Max, *Untersuchungen über optisch wahrnehmbare Phänomene* an Punktionsflüssigkeiten bei Carcinom- und anderen Krankheiten, in: Zeitschrift für Krebsforschung 32, 1930, pp. 171–181.

Berliner Gesellschaft für Rassenhygiene (ed.), Was will die *Rassenhygiene*?, Berlin 1917.

Binding, Karl; Alfred E. Hoche, Die *Freigabe* der Vernichtung lebensunwerten Lebens. Ihr Maß und ihre Form, Leipzig 1920, 2nd edition, 1922.

Binswanger, Ludwig; Stavros Zurukzoglu (eds), *Verhütung* erbkranken Nachwuchses. Eine kritische Betrachtung und Würdigung, Basel 1938.

Biswas, Prophulla Chandra, Über *Hand- und Fingerleisten* von Indern, in: ZMA 35, 1936, pp. 519–550.

Blasbalg, Jenny, Ausländische und deutsche *Gesetze* und Gesetzentwürfe über Unfruchtbarmachung, in: Zeitschrift für die gesamte Strafrechtswissenschaft 52, 1932, pp. 477–496.

Boeters, Heinz, Das *Hypophysen-Vorderlappenhormon* (Prolan) und die männliche Keimdrüse, Diss. Berlin 1931.

Boeters, Heinz, *Untersuchungen* über Familienaufbau und Fruchtbarkeitsziffern bei rußlanddeutschen Bauern, in: AfRGB 30, 1936, pp. 36–42.

Bohn, Wolf, Die *Deutsche Gesellschaft für Rassenhygiene* seit der Machtübernahme, in: Allgemeine Zeitschrift für Psychiatrie und ihre Grenzgebiete 112, 1939, pp. 463–469.

Bonhoeffer, Karl (ed.), Die psychiatrischen *Aufgaben* bei der Ausführung des Gesetzes zur Verhütung erbkranken Nachwuchses, Berlin 1934.

Bonhoeffer, Karl (ed.), Die *Erbkrankheiten*. Klinische Vorträge im 2. erbbiologischen Kursus in Berlin, Berlin 1936.

Bonhoeffer, Karl, *Lebenserinnerungen*, in: Jürg Zutt et al. (eds), Karl Bonhoeffer. Zum hundertsten Geburtstag am 31. März 1968, Berlin 1969, pp. 8–107.

Braeuning, Hermann; Franz Redeker, Die haematogene *Lungentuberkulose* des Erwachsenen, Leipzig 1931.

Bratz, Emil, Kann die *Versorgung* der Geisteskranken billiger gestaltet werden und wie?, in: Allgemeine Zeitschrift für Psychiatrie 95, 1932, pp. 1–40.

Brauns, Luise, *Studien* an Zwillingen im Säuglings- und Kleinkindalter. Ein Beitrag zur Zwillingsbiologie, in: Zeitschrift für Kinderforschung 43, 1934, pp. 86–129.

Breymann, Hans, Über die *Notwendigkeit* eines Zusammengehens von Genealogen und Medizinern in der Familienforschung, in: AfRGB 9, 1912, pp. 18–29.

Bühler, Engelhard, *Experimentelle Untersuchungen* über Diphterie-Immunität, in: Verhandlungen der Deutschen Gesellschaft für innere Medizin 49, 1937, pp. 368–373.

Bühler, Engelhard, Die *Oberlidfalte* am Europäerauge, in: ZMA 38, 1939, pp. 56–62.

Bühler, Engelhard, *Untersuchungen über die Erblichkeit des Isoagglutinationstiters*, in: ZIAVL 70, 1935, pp. 463–467.

Bühler, Engelhard, *Zwillingsstudien* über Falten und Furchen des Antlitzes, in: Verhandlungen der Deutschen Gesellschaft für Rassenforschung 9, 1938, pp. 54–61.

Bühler, Engelhard; Fritz Lenz, Über die *Frage* der Erblichkeit der Disposition bzw. Immunität bei Kinderkrankheiten, in: ZIAVL 73, 1936, pp. 536–541.

Büscher, Karl-Georg, *Wandel* der Gesichtspunkte der Gattenwahl im Spiegel der privaten Heiratsanzeigen, Diss. Berlin 1941.

Burkert, Günter, *Auslesevorgänge* durch Ab- und Zuwanderung in einer hessischen Landbevölkerung, in: AfRGB 32, 1938, pp. 407–427.

Busse, Hertha, Über normale *Asymmetrien* des Gesichts und im Körperbau des Menschen, in: ZMA 35, 1936, pp. 412–445.

Busse, Hertha, Altslawische *Skelettreste* im Potsdamer Havelland, in: Zeitschrift für Ethnologie 66, 1934, pp. 111–128.

Carmena, Miguel, Ist die persönliche *Affektlage* oder "Nervosität" eine vererbte Eigenschaft? In: Zeitschrift für die gesamte Neurologie und Psychiatrie 150, 1934, pp. 434–445.

Carmena, Miguel, *Schreibdruck* bei Zwillingen, in: Zeitschrift für die gesamte Neurologie und Psychiatrie 152, 1935, pp. 19–24.

Claußen, Ferdinand, *Phänogenetik* vom Menschen, in: ZIAVL 76, 1939, pp. 14–46.

Conitzer, Harry, Die *Rothaarigkeit*, in: ZMA 29, 1931, pp. 84–147.

Curtius, Friedrich, *Nachgeburtsbefunde* bei Zwillingen und Ähnlichkeitsdiagnose, in: Archiv für Gynäkologie 140, 1930, pp. 361–366.

Curtius, Friedrich, Familiäre diffuse *Sklerose* und familiäre spastische Spinalparalyse in einer Sippe. Ein Beitrag zur Genealogie der Heredodegenerationen, in: Zeitschrift für die gesamte Neurologie und Psychiatrie 126, 1930, pp. 209–227.

Curtius, Friedrich; Otmar v. Verschuer, Die *Anlage zur Entstehung von Zwillingen* und ihre Vererbung, in: AfRGB 26, 1932, pp. 361–387.

Czapnik, Carl Role, Über die *Erbbedingtheit* der Intersexualität, in: AfRGB 36, 1942, pp. 163–221.

Deeg, Peter; Julius Streicher (eds), *Hofjuden*, 8th/9th edition, Nuremberg 1939.

Der neue Entwurf des allgemeinen Deutschen Strafgesetzbuches vom ärztlichen Standpunkte. Bericht über die Sitzung des verstärkten Ausschusses (für gerichtliche Medizin pp.) des Landesgesundheitsrates am 30. und 31. Oktober 1925, in: Veröffentlichungen aus dem Gebiete der Medizinalverwaltung 21, 1926, pp. 35–172.

Die Eugenik im Dienste der Volkswohlfahrt. Bericht über die Verhandlungen eines zusammengesetzten Ausschusses des preußischen Landesgesundheitsrates vom 2. Juli 1932 (=Veröffentlichungen auf dem Gebiete der Medizinalverwaltung; 38, H. 5), Berlin 1932, pp. 1–112.

Diedrich, Heinz, *Erhebungen* an Stettiner Grundschülern über Schulleistung, Begabung und Geschwisterzahl, in: AfRGB 35, 1941, pp. 1–17.

Diehl, Karl, Das *Erbe* als Gestaltungsfaktor der Tuberkulose, in: Die Tuberkulose, vol. I: Allgemeine Biologie und Pathologie der Tuberkulose, Leipzig 1943, pp. 633–695.

Diehl, Karl, *Tierexperimentelle Erbforschung* bei der Tuberkulose, in: Beiträge zur Klinik der Tuberkulose 97, 1942, pp. 331–349.

Diehl, Karl; Otmar von Verschuer, Der *Erbeinfluß* bei der Tuberkulose (Zwillingstuberkulose II), Jena 1936.

Diehl, Karl, *Erbuntersuchungen* an tuberkulösen Zwillingen, in: Beiträge zur Klinik der Tuberkulose 75, 1930, pp. 206–215.

Diehl, Karl, *Tuberkulose und Eugenik*, in: Eugenik, Erblehre, Erbpflege 3, 1933, pp. 8–14.

Diehl, Karl, *Zwillingstuberkulose*. Zwillingsforschung und erbliche Tuberkulosedisposition, Jena 1933.

Dönges, Karl, *Fragen* zur erblichen Bedingtheit der sog. Luxatio coxae congenita, Diss. Frankfurt 1939/40.

Dokumente aus der eugenischen Bewegung: I. Anträge für das Reichsstrafgesetzbuch, in: Das kommende Geschlecht 5, 1929, pp. 20–24.

Dornfeldt, Walter, *Studien* über Schädelform und Schädelveränderung von Berliner Ostjuden und ihren Kindern, in: ZMA 39, 1941, pp. 290–372.

Dostojewskij, Fjodor Michaijlowitsch, *Tagebuch* eines Schriftstellers (=Sämtliche Werke in zehn Bänden; 5), Munich 1977.

Dürre, Konrad, Die *eugenische Bewegung* in Deutschland und in anderen Ländern, in: Erblehre – Erbpflege, pp. 80–91.

Duis, Bernhard T., *Fingerleisten* bei Schizophrenen, in: ZMA 34, 1936, pp. 391–417.

Eggert, Horst, *Ehescheidungen* und ihre gesellschaftsbiologischen Ursachen, in: AfRGB 34, 1940, pp. 339–382.

Eichstädt, Volkmar, *Bibliographie* zur Geschichte der Judenfrage, Hamburg 1938.

Erblehre – Erbpflege, Zentralinstitut für Erziehung und Unterricht (ed.), Berlin 1933.

Eugenische Maßnahmen als Aufgabe des Staates, in: Deutsches Ärzteblatt 61, 1932, pp. 119–121.

Fetzer, Christian, Rassenanatomische *Untersuchungen* an 17 Hottentottenköpfen, in: ZMA 16, 1914, pp. 95–156.

Fischer, Eugen, Sind die *alten Kanarier* ausgestorben? in: Zeitschrift für Ethnologie 62, 1930, pp. 258–281.

Fischer, Eugen, *Anthropologische Erhebungen* an der deutschen Bevölkerung, in: Verhandlungen der Gesellschaft für Physische Anthropologie 4, 1930, pp. 21 f.

Fischer, Eugen, *Aufgaben* der Anthropologie, menschlichen Erblehre und Eugenik, in: Die Naturwissenschaften 14, 1926, pp. 749–755.

Fischer, Eugen, Der *Begriff des völkischen Staates*, biologisch betrachtet. Rede bei der Feier der Erinnerung an den Stifter der Berliner Universität, König Friedrich Wilhelm III. in der Alten Aula am 29. Juli 1933, Berlin 1933.

Fischer, Eugen, Anthropologische *Bemerkungen* zu den Masken, in: Georg Karo, Schachtgräber von Mykenai, 1933, pp. 320–331.

Fischer, Eugen, *Einleitender Vortrag* auf dem Internationalen Kongreß für Bevölkerungswissenschaft, in: Hans Harmsen, Franz Lohse (eds), Bevölkerungsfragen. Bericht des Internationalen Kongresses für Bevölkerungswissenschaft, Berlin 26.8.1935–1.9.1935, Munich 1936, pp. 39–43.

Fischer, Eugen, Die *Entwicklungsgeschichte des Dachses* und die Frage der Zwillingsbildung, in: Verhandlungen der Anatomischen Gesellschaft, 40. Tagung in Breslau, 10–13 April 1931, pp. 22–34.

Fischer, Eugen, *Erbschädigung* beim Menschen, in: Das kommende Geschlecht 5, 1930, pp. 1–19.
Fischer, Eugen, *Europäer-Polynesier-Kreuzung*, in: ZMA 28, 1930, pp. 205–209.
Fischer, Eugen, Die *Fortschritte der menschlichen Erblehre* als Grundlage eugenischer Bevölkerungspolitik, in: Deutsche Forschung 20, 1933, pp. 55–71.
Fischer, Eugen, Zur *Frage der Zwillingsbildung* beim Nagetier, in: Wilhelm Roux' Archiv für Entwicklungsmechanik der Organismen 118 [Fs. für Hans Spemann], 1929, pp. 352–358.
Fischer, Eugen, Zur *Frage einer äthiopischen Rasse*, in: ZMA 27, 1929, pp. 339–341.
Fischer, Eugen, Zur *Frage "Rassenmischehe"* in den Kolonien, in: Volk und Rasse 11, 1936, pp. 460 f.
Fischer, Eugen, *Geistige Rassenunterschiede*, in: Rassenpolitische Auslands-Korrespondenz 1939, No. 4.
Fischer, Eugen, Versuch einer *Genanalyse* des Menschen, in: ZIAVL 54, 1930, pp. 128–234.
Fischer, Eugen, *Genetik* und Stammesgeschichte der menschlichen Wirbelsäule, in: Biologisches Zentralblatt 53, 1933, pp. 203–220.
Fischer, Eugen, Aus der *Geschichte* der Deutschen Gesellschaft für Rassenhygiene, in: AfRGB 24, 1930, pp. 1–5.
Fischer, Eugen, Das *Kaiser-Wilhelm-Institut für Anthropologie*, menschliche Erblehre und Eugenik, in: Adolf von Harnack (ed.), Handbuch der Kaiser-Wilhelm-Gesellschaft zur Förderung der Wissenschaften, Berlin 1928, pp. 116–121.
Fischer, Eugen, *Menschliche Erblehre* und ärztliche Praxis, in: Deutsche Medizinische Wochenschrift 65, 1939, pp. 485–487.
Fischer, Eugen, *Menschliche Erblehre* und ärztliche Praxis, in: Wiener Klinische Wochenschrift 52, 1939, p. 601.
Fischer, Eugen, *Neue Rehobother Bastardstudien, I*: Antlitzveränderungen verschiedener Altersstufen bei Bastarden, in: ZMA 37, 1938, pp. 127–139.
Fischer, Eugen, *Neue Rehobother Bastardstudien, II*: Fortführung und Ergänzung der Sippentafeln, in: ZMA 40, 1942, pp. 1–33.
Fischer, Eugen, Versuch einer *Phänogenetik* der normalen Eigenschaften des Menschen, in: ZIAVL 76, 1939, pp. 47–117.
Fischer, Eugen, *Rasse und Vererbung geistiger Eigenschaften*, in: ZMA 38, 1939, pp. 1–9.
Fischer, Eugen, Die menschlichen *Rassen als Gruppen* mit gleichen Gen-Sätzen (=Abhandlungen der Preußischen Akademie der Wissenschaften, Mathematisch-naturwissenschaftliche Klasse; ed. 1940, No. 3), Berlin 1940.
Fischer, Eugen, *Rassenentstehung* und älteste Rassengeschichte der Hebräer, in: Sitzungsberichte der Dritten Münchner Arbeitstagung des Reichsinstituts für Geschichte des neuen Deutschlands vom 5.–7. Juli 1938 (=Forschungen zur Judenfrage; 3), Hamburg 1938, pp. 121–136.
Fischer, Eugen, Zur *Rassenfrage der Etrusker* (=Sonderausgabe aus den Sitzungsberichten der Preußischen Akademie der Wissenschaften, Physikalisch-mathematische Klasse; ed. 1938, No. 25), Berlin 1938.
Fischer, Eugen, Die *Rehobother Bastards* und das Bastardisierungsproblem beim Menschen. Anthropologische und ethnologische Studien am Rehobother Bastardvolk in Deutsch-Südwestafrika, Jena 1913, reprint Graz 1961.
Fischer, Eugen, *Rez.: Hans Friedrich Karl Günther, Rassenkunde des deutschen Volkes* [1926], in: ZMA 26, 1926/27, p. 190.
Fischer, Eugen, *Rez.: Hans Friedrich Karl Günther, Rassenkunde des deutschen Volkes* [1928], in: ZMA 27, 1928/30, p. 158.
Fischer, Eugen, *Rez.: Hans Friedrich Karl Günther, Rassenkunde des jüdischen Volkes* [1930], in: ZMA 29, 1931, p. 519.
Fischer, Eugen, *Rez.: Houston Stewart Chamberlain, Rasse und Persönlichkeit* [1925], in: ZMA 25, 1925, p. 534.
Fischer, Eugen, *Rez.: Ludwig Schemann, Die Rasse in den Geisteswissenschaften* [1928], in: ZMA 27, 1928/30, p. 346.
Fischer, Eugen, *Rez. Walter Darré, Neuadel aus Blut und Boden*, in: ZMA 29, 1931, p. 520.

Fischer, Eugen, *Strahlenbehandlung* und Nachkommenschaft, in: Deutsche Medizinische Wochenschrift 55, 1929, pp. 89–91.

Fischer, Eugen, *Taubstummheit* und Eugenik, in: Eugenik, Erblehre, Erbpflege 3, 1933, pp. 121–124.

Fischer, Eugen, *Untersuchungen* über die süddeutsche Brachykephalie, III: Die Gebeine aus dem karolingischen Kloster Lorsch, in: ZMA 31, 1933, pp. 283–298.

Fischer, Eugen, Über *Varietätenforschung*, in: Verhandlungen der Gesellschaft für Physische Anthropologie 3, 1929, pp. 16–22.

Fischer, Eugen, Rassenkundliche Probleme in *Weißafrika*, in: Koloniale Völkerkunde, pp. 130–139.

Fischer, Eugen; Karl Diehl, *Experimente* über die Tuberkulose bei Kaninchen, in: Der Erbarzt 8, 1940, pp. 93–99.

Fischer, Eugen; Hans Friedrich Karl Günther, *Deutsche Köpfe* nordischer Rasse, Munich 1927.

Fischer, Eugen; Gerhard Kittel, Das antike *Weltjudentum*. Tatsachen, Texte, Bilder (=Forschungen zur Judenfrage; 7), Hamburg 1943 [consigned 1944].

Fischer, Max, Der *Alkoholmißbrauch*, in: Das kommende Geschlecht 4, 1929, pp. 1–72.

Fischer, Max, Die *Formung* der menschlichen Nase in der Pubertät, in: Archiv für Frauenkunde und Konstitutionsforschung 16, 1930, pp. 113–117.

Fischer, Max, *Hämophilie* und Blutsverwandtschaft, in: Zeitschrift für Konstitutionsforschung 16, 1932, pp. 502–512.

Fischer, Max; Heinz Lemser, Zur Frage der Erkennungsmöglichkeit heterozygoter und homozygoter Erbanlagen für Diabetes mit Hilfe der Kupferreaktion, in: Der Erbarzt 5, 1938, p. 73.

Fischer, Werner; Benno Raquet, *Beitrag* zur Frage des Nachweises einer serologischen Differenzierung der menschlichen Rassen, in: Zeitschrift für Immunitätsforschung 94, 1938, pp. 104–121.

Fishberg, Maurice, Die *Rassenmerkmale* der Juden. Eine Einführung in ihre Anthropologie, Munich 1913.

Fleury Cuello, Eduardo, *Untersuchungen* über die süddeutsche Brachykephalie, in: ZMA 30, 1932, pp. 406–428.

Francke, Ernst, *Körperbau* und Refraktion, in: Klinisches Mitteilungsblatt für Augenheilkunde 101, 1938, pp. 184–204.

Frede, Maria, *Untersuchungen* an der Wirbelsäule und dem Extremitätenplexus der Ratte, in: ZMA 33, 1934, pp. 96–150.

Fried, Ferdinand, Der *Aufstieg* der Juden, Goslar 1937.

Frischeisen-Köhler, Ida, Über die *Empfindlichkeit* für Schnelligkeitsunterschiede, in: Psychologische Forschungen 18, 1933, pp. 286–290.

Frischeisen-Köhler, Ida, Das persönliche *Tempo*. Eine erbbiologische Untersuchung, Leipzig 1933.

Frischeisen-Köhler, Ida, *Untersuchungen* an Schulzeugnissen von Zwillingen, in: Zeitschrift für angewandte Psychologie 37, 1930, pp. 385–416.

Fritsch, Theodor, *Handbuch* der Judenfrage. Die wichtigsten Tatsachen zur Beurteilung des jüdischen Volkes, 49th edition, Leipzig 1943.

Galton, Francis, The *History of Twins* as a Criterion of the Relative Powers of Nature and Nurture, in: Journal of the Anthropological Institute of Great Britain and Ireland 5, 1876, pp. 391–406.

Geipel, Georg, *Anleitung* zur erbbiologischen Beurteilung der Finger- und Handleisten, Munich 1935.

Geipel, Georg, Der *Formindex* der Fingerleistenmuster, in: ZMA 36, 1937, pp. 330–361.

Geipel, Georg, Die *Gesamtzahl* der Fingerleisten als neues Merkmal zur Zwillingsdiagnose, in: ZMA 39, 1941, pp. 414–419.

Geipel, Georg, *Methode der Auswertung* der Fingerleistenmuster für Vaterschaftsgutachten und Gefahren falscher Anwendung, in: Der Erbarzt 4, 1937, p. 137.

Geipel, Georg, Zur *Methode der Ermittlung* der Fingerleisten im Vaterschaftsnachweis, in: Der Erbarzt 11, 1943, pp. 25–31.

Geipel, Georg, Die *Verteilung* der Fingerleistenmuster und die homologe Konkordanz bei ein- und zweieiigen Zwillingen, in: ZMA 40, 1942, pp. 51–79.

Gerhardt, Kurt, Zur *Frage* Brachykephalie und Schädelform, in: ZMA 37, 1938, pp. 277–489.
Gerstenmaier, Eugen, *Streit* und Friede hat seine Zeit. Ein Lebensbericht, Frankfurt/Main 1981.
Geyer, Horst, Zur *Ätiologie* der Mongoloiden Idiotie, Leipzig 1939.
Geyer, Horst, Die *Epilepsien*, in: Fortschritte der Erbpathologie, Rassenhygiene und ihrer Grenzgebiete 1, 1937, pp. 78–109.
Geyer, Horst, Die *erbliche Fallsucht* und andere Anfallskrankheiten, in: Fortschritte der Erbpathologie, Rassenhygiene und ihrer Grenzgebiete 3, 1939, pp. 259–294.
Geyer, Horst, *Gegensätzliche Äußerung* seelischer Anlagen bei erbgleichen Zwillingen, in: ZMVKL 24, 1940, pp. 536–546.
Geyer, Horst, Über *Hirnwindungen* bei Zwillingen, in: ZMA 38, 1939, pp. 51–55.
Geyer, Horst, *Rassenhygiene* und Littlesche Krankheit, in: Der Erbarzt 2, 1935, pp. 83–86.
Geyer, Horst, Über den *Schlaf* von Zwillingen, in: ZIAVL 73, 1936, pp. 524–527.
Geyer, Horst, Die angeborenen und früh erworbenen *Schwachsinnszustände*. I. Allgemeines über Ätiologie und Wesen der unkomplizierten Oligophrenien, in: Fortschritte der Neurologie, Psychiatrie und ihrer Grenzgebiete 10, 1938, pp. 289–323.
Geyer, Horst, *Subcorticale Mechanismen* bei schlafenden Zwillingen, in: Zeitschrift für die gesamte Neurologie und Psychiatrie 161, 1938, pp. 378–383.
Geyer, Horst, Der *Würzburger Vererbungskongreß* 1938, in: Ziel und Weg 8, 1938, pp. 648–651.
Geyer, Horst; Ole Pedersen, Zur *Erblichkeit* der Neubildungen des Zentralnervensystems, in: Zeitschrift für die gesamte Neurologie und Psychiatrie 165, 1939, pp. 284–294.
Gigas, Heinz, *Untersuchungen* über Muskelvarietäten, in: ZMA 39, 1941, pp. 480–541.
Glatzel, Hans, Der *Anteil* der Erbanlage und Umwelt an der Variabilität des normalen Blutbildes, in: Deutsches Archiv für klinische Medizin 170, 1931, pp. 470–489.
Glatzel, Hans, *Beiträge zur Zwillingspathologie*, in: Zeitschrift für klinische Medizin 116, 1931, pp. 632–668.
Glatzel, Hans, Die *Erbanlage* in ihrer Bedeutung für die normale Magenfunktion, in: Zeitschrift für klinische Medizin 118, 1931, pp. 242–260.
Glum, Friedrich, Zwischen *Wissenschaft*, Wirtschaft und Politik. Erlebtes und Erdachtes in vier Reichen, Bonn 1964.
Göllner, Herbert, *Volks- und Rassenkunde* der Bevölkerung von Friedersdorf (=Deutsche Rassenkunde; 9), Jena 1932.
Goldschmidt, Richard B., In and Out of the *Ivory Tower*, Seattle 1960.
Goldschmidt, Richard B, *Physiologie* der Vererbung, Berlin 1927.
Gottschaldt, Kurt, Der *Aufbau* des kindlichen Handelns (=Zeitschrift für angewandte Psychologie, Beihefte; 68), Leipzig 1933.
Gottschaldt, Kurt, *Bemerkung* zu dem Aufsatz von Fritz Lenz "Zur Problematik der psychologischen Erbforschung usw.", in: AfRGB 37, 1943/44, pp. 21 f.
Gottschaldt, Kurt, *Erbe und Umwelt* in der Entwicklung der geistigen Persönlichkeit, in: ZMA 38, 1939, pp. 10–17.
Gottschaldt, Kurt, *Erbpsychologie* der Elementarfunktionen der Begabung, in: Just (ed.), Handbuch der Erbbiologie, vol. V/1, pp. 445–537.
Gottschaldt, Kurt, Die *Methodik der Persönlichkeitsforschung* in der Erbpsychologie, Leipzig 1942.
Gottschaldt, Kurt, Zur *Methodik erbpsychologischer Untersuchungen* in einem Zwillingslager, in: ZIAVL 73, 1936, pp. 518–523.
Gottschaldt, Kurt, *Phänogenetische Fragestellungen* im Bereich der Erbpsychologie, in: ZIAVL 76, 1939, pp. 118–156.
Gottschaldt, Kurt, *Psychologische Probleme* und Methoden *in der Kolonialforschung*, in: Hugo Adolf Bernatzik (ed.), Afrika. Handbuch der angewandten Völkerkunde, vol. I, Munich 1947, pp. 161–187.
Gottschaldt, Kurt, Zur *Problematik* der psychologischen Erbforschung, in: AfRGB 36, 1942, pp. 28–56.
Gottschaldt, Kurt, *Umwelterscheinungen* im erbpsychologischen Bild, in: Die Naturwissenschaften 25, 1937, pp. 431–434.

Gottschaldt, Kurt, Über die *Vererbung* von Intelligenz und Charakter, in: Fortschritte der Erbpathologie, Rassenhygiene und ihrer Grenzgebiete 1, 1937, pp. 1–21.
Gottschick, Johann, Die beiden *Hauptfragen* der Zwillingsbiologie, in: AfRGB 31, 1937, pp. 377–394.
Gottschick, Johann, *Psychiatrie der Kriegsgefangenschaft*. Dargestellt auf Grund von Beobachtungen in den USA an deutschen Kriegsgefangenen aus dem letzten Weltkrieg, Stuttgart 1963.
Gottschick, Johann, *Zwillingsbefunde* und Reinrassigkeit, in: AfRGB 33, 1939, pp. 102–110.
Grebe, Hans, Zur *Ätiologie* der Arhincephalie, in: Der Erbarzt 12, 1944, pp. 138–145.
Grebe, Hans, Die *Akrocephalosyndaktylie*. Eine klinisch-ätiologische Studie, in: ZMVKL 28, 1944, pp. 209–261.
Grebe, Hans, Über erbpathologische *Befunde* bei Vaterschaftsbegutachtungen, in: Der Erbarzt 12, 1944, pp. 17–22.
Grebe, Hans, *Chondrodysplasie*, Rome 1955.
Grebe, Hans, Über *Differentialdiagnose* und Erbverhältnisse bei primordialem Zwergwuchs, in: Der Erbarzt 10, 1942, pp. 195–210.
Grebe, Hans, *Dysplasie* der rechten Körperhälfte bei einem Paarling von eineiigen Zwillingsschwestern, in: Der Erbarzt 10, 1942, pp. 99–109.
Grebe, Hans, Zur *Erblichkeit* von Darmmißbildungen, in: Der Erbarzt 11, 1943, pp. 104–110.
Grebe, Hans, Die *Erbpathologie* in ihrer Bedeutung für die Gesamtmedizin und das Werk von Otmar Freiherr von Verschuer, in: Die Umschau in Wissenschaft und Technik 48, 1944, pp. 105 f.
Grebe, Hans, *Erbpathologische Arbeitsgemeinschaft*, in: Der Erbarzt 11, 1943, pp. 151–153.
Grebe, Hans, Die *Fistula* sacrococcyga, ein Erbmerkmal, in: Der Erbarzt 10, 1942, pp. 123–126.
Grebe, Hans, Die *Häufigkeit* der erblichen und nichterblichen Blindheitsursachen, in: Der Erbarzt 5, 1938, p. 22.
Grebe, Hans, Über *Hernien* und Erbanlage, in: Der Erbarzt 12, 1944, pp. 65–72.
Grebe, Hans, *Hydrophtalmus* congenitus, in: Der Erbarzt 11, 1943, pp. 92–94.
Grebe, Hans, *Lipomatosis*, psychische Anomalien und Mißbildungen in einer Sippe, in: Der Erbarzt 11, 1943, pp. 55–63.
Grebe, Hans, Angeborene *Mißbildung* und weitere Kinder, in: Medizinische Klinik 39, 1943, pp. 484–490.
Grebe, Hans, Der *Nachweis* der Heterozygoten bei rezessiven Erbleiden, in: Der Erbarzt 11, 1943, pp. 1–9.
Grebe, Hans, Gibt es eine erbliche *Struma*?, in: Der Erbarzt 10, 1942, pp. 278–282.
Grebe, Hans, Über die *Todesursache* bei Totgeburten und Frühverstorbenen insbesondere durch Mißbildungen, in: Der Erbarzt 10, 1942, pp. 110–119, 126–143.
Grebe, Hans, *Untersuchungen* über Papillarlinienveränderungen bei Syndaktylie und Polydaktylie, in: ZMA 39, 1941, pp. 62–78.
Grebe, Hans; Wolf Weißwange, Die *Chondrodysplasie* und verwandte Systemerkrankungen im Röntgenbild, in: Fortschritte auf dem Gebiet der Röntgenstrahlen 67, 1943, pp. 99–116 and 233–246.
Gremmler, J., Die *Beziehungen* der Hypoxämie zum epileptischen Anfall und zu den Höhenkrämpfen, in: Der Nervenarzt 15, 1942, pp. 467–476.
Grohmann, Herbert, Zur *Erbpathologie* der Recklinghausenschen Krankheit, in: Der Erbarzt 5, 1939, p. 20.
Grohmann, Herbert, *Heterogenie* der rezessiven Taubstummheit, in: Der Erbarzt 7, 1939, pp. 12–20.
Grotjahn, Alfred, *Soziale Pathologie*. Versuch einer Lehre von den sozialen Beziehungen der Krankheiten als Grundlage der sozialen Hygiene Berlin 1912.
Gruber, Max von, *Organisation* der Forschung und Sammlung von Materialien über die Entartungsfrage, in: Concordia 17, 1910, pp. 225–228.
Gruber, Max von; Ernst Rüdin (eds), *Fortpflanzung*, Vererbung, Rassenhygiene. Katalog der Gruppe Rassenhygiene der Internationalen Hygiene-Ausstellung 1911 in Dresden, Munich 1911.

Günder, Richard, *Beiträge* zur Frage der Pseudohämophilie, in: AfRG 33, 1939, pp. 412–417.
Günder, Richard, *Gerinnungsprüfungen* in einer großen, bisher nicht beschriebenen Blutersippe, in: AfRGB 33, 1939, pp. 490–506.
Günder, Richard, *Mitteilungen* über neue Blutersippen, in: AfRGB 33, 1939, pp. 355–364.
Günther, Eckhard, Das *Judentum*, in: Mainfranken, 1789–1816, n.p. 1943.
Günther, Hans F. K., *Rassenkunde des jüdischen Volkes*, 2nd edition Munich 1930.
Gütt, Arthur, Zum *Geleit*, in: Der Erbarzt 1, 1934, p. 1.
Gütt, Arthur; Ernst Rüdin; Falk Ruttke, *Gesetz* zur Verhütung erbkranken Nachwuchses vom 14. Juli 1933 mit Auszug aus dem Gesetz gegen gefährliche Gewohnheitsverbrecher und über Maßregeln der Sicherung und Besserung vom 24 November 1933, Munich 1934.
Haase, Friedrich Hermann, Die *Übersterblichkeit* der Knaben als Folge rezessiver geschlechtsgebundener Erbanlagen, in: ZMVKL 22, 1938, pp. 105–126.
Haecker, Valentin, Einige *Aufgaben der Phänogenetik*, in: Studia Mendeliana. Ad centesimum diem natalem Gregorii Mendelii a grata patria celebrandum, Brno 1923, pp. 78–91.
Haecker, Valentin, *Aufgaben und Ergebnisse* der Phänogenetik, in: Bibliographia Genetica 1, 1925, pp. 93–314.
Haecker, Valentin, Phänogenetisch gerichtete *Bestrebungen* in Amerika, in: ZIAVL 40, 1926, pp. 232–238.
Haecker, Valentin, Entwicklungsgeschichtliche *Eigenschaftsanalyse* (Phänogenetik), Jena 1918.
Hässler, Johann Nepomuk, *Untersuchungen* über Ehe- und Fruchtbarkeitsverhältnisse in einer Bauern- und Bürgersippe der Baar, Freiburg 1941.
Haetinger, Martin, Zur anthropologischen *Stellung* der Mokén des Mergui-Archipels, in: ZMA 40, 1943, pp. 193–273.
Hara, Sei, *Untersuchung* der Fingerleisten von Zwillingen, in: ZMA 30, 1932, pp. 564–570.
Harmsen, Hans, *Bevölkerungspolitische Neuorientierung* unsrer Gesundheitsfürsorge, in: Gesundheitsfürsorge. Zeitschrift der evangelischen Kranken- und Pflegeanstalten 5, 1931, pp. 1–6.
Harmsen, Hans, *Eugenetische Neuorientierung* unsrer Gesundheitsfürsorge, in: Gesundheitsfürsorge. Zeitschrift der evangelischen Kranken- und Pflegeanstalten 5, 1931, pp. 127–131.
Harmsen, Hans, *Verminderung* der Kosten für die geistig und körperlich Minderwertigen. Wichtige Beschlüsse des preußischen Staatsrates zu eugenetischen Forderungen, in: Gesundheitsfürsorge 6, 1932, pp. 41–49.
Hassell, Ulrich von, Die *Hassell-Tagebücher* 1938–1944. Aufzeichnungen vom Anderen Deutschland, Baron Friedrich Hiller von Gaertringen (ed.), Berlin 1988.
Hauschild, Rita, *Bastardstudien* an Chinesen, Negern, Indianern in Trinidad und Venezuela, in: ZMA 39, 1941, pp. 181–289.
Hauschild, Rita, *Rassenunterschiede* zwischen negriden und europiden Primordialcranien des 3. Fetalmonats, in: ZMA 36, 1937, pp. 215–280.
Hecht, Günther, Die *Bedeutung* des Rassengedankens in der Kolonialpolitik, in: Deutscher Kolonialdienst. Ausbildungsblätter des Kolonialpolitischen Amtes der NSDAP (Reichsleitung) 2, 1937, No. 11, pp. 4–8, No. 12, pp. 1–5.
Hecht, Günther, *Kolonialfrage* und Rassengedanke (=Schriftenreihe des Rassenpolitischen Amtes der NSDAP; 16), Berlin n.d. [1939].
Helmersen, Erwin von, Die *Nachkommenschaft* einer armenischen Familie in einem deutschen Siedlungsdorf der Bukowina, Diss. Berlin 1943.
Heman, Carl Friedrich, *Geschichte* des jüdischen Volkes seit der Zerstörung Jerusalems, Calw 1908.
Hermann, Albert, Die *deutschen Bauern* des Burzenlandes (=Deutsche Rassenkunde. 15/16), Jena 1937.
Hesch, Michael, *Otto Reche* als Rassenforscher, in: Hesch Michael, Günther Spannaus (eds), Kultur und Rasse. Otto Reche zum 60. Geburtstag, Munich 1939, pp. 9–16.
Hinkel, Hans, *Judenviertel* Europas. Die Juden zwischen Ostsee und Schwarzem Meer, Berlin 1939.
Hoffmann, Géza von, Die *Rassenhygiene* in den Vereinigten Staaten von Nordamerika, Munich 1913.

Horneck, Karl G., Über den *Nachweis* serologischer Verschiedenheiten der menschlichen Rassen, in: ZMVKL 26, 1942, pp. 309–319.

Idelberger, Annemarie, *Zwillingsforschung*, in: Zeitschrift für Kinderforschung 47, 1938, pp. 497–511.

Idelberger, Karlheinz, Zur *Frage der anlagemäßigen Entstehung* des angeborenen Klumpfußes und seiner Beziehungen zu intellektuellen Störungen, in: AfRGB 33, 1939, pp. 304–333.

Idelberger, Karlheinz, Zur *Frage der exogenen Entstehung* der angeborenen Hüftverrenkung (Zwillingshäufigkeit und Geschlechtsverhältnis), in: AfRGB 35, 1941, pp. 314–324.

Ivanicek, Franjo, *Beiträge* zur Anthropologie und Rassengeschichte der Kroaten, in: ZMA 40, 1942, pp. 177–192.

Just, Günther (ed.), *Handbuch der Erbbiologie* des Menschen, 5 vols, Berlin 1940.

Just, Günther, *Probleme des höheren Mendelismus* beim Menschen, in: ZIAVL 67, 1934, pp. 263–268.

Karvé, Irawati, Normale *Asymmetrie* des menschlichen Schädels, Diss. Berlin 1931.

Karvé, Irawati, *Beobachtungen* über die Augenfarben an Chitpavan-Brahmanen, in: ZMA 29, 1931, pp. 498–501.

Kattentidt, Balder, *Ceterum censeo*, tuberculosem esse delendam, in: Zeitschrift für Tuberkulose 74, 1936, pp. 251–254.

Kayser-Petersen, Julius, Hie *praktischer Arzt* – hie Tuberkulosefürsorgearzt! Ein streitbares Gespräch mit friedlichem Ausgang, Reinbek 1937.

Kehl, Heinz, Über die *Erblichkeit* der Myopie, Diss. Berlin 1938.

Keller, Josef; Hanns Andersen, Der *Jude* als Verbrecher, Berlin 1937.

Kessler, Gerhard, Die *Familiennamen* der Juden in Deutschland, Leipzig 1935.

Kiffner, Fritz, Ein *Beitrag* zur Morphologie der Sakai, in: ZMA 27, 1930, pp. 179–198.

Kiffner, Fritz, *Stereoröntgenbefunde* an Zwillingsplacenten, in: Archiv für Gynäkologie 136, 1929, pp. 111–121.

Kim, Baeckpyeng, *Rassenunterschiede* am embryonalen Schweineschädel und ihre Entstehung, in: ZMA 32, 1933, pp. 486–523.

King, Wuhou Wayne, Die *Hautleisten* am Mittel- und Grundglied von Chinesenhänden und deren übriges Leistensystem, in: ZMA 38, 1939, pp. 309–342.

Klein, Hans, Die *Pelger-Anomalie* der Leucocyten und die pathologische Anatomie des neugeborenen homozygoten Pelger-Kaninchens. Ein Beitrag zum Formenkreis der fetalen Chondrodystrophie, in: ZMVKL 29, 1949, pp. 551–620.

Klenck, Willy; Walter Scheidt, *Niedersächsische Bauern* I, Geestbauern im Elbe-Wesermündegebiet (=Deutsche Rassenkunde; 1), Jena 1929.

Koch, Gerhard, *Humangenetik* und Neuro-Psychiatrie in meiner Zeit (1932–1978). Jahre der Entscheidung, Erlangen 1993.

Koch, Gerhard, *Paramyotonia congenita*, in: Der Erbarzt 11, 1943, pp. 167–174.

Kolb, Gustav, *Reform* der Irrenfürsorge, in: Zeitschrift für die gesamte Neurologie und Psychiatrie 47, 1919, pp. 137–172.

Koloniale Völkerkunde, Koloniale Sprachforschung, Koloniale Rassenforschung. Berichte über die Arbeitstagung im Januar 1943 in Leipzig (=Beiträge zur Kolonialforschung, Tagungsbd.; 1), Berlin 1943.

Koslowski, Heinz, Die *Einfügung* französischer (hugenottischer) Flüchtlinge in das deutsche Volk, in: ZMA 39, 1940, p. 117–176.

Kranz, Heinrich, Zur *Frage der Konkordanz* bei kriminellen Zwillingspaaren, in: Forschungen und Fortschritte 9, 1933, pp. 495 f.

Kranz, Heinrich, Die *Haare* von Ostgrönländern und westgrönländischen Eskimo-Dänen-Mischlingen, in: Wegener (ed.), Wissenschaftliche Ergebnisse der Deutschen Grönlandexpedition, pp. 65–84.

Kranz, Heinrich, Die *Kriminalität bei Zwillingen*, in: ZIAVL 67, 1934, pp. 308–313.

Kranz, Heinrich, *Tumoren* bei Zwillingen, in: Bericht über die Jahresversammlung (13.–17. September 1931) der Deutschen Gesellschaft für Vererbungswissenschaft 9, 1931, pp. 353–361.

Kranz, Heinrich, *Vererbung* der menschlichen Haarform, in: Eugenik, Erblehre, Erbpflege 1, 1931, pp. 134 f.
Krehl, Ludolf von, *Der Arzt*, Stuttgart 1937.
Kühne, Konrad, Die *Vererbung* der Variationen der menschlichen Wirbelsäule, in: ZMA 30, 1931, pp. 1–221.
Kühne, Konrad, Die *Zwillingswirbelsäule* (Eine erbgenetische Forschung), in: ZMA 35, 1936, pp. 1–376.
Künneth, Walter, *Lebensführungen*. Der Wahrheit verpflichtet, Wuppertal 1979.
Künneth, Walter, Die biblische *Offenbarung* und die Ordnungen Gottes, in: Walter Künneth, Helmuth Schreiner (eds), Nation, pp. 21–44.
Künneth, Walter; Helmuth Schreiner (eds), Die *Nation* vor Gott. Zur Botschaft der Kirche im Dritten Reich, 2nd edition, Berlin 1937.
Lange, Bruno, Die *Bedeutung* von Erbfaktoren für Entstehung und Verlauf der Tuberkulose, in: Zeitschrift für Tuberkulose 72, 1935, pp. 241–262.
Lassen, Marie-Thérèse, Zur *Frage* der Vererbung "sozialer und sittlicher Charakteranlagen" (auf Grund von Fragebögen über Zwillinge), in: AfRGB 25, 1931, pp. 268–278.
Lassen, Marie-Thérèse, *Nachgeburtsbefunde* bei Zwillingen und Ähnlichkeitsdiagnose, II. Mitteilung, in: Archiv für Gynäkologie 147, 1931, pp. 48–64.
Laughlin, Harry Hamilton, Eugenical *Sterilization* in the United States, Chicago 1922.
Lehmann, Melanie, Verleger J. *F. Lehmann*. Ein Leben im Kampf um Deutschland, Munich 1935.
Lehmann, Wolfgang, Die *Bedeutung* der Erbveranlagung bei der Entstehung der Rachitis, in: Zeitschrift für Kinderheilkunde 57, 1936, pp. 603–643.
Lehmann, Wolfgang, Anthropologische *Beobachtungen* auf den kleinen Sundainseln, in: Zeitschrift für Ethnologie 66, 1934, pp. 268–276.
Lehmann, Wolfgang, *Erbuntersuchung* an rachitischen Zwillingen, in: Monatsschrift für Kinderheilkunde 62, 1934, pp. 205–212.
Lehmann, Wolfgang, *Zwillingspathologische Untersuchungen* über die dystrophische Diathese, in: ZIAVL 70, 1935, pp 472–476.
Lehmann, Wolfgang, Eduard A. Witteler, *Zwillingsbeobachtung* zur Erbpathologie der Polydaktylie, in: Zentralblatt für Chirurgie 62, 1935, pp. 2844–2852.
Lemser, Heinz, Zur *Eiigkeitsdiagnose* bei Zwillingen und über Grenzen ihrer Sicherheit, in: Der Erbarzt 4, 1937, pp. 118–122.
Lemser, Heinz, Zur *Erb- und Rassenpathologie* des Diabetes mellitus, in: AfRGB 32, 1938, pp. 481–516.
Lemser, Heinz, Zur *Erb- und Rassenpathologie* des Diabetes mellitus, *II*: Die Frage einer Rassenpathologie beim Diabetes mellitus, in: AfRGB 33, 1939, pp. 193–224.
Lemser, Heinz, *Kann eine Erbanlage für Diabetes latent bleiben*? in: Der Erbarzt 5, 1938, S. 33.
Lemser, Heinz, Inwieweit läßt sich eine *nicht manifestierte Erbanlage* für Diabetes mit Hilfe von Belastungsproben erkennen?, in: Münchner Medizinische Wochenschrift 85, 1938, p. 1657.
Lenz, Fritz, Menschliche *Auslese* und Rassenhygiene (Eugenik) (=Erwin Baur, Eugen Fischer, Fritz Lenz, Menschliche Erblichkeitslehre und Rassenhygiene, Bd. II), 3rd edition, Munich 1931, 4th edition, Munich 1938.
Lenz, Fritz, *Bund zur Erhaltung* und Mehrung der deutschen Volkskraft, in: AfRGB 11, 1914/15, pp. 554 f.
Lenz, Fritz, Die *Denkschrift* Muckermanns an die Regierung, in: AfRGB 26, 1932, pp. 231–233.
Lenz, Fritz, Ein "*Deutscher Bund für Volksaufartung* und Erbkunde", in: AfRGB 17, 1925, pp. 349 f.
Lenz, Fritz, Zur *Frage der Fortpflanzung der Hilfsschüler*, in: AfRGB 35, 1941, pp. 54 f.
Lenz, Fritz, Zur *Frage der Ursachen von Zwillingsgeburten*, in: AfRGB 27, 1933, pp. 294–318.
Lenz, Fritz, *Gedanken zur Rassenhygiene* (Eugenik), in: AfRGB 37, 1943/44, pp. 84–109.
Lenz, Fritz, Die *Häufigkeit* der Verwandtenehen und ihr Rückgang, in: Der Erbarzt 5, 1938, p. 97.
Lenz, Fritz, *Mendeln die Geisteskrankheiten?*, in: ZIAVL 73, 1936, pp. 559–571.
Lenz, Fritz, Zur *Problematik der psychologischen Erbforschung*, in: AfRGB 35, 1941, pp. 345–368.
Lenz, Fritz, Zur *Problematik der psychologischen Erbforschung und der Lehre vom Schichtenbau der Seele*, in: AfRGB 37, 1943/44, pp. 6–21.

Lenz, Fritz, *Rasse und Klima*, in: Verhandlungen der Gesellschaft deutscher Naturforscher und Ärzte 95, 1938, pp. 34–38.

Lenz, Fritz, Über *Rassen und Rassenbildung*, in: Unterrichtsblätter Mathematik und Naturwissenschaft 40, 1934, pp. 177–190.

Lenz, Fritz., *Rez. R.W. Darré, Neuadel aus Blut und Boden*, in: AfRGB 26, 1932, pp. 444–447.

Lenz, Fritz, Die *Stellung* des Nationalsozialismus zur Rassenhygiene, in: AfRGB 25, 1931, pp. 300–308.

Lenz, Fritz, Zur *Sterilisierungsfrage*, in: Klinische Wochenschrift 13, 1934, pp. 294 f.

Lenz, Fritz, Über *Wege und Irrwege* rassenkundlicher Untersuchungen, in: ZMA 39, 1941, pp. 385–413.

Lenz, Fritz, *Wer wird schizophren?*, in: Der Erbarzt 4, 1937, p. 154.

Lenz, Fritz, Inwieweit kann man aus *Zwillingsbefunden* auf Erbbedingtheit oder Umwelteinfluß schließen, in: Deutsche Medizinische Wochenschrift 61, 1935, pp. 873–875.

Lenz, Fritz; Otmar v. Verschuer, Zur *Bestimmung* des Anteils von Erbanlage und Umwelt an der Variabilität, in: AfRGB 20, 1928, pp. 425–428.

Liek, Erwin, *Der Arzt und seine Sendung*, Munich 1926.

Löffler, Lothar, *Röntgenschädigungen* der menschlichen Keimzelle und Nachkommenschaft, in: Strahlentherapie 34, 1929, pp. 736–766.

Lüth, Karl-Friedrich, *Endokrine Störungen* bei eineiigen Zwillingen, in: ZMVKL 21, 1937, pp. 55–67.

Lüth, Karl-Friedrich, Über *erbliche Disposition* zu Appendizitis, in: Der Erbarzt 5, 1938, p. 88.

Lüth, Karl-Friedrich, Eine *Sippe* mit gehäuftem Albinismus totalis, in: Der Erbarzt 4, 1937, pp. 31 f.

Lüth, Karl-Friedrich, *Untersuchungen* über die Alkoholblutkonzentration nach Alkoholgaben bei 10 eineiigen und 10 zweieiigen Zwillingspaaren, in: Deutsche Zeitschrift für Gerichtliche Medizin 32, 1939, pp. 145–164.

Lüth, Karl-Friedrich, *Uterusmyonom* bei Zwillingen, in: Der Erbarzt 5, 1938, p. 38.

Luxenburger, Hans, Die *Zwillingsforschung* als Methode der Erbforschung beim Menschen, in: Just (ed.), Handbuch der Erbbiologie, vol. II, pp. 213–248.

Maas, Gertrud, Die *Kinderzahl* in Ehen mit und ohne Ehestandsdarlehen, in: AfRGB 37, 1943/44, pp. 227–275.

Macco, Hans, *Rasseprobleme* im Dritten Reich, Berlin 1933.

Magnussen, Karin, Erbbiologischer *Beitrag* zum Nachweis der ektodermalen Herkunft des Musc. sphincter iridis, in: Zeitschrift für Zellforschung (Dept. A) 33, 1943/45, pp. 408–411.

Magnussen, Karin, Zur *Bestimmung der Farbe* und Pigmentverteilung der menschlichen Iris, in: Der Erbarzt 11, 1943, pp. 174–180.

Magnussen, Karin, Zur *Bestimmung der Irisstruktur* im menschlichen Auge, in: Der Erbarzt 11, 1943, pp. 86–92.

Magnussen, Karin, Über die *Beziehung* zwischen Irisfarbe, histologischer Pigmentverteilung und Pigmentierung des Bulbus beim menschlichen Auge, in: ZMA 41, 1949, pp. 295–312.

Magnussen, Karin, Über eine sichelförmige *Hornhautüberwachsung* am Kaninchenauge, in: Der Erbarzt 12, 1944, pp. 60–62.

Magnussen, Karin, Rassen- und bevölkerungspolitisches *Rüstzeug*. Statistik, Gesetzgebung und Kriegsaufgaben, 3rd edition, Munich 1943.

Magnussen, Karin, *Untersuchungen zur Entwicklungsphysiologie* des Schmetterlingsflügels, in: Wilhelm Roux' Archiv für Entwicklungsmechanik der Organismen 128, 1933, pp. 447–479.

Malán, Mihali, Zur *Erblichkeit* der Orientierungsfähigkeit im Raum, in: ZMA 39, 1940, pp. 1–23.

Mengele, Josef, *Rez.: Georg v. Knorre, Über Vererbung angeborener Herzfehler*, in: Der Erbarzt 9, 1941, pp. 213 f.

Mengele, Josef, *Rez.: Gerhard Venzmer, Erbmasse und Krankheit*, in: Der Erbarzt 8, 1940, p. 214.

Mengele, Josef, *Rez.: Gottfried Pressler, Untersuchungen über den Einfluß der Großstadt* auf die Kopfform sowie Beiträge zur Anthropologie und Stammeskunde Hannovers, in: Der Erbarzt 8, 1940, p. 47.

Mengele, Josef, *Rez.: Lothar Stengel v. Rutkowski, Grundzüge der Erbkunde und Rassenpflege*, in: Der Erbarzt 8, 1940, p. 116.

Mengele, Josef, *Sippenuntersuchungen* bei Lippen-Kiefer-Gaumenspalte, in: ZMVKL 23, 1939, pp. 17–42 [at the same time Diss. Frankfurt/Main 1938].

Mengele, Josef, *Tagung der Gesellschaft für physische Anthropologie*, in: Der Erbarzt 4, 1937, pp. 140 f.

Mengele, Josef, Rassenmorphologische *Untersuchungen des vorderen Unterkieferabschnittes* bei vier rassischen Gruppen, in: Gegenbaurs Morphologisches Jahrbuch 79, 1937, S. 60–116 [at the same time Diss. Munich 1935].

Mengele, Josef, Zur *Vererbung der Ohrfisteln*, in: Der Erbarzt 8, 1940, pp. 59 f.

Meyer-Heydenhagen, Gisela, Die palmaren *Hautleisten* bei Zwillingen, Diss. Berlin 1934.

Mollison, Theodor, Das *Anthropologische Institut der Universität München*, in: Zeitschrift für Rassenkunde und ihre Nachbargebiete 9, 1939, pp. 275–277.

Mollison, Theodor, *Rassenkunde und Rassenhygiene*, in: Rüdin (ed.), Erblehre, pp. 34–48.

Mollison, Theodor, *Serodiagnostik* als Methode der Tiersystematik und Anthropologie, in: Emil Abderhalden (ed.), Handbuch der biologischen Arbeitsmethoden, Bd. IX/1, Munich 1923, pp. 553–584.

Mollison, Theodor, Serologische *Verwandtschaftsforschung* am Menschen und anderen Primaten, in: Tagungsberichte der Deutschen Anthropologischen Gesellschaft. Bericht über die allgemeine Versammlung der Deutschen Anthropologischen Gesellschaft, Augsburg 1926, pp. 88–92.

Morgan, Thomas Hunt, The *Relation* of Genetics to Physiology and Medicine. Nobel Lecture. Les Prix Nobel en 1934, Stockholm 1935.

Mrugowsky, Joachim, Das *ärztliche Ethos*. Christoph Wilhelm Hufelands Vermächtnis einer fünfzigjährigen Erfahrung, Munich 1939.

Muckermann, Hermann, *Aus der Hauptversammlung* der Deutschen Gesellschaft für Rassenhygiene (Eugenik) zu München am 18. September 1931, in: AfRGB 26, 1932, pp. 94–108.

Muckermann, Hermann, *Differenzierte Fortpflanzung*, in: AfRGB 24, 1930, pp. 269–290.

Muckermann, Hermann, Eugenik und christliche *Ehe*, in: Schönere Zukunft 6, 1930, pp. 718–720.

Muckermann, Hermann, Die *Enzyklika Casti conubii* und die Eugenik, in: Ethik 7, 1930/1931, pp. 393–400.

Muckermann, Hermann, *Eugenik*, Berlin 1934.

Muckermann, Hermann, *Eugenik und Katholizismus*, 2nd edition, Berlin 1934.

Muckermann, Hermann, *Kind und Volk*, I: Vererbung und Auslese, II: Gestaltung der Lebenslage, Freiburg 1933–1934.

Muckermann, Hermann, Um das *Leben der Ungeborenen*, 4th edition, Berlin 1934.

Muckermann, Hermann, *Neue Forschungen* über das Problem der differenzierten Volksvermehrung, in: ZIAVL 54, 1930, pp. 287–295.

Muckermann, Hermann, *Rassenforschung und Volk der Zukunft*, 3rd edition, Bonn 1934.

Muckermann, Hermann, *Volkstum, Staat und Nation*, eugenisch gesehen, 2nd ed. Essen 1934.

Muckermann, Hermann, *Wesen der Eugenik* und Aufgaben der Gegenwart, in: Das kommende Geschlecht 5, 1929, pp. 1–48.

Muckermann, Hermann, *Wirkungen des Alkoholgenusses* auf die Nachkommenschaft (=Die Alkoholfrage in Wohlfahrtspflege und Sozialpolitik; 1), Berlin 1928.

Mühlmann, Wilhelm Emil, *Rassen- und Völkerkunde*. Lebensprobleme der Rassen, Gesellschaften und Völker, Braunschweig 1936.

Mühlmann, Wilhelm Emil, *Untersuchungen* über die süddeutsche Brachykephalie I: Badische Schädel aus dem 16.–18. Jahrhundert, in: ZMA 30, 1932, pp. 382–405.

Münter, Heinrich, *Lungentuberkulose* und Erblichkeit, in: Beiträge zur Klinik der Tuberkulose 76, 1930, pp. 257–414.

Nachtsheim, Hans, *Krampfbereitschaft und Genotypus. I*: Die Epilepsie der Weißen Wiener-Kaninchen, in: ZMVKL 22, 1939, pp. 791–810.

Nachtsheim, Hans, *Krampfbereitschaft und Genotypus. II*: Weitere Untersuchungen zur Epilepsie der Weißen Wiener-Kaninchen, in: ZMVKL 25, 1941, pp. 229–244.

Nachtsheim, Hans, *Krampfbereitschaft und Genotypus. III*: Das Verhalten epileptischer und nichtepileptischer Kaninchen im Cardiazolkrampf, in: ZMVKL 26, 1942, pp. 22–74.
Nachtsheim, Hans, *Krampfbereitschaft und Lebensalter*. Nach Beobachtungen und Versuchen am Kaninchen, in: Zeitschrift für Altersforschung 3, 1941, pp. 1–21.
Nachtsheim, Hans, *Modelle menschlicher Erbleiden* beim Tier, in: Forschungen und Fortschritte 20, 1944, pp. 62 f.
Nachtsheim, Hans, Die *Notwendigkeit* einer aktiven Erbgesundheitspflege, in: Der Landarzt 40, 1964, pp. 1383–1385.
Nachtsheim, Hans, Die *Pelger-Anomalie* und ihre Vererbung bei Mensch und Tier, in: Der Erbarzt 10, 1942, pp. 175–188.
Nachtsheim, Hans, Die *Pelger-Anomalie* und ihre Vererbung bei Mensch und Tier. *II*: Die homozygoten Pelger und ihr Schicksal, in: Der Erbarzt 11, 1943, pp. 129–142.
Nachtsheim, Hans; Hans Klein, *Hydrops congenitus universalis* beim Kaninchen, eine erbliche fetale Erythroblastose (=Abhandlungen der Deutschen Akademie der Wissenschaften zu Berlin, Mathematisch-naturwissenschaftliche Klasse; ed. 1947, No. 5), Berlin 1948.
Naujoks, Harry, Mein *Leben* im KZ Sachsenhausen 1936–1942. Erinnerungen des ehemaligen Lagerältesten, Cologne 1987.
Nehls, Gerhard, *Caries* und Paradentose bei Zwillingen, in: ZMVKL 24, 1940, pp. 235–247.
Nehse, Erich, *Beiträge* zur Morphologie, Variabilität und Vererbung der menschlichen Kopfbehaarung, Diss. Berlin 1936.
Newman, Horatio H. et al., *Twins*. A Study of Heredity and Environment, Chicago 1937.
Nicolai, Georg Friedrich, Die *Biologie* des Krieges, 2 vols, Zurich 1919.
Nyiszli, Miklós, Im *Jenseits* der Menschlichkeit. Ein Gerichtsmediziner in Auschwitz, Berlin 1992.
Oetting, Margot Irmgard, *Hand- und Fingerleisten* einiger Guayaki-Indianer, in: ZMA 41, 1944, pp. 275–283.
Ohm, August, Die *Entwicklung* der sozialen Persönlichkeit während der Untersuchungshaft, Diss. Berlin 1938.
Ostermann, Arthur, Negative *Eugenik*, in: Erblehre – Erbpflege, pp. 66–79.
Passarge, Siegfried, Das *Judentum* als landschaftskundlich-ethnologisches Problem, Munich 1929.
Pedersen, Ole; Horst Geyer, Diskordantes *Auftreten von Hirntumoren* bei erbgleichen Zwillingen, in: Zentralblatt für Neurochirurgie 3, 1938, pp. 53–63.
Perret, Gustav, *Cro-Magnon-Typen* vom Neolithikum bis heute, in: ZMA 37, 1937, pp. 1–100.
Pessler, Gottfried, *Untersuchung* über den Einfluß der Großstadt auf die Kopfform sowie Beiträge zur Anthropologie und Stammeskunde Hannovers, in: ZMA 38, 1939, pp. 210–215.
Petri, Elsa, *Untersuchungen* zur Erbbedingtheit der Menarche, in: ZMA 33, 1934, pp. 43–48.
Piebenga, Haring Tjittes, Über das *Hautleistensystem* der Bevölkerung der Insel Urk, in: ZMA 40, 1942, pp. 149–177.
Piebenga, Haring Tjittes, Systematische und erbbiologische *Untersuchungen* über das Hautleistensystem der Friesen, Flamen und Wallonen, in: ZMA 37, 1938, pp. 140–165.
Ploetz-Radmann, Maria, Die *Hautleistenmuster* der unteren beiden Fingerglieder der menschlichen Hand, in: ZMA 36, 1937, pp. 281–310.
Poll, Heinrich, Über *Zwillingsforschung* als Hilfsmittel menschlicher Erbforschung, in: Zeitschrift für Ethnologie 1, 1914, pp. 87–105.
Popenoe, Paul, *Rassenhygiene* (Eugenik) in den Vereinigten Staaten, in: AfRGB 15, 1923/24, pp. 182–187.
Prigge, Richard; Otmar v. Verschuer, Gibt es erbbedingte *Resistenzunterschiede* gegen Diphterietoxin beim Meerschweinchen?, in: Der Erbarzt 11, 1943, pp. 157–166.
Pünder, Hermann, *Politik* in der Reichskanzlei. Aufzeichnungen aus den Jahren 1929–1932, Thilo Vogelsang (ed.), Stuttgart 1961.
Pugel, Theodor; Robert Körber, *Antisemitismus* der Welt in Wort und Bild. Der Wettstreit um die Judenfrage, Dresden 1936.
Quelprud, Thordar, Zur *Erblichkeit* des Darwinschen Höckerchens, in: ZMA 34, 1934, pp. 343–363.
Quelprud, Thordar, *Familienforschungen* über Merkmale des äußeren Ohres, in: ZIAVL 67, 1934, pp. 296–299.

Quelprud, Thordar, Die *Ohrmuschel* und ihre Bedeutung für die erbbiologische Abstammungsprüfung, in: Der Erbarzt 2, 1935, p. 121.

Quelprud, Thordar, *Über Zwillingsohren*, in: Zeitschrift für Ethnologie 64, 1932, pp. 130–133.

Quelprud, Thordar, *Untersuchungen* der Ohrmuschel von Zwillingen, in: ZIAVL 62, 1929, pp. 160–165.

Quelprud, Thordar, *Zwillingsohren*, in: Eugenik, Erblehre, Erbpflege 2, 1932.

Rath, Bruno, *Rotgründblindheit* in der Calmbacher Blutersippe. Nachweis des Faktorenaustausches beim Menschen, in: AfRGB 32, 1938, pp. 397–407.

Redeker, Franz, *Tuberkulosevererbung* und Eugenik, in: Zeitschrift für Tuberkulose 62, 1931, pp. 25–34.

Reichert, Ottwil; Josef Mengele, Zur *Erbbedingtheit* der Thrombangitis obliterans, in: ZMVKL 23, 1939, pp. 53–66.

Reventlow, Ernst zu, *Judas Kampf* und Niederlage in Deutschland, Berlin 1937.

Richter, Brigitte, *Burkhards und Kaulstoß* (=Deutsche Rassenkunde; 14), Jena 1936.

Ried, Hans August, *Miesbacher Landbevölkerung*. Eine rassenkundliche und volkskundliche Untersuchung aus Oberbayern (=Deutsche Rassenkunde; 3), Jena 1930.

Riemann, Hans, *Erwiderung* auf die Arbeit Gottschicks, in: AfRGB 34, 1940, pp. 70 f.

Ritter, Hans, *Cro-Magnon-Merkmale* an den Gliedmaßenknochen der Guanchen und der fälischen Rasse, in: ZMA 41, 1944, pp. 1–48.

Rodenwaldt, Ernst, Vom *Seelenkonflikt* des Mischlings, in: ZMA 34, 1934, pp. 364–375.

Roemer, Hans; Gustav Kolb; Valentin Faltlhauser (eds), Die *offene Fürsorge* in der Psychiatrie und ihren Grenzgebieten, Berlin 1927.

Rößle, Robert, Das *Verhalten* von Syphilis und Tuberkulose in Familien, in: Schweizerische Medizinische Wochenschrift 19, 1938, pp. 3–9.

Roth, Otto, *Wachstumsversuche* an Ratten, in: ZMA 33, 1935, pp. 409–438.

Rüdin, Ernst (ed.), *Erblehre* und Rassenhygiene im völkischen Staat, Munich 1934.

Rüdin, Ernst, *Psychiatrische Indikation* zur Sterilisierung, in: Das kommende Geschlecht 5, 1929, pp. 1–19.

Rüdin, Ernst, Einige *Wege und Ziele* der Familienforschung, mit Rücksicht auf die Psychiatrie, in: Zeitschrift für die gesamte Neurologie und Psychiatrie 7, 1911, pp. 487–585.

Ruhenstroth-Bauer, Gerhard; Hans Nachtsheim, Die *Bedeutung des Sauerstoffmangels* für die Auslösung des epileptischen Anfalls. Nach Versuchen an epileptischen Kaninchen, in: Klinische Wochenschrift 23, 1944, pp. 18–21.

Ruppin, Arthur, Les *juifs* dans le monde moderne, Paris 1934.

Saller, Karl, Die *Fehmarner*. Eine anthropologische Untersuchung aus Ostholstein (=Deutsche Rassenkunde; 4), Jena 1930.

Saller, Karl, Die *Keuperfranken*. Eine anthropologische Untersuchung aus Mittelfranken (=Deutsche Rassenkunde; 2), Jena 1930.

Saller, Karl, Die *Rassenlehre* des Nationalsozialismus in Wissenschaft und Propaganda, Darmstadt 1961.

Schade, Heinrich, Anthropologische *Befunde* bei einer erbbiologischen Bestandsaufnahme in der Schwalm, in: Verhandlungen der Deutschen Gesellschaft für Rassenforschung 10, 1940, pp. 36–39.

Schade, Heinrich, *Beitrag* zur Feststellung der Häufigkeit von Erbkrankheiten, in: Der Erbarzt 8, 1940, pp. 126–128.

Schade, Heinrich, Zur endogenen *Entstehung* von Gliedmaßendefekten, in: ZMA 36, 1937, pp. 375–381.

Schade, Heinrich, *Erbbiologische Bestandsaufnahme*, in: Fortschritte der Erbpathologie, Rassenhygiene und ihrer Grenzgebiete 1, 1937, pp. 37–48.

Schade, Heinrich, *Ergebnisse* einer Bevölkerungsuntersuchung in der Schwalm, in: Abhandlungen der Akademie der Wissenschaften und der Literatur Mainz, Mathematisch-naturwissenschaftliche Klasse, ed. 1950, No. 16, pp. 417–491.

Schade, Heinrich, Die *Häufigkeit* des Schwachsinns in einer geschlossenen bäuerlichen Bevölkerung, erhoben bei einer erbbiologischen Bestandsaufnahme, in: ZIAVL 73, 1936, pp. 577–579.

Schade, Heinrich, Der *Internationale Kongreß für Bevölkerungswissenschaften* in Berlin, in: Der Erbarzt 1, 1935, pp. 140–142.

Schade, Heinrich, Ausländische *Stimmen* zur deutschen Erb- und Rassenpflege, in: Rassenpolitische Auslands-Korrespondenz 1936, No. 5, pp. 3 f.

Schade, Heinrich, *Untersuchung* zur Auflösung eines kleinen sozialen, großbäuerlichen Isolates, in: Abhandlungen der Akademie der Wissenschaften und der Literatur Mainz, Mathematisch-naturwissenschaftliche Klasse, ed. 1959, No. 11, pp. 841–870.

Schade, Heinrich, *Untersuchungen* zur Frage der Erblichkeit von Mangel- und Fehlbildungen der Gliedmaßen, in: Der Erbarzt 8, 1940, pp. 239–256.

Schade, Heinrich; Maria Küper, Der angeborene *Schwachsinn* in der Rechtsprechung der Erbgesundheitsobergerichte II, in: Der Erbarzt 5, 1938, pp. 66–77.

Schaeuble, Johannes, Einige anthropologische *Beobachtungen* an chilenischen Mischlingen, in: Zeitschrift für Ethnologie 68, 1936, pp. 251–256.

Schaeuble, Johannes, Die *Entstehung* der palmaren digitalen Triradien, in: ZMA 31, 1933, pp. 403–438.

Schaeuble, Johannes, *Indianer* und Mischlinge in Südchile, in: ZMA 38, 1939, pp. 63–66.

Schallmayer, Wilhelm, *Vererbung* und Auslese im Lebenslauf der Völker, Jena 1903.

Scheidt, Walter; Hinrich Wriede, Die *Elbinsel Finkenwärder*, Munich 1927.

Schiff, Fritz, *Serologische Untersuchungen* an Zwillingen, II. Mitteilung, in: ZMA 32, 1933, pp. 244–249.

Schiff, Fritz; Otmar von Verschuer, *Serologische Untersuchungen* an Zwillingen, in: Klinische Wochenschrift 10, 1931, pp. 723–726.

Schmidt, Ilse, Über *Beziehungen* zwischen Landflucht und Intelligenz, in: AfRGB 32, 1938, pp. 358–370.

Schnecke, Christian, *Zwergwuchs* beim Kaninchen und seine Vererbung, in: ZMVKL 25, 1941, pp. 425–457.

Schreiber, Georg, *Deutsches Reich* und deutsche Medizin. Studien zur Medizinalpolitik des Reiches in der Nachkriegszeit (1918–1926), Leipzig 1926.

Schrijver, Fritz, Über die *Erforschung* erblicher Abweichungen beim Geschmackssinn, in: Zeitschrift für Rassenphysiologie 6, 1933, pp. 177–179.

Schröder, Lore, Die *Frage* der Entstehung von Erbschädigung beim Menschen durch Gifte, Diss. Berlin 1933.

Schulze, Gerhard, Über die *Frage* des Einflusses der Rachitis auf einige Kopfmaße und den Längenbreitenindex, in: AfRGB 35, 1941, pp. 18–28.

Schwabe, Karl-Heinz, Rassenbiologische *Erhebungen* in Hennickendorf, einem Dorfe der Mark Brandenburg, in: AfRGB 35, 1941, pp. 293–313.

Schwarzweller, Franz, Ein *Beitrag* zur Genese des angeborenen Klumpfußes, in: Der Erbarzt 3, 1936, pp. 182–186.

Selchow, Bogislav von, *Hundert Tage* aus meinem Leben, Leipzig 1936.

Semigothaisches genealogisches Taschenbuch aristokratisch-jüdischer Heiraten, Munich 1912.

Siemens, Hermann Werner, Einige *Bemerkungen über die Ähnlichkeitsdiagnose* der Eineiigkeit, in: AfRGB 31, 1937, pp. 211–214.

Siemens, Hermann Werner, Über die *Eineiigkeitsdiagnose* der Zwillinge aus den Eihäuten und aus dem dermatologischen Befund, in: ZIAVL 37, 1925, pp. 122–124.

Siemens, Hermann Werner, Zur *Geschichte der Zwillingsmethode*, in: ZMVKL31, 1952, pp. 171–173.

Siemens, Hermann Werner, Die *Zwillingspathologie*, Berlin 1924.

Skerlj, Bozo, *Menarche* und Klima in Europa, in: Zeitschrift für Ethnologie 63, 1931, pp. 413–415.

Söhner, Ilse, Das *Recht* der Sterilisation, jur. Diss. Freiburg 1933.

Sombart, Werner, Die *Juden* und das Wirtschaftsleben, Leipzig 1911.

Sommer, Robert, *Organisation* und Aufgaben eines Reichsinstitutes für Familienforschung und Vererbungslehre, in: Deutsche Medizinische Wochenschrift 46, 1914, pp. 708–711.

Steffens, Christel, Über *Zehenleisten* bei Zwillingen, in: ZMA 37, 1938, pp. 218–258.
Stein, Gerhard, Zur Psychologie und Anthropologie der Zigeuner in Deutschland, Gräfenhainichen 1941 [at the same time Diss. Frankfurt/Main 1938].
Steiner, Franz, *Nachgeburtsbefunde* bei Mehrlingen und Ähnlichkeitsdiagnose, in: Archiv für Gynäkologie 159, 1935, pp. 509–523.
Steiner, Franz, *Untersuchungen* zur Frage der Erblichkeit des Diabetes mellitus, in: Deutsches Archiv für klinische Medizin 178, 1936, pp. 497–510.
Steinwallner, Bruno, Rassenhygienische *Gesetzgebung* und Maßnahmen im Ausland, in: Fortschritte der Erbpathologie, Rassenhygiene und ihrer Grenzgebiete 1, 1937/38, pp. 193–260.
Sterz, Georg; Horst Geyer, Zur *Erbpathologie* der spinalen Ataxie unter besonderer Berücksichtigung des "Status dysraphicus", in: Zeitschrift für die gesamte Neurologie und Psychiatrie 157, 1937, pp. 795–806.
Stigler, Robert, Rassenphysiologische *Untersuchungen* an farbigen Kriegsgefangenen in einem Kriegsgefangenenlager, in: Zeitschrift für Rassenphysiologie 13, 1943, pp. 26–57.
Störring, Ferdinand K.; Heinz Lemser, Über die *Beziehungen* von Akromegalie und Diabetes, in: Münchner Medizinische Wochenschrift 87, 1940, p. 338.
Stroothenke, Wolfgang, *Erbpflege* und Christentum, Leipzig 1940.
Suchalla, Harry, *Variabilität* und Erblichkeit von Schädelmerkmalen bei Zwerg- und Riesenkaninchen; dargestellt an Hermelin- und Widderkaninchen, in: ZMA 40, 1943, pp. 274–333.
Tao, Yun-kuei, *Chinesen-Europäer-Mischlinge*, in: Eugenik, Erblehre, Erbpflege 1, 1931, pp. 247–250.
Tao, Yun-kuei, *Chinesen-Europäerinnen-Kreuzung*, Diss. Berlin 1935.
Theilhaber, Felix A., Der *Untergang* der deutschen Juden. Eine volkswirtschaftliche Studie, Munich 1911.
Thieme, Hellmuth, Die *Neubauernauslese* und ihre Bedeutung für eine rassenhygienische Bevölkerungspolitik, Diss. Berlin 1943.
Tübinger Gesellschaft für Rassenhygiene, in: Zeitschrift für kulturgeschichtliche und biologische Familienkunde 1, 1924, pp. 34 f.
Uhlebach, Rudolf, *Messungen* an Hand- und Fußskeletten von Hottentotten, in: ZMA 16, 1914, pp. 449–464.
Unseres Heiligen Vaters Pius' XI. durch göttliche Vorsehung Papst *Rundschreiben* über die christliche Ehe in Hinsicht auf die gegenwärtigen Verhältnisse, Bedrängnisse, Irrtümer und Verfehlungen in Familie und Gesellschaft. Autorisierte Ausgabe, Freiburg 1931.
Veit, Gertrud, *Über Dornfortsatzbrüche*, Bottrop 1936.
Verhandlungen des Reichstags, Wahlperiode 3, 1927.
Verhandlungen des Ersten Deutschen Soziologentages vom 19.–20. Oktober 1910 in Frankfurt am Main. Reden und Vorträge und Debatten, Tübingen 1911, ND Frankfurt/Main 1969.
Verschuer, Otmar von, Die *Ähnlichkeitsdiagnose* der Eineiigkeit von Zwillingen, in: Anthropologischer Anzeiger 5, 1928, pp. 244–248.
Verschuer, Otmar von, Der *Anteil von Erbanlage und Umwelt* an den Ursachen der Verschiedenheiten zwischen zweieiigen Zwillingen (Methoden der zwillingsanthropologischen Forschung), in: Sitzungsberichte der Anthropologischen Gesellschaft in Wien (Bericht über die Tagung in Salzburg vom September 1926), 1926/27, pp. 36–38.
Verschuer, Otmar von, *Anthropologische Studien* an ein- und zweieiigen Zwillingen. Aus den Verhandlungen der Deutschen Gesellschaft für Vererbungswissenschaft, Hamburg, in: ZIAVL 41, 1926, pp. 115–119.
Verschuer, Otmar von, *Aufgaben und Ziele* der menschlichen Erblichkeitslehre, in: Münchner Medizinische Wochenschrift 74, 1927, p. 999.
Verschuer, Otmar von, *Auslese und Rassenhygiene*, in: Der Deutsche Gedanke 2, 1925, pp. 744–751.
Verschuer, Otmar von, *Bemerkungen zur Genanalyse* beim Menschen, in: Der Erbarzt 7, 1939, pp. 65–69.
Verschuer, Otmar von, Die *biologischen Grundlagen* der menschlichen Mehrlingsforschung, in: ZIAVL 59, 1932, pp. 147–205.

Verschuer, Otmar von, *Erbanlage als Schicksal und Aufgabe* (=Preußische Akademie der Wissenschaften, Vorträge und Schriften; 18), Berlin 1944.
Verschuer, Otmar von, Der *Erbarzt an der Jahreswende*, in: Der Erbarzt 10, 1942, pp. 1–3.
Verschuer, Otmar von, Das *Erbbild vom Menschen*, in: Der Erbarzt 7, 1939, pp. 1–12.
Verschuer, Otmar von., Zur *Erbbiologie der Fingerleisten*, zugleich ein Beitrag zur Zwillingsforschung, in: Forschungen und Fortschritte 9, 1933, pp. 477 f.
Verschuer, Otmar von, *Erbbiologische Erkenntnisse* zur Begründung der deutschen Bevölkerungs- und Rassenpolitik, in: Eugen Gerstenmeier (ed.), Kirche, Volk und Staat, Berlin 1937, pp. 63–75.
Verschuer, Otmar von, Ein *erbgleiches Zwillingspaar* mit hervorragender Begabung für Schachspiel, in: Eugenik, Erblehre, Erbpflege 1, 1931, pp. 177–179.
Verschuer, Otmar von, *Erbpathologie*, 2nd edition, Dresden 1934, 3rd edition, Dresden 1945.
Verschuer, Otmar von, *Erbpathologische Arbeitsgemeinschaft*, in: Der Erbarzt 11, 1943, p. 91.
Verschuer, Otmar von, *Erbpsychologische Untersuchungen* an Zwillingen, in: ZIAVL 54, 1930, pp. 280–285.
Verschuer, Otmar von, *Erbuntersuchungen an tuberkulösen Zwillingen*, in: Beiträge zur Klinik der Tuberkulose 81, 1932, pp. 223–233.
Verschuer, Otmar von, Der *erste Nachweis von Faktorenaustausch* (crossing-over) beim Menschen, in: Der Erbarzt 5, 1938, p. 3.
Verschuer, Otmar von, *Eugenik und Hermann Muckermann*, in: Deutschlands Erneuerung 16, 1932, pp. 463–466.
Verschuer, Otmar von, *Eugenische Fragen*, in: Monatsheft der Sozialen Arbeitsgemeinschaft Evangelischer Männer und Frauen Thüringens 8, 1933, pp. 42–44.
Verschuer, Otmar von, Ein *Fall von Monochorie* bei zweieiigen Zwillingen, in: Münchner Medizinische Wochenschrift 72, 1925, p. 184.
Verschuer, Otmar von, Zur *Frage des Faktorenaustausches* beim Menschen, in: Der Erbarzt 5, 1938, p. 29.
Verschuer, Otmar von, Grundlegende *Fragen der vererbungsbiologischen Zwillingsforschung*, in: Münchner Medizinische Wochenschrift 73, 1926, pp. 1562–1565.
Verschuer, Otmar von, Zur *Frage der Zwillingsdiagnose*, in: AfRGB 32, 1938, S. 69–74.
Verschuer, Otmar von, Die häufigste *Geburtsstunde* von Zwillingen, in: Eugenik, Erblehre, Erbpflege 3, 1933, p. 40.
Verschuer, Otmar von, Das *Gesetz zur Verhütung erbkranken Nachwuchses* – ein Weg zur erblichen Gesundung des Volkes, in: Deutsches Adelsblatt, No. 34, 19/8/1933.
Verschuer, Otmar von, Was kann der *Historiker*, der Genealoge und der Statistiker zur Erforschung des biologischen Problems der Judenfrage beitragen? in: Sitzungsberichte der Zweiten Arbeitstagung der Forschungsabteilung Judenfrage des Reichsinstituts für Geschichte des neuen Deutschlands vom 12.–14. May 1937 (=Forschungen zur Judenfrage; 2), Hamburg 1937, pp. 216–222.
Verschuer, Otmar von, *Intellektuelle Entwicklung* und Vererbung, in: Günther Just (Hg.), Vererbung und Erziehung, Berlin 1930, pp. 176–207.
Verschuer, Otmar von, Das ehemalige *Kaiser-Wilhelm-Institut für Anthropologie*, menschliche Erblehre und Eugenik. Bericht über die wissenschaftliche Forschung 1927–1945, in: ZMA 55, 1964, pp. 127–174.
Verschuer, Otmar von, Eine *Kartei* der Gemeinschaftsunfähigen, in: Der Erbarzt, 8, 1940, p. 235.
Verschuer, Otmar von, *Leitfaden* der Rassenhygiene, 2nd edition, Leipzig 1944.
Verschuer, Otmar von, *Mitarbeiter* der Krankenhäuser bei den Aufgaben der Eugenik, in: Zeitschrift für das gesamte Krankenhauswesen 35, 1939, pp. 156 f.
Verschuer, Otmar von, *Physiologie und Pathologie* in der Anthropologie, in: Forschungen und Fortschritte 6, 1930, pp. 351 f.
Verschuer, Otmar von, Die *Rasse als biologische Größe*, in: Walter Künneth, Helmuth Schreiner (eds), Nation, pp. 24–37.

Verschuer, Otmar von, *Rassenbiologie der Juden*, in: Sitzungsberichte der Dritten Münchner Arbeitstagung des Reichsinstituts für Geschichte des neuen Deutschlands vom 5.-7. Juli 1938 (=Forschungen zur Judenfrage; 3), Hamburg 1938, pp. 137–151.

Verschuer, Otmar von, *Rassenhygiene*, in: Deutsche Politik. Ein völkisches Handbuch, 15. Teil, Frankfurt/Main 1925.

Verschuer, Otmar von, *Rassenhygiene als Wissenschaft und Staatsaufgabe* (=Frankfurter Akademische Reden; 7), Frankfurt/Main 1936.

Verschuer, Otmar von, *Rez.: Deutsche Rassenkunde*, Bde. 1–4, in: ZMA 28, 1930, pp. 378–381.

Verschuer, Otmar von, *Rez. Emil Abderhalden, Rasse und Vererbung* vom Standpunkt der Feinstruktur von Blut- und zelleigenen Eiweißstoffen aus betrachtet, in: Der Erbarzt 8, 1940, pp. 91 f.

Verschuer, Otmar von, *Rez.: Gunnar Dahlberg, Twin Births* and Twins from a Hereditary Point of View, in: AfRGB 19, 1927, pp. 88–92.

Verschuer, Otmar von, *Rez.: Hans F. K. Günther, Der nordische Gedanke* unter den Deutschen, in: AfRGB 18, 1926, pp. 325–328.

Verschuer, Otmar von, *Sozialpolitik und Rassenhygiene*, Langensalza 1928.

Verschuer, Otmar von, Der gegenwärtige *Stand der Zwillingsforschung*, in: Archiv für Soziale Hygiene und Demographie 1, 1925, pp. 1–4.

Verschuer, Otmar von, Chemisch-physikalische *Studien über den Blutserumeiweißgehalt* an Hand vergleichender Eiweißbestimmungen im Blutserum nach der Kjeldahl-, Refraktometer- und Viskosimetermethode und im Coffeinversuch, med. Diss. Munich 1923.

Verschuer, Otmar von, Vom *Umfang der erblichen Belastung* im deutschen Volke, in: AfRGB 24, 1930, pp. 238–268.

Verschuer, Otmar von, Die *Umweltwirkung* auf die anthropologischen Merkmale nach Untersuchungen an eineiigen Zwillingen. Aus den Verhandlungen der deutschen Gesellschaft für Vererbungswissenschaft, Innsbruck, in: ZIAVL 37, 1925, pp. 119–122.

Verschuer, Otmar von, Die *Unfruchtbarmachung* bei schwerer erblicher geistiger Störung, in: Der Erbarzt 5, 1938, pp. 125–127.

Verschuer, Otmar von, Die *Variabilität* des menschlichen Körpers an Hand von Wachstumsstudien an ein- und zweieiigen Zwillingen, in: Verhandlungen des V. Internationalen Kongresses für Vererbungswissenschaft, Berlin 1927 (=Zeitschrift für induktive Abstammungs- und Vererbungslehre, Suppl.Bd.; 1), Leipzig 1928, pp. 1508–1516.

Verschuer, Otmar von, *Vaterschaftsbestimmung*, in: Der Erbarzt 12, 1944, pp. 6–17.

Verschuer, Otmar von, Die *Vaterschaftsgutachten* des Frankfurter Universitätsinstituts für Erbbiologie und Rassenhygiene. Ein vorläufiger Überblick, in: Der Erbarzt 9, 1941, pp. 25–31.

Verschuer, Otmar von, *Vererbung, Auslese und Rassenhygiene*, in: Der Deutsche Gedanke 2, 1925, pp. 744–751.

Verschuer, Otmar von, Die *vererbungsbiologische Zwillingsforschung*. Ihre biologischen Grundlagen. Studien an 102 eineiigen und 45 gleichgeschlechtlichen zweieiigen Zwillings- und an 2 Drillingspaaren, in: Ergebnisse der Inneren Medizin und Kinderheilkunde 31, 1927, pp. 35–102.

Verschuer, Otmar von, *Vier Jahre Frankfurter Universitätsinstitut* für Rassenhygiene, in: Der Erbarzt 6, 1939, pp. 57–64.

Verschuer, Otmar von, Die *Wirkung der Umwelt* auf die anthropologischen Merkmale nach Untersuchungen an eineiigen Zwillingen, in: AfRGB 17, 1925, pp. 149–164.

Verschuer, Otmar von, Die *Wirkung von Genen* und Parasiten im Körper des Menschen, in: Ärztliche Forschung 2, 1948, pp. 378–388.

Verschuer, Otmar von, Die *Zwillingsforschung im Dienste der Erblehre*, in: Zeitschrift für ärztliche Fortbildung 31, 1934, pp. 189 f.

Verschuer, Otmar von; Georg Geipel, Zur *Frage* der Erblichkeit des Formindex der Fingerleistenmuster, in: ZIAVL 70, 1935, pp. 460–463.

Verschuer, Otmar von; V. Zipperlein, Die erb- und umweltbedingte *Variabilität* der Herzform (Nach Röntgenaufnahmen des Herzens bei Zwillingen), in: Zeitschrift für klinische Medizin 112, 1929, pp. 69–92.

Von wissenschaftlichen Instituten, in: Kultur und Leben 4, 1927, pp. 315–316.

Waddington, Conrad H., The *Epigenotype*, in: Endeavour 1, 1942, pp. 18–20.

Wagner, Georg, Rassenbiologische *Beobachtungen* an Zigeunern und Zigeunerzwillingen, Diss. Berlin 1943.

Wagner, Georg, *Partielle Irisfärbung* (Hornhautüberwachsung), ein neues Erbmerkmal, in: Der Erbarzt 12, 1944, pp. 62–64.

Walter, Otto, Die *Tuberkulosebekämpfung* als politische Aufgabe, in: Beiträge zur Klinik der Tuberkulose 86, 1935, pp. 414–424.

Wegener, Kurt (ed.), *Wissenschaftliche Ergebnisse der Deutschen Grönland-Expedition* Alfred Wegener 1929 und 1930/31, vol. VI: Anthropologie und Zoologie, Leipzig 1934.

Weinert, Hans, *Blutgruppenuntersuchungen* an Menschenaffen und ihre stammesgeschichtliche Bewertung, in: Zeitschrift für Rassenphysiologie 4, 1931, pp. 8–23.

Weinert, Hans, Die *fossilen Menschenreste*, in: Fritz Wiegers (Hg.), Diluviale Vorgeschichte des Menschen, vol. I: Allgemeine Diluvialprähistorie, Stuttgart 1928, pp. 199–289.

Weinert, Hans, *Kreuzungsmöglichkeit* zwischen Affe und Mensch, in: Volksaufartung, Erbkunde, Eheberatung 4, 1929, pp. 219–222.

Weinert, Hans, *Neue Blutgruppenuntersuchungen* an Affen im Jahre 1932, in: Zeitschrift für Rassenphysiologie 6, 1933, pp. 75–81.

Weinert, Hans, *Pithecanthropus Erectus*, in: Zeitschrift für Anatomie und Entwicklungsgeschichte 87, 1928, pp. 429–547.

Weinert, Hans, Der "*Sinanthropus pekinensis*" als Bestätigung des Pithecanthropus erectus, in: ZMA 29, 1931, pp. 159–187.

Weinert, Hans, *Ursprung der Menschheit*, Stuttgart 1932.

Weinert, Hans, *Weitere Blutgruppenuntersuchungen* an Affen. 2. Mitteilung, in: Zeitschrift für Rassenphysiologie 5, 1932, pp. 59–68.

Weitz, Wilhelm, *Studien* an eineiigen Zwillingen, in: Zeitschrift für klinische Medizin 101, 1925, pp. 115–154.

Weitz, Wilhelm, Über die *Vererbung* bei der Muskeldystrophie, in: Deutsche Zeitschrift für Nervenheilkunde 72, 1921, pp. 143–204.

Werner, Martin, Die *Erb- und Umweltbedingtheit* der Unterschiede bei der vitalen Lungenkapazität und einigen zugehörigen Körpermaßen und Indices, in: ZMVKL 21, 1937, pp. 293–305.

Werner, Martin, *Erbunterschiede* bei einigen Funktionen des vegetativen Systems nach experimentellen Untersuchungen an 30 Zwillingspaaren, in: Verhandlungen der Deutschen Gesellschaft für innere Medizin 47, 1935, pp. 444–449.

Werner, Martin, Zwillingsphysiologische *Untersuchungen* über den Grundumsatz und die spezifisch-dynamische Eiweißwirkung, in: ZIAVL 70, 1935, pp. 467–471.

Wessel, Helene, *Lebenshaltung* aus Fürsorge und aus Erwerbstätigkeit. Eine Untersuchung des Kostenaufwandes für Sozialversicherung, Fürsorge und Versorgung im Vergleich zum Familieneinkommen aus Erwerbstätigkeit, Eberswalde 1931.

Westenhöfer, Max, Die *Aufgaben* der Rassenhygiene (des Nachkommenschutzes) im neuen Deutschland, Berlin 1920.

Wilde, Kurt, *Erbpsychologische Untersuchungen* über die Übungsfähigkeit, in: Archiv für die gesamte Psychologie 109, 1941, pp. 82–119.

Wilde, Kurt, Über *Intelligenzuntersuchungen* an Zwillingen, in: ZIAVL 73, 1936, pp. 512–517.

Wilde, Kurt, *Meß- und Auswertungsmethoden* in erbpsychologischen Zwillingsuntersuchungen, in: Archiv für Psychologie 109, 1941, pp. 1–81.

Winkler, Hans, Über die *Rolle* von Kern und Protoplasma bei der Vererbung, in: ZIAVL 33, 1924, pp. 238–253.

Witte, Peter et al. (compil.), Der *Dienstkalender Heinrich Himmlers* 1941/42, Hamburg 1999.

Wolfslast, Werner, Eine *Sippe* mit recessiver geschlechtsgebundener spastischer Diplegie, in: ZMVKL 27, 1943, pp. 189–198.

Würth, Adolf, *Bemerkungen* zur Zigeunerfrage und Zigeunerforschung in Deutschland, in: Anthropologischer Anzeiger. Mitteilungsblatt der Deutschen Gesellschaft für Rassenforschung 15, 1938, pp. 95–98.

Würth, Adolf, Die *Entstehung* der Beugefurchen der menschlichen Hohlhand, in: ZMA 36, 1937, pp. 187–214.

Zollschan, Ignaz, Das *Rassenproblem* unter besonderer Berücksichtigung der theoretischen Grundlagen der jüdischen Rassenfrage, Vienna 1910.

Literature

Abelshauser, Werner, Die *Weimarer Republik* – ein Wohlfahrtsstaat? in: Werner Abelshauser (ed.), Die Weimarer Republik als Wohlfahrtsstaat. Zum Verhältnis von Wirtschafts- und Sozialpolitik in der Industriegesellschaft, Stuttgart 1987, pp. 9–31.

Adams, Mark B. (ed.), The Wellborn *Science*: Eugenics in Germany, France, Brazil, and Russia, New York 1990.

Allen, Garland E., *Thomas Hunt Morgan*. The Man and His Science, Princeton 1978.

Allen, Garland E, *Thomas Hunt Morgan and the Split* between Embryology and Genetics 1910–1935, in: Thomas J. Horder et al. (eds), A History of Embryology, Cambridge 1985, pp. 113–146.

"*Alles für Deutschland*, Deutschland für Christus". Evangelische Kirche in Frankfurt am Main 1929 bis 1945, Matthias Benad et al. (eds), 2nd edition, Frankfurt/Main 1985.

Aly, Götz; Susanne Heim, *Vordenker* der Vernichtung. Auschwitz und die deutschen Pläne für eine europäische Ordnung, Frankfurt 1993.

Ash, Mitchell G., Die *erbpsychologische Abteilung* am Kaiser-Wilhelm-Institut für Anthropologie, menschliche Erblehre und Eugenik 1935–1945, in: Wolfgang Schönpflug (ed.), Geschichte der Psychologie in Berlin. Vorträge einer Arbeitsgruppe des 36. Kongresses der Deutschen Gesellschaft für Psychologie in Berlin am 4. Oktober 1988, Berlin 1990, pp. 115–129.

Ash, Mitchell G, Ein *Institut* und eine Zeitschrift. Zur Geschichte des Berliner Psychologischen Instituts und der Zeitschrift "Psychologische Forschung" vor und nach 1933, in: Carl Friedrich Graumann (ed.), Psychologie im Nationalsozialismus, Berlin 1985, pp. 113–137.

Ash, Mitchell G, From "*Positive Eugenics*" to Behavioral Genetics: Psychological Twin Research under Nazism and Since, in: Peter Drewek, Christoph Lüth (eds), Geschichte der Erziehungswissenschaft – History of Educational Studies – Histoire des Sciences de l'Education (=Paedagogica Historica. International Journal of the History of Education, Supplementary Series; 3), Gent 1998, 336–357.

Ash, Mitchell G, *Wissenschaft und Politik* als Ressourcen für einander, in: Rüdiger vom Bruch, Brigitte Kaderas (eds), Wissenschaften und Wissenschaftspolitik. Bestandsaufnahmen zu Formationen, Brüchen und Kontinuitäten im Deutschland des 20. Jahrhunderts, Stuttgart 2002, pp. 32–49.

Ayass, Wolfgang, "*Asoziale*" im Nationalsozialismus, Stuttgart 1995.

Bär, Gesine, "Wir stehen nicht allein". *Schwedische Eugenik* im Spiegel der deutschen nationalsozialistischen Rassenforschung, in: Nordeuropa-Forum 2, 2002, pp. 25–41.

Barkan, Elazar, The *Retreat* of Scientific Racism. Changing Concepts of Race in Britain and the United States between the World Wars, Cambridge 1992.

Bauman, Zygmunt, *Dialektik* der Ordnung. Die Moderne und der Holocaust, Hamburg 1992.

Bauman, Zygmunt, *Moderne* und Ambivalenz. Das Ende der Eindeutigkeit, Frankfurt/Main 1995.

Beddies, Thomas, Zur *Geschichte* der Karl-Bonhoeffer-Nervenklinik, ehem. Wittenauer Heilstätten, ehem. Irrenanstalt der Stadt Berlin zu Dalldorf, in: Thomas Beddies, Andrea Dörries (eds), Patienten, pp. 37–187.

Beddies, Thomas; Andrea Dörries (eds), Die *Patienten* der Wittenauer Heistätten in Berlin 1919–1960, Husum 1999.

Benzenhöfer, Udo, *Der gute Tod*? Euthanasie und Sterbehilfe in Geschichte und Gegenwart, Munich 1999.

Benzenhöfer, Udo, *Hans Heinze*: Kinder- und Jugendpsychiatrie und "Euthanasie", in: Arbeitskreis zur Erforschung der nationalsozialistischen "Euthanasie" und Zwangssterilisation (ed.), Beiträge zur NS-"Euthanasie"-Forschung 2002. Fachtagungen vom 24. bis 26. Mai 2002 in Linz und Hartheim/Alkoven und vom 15. bis 17. November 2002 in Potsdam, Ulm 2003, pp. 9–51.

Bergemann, Claudia, *Mitgliederverzeichnis* der Kaiser-Wilhelm-Gesellschaft zur Förderung der Wissenschaften, 2 vols., Berlin 1990/91.

Bergmann, Anna, Der entseelte *Patient*. Die moderne Medizin und der Tod, Berlin 2004.

Bergmann, Anna; Gabriele Czarnowski; Annegret Ehmann, *Menschen* als Objekte humangenetischer Forschung und Politik im 20. Jahrhundert. Zur Geschichte des Kaiser-Wilhelm-Instituts für Anthropologie, menschliche Erblehre und Eugenik in Berlin-Dahlem (1927–1945), in: Pross, Götz Aly (eds), Wert, pp. 120–142.

Blacker, Carlos P., *Eugenics*. Galton and After, London 1952.

Bock, Gisela, *Zwangssterilisation* im Nationalsozialismus. Studien zur Rassenpolitik und Frauenpolitik, Opladen 1986.

Bräutigam, Helmut, "Wir beherbergten *Angehörige der Ostvölker*, Männer vom Balkan ...". Fremd- und Zwangsarbeit im Evangelischen Johannesstift 1939–1945, Berlin 2001.

Broberg, Gunnar; Nils Roll-Hansen (eds), *Eugenics and the Welfare State*. Sterilization Policy in Denmark, Sweden, Norway, and Finland, East Lansing 1996.

Broberg, Gunnar; Nils Roll-Hansen; Mattias Tydén, *Eugenics in Sweden*: Efficient Care, in: Gunnar Broberg, Nils Roll-Hansen (eds), Eugenics and the Welfare State, pp. 77–149.

Brucker-Boroujerdi, Ute; Wolfgang Wippermann, "Das *Zigeunerlager*" *Berlin-Marzahn* 1936–1945, in: Pogrom. Zeitschrift für bedrohte Völker 130, 1987, pp. 77–80.

Brunck, Helma, Die *Deutsche Burschenschaft* in der Weimarer Republik und im Nationalsozialismus, Munich 1999.

Burkhardt, Claudia, *Euthanasie* – "Vernichtung lebensunwerten Lebens" im Spiegel der Diskussionen zwischen Juristen und Medizinern von 1900 bis 1940, med. Diss. Mainz 1981.

Byer, Doris, Der *Fall Hugo A. Bernatzik*. Ein Leben zwischen Ethnologie und Öffentlichkeit, 1897–1953, Cologne 1999.

Byer, Doris, *Rassenhygiene* und Wohlfahrtspflege. Zur Entstehung eines sozialdemokratischen Machtdispositivs in Österreich bis 1934, Frankfurt/Main 1988.

Dahms, Hans-Joachim, Die *Universität Göttingen* 1918 bis 1989, in: Rudolf v. Thadden, Günter J. Trittel (eds), Göttingen. Geschichte einer Universitätsstadt, vol. III: Von der preußischen Mittelstadt zur südniedersächsischen Großstadt 1866–1989, Göttingen 1999, pp. 395–456.

Damm, Sabine; Norbert Emmerich, Die *Irrenanstalt Dalldorf-Wittenau* bis 1933, in: Totgeschwiegen, pp. 11–47.

Daum, Monika; Hans-Ulrich Deppe, *Zwangssterilisation* in Frankfurt am Main 1933–1945, Frankfurt/Main 1991.

Deichmann, Ute, *Biologen* unter Hitler. Vertreibung, Karrieren, Forschung, Frankfurt/Main 1992.

Deichmann, Ute, *Flüchten*, Mitmachen, Vergessen. Chemiker und Biochemiker in der NS-Zeit, Weinheim 2001.

Deichmann, Ute, *Hans Nachtsheim*, a Human Geneticist under National Socialism and the Question of Freedom of Science, in: Michael Fortun, Everett Mendelsohn (eds), The Practices of Human Genetics, Dordrecht 1999, pp. 143–153.

Dicke, Jan Nikolas, *Eugenik* und Rassenhygiene in Münster zwischen 1918 und 1939, Berlin 2004.

Dieckhöfer, Hans-Ulrich; Christoph Kaspari, Die *Tätigkeit* des Sozialhygienikers und Eugenikers Alfred Grotjahn (1869–1931) als Reichstagsabgeordneter der SPD 1921–1924, in: Medizinhistorisches Journal 21, 1986, pp. 308–331.

Dierks, Klaus, *Chronologie* der namibischen Geschichte von der vorgeschichtlichen Zeit zum unabhängigen Namibia, Windhuk 2000.

Ditt, Karl, Die *Kulturraumforschung* zwischen Wissenschaft und Politik. Das Beispiel Franz Petri (1903–1993), in: Westfälische Forschungen 46, 1996, pp. 73–176.

Ditt, Karl, *Raum* und Volkstum. Die Kulturpolitik des Provinzialverbandes Westfalen 1923–1945, Münster 1988.

Ditt, Karl, *Was ist "westfälisch"*? Zur Geschichte eines Stereotyps, in: Westfälische Forschungen 52, 2002, pp. 45–94.

Doeleke, Werner, *Alfred Ploetz* (1860–1940). Sozialdarwinist und Gesellschaftsbiologe, med. Diss. Frankfurt 1975.

Dowbiggin, Ian Robert, *Keeping America Sane*. Psychiatry and Eugenics in the United States and Canada, 1880–1940, Ithaca 1997.

Drechsler, Horst, *Südwestafrika* unter deutscher Kolonialherrschaft. Der Kampf der Herero und Nama gegen den deutschen Imperialismus (1884–1915), Berlin 1966.

Ebbinghaus, Angelika; Klaus Dörner (eds), *Vernichten* und Heilen. Der Nürnberger Ärzteprozeß und seine Folgen, Berlin 2002.

Eckart, Wolfgang U., *Medizin* und Kolonialimperialismus. Deutschland 1884–1945, Paderborn 1997.

Eckart, Wolfgang U., "Der größte *Versuch*, den die Einbildungskraft ersinnen kann" – Der Krieg als hygienisch-bakteriologisches Laboratorium und Erfahrungsfeld, in: Wolfgang U. Eckart, Christoph Gradmann (eds), Die Medizin und der Erste Weltkrieg, Pfaffenweiler 1996, pp. 299–319.

Falter, Jürgen W., *Hitlers Wähler*, Munich 1991.

Fangerau, Heiner, *Etablierung* eines rassenhygienischen Standardwerkes, 1921–1941. Der Baur-Fischer-Lenz im Spiegel der zeitgenössischen Rezensionsliteratur, Frankfurt/Main 2001.

Faulstich, Heinz, *Hungersterben* in der Psychiatrie 1914–1949. Mit einer Topographie der NS-Psychiatrie, Freiburg 1998.

Ferdinand, Ursula, *Bevölkerungswissenschaft* und Rassismus. Die internationalen Bevölkerungskongresse der International Union of the Scientific Investigation of Population Problems (IUSIPP) als paradigmatische Foren, in: Rainer Mackensen (ed.), Bevölkerungslehre und Bevölkerungspolitik im "Dritten Reich", Opladen 2004, pp. 61–98.

Fichtner, Gerhard, Die *Euthanasiediskussion* in der Zeit der Weimarer Republik, in: Albin Eser (ed.), Suizid und Euthanasie als human- und sozialwissenschaftliches Problem, Stuttgart 1976, pp. 24–40.

Field, Geoffrey G., *Evangelist* of Race. The Germanic Visions of Houston Stewart Chamberlain, New York 1981.

Foucault, Michel, Der *Wille* zum Wissen, Frankfurt/Main 1983.

Fraenkel, Ernst, Der *Doppelstaat*. Recht und Justiz im "Dritten Reich" (1941), Frankfurt/Main 1974.

Freund, Wolfgang, *Volk*, Reich und Westgrenze. Wissenschaften und Politik in der Pfalz, im Saarland und im annektierten Lothringen 1925–1945, Diss. Saarbrücken 2002.

Frewer, Andreas, *Medizin und Moral* in Weimarer Republik und Nationalsozialismus. Die Zeitschrift "Ethik" unter Emil Abderhalden, Frankfurt/Main 2000.

Fuchs, Petra, "*Körperbehinderte*" zwischen Selbstaufgabe und Emanzipation. Selbsthilfe – Integration – Aussonderung, Neuwied 2001.

Gailus, Manfred, *Antisemitismus* und protestantisches Sozialmilieu Berlins 1930 bis 1945, in: Michael Grüttner et al. (eds), Geschichte und Emanzipation. Fs. für Reinhard Rürup, Frankfurt 1999, 333–358.

Gailus, Manfred, *Protestantismus* und Nationalsozialismus. Studien zur nationalsozialistischen Durchdringung des protestantischen Sozialmilieus in Berlin, Cologne 2001.

Gailus, Manfred, Vom evangelischen *Sozialpfarrer* zum nationalsozialistischen Sippenforscher. Die merkwürdigen Lebensläufe des Berliner Theologen Karl Themel (1890–1973), in: Zeitschrift für Geschichtswissenschaft 49, 2001, pp. 796–826.

Gausemeier, Bernd, Rassenhygienische *Radikalisierung* und kollegialer Konsens, in: Carola Sachse (ed.), Verbindung, 178–198.

Gausemeier, Bernd, *Walter Scheidt* und die "Bevölkerungsbiologie". Ein Beitrag zur Geschichte der "Rassenbiologie" in der Weimarer Republik und im "Dritten Reich", Magisterarbeit im Fachbereich Geschichte, Freie Universität Berlin 1998.

Gay, Peter, Die *Republik der Außenseiter*. Geist und Kultur in der Weimarer Zeit 1918–1933, Frankfurt/Main 1987.

Gedenkbuch. Die Sinti und Roma im Konzentrationslager Auschwitz-Birkenau, Staatliches Museum Auschwitz-Birkenau (ed.), Munich 1993.

Geisenhainer, Katja, "*Rasse* ist Schicksal". Otto Reche (1879–1966) – ein Leben als Anthropologe und Völkerkundler, Leipzig 2002.

Gerrens, Uwe, *Medizinisches Ethos* und theologische Ethik. Karl und Dietrich Bonhoeffer in der Auseinandersetzung um Zwangssterilisation und "Euthanasie" im Nationalsozialismus, Munich 1996.

Gessler, Bernhard, *Eugen Fischer* (1874–1967). Leben und Werk des Freiburger Anatomen, Anthropologen und Rassenhygienikers bis 1927, Frankfurt/Main 2000.

Gilsenbach, Reimar, *Erwin Baur*. Eine deutsche Chronik, in: Arbeitsmarkt und Sondererlaß. Menschenverwertung, Rassenpolitik und Arbeitsamt (=Beiträge zur nationalsozialistischen Gesundheits- und Sozialpolitik; 8), Berlin 1990, pp. 184–197.

Gilsenbach, Reimar, Wie *Lolitschai* zur Doktorwürde kam, in: Feinderklärung, Zigeunerforschung und Asozialenpolitik (=Beiträge zur nationalsozialistischen Gesundheits- und Sozialpolitik; 6), Berlin 1988, pp. 101–134.

Grell, Ursula, *Karl Bonhoeffer* und die Rassenhygiene, in: Totgeschwiegen, pp. 207–218.

Grimm, Hans, *Felix v. Luschan* als Anthropologe. Von der Kraniologie zur Humanbiologie, in: Ethnographisch-Anthropologische Zeitschrift 27, 1986, pp. 415–425.

Grosch-Obenauer, Dagmar, *Hermann Muckermann* und die Eugenik, Diss. Mainz 1986.

Grosse, Pascal, *Kolonialismus*, Eugenik und bürgerliche Gesellschaft in Deutschland, 1850–1918, Frankfurt/Main 2000.

Gruchmann, Lothar, *Euthanasie* und Justiz im Dritten Reich, in: VfZ 20, 1972, pp. 235–279.

Grüber, Katrin, *Plädoyer* für eine verantwortbare medizinische Forschung – Abschied vom Gendogma, in: Ulrich Bach, Andreas de Kleine (eds), Auf dem Weg in die Totale Medizin? Eine Handreichung zur "Bioethik"-Debatte, Neukirchen-Vluyn 1999, pp. 50–60.

Günther, Maria, Die *Institutionalisierung* der Rassenhygiene an den deutschen Hochschulen vor 1933, med. Diss. Mainz 1982.

Hafner, Karl Heinz; Rolf Winau, Die *Freigabe* der Vernichtung lebensunwerten Lebens, in: Medizinhistorisches Journal 9, 1974, pp. 227–254.

Hagemann, Rudolf, *Erwin Baur* (1875–1933), Pionier der Genetik und Züchtungsforschung. Seine wissenschaftlichen Leistungen und ihre Ausstrahlung auf Genetik, Biologie und Züchtungsforschung von heute, Eichenau 2000.

Hahn, Daphne, *Modernisierung* und Biopolitik. Sterilisierung und Schwangerschaftsabbruch in Deutschland nach 1945, Frankfurt/Main 2000.

Hahn, Susanne, *Alternsforschung* und Altenpflege im Nationalsozialismus, in: Christoph Meinel, Peter Voswinckel (eds), Medizin, Naturwissenschaft, Technik und Nationalsozialismus. Kontinuitäten und Diskontinuitäten, Stuttgart 1994, pp. 221–229.

Haller, Mark H., *Eugenics*. Hereditarian Attitudes in American Thought, New Brunswick 1984.

Hanrath, Sabine, Zwischen "*Euthanasie*" und Psychiatriereform. Anstaltspsychiatrie in Westfalen und Brandenburg: Ein deutsch-deutscher Vergleich (1945–1964), Paderborn 2002.

Hansen, Bent Sigurd, Something Rotten in the State of Denmark. Eugenics and the Ascent of the Welfare State, in: Gunnar Broberg, Nils Roll-Hansen (eds), Eugenics and the Welfare State, pp. 195–258.

Harwood, Jonathan, Eine vergleichende *Analyse* zweier genetischer Forschungsinstitute: die Kaiser-Wilhelm-Institute für Biologie und für Züchtungsforschung, in: Bernhard vom Brocke, Hubert Laitko (eds), Kaiser-Wilhelm-/Max-Planck-Gesellschaft, pp. 331–348.

Harwood, Jonathan, *Styles* of Scientific Thought. The German Genetics Community 1900–1933, Chicago 1993.

Hehemann, Rainer, Die "*Bekämpfung des Zigeunerunwesens*" im Wilhelminischen Deutschland und in der Weimarer Republik 1871–1933, Frankfurt/Main 1987.

Heiber, Helmut, Der *Generalplan Ost*, in: VfZ 6, 1958, pp. 280–325.

Heiber, Helmut, *Walter Frank* und sein Reichsinstitut für Geschichte des neuen Deutschlands, Stuttgart 1966.

Heinemann, Isabel, "*Rasse*, Siedlung, deutsches Blut". Das Rasse- und Siedlungshauptamt der SS und die rassenpolitische Neuordnung Europas, Göttingen 2003.

Henning, Eckart; Marion Kazemi, *Chronik* der Kaiser-Wilhelm-Gesellschaft zur Förderung der Wissenschaften, Berlin 1988.

Herlitzius, Anette, *Frauenbefreiung* und Rassenideologie. Rassenhygiene und Eugenik im politischen Programm der "Radikalen Frauenbewegung" (1900–1933), Wiesbaden 1995.

Hesse, Hans, *Augen* aus Auschwitz. Ein Lehrstück über nationalsozialistischen Rassenwahn und medizinische Forschung – Der Fall Dr. Karin Magnussen, Essen 2001.

Heuer, Bernd; Peter Propping, *Vergleich* des "Archivs für Rassen- und Gesellschaftsbiologie" (1904–1933) und des "Journals of Heredity" (1910–1939), in: Medizinhistorisches Journal 26, 1991, pp. 78–93.

Hildebrand, Klaus, Vom *Reich* zum Weltreich. Hitler, NSDAP und koloniale Frage 1919–1945, Munich 1969.

Hoffmann, Christoph, Die *Inhalte* des Begriffs "Euthanasie" im 19. Jahrhundert und seine Wandlungen in der Zeit bis 1920, med. Diss. Berlin 1969.

Hohmann, Joachim S., *Robert Ritter* und die Erben der Kriminalbiologie. "Zigeunerforschung" im Nationalsozialismus und in Westdeutschland im Zeichen des Rassismus, Frankfurt/Main 1991.

Horst, Ferdinand; Kurt-Erich Maier, *Eugen Fischer* (1874–1962), in: Badische Heimat 79, 1999, pp. 698–705.

Hoßfeld, Uwe, Die *Jenaer Jahre* des "Rasse-Günther" von 1930 bis 1935, in: Medizinhistorisches Journal 34, 1999, pp. 47–103.

Hoßfeld, Uwe, Von der *Rassenkunde*, Rassenhygiene und biologischen Erbstatistik zur Synthetischen Theorie der Evolution: Eine Skizze der Biowissenschaften, in: Uwe Hoßfeld et al. (eds.), "Kämpferische Wissenschaft". Studien zur Universität Jena im Nationalsozialismus, Cologne 2003, pp. 519–574.

Hrdlicka, Manuela R., *Alltag* im KZ. Das Lager Sachsenhausen bei Berlin, Opladen 1992.

Huonker, Thomas, *Anstaltseinweisungen*, Kindswegnahmen, Eheverbote, Sterilisationen, Kastrationen. Fürsorge, Zwangsmaßnahmen, "Eugenik" und Psychiatrie in Zürich zwischen 1890 und 1970, Zurich 2002.

Ipsen, Carl, *Dictating Demography*: The Problem of Population in Fascist Italy, Cambridge 1996.

Jakobi, Helga et al. (eds), *Aeskulap & Hakenkreuz*. Zur Geschichte der Medizinischen Fakultät in Gießen zwischen 1933 und 1945, 2nd edition, Frankfurt/Main 1989.

Jasper, Hinrich, *Maximinian de Crinis* (1889–1945). Eine Studie zur Psychiatrie im Nationalsozialismus, Husum 1991.

Kaasch, Michael, *Sensation*, Irrtum, Betrug? Emil Abderhalden und die Geschichte der Abwehrfermente, in: Wieland Berg et al. (eds), Vorträge und Abhandlungen zur Wissenschaftsgeschichte 1999/2000, Halle 2000, pp. 145–210.

Kaiser, Jochen-Christoph, *Innere Mission* und Rassenhygiene. Zur Diskussion im Centralausschuß für Innere Mission 1930–1938, in: Lippische Mitteilungen aus Geschichte und Landeskunde 55, 1986, pp. 197–217.

Kaiser, Jochen-Christoph, Sozialer *Protestantismus* im 20. Jahrhundert. Beiträge zur Geschichte der Inneren Mission 1914–1945, Munich 1989.

Kaiser, Jochen-Christoph et al. (eds.), *Eugenik*, Sterilisation, "Euthanasie". Politische Biologie in Deutschland 1895–1945. Eine Dokumentation, Berlin 1992.

Kaspari, Christoph, Der *Eugeniker* Alfred Grotjahn (1869–1931) und die "Münchner Rassenhygiene", in: Medizinhistorisches Journal 24, 1989, pp. 306–332.

Kater, Michael H., Das "*Ahnenerbe*" der SS, 1935–1945. Ein Beitrag zur Kulturpolitik des Dritten Reiches, 2nd edition, Munich 1997.

Kater, Michael H, Doctor Leonardo *Conti* and his Nemesis: The Failure of Centralized Medicine in the Third Reich, in: Central European History 18, 1985, pp. 299–325.

Kaufmann, Doris, *Eugenik* – Rassenhygiene – Humangenetik. Zur lebenswissenschaftlichen Neuordnung der Wirklichkeit in der ersten Hälfte des 20. Jahrhunderts, in: Richard van

Dülmen (ed.), Erfindung des Menschen. Schöpfungsbilder und Körperbilder 1500–2000, Vienna 1998, pp. 347–365.

Kaufmann, Doris, *Eugenische Utopie* und wissenschaftliche Praxis im Nationalsozialismus. Zur Wissenschaftsgeschichte der Schizophrenieforschung, in: Wolfgang Hardtwig (ed.), Utopie und politische Herrschaft im Europa der Zwischenkriegszeit, Munich 2003, pp. 309–325.

Kaufmann, Doris, "*Rasse und Kultur*". Die amerikanische Kulturanthropologie um Franz Boas (1858–1942) in der ersten Hälfte des 20. Jahrhunderts – ein Gegenentwurf zur Rassenforschung in Deutschland, in: Hans-Walter Schmuhl (ed.), Rassenforschung, pp. 309–327.

Keller, Christoph, Der *Schädelvermesser*. Otto Schlaginhaufen – Anthropologe und Rassenhygieniker. Eine biographische Reportage, Zurich 1995.

Kelting, Kerstin, Das *Tuberkuloseproblem* im Nationalsozialismus, Diss. Kiel 1974.

Kersting, Franz-Werner; Hans-Walter Schmuhl, *Einleitung*, Quellen, pp. 1–64.

Kersting, Franz-Werner; Hans-Walter Schmuhl (eds), *Quellen* zur Geschichte der Anstaltspsychiatrie in Westfalen, Bd. II: 1914–1955, Paderborn 2004.

Kevles, Daniel J., *In the Name* of Eugenics. Genetics and the Uses of Human Heredity, reprint Cambridge 1995.

Kieta, Mieczyslaw, Das *Hygiene-Institut* der Waffen-SS und Polizei in Auschwitz, in: Auschwitz-Hefte 1, 1987, pp. 213–218.

Klee, Ernst, *Auschwitz*, die NS-Medizin und ihre Opfer, Frankfurt/Main 2001.

Klee, Ernst, "*Euthanasie*" im NS-Staat, Frankfurt/Main 1983.

Klee, Ernst, Deutsche *Medizin* im Dritten Reich. Karrieren vor und nach 1945, Frankfurt/Main 2001.

Klee, Ernst, Das *Personenlexikon* zum Dritten Reich. Wer war was vor und nach 1945, Frankfurt/Main 2003.

Kline, Wendy, Building a Better *Race*. Gender, Sexuality, and Eugenics from the Turn of the Century to the Baby Boom, Berkeley 2001.

Knaape, Hans-Heinrich, Die *medizinische Forschung* an geistig behinderten Kindern in Brandenburg-Görden in der Zeit des Faschismus, in: Achim Thom, Samuel Mitja Rapoport (eds), Das Schicksal der Medizin im Faschismus. Internationales wissenschaftliches Symposium europäischer Sektionen der IPPNW 17.-20. November 1988, Erfurt/Weimar, Neckarsulm 1989, pp. 224–227.

Koch, Peter-Ferdinand, *Menschen-Versuche*. Die tödlichen Experimente deutscher Ärzte, Munich 1996.

Kröner, Hans-Peter, Die *Eugenik* in Deutschland von 1891 bis 1934, med. Diss. Münster 1980.

Kröner, Hans-Peter, *Von der Rassenhygiene zur Humangenetik*. Das Kaiser-Wilhelm-Institut für Anthropologie, menschliche Erblehre und Eugenik nach dem Kriege, Stuttgart 1997.

Kröner, Hans-Peter, Von der *Vaterschaftsbestimmung* zum Rassegutachten. Der erbbiologische Ähnlichkeitsvergleich als "österreichisch-deutsches Projekt" 1926–1945, in: Berichte zur Wissenschaftsgeschichte 22, 1999, pp. 257–264.

Kröner, Hans-Peter et al., *Erwin Baur*. Naturwissenschaft und Politik, Cologne 1994.

Kroll, Jürgen, Zur *Entstehung* und Institutionalisierung einer naturwissenschaftlichen und sozialpolitischen Bewegung. Die Entwicklung der Eugenik/Rassenhygiene bis zum Jahr 1933, med. Diss. Tübingen 1983.

Krüger, Martina, *Kinderfachabteilung Wiesengrund*. Die Tötung behinderter Kinder in Wittenau, in: Totgeschwiegen, pp. 151–176.

Kubica, Helena, Dr. *Mengele* und seine Verbrechen im Konzentrationslager Auschwitz-Birkenau, in: Hefte von Auschwitz 20, 1997, pp. 369–455.

Kudlien, *Max v. Gruber* und die frühe Hitlerbewegung, in: Medizinhistorisches Journal 17, 1982, pp. 373–389.

Küchenhoff, Bernhard, Eugenisch motiviertes *Denken* und Handeln im "Burghölzli" am Anfang des 20. Jahrhunderts, in: Schweizer Archiv für Neurologie und Psychiatrie 154, 2003, pp. 11–19.

Kühl, Stefan, Die *Internationale* der Rassisten. Aufstieg und Niedergang der internationalen Bewegung für Eugenik und Rassenhygiene im 20. Jahrhundert, Frankfurt/Main 1997.

Kuhn, Thomas S., Die *Struktur wissenschaftlicher Revolutionen*, 2nd edition, Frankfurt/Main 1976. [The Structure of Scientific Revolutions]

Kum'a N'Dumbè III, Alexandre, *Pläne* zu einer nationalsozialistischen Kolonialherrschaft in Afrika, in: Wolfgang Benz, Hermann Graml (eds), Aspekte deutscher Außenpolitik im 20. Jahrhundert. Aufsätze Hans Rothfels zum Gedächtnis, Stuttgart 1976, pp. 165–192.

Kundrus, Birthe (ed.), *Phantasiereiche*. Zur Kulturgeschichte des deutschen Kolonialismus, Frankfurt/Main 2003.

Labisch, Alfons, *Alfred Grotjahn* (1869–1931) und das gesundheitspolitische Programm der Mehrheitssozialdemokraten von 1922, in: Medizin – Mensch – Gesellschaft 8, 1983, pp. 192–197.

Labisch, Alfons; Florian Tennstedt, Der *Weg* zum "Gesetz über die Vereinheitlichung des Gesundheitswesens" vom 3 July 1934. Entwicklungslinien und -momente des staatlichen und kommunalen Gesundheitswesens in Deutschland, 2 vols., Düsseldorf 1985.

Lang, Hans-Joachim, *Grab* Nr. 27, Grafenhausen, August Hirt. Über die Verbrechen und das Lebensende eines weltweit gesuchten Anatomieprofessors, in: Land zwischen Hochrhein und Schwarzwald. Beiträge zur Geschichte des Landkreises Waldshut, hg. vom Geschichtsverein Hochrhein, Waldshut 1998, pp. 291–301.

Langbein, Hermann, *Menschen in Auschwitz*, Vienna 1995.

Lasik, Alexander, Die *Personalbesetzung* des Gesundheitsdienstes der SS im Konzentrationslager Auschwitz-Birkenau in den Jahren 1939–1945, in: Hefte von Auschwitz 20, 1997, pp. 290–368.

Lebzelter, Gisela, Die "*Schwarze Schmach*". Vorurteile – Propaganda – Mythos, in: Geschichte und Gesellschaft 11, 1985, pp. 37–58.

Lifton, Robert J., *Ärzte* im Dritten Reich, Stuttgart 1988.

Lilienthal, Georg, *Anthropologie* und Nationalsozialismus: Das erb- und rassenkundliche Abstammungsgutachten, in: Jahrbuch des Instituts für Geschichte der Medizin der Robert Bosch Stiftung 6, 1987, pp. 71–91.

Lilienthal, Georg, *Rassenhygiene* im Dritten Reich. Krise und Wende, in: Medizinhistorisches Journal 14, 1979, pp. 114–134.

Lilienthal, Georg, "*Rheinlandbastarde*", Rassenhygiene und das Problem der rassenideologischen Kontinuität. Zur Untersuchung von Reiner Pommerin, in: Medizinhistorisches Journal 15, 1980, pp. 426–436.

Lindemann, Gerhard, *Antijudaismus* und Antisemitismus in den evangelischen Landeskirchen während der NS-Zeit, in: Geschichte und Gesellschaft 29, 2003, pp. 575–607.

Lindemann, Gerhard, "*Typisch jüdisch*". Die Stellung der Evangelisch-Lutherischen Landeskirche Hannovers zu Antijudaismus, Judenfeindschaft und Antisemitismus 1919–1949, Berlin 1998.

Lindemann, Jean, Emil *Abderhaldens Abwehrenzyme*, in: Naturwissenschaftliche Rundschau 52, 1999, pp. 92–94.

Lösch, Niels C., *Rasse* als Konstrukt. Leben und Werk Eugen Fischers, Frankfurt/Main 1997.

Lohalm, Uwe, Die *Wohlfahrtskrise* 1930–1933. Vom ökonomischen Notprogramm zur rassenhygienischen Neubestimmung, in: Frank Bajohr et al. (eds), Zivilisation und Barbarei. Die widersprüchlichen Potentiale der Moderne, Hamburg 1991, pp. 193–225.

McLaren, Angus, Our Own *Master Race*. Eugenics in Canada, 1885–1945, Toronto 1990.

Macrakis, Kristie, *Surviving the Swastika*. Scientific Research in Nazi Germany, New York 1993.

Madajczyk, Czesław, Vom *Generalplan Ost* zum Generalsiedlungsplan, Munich 1994.

Mai, Christoph, *Humangenetik* im Dienste der "Rassenhygiene". Zwillingsforschung in Deutschland bis 1945, Aachen 1997.

Maitra, Robin T., "...*wer imstande und gewillt ist*, dem Staate mit Höchstleistungen zu dienen!" Hans Reiter und der Wandel der Gesundheitskonzeption im Spiegel der Lehr- und Handbücher der Hygiene zwischen 1920 und 1960, Husum 2001.

Massin, Benoît, *Mengele*, die Zwillingsforschung und die "Auschwitz-Dahlem-Connection", in: Carola Sachse (ed.), Verbindung, pp. 201–254.

Massin, Benoît, *Rasse und Vererbung* als Beruf. Die Hauptforschungsrichtungen am Kaiser-Wilhelm-Institut für Anthropologie, menschliche Erblehre und Eugenik im Nationalsozialismus, in: Hans-Walter Schmuhl (ed.), Rassenforschung, pp. 190–244.

Mazumdar, Pauline Margaret Hodgson, *Blood and Soil*: The Serology of Aryan Racial State, in: Bulletin of the History of Medicine 64, 1990, pp. 187–219.

Mazumdar, Pauline Margaret Hodgson, *Eugenics*, Human Genetics and Human Failings. The Eugenics Society, its Sources and its Critics in Britain, London 1992.

Mazumdar, Pauline Margaret Hodgson, *Species* and Specificity. An Interpretation of the History of Immunology, Cambridge 1995.

Meister, Johannes, Die "*Zigeunerkinder*" von der St. Josefspflege in Mulfingen, in: 1999. Zeitschrift für Sozialgeschichte des 20. und 21. Jahrhunderts 2, 1987, V. 2, pp. 14–51.

Meyer, Beate, "*Jüdische Mischlinge*". Rassenpolitik und Verfolgungserfahrung 1933–1945, Hamburg 1999.

Milton, Sybil, *Vorstufe* zur Vernichtung. Die Zigeunerlager nach 1933, in: VfZ 43, 1995, pp. 115–130.

Mitscherlich, Alexander; Fred Mielke (eds), *Medizin* ohne Menschlichkeit. Dokumente des Nürnberger Ärzteprozesses, Frankfurt/Main 1978.

Müller, Joachim, *Sterilisation* und Gesetzgebung bis 1933, Husum 1985.

Müller-Hill, Benno, Das *Blut* von Auschwitz und das Schweigen der Gelehrten, in: Doris Kaufmann (ed.), Geschichte, pp. 189–227.

Müller-Hill, Benno, *Genetics after Auschwitz*, in: Holocaust and Genocide Studies 2, 1987, pp. 3–20.

Müller-Hill, Benno, *Tödliche Wissenschaft*. Die Aussonderung von Juden, Zigeunern und Geisteskranken 1933–1945, Reinbek 1984.

Müller-Hill, Benno; Ute Deichmann, The *Fraud* of Abderhalden's Enzymes, in: Nature 393, 1998, pp. 109–111.

Münzel, Mark; Bernhard Streck (eds), *Kumpania* und Kontrolle. Moderne Behinderungen zigeunerischen Lebens, Giessen 1981.

Nadav, Daniel S., *Julius Moses* (1868–1942) und die Politik der Sozialhygiene in Deutschland, Gerlingen 1985.

Nadav, Daniel S, *Julius Moses und Alfred Grotjahn*. Das Verhalten zweier sozialdemokratischer Ärzte zu Fragen der Eugenik und Bevölkerungspolitik, in: Pross; Götz Aly (eds), Wert, pp. 143–152.

Nemitz, Kurt, *Antisemitismus* in der Wissenschaftspolitik der Weimarer Republik. Der "Fall Ludwig Schemann", in: Jahrbuch des Instituts für Deutsche Geschichte 12, 1983, pp. 377–407.

Neumärker, Klaus; Michael Seidel, *Karl Bonhoeffer* und seine Stellung zur Sterilisierungsgesetzg ebung, in: Totgeschwiegen, pp. 269–286.

Nitschke, Asmus, Die "*Erbpolizei*" im Nationalsozialismus. Zur Alltagsgeschichte der Gesundheitsämter im Dritten Reich, Opladen 1999.

Nowak, Kurt, "*Euthanasie*" und Sterilisierung im "Dritten Reich". Die Konfrontation der evangelischen und katholischen Kirche mit dem "Gesetz zur Verhütung erbkranken Nachwuchses" und der "Euthanasie"-Aktion, 3rd edition, Göttingen 1984.

Peukert, Detlev J. K., Die *Weimarer Republik*. Krisenjahre der Klassischen Moderne, Frankfurt 1987(*The Weimar Republic: The Crisis of Classical Modernity*. Translated by Richard Deveson. Hill and Wang, New York, 1989).

Plumpe, Gottfried, Die *I.G. Farbenindustrie* AG. Wirtschaft, Technik und Politik 1914–1945, Berlin 1990.

Pommerin, Reiner, "*Sterilisierung der Rheinlandbastarde*". Das Schicksal einer farbigen deutschen Minderheit 1918–1937, Düsseldorf 1979.

Popplow, Ullrich, Die *Machtergreifung* in Augenzeugenberichten. Göttingen 1932–1935, in: Göttinger Jahrbuch 1977, pp. 157–200.

Posner, Gerald L.; John Ware, *Mengele*. Die Jagd nach dem Todesengel, Berlin 1998.

Potthast, Thomas, "*Rassenkreise*" und die Bedeutung des "Lebensraums". Zur Tier-Rassenforschung in der Evolutionsbiologie, in: Hans-Walter Schmuhl (ed.), Rassenforschung, pp. 275–308.

Proctor, Robert N, *Adolf Butenandt* (1903–1995). Nobelpreisträger, Nationalsozialist und MPG-Präsident. Ein erster Blick in den Nachlaß (=Ergebnisse. Vorabdrucke aus dem Forschungsprogramm "Geschichte der Kaiser-Wilhelm-Gesellschaft im Nationalsozialismus"; 2), Berlin 2000.

Proctor, Robert N, From *Anthropologie* to Rassenkunde in the German Anthropological Tradition, in: George W. Stocking (ed.), Bones, Bodies, Behavior. Essays on Biological Anthropology, Madison 1988, pp. 138–179.

Proctor, Robert N, *Blitzkrieg* gegen den Krebs. Gesundheit und Propaganda im Dritten Reich, Stuttgart 2002.

Proctor, Robert N, *Racial Hygiene*: Medicine under the Nazis, Cambridge 1988.

Raphael, Lutz, Die *Verwissenschaftlichung* des Sozialen als methodische und konzeptionelle Herausforderung für eine Sozialgeschichte des 20. Jahrhunderts, in: Geschichte und Gesellschaft 22, 1996, pp. 165–193.

Rehse, Helga, *Euthanasie*, Vernichtung lebensunwerten Lebens und Rassenhygiene in Programmschriften vor dem Ersten Weltkrieg, med. Diss. Heidelberg 1969.

Reilly, Philip R., The Surgical *Solution*. A History of Involuntary Sterilization in the United States, Baltimore 1991.

Reyer, Jürgen, Alte *Eugenik* und neue Wohlfahrtspflege. Entwertung und Funktionalisierung der Fürsorge vom Ende des 19. Jahrhunderts bis zur Gegenwart, Freiburg 1991.

Richter, Ingrid, *Katholizismus* und Eugenik in der Weimarer Republik und im Dritten Reich. Zwischen Sittlichkeitsreform und Rassenhygiene, Paderborn 2001.

Ringer, Fritz K., The *Decline* of the German Mandarins. The German Academic Community, 1890–1933, Cambridge 1969.

Rissom, Renate, *Fritz Lenz* und die Rassenhygiene, Husum 1983.

Ritter, Hans Jakob, "Nicht unbeeinflusst durch nördliche *Winde*"? Schweizer Psychiatrie und Eugenik in der Zwischenkriegszeit, in: Psychiatrische Praxis 27, 2000, pp. 127–133.

Roelcke, Volker, *Programm* und Praxis der psychiatrischen Genetik an der Deutschen Forschungsanstalt für Psychiatrie unter Ernst Rüdin. Zum Verhältnis von Wissenschaft, Politik und Rasse-Begriff vor und nach 1933, in: Hans-Walter Schmuhl (ed.), Rassenforschung.

Rössler, Mechthild; Sabine Schleiermacher (eds), Der "*Generalplan Ost*". Aspekte der nationalsozialistischen Planungs- und Vernichtungspolitik, Cologne 1990.

Roll-Hansen, Nils, *Norwegian Eugenics*. Sterilization as Social Reform, in: Gunnar Broberg, Nils Roll-Hansen (eds), Eugenics and the Welfare State, pp. 151–258.

Rossijanow, Kirill, Gefährliche *Beziehungen*: Experimentelle Biologie und ihre Protektoren, in: Dietrich Beyrau (ed.), Im Dschungel der Macht. Intellektuelle Professionen unter Stalin und Hitler, Göttingen 2000, pp. 340–359.

Roth, Karl Heinz, "*Erbbiologische Bestandsaufnahme*" – ein Aspekt "ausmerzender" Erfassung vor der Entfesselung des Zweiten Weltkrieges, in: Karl Heinz Roth (ed.), Erfassung, pp. 57–100.

Roth, Karl Heinz (ed.), *Erfassung* zur Vernichtung. Von der Sozialhygiene zum "Gesetz über Sterbehilfe", Berlin 1984.

Roth, Karl Heinz, *Josef Mengele* als Anthropologe, in: Dokumentationsstelle zur NS-Sozialpolitik, Mitteilungen 1, 1985, pp. 9 f.

Roth, Karl Heinz, Die wissenschaftliche *Normalität* des Schlächters, in: Dokumentationsstelle zur NS-Sozialpolitik, Mitteilungen 1, 1985, pp. 1–9.

Roth, Karl Heinz, *(Schein-)Alternativen* im Gesundheitswesen: Alfred Grotjahn (1869–1931) – Integrationsfigur etablierter Sozialmedizin und nationalsozialistischer "Rassenhygiene", in: Karl Heinz Roth (ed.), Erfassung, pp. 31–56.

Roth, Karl Heinz, *Schöner neuer Mensch*. Der Paradigmenwechsel der klassischen Genetik und seine Auswirkungen auf die Bevölkerungsbiologie des "Dritten Reichs", in: Heidrun Kaupen-Haas (ed.), Der Griff nach der Bevölkerung. Aktualität und Kontinuität nazistischer Bevölkerungspolitik, Nördlingen 1986, pp. 11–63.

Roth, Karl Heinz; Götz Aly, Das "*Gesetz über die Sterbehilfe* bei unheilbar Kranken". Protokolle der Diskussion über die Legalisierung der nationalsozialistischen Anstaltsmorde in den Jahren 1938–1941, in: Karl Heinz Roth, Götz Aly (eds), Erfassung, pp. 101–179.

Sachse, Carola, *Adolf Butenandt* und Otmar von Verschuer. Eine Freundschaft unter Wissenschaftlern (1942–1969), in: Wolfgang Schieder, Achim Trunk (eds), Adolf Butenandt, pp. 286–319.

Sachse, Carola (ed.), Die *Verbindung nach Auschwitz*. Biowissenschaften und Menschenversuche an Kaiser-Wilhelm-Instituten. Dokumentation eines Symposiums (=Geschichte der Kaiser-Wilhelm-Gesellschaft im Nationalsozialismus; 6), Göttingen 2003.

Sachse, Carola; Benoît Massin, Biowissenschaftliche *Forschung* an Kaiser-Wilhelm-Instituten und die Verbrechen des NS-Regimes. Informationen über den gegenwärtigen Wissensstand (=Ergebnisse. Vorabdrucke aus dem Forschungsprogramm "Geschichte der Kaiser-Wilhelm-Gesellschaft im Nationalsozialismus"; 3), Berlin 2000.

Sachße, Christoph; Florian Tennstedt, Der *Wohlfahrtsstaat* im Nationalsozialismus (=Geschichte der Armenfürsorge in Deutschland; 3), Stuttgart 1992.

Sandner, Peter, *Frankfurt. Auschwitz.* Die nationalsozialistische Verfolgung der Sinti und Roma in Frankfurt am Main, Frankfurt/Main 1998.

Saretzki, Thomas, *Reichsgesundheitsrat* und Preußischer Landesgesundheitsrat in der Weimarer Republik, Berlin 2000.

Schieder, Wolfgang; Achim Trunk (eds), *Adolf Butenandt* und die Kaiser-Wilhelm-Gesellschaft. Wissenschaft, Industrie und Politik im "Dritten Reich" (=Geschichte der Kaiser-Wilhelm-Gesellschaft im Nationalsozialismus; 7), Göttingen 2004.

Schleiermacher, Sabine, *Sozialethik* im Spannungsfeld von Sozial- und Rassenhygiene. Der Mediziner Hans Harmsen im Centralausschuß für die Innere Mission, Husum 1998.

Schlich, Thomas, *Wissenschaft.* Die Herstellung wissenschaftlicher Fakten als Thema der Geschichtsforschung, in: Thomas Schlich, Norbert Paul (eds), Medizingeschichte: Aufgaben, Probleme. Perspektiven, Frankfurt/Main 1998, pp. 107–129.

Schmaltz, Florian, *Kampfstoff-Forschung* im Nationalsozialismus. Zur Kooperation von Kaiser-Wilhelm-Instituten, Militär und Industrie, phil. Diss. Bremen 2004.

Schmiedebach, Heinz-Peter, *Sozialdarwinismus*, Biologismus, Pazifismus – Ärztestimmen zum Ersten Weltkrieg, in: Johanna Bleker, Heinz-Peter Schmiedebach (eds), Medizin und Krieg. Vom Dilemma der Heilberufe 1865 bis 1985, Frankfurt/Main 1987, pp. 93–121.

Schmokel, Wolfe W., Der *Traum* vom Reich. Deutscher Kolonialismus 1919–1945, Gütersloh 1967.

Schmuhl, Hans-Walter, *Ärzte in der Anstalt Bethel*, 1870–1945, Bielefeld 1998.

Schmuhl, Hans-Walter, *Ärzte in der Westfälischen Diakonissenanstalt Sarepta*, 1890–1970, Bielefeld 2001.

Schmuhl, Hans-Walter, *Arbeitsmarktpolitik* und Arbeitsverwaltung in Deutschland 1871–2002. Zwischen Fürsorge, Hoheit und Markt, Nuremberg 2003.

Schmuhl, Hans-Walter, *Hirnforschung* und Krankenmord. Das Kaiser-Wilhelm-Institut für Hirnforschung 1937–1945, in: VfZ 50, 2002, pp. 559–609 [longer version in: Ergebnisse. Vorabdrucke aus dem Forschungsprogramm "Geschichte der Kaiser-Wilhelm-Gesellschaft im Nationalsozialismus"; 1, Berlin 2000].

Schmuhl, Hans-Walter, Evangelische *Krankenhäuser* und die Herausforderung der Moderne. 75 Jahre Deutscher Evangelischer Krankenhausverband (1926–2001), Leipzig 2002.

Schmuhl, Hans-Walter, *Max Weber* und das Rassenproblem, in: Manfred Hettling et al. (eds), Was ist Gesellschaftsgeschichte? Positionen, Themen, Analysen (Hans-Ulrich Wehler zum 60. Geburtstag), Munich 1991, pp. 331–42.

Schmuhl, Hans-Walter, "*Neue Rehobother Bastardstudien*". Eugen Fischer und die Anthropometrie zwischen Kolonialforschung und nationalsozialistischer Rassenpolitik, in: Gert Theile (ed.), Anthropometrie. Vermessung des Menschen von Lavater bis Avatar, expected publication Paderborn 2005.

Schmuhl, Hans-Walter, *Rasse*, Rassenforschung, Rassenpolitik. Annäherungen an das Thema, in: Hans-Walter Schmuhl (ed.), Rassenforschung, pp. 7–37.

Schmuhl, Hans-Walter (ed.), *Rassenforschung* an Kaiser-Wilhelm-Instituten vor und nach 1933 (=Geschichte der Kaiser-Wilhelm-Gesellschaft im Nationalsozialismus; 4), Göttingen 2003.

Schmuhl, Hans-Walter, *Rassenhygiene in Deutschland* – Eugenik in der Sowjetunion: Ein Vergleich, in: Dietrich Beyrau (ed.), Im Dschungel der Macht. Intellektuelle Professionen unter Stalin und Hitler, Göttingen 2000, pp. 360–377.

Schmuhl, Hans-Walter, *Rassenhygiene, Nationalsozialismus, Euthanasie*. Von der Verhütung zur Vernichtung "lebensunwerten Lebens", 2nd edition, Göttingen 1992.

Schmuhl, Hans-Walter, *Rassismus* unter den Bedingungen charismatischer Herrschaft. Zum Übergang von der Verfolgung zur Vernichtung gesellschaftlicher Minderheiten im Dritten

Reich, in: Karl Dietrich Bracher et al. (eds), Deutschland 1933–1945. Neue Studien zur nationalsozialistischen Herrschaft, 2nd edition, Bonn 1993, pp. 182–97.

Schmuhl, Hans-Walter, *Zwangsarbeit* in Kirche und Diakonie, in: Jochen-Christoph Kaiser (ed.), Zwangsarbeit in Kirche und Diakonie, Stuttgart 2005.

Schneck, Peter, Die *Entwicklung* der Eugenik als soziale Bewegung in der Epoche des Imperialismus, in: Hans-Martin Dietl et al. (eds), Eugenik. Entstehung und gesellschaftliche Bedingtheit, Jena 1984, pp. 24–58.

Schneider, Gabriele, *Mussolini* in Afrika. Die faschistische Rassenpolitik in den italienischen Kolonien, Cologne 2000.

Scholder, Klaus (ed.), Die *Mittwochs-Gesellschaft*. Protokolle aus dem geistigen Deutschland 1932 bis 1944, Berlin 1982.

Schulle, Diana, Das *Reichssippenamt*. Eine Institution nationalsozialistischer Rassenpolitik, Berlin 2001.

Schulze, Winfried, Der *Stifterverband* für die Deutsche Wissenschaft 1920–1995, Berlin 1995.

Schwartz, Michael, Sozialistische *Eugenik*. Eugenische Sozialtechnologien in Debatten und Politik der deutschen Sozialdemokratie 1890–1933, Bonn 1995.

Schwartz, Michael, Konfessionelle *Milieus* und Weimarer Eugenik, in: Historische Zeitschrift 261, 1995, pp. 403–448.

Schwartz, Michael, "*Proletarier*" und "Lumpen". Sozialistische Ursprünge eugenischen Denkens, in: VfZ 42, 1994, pp. 437–470.

Schwartz, Michael, *Wissen* und Macht. Metamorphosen eugenischer Biopolitik, in: Frank Becker et al. (eds), Politische Gewalt in der Moderne. Fs. für Hans-Ulrich Thamer, Münster 2003, pp. 165–192.

Schweizer, Magdalena, Die psychiatrische *Eugenik* in Deutschland und in der Schweiz zur Zeit des Nationalsozialismus, Bern 2002.

Schwerin, Alexander von, *Experimentalisierung* des Menschen. Der Genetiker Hans Nachtsheim und die vergleichende Erbpathologie, 1920–1945 (=Geschichte der Kaiser-Wilhelm-Gesellschaft im Nationalsozialismus; 10), Göttingen 2004.

Seidelman, William E., *Mengele Medicus*. Medicine's Nazi Heritage, in: International Journal of Health Service 19, 1989, pp. 599–610.

Seidler, Horst; Andreas Rett, Das *Reichssippenamt* entscheidet. Rassenbiologie im Nationalsozialismus, Vienna 1982.

Siemen, Hans Ludwig, *Menschen* blieben auf der Strecke ... Psychiatrie zwischen Reform und Nationalsozialismus, Gütersloh 1987.

Simon, Gerd; Joachim Zahn, *Nahtstellen* zwischen sprachstrukturalistischem und rassistischem Diskurs. Eberhard Zwirner und das "Deutsche Spracharchiv" im Dritten Reich, Ms. (accessible at: www.http://homepages.uni-tuebingen.de/gerd.simon).

Sparing, Frank, Von der *Rassenhygiene* zur Humangenetik – Heinrich Schade, in: Michael G. Esch (ed.), Die Medizinische Akademie Düsseldorf im Nationalsozialismus, Essen 1997, pp. 341–363.

Stadler, Michael, Das *Schicksal* der nichtemigrierten Gestaltpsychologen im Nationalsozialismus, in: Carl Friedrich Graumann (ed.), Psychologie im Nationalsozialismus, Berlin 1985, pp. 139–164.

Stepan, Nancy Ley, "The *Hour* of Eugenics". Race, Gender, and Nation in Latin America, Ithaca 1991.

Stingerlin, Martin (ed.), *Biopolitik* und Rassismus, Frankfurt/Main 2003.

Strebel, Bernhard; Jens-Christian Wagner, *Zwangsarbeit* für Forschungseinrichtungen der Kaiser-Wilhelm-Gesellschaft 1939–1945. Ein Überblick (=Ergebnisse. Vorabdrucke aus dem Forschungsprogramm "Geschichte der Kaiser-Wilhelm-Gesellschaft im Nationalsozialismus"; 11), Berlin 2003.

Stuchlik, Gerda, Das *Frankfurter Institut* für Erbbiologie und Rassenhygiene, in: Christoph Dorner et al. (eds), Die braune Machtergreifung. Universität Frankfurt 1930–1945, Frankfurt/Main n.d. [1989], pp. 161–203.

Stürzbecher, Manfred, *Otto Krohne*, in: Berliner Ärzteblatt 92, 1979, pp. 697 f.

Süß, Winfried, Der beinahe unaufhaltsame *Aufstieg* des Karl Brandt. Zur Stellung des "Reichskommissars für das Sanitäts- und Gesundheitswesen" im gesundheitspolitischen

Machtgefüge des "Dritten Reiches", in: Wolfgang Woelk, Jörg Vögele (eds), Geschichte der Gesundheitspolitik in Deutschland. Von der Weimarer Republik bis in die Frühgeschichte der "doppelten Staatsgründung", Berlin 2002, pp. 197–224.

Süß, Winfried, Der "*Volkskörper*" im Krieg. Gesundheitspolitik, medizinische Versorgung und Krankenmord im nationalsozialistischen Deutschland 1939–1945, Munich 2003.

Thomann, Klaus-Dieter, *Otmar Freiherr von Verschuer* – ein Hauptvertreter der faschistischen Rassenhygiene, in: Achim Thom, Horst Spaar (eds), Medizin im Faschismus. Symposium über das Schicksal der Medizin in der Zeit des Faschismus in Deutschland 1933–1945, Leipzig 31.1.-2.2.1983, Berlin 1985, pp. 57–67.

Tollmien, Cordula, *Nationalsozialismus in Göttingen* (1933–1945), Göttingen 1999.

Totgeschwiegen 1933–1945. Zur Geschichte der Wittenauer Heilstätten, seit 1957 Karl-Bonhoeffer-Nervenklinik, Arbeitsgruppe zur Erforschung der Geschichte der Karl-Bonhoeffer-Nervenklinik (ed.), 2nd edition, Berlin 1989.

Trombley, Stephen, The *Right* to Reproduce: A History of Coercive Sterilization, London 1988.

Trunk, Achim, *Rassenforschung* und Biochemie. Ein Projekt – und die Frage nach dem Beitrag Butenandts, in: Wolfgang Schieder, Achim Trunk (eds), Adolf Butenandt, pp. 247–285 [heavily abridged and reworked version of idem., Zweihundert Blutproben].

Trunk, Achim, *Zweihundert Blutproben* aus Auschwitz. Ein Forschungsvorhaben zwischen Anthropologie und Biochemie (1943–1945) (=Ergebnisse. Vorabdrucke aus dem Forschungsprogramm "Geschichte der Kaiser-Wilhelm-Gesellschaft im Nationalsozialismus"; 12) Berlin 2003.

Trus, Armin, Der "*Heilige Krieg*" der Eugeniker, in: Gerhard Freiling, Günter Schärer-Pohlmann (eds), Geschichte und Kritik. Beiträge zu Gesellschaft, Politik und Ideologie in Deutschland. Heinrich Brinkmann zum 60. Geburtstag, Giessen 2002, pp. 245–286.

Tutzke, Dietrich, *Alfred Grotjahn*, Leipzig 1979.

Usborne, Cornelie, *Frauenkörper*, Volkskörper, Staatskörper. Geburtenkontrolle und Bevölkerungspolitik in der Weimarer Republik, Münster 1994.

Völklein, Ulrich, *Josef Mengele*. Der Arzt von Auschwitz, Göttingen 2002.

Vogt, Annette, *Wissenschaftlerinnen* in Kaiser-Wilhelm-Instituten A-Z, Berlin 1999.

Volkov, Shulamit, *Antisemitismus* als kultureller Code. in: idem, Jüdisches Leben und Antisemitismus im 19. und 20. Jahrhundert. Zehn Essays, 2nd edition, Munich 2000, pp. 13–36.

Vom Brocke, Bernhard; Hubert Laitko (eds), Die *Kaiser-Wilhelm-/Max-Planck-Gesellschaft* und ihre Institute. Studien zu ihrer Geschichte: Das Harnack-Prinzip, Berlin 1996.

Waibel, Annette, Die *Provinzialkinderanstalt* für seelisch Abnorme in Bonn, in: Folgen der Ausgrenzung. Studien zur Geschichte der NS-Psychiatrie in der Rheinprovinz, Archivberatungsstelle Rheinland (ed.), Pulheim-Brauweiler, Cologne 1995, pp. 67–88.

Walter, Bernd, *Psychiatrie* und Gesellschaft in der Moderne. Geisteskrankenfürsorge in der Provinz Westfalen zwischen Kaiserreich und NS-Regime, Paderborn 1996.

Weber, Matthias M., *Ernst Rüdin*. Eine kritische Biographie, Berlin 1993.

Weber, Matthias M, Rassenhygienische und genetische *Forschungen* an der Deutschen Forschungsanstalt für Psychiatrie/Kaiser-Wilhelm-Institut in München vor und nach 1933, in: Doris Kaufmann (ed.), Geschichte, pp. 95–111.

Weber, Matthias M, "Ein *Forschungsinstitut* für Psychiatrie ...". Die Entwicklung der Deutschen Forschungsanstalt für Psychiatrie in München zwischen 1917 und 1945, in: Sudhoffs Archiv 75, 1991, pp. 74–89.

Weber, Matthias M, *Harnack-Prinzip* oder Führerprinzip? Erbbiologie unter Ernst Rüdin an der Deutschen Forschungsanstalt für Psychiatrie in München, in: Bernhard vom Brocke, Hubert Laitko (eds), Kaiser-Wilhelm-/Max-Planck-Gesellschaft, pp. 411–422.

Weber, Matthias M, *Psychiatrie* als Rassenhygiene. Ernst Rüdin und die Deutsche Forschungsanstalt für Psychiatrie in München, in: Medizin, Gesellschaft und Geschichte 10, 1991, pp. 149–169.

Weindling, Paul J., *Genetik und Menschenversuche* in Deutschland, 1940–1950. Hans Nachtsheim, die Kaninchen von Dahlem und die Kinder vom Bullenhuser Damm, in: Hans-Walter Schmuhl (ed.), Rassenforschung, pp. 245–274.

Weindling, Paul J, *Health*, Race and German Politics between National Unification and Nazism 1870–1945, Oxford 1989.

Weindling, Paul J., *International Eugenics*. Swedish Sterilization in Context, in: Scandinavian Journal of History 24, 1999, pp. 179–197.

Weindling, Paul J, Die *Preußische Medizinalverwaltung* und die "Rassenhygiene". Anmerkungen zur Gesundheitspolitik der Jahre 1905–1933, in: Zeitschrift für Sozialreform 30, 1984, pp. 675–687.

Weindling, Paul J, *Soziale Hygiene*, Eugenik und medizinische Praxis. Der Fall Alfred Grotjahn, in: Heinz-Harald Abholz u. a., Krankheit und Ursachen (=Das Argument. Sonderband; 119), Berlin 1984, pp. 6–20.

Weindling, Paul J, *Weimar Eugenics*: The Kaiser Wilhelm Institute for Anthropology, Human Heredity and Eugenics in Social Context, in: Annals of Science 42, 1985, pp. 303–318.

Weingart, Peter, *Doppel-Leben*. Ludwig Ferdinand Clauß: zwischen Rassenforschung und Widerstand, Frankfurt/Main 1995.

Weingart, Peter, *Science* and Political Culture: Eugenics in Comparative Perspective, in: Scandinavian Journal of History 24, 1999, pp. 163–177.

Weingart, Peter; Jürgen Kroll; Kurt Bayertz, *Rasse*, Blut und Gene. Geschichte der Eugenik und Rassenhygiene in Deutschland, Frankfurt/Main 1988.

Weingartner, James J., *Massacre* at Mechterstädt – The Case of the Marburger Studentencorps 1920, in: The Historian 37, 1975, pp. 598–618.

Weinreich, Max, *Hitler's Professors*. The Part of Scholarship in German's Crimes Against the Jewish People, New York 1946, ND New Haven 1999.

Weiss, Sheila F., *Humangenetik* und Politik als wechselseitige Ressourcen. Das Kaiser-Wilhelm-Institut für Anthropologie, menschliche Erblehre und Eugenik im "Dritten Reich" (=Ergebnisse. Vorabdrucke aus dem Forschungsprogramm "Geschichte der Kaiser-Wilhelm-Gesellschaft im Nationalsozialismus"; 17), Berlin 2004.

Weiss, Sheila F, *Race and Class* in Fritz Lenz's Eugenics, in: Medizinhistorisches Journal 27, 1992, pp. 5–25.

Weiss, Sheila F, *Race Hygiene* and National Efficiency. The Eugenics of Wilhelm Schallmayer, Berkeley 1987.

Weiss, Sheila F, 'The *Sword* of Our Science' as a Foreign Policy Weapon: The Political Function of German Human Geneticists in the International Arena During the Third Reich (=Ergebnisse. Vorabdrucke aus dem Forschungsprogramm "Geschichte der Kaiser-Wilhelm-Gesellschaft im Nationalsozialismus"; 22), Berlin 2005.

Wendel, Günter, Die *Kaiser-Wilhelm-Gesellschaft* 1911–1914. Zur Anatomie einer imperialistischen Forschungsgesellschaft, Berlin 1975.

Weß, Ludger, *Humangenetik* zwischen Wissenschaft und Rassenideologie. Das Beispiel Otmar von Verschuer (1896–1969), in: Karsten Linne, Thomas Wohlleben (eds), Patient Geschichte. Fs. für Karl Heinz Roth, Frankfurt/Main 1993, pp. 166–184.

Wettley, Annemarie, *August Forel*. Ein Arztleben im Zwiespalt seiner Zeit, Salzburg 1953.

Wetzell, Richard, *Kriminalbiologische Forschung* an der Deutschen Forschungsanstalt für Psychiatrie in der Weimarer Republik und im Nationalsozialismus, in: Hans-Walter Schmuhl (ed.), Rassenforschung, pp. 68–98.

Winau, Rolf, *Versuche* mit Menschen. Historische Entwicklung und ethischer Diskurs, in: Carola Sachse (ed.), Verbindung, pp. 158–177.

Winkler, Ulrike, "*Hauswirtschaftliche Ostarbeiterinnen*" – Zwangsarbeit in deutschen Haushalten, in: Ulrike Winkler (ed.), Stiften gehen. NS-Zwangsarbeit und Entschädigungsdebatte, Cologne 2000, pp. 148–168.

Wojak, Irmtrud, *Das "irrende Gewissen"* der NS-Verbrecher und die deutsche Rechtsprechung. Die "jüdische Skelettsammlung" am Anatomischen Institut der "Reichsuniversität Straßburg" in: Fritz-Bauer-Institut (ed.), "Beseitigung des jüdischen Einflusses...". Antisemitische Forschung, Eliten und Karrieren im Nationalsozialismus, Jahrbuch 1998/1999 zur Geschichte und Wirkung des Holocaust, Frankfurt/Main 1999, pp. 101–130.

Zierold, Kurt, *Forschungsförderung* in drei Epochen. Deutsche Forschungsgemeinschaft: Geschichte, Arbeitsweise, Kommentar, Wiesbaden 1968.

Zimmermann, Michael, *Rassenutopie* und Genozid. Die nationalsozialistische "Lösung der Zigeunerfrage" Hamburg 1996.

Zimmermann, Susanne, Die *Berufung* von Hans F. K. Günther zum Professor für Sozialanthropologie an der Universität Jena im Jahre 1930, in: Würzburger medizinhistorische Mitteilungen 14, 1996, pp. 489–497.

Zofka, Zdenek, Der KZ-Arzt *Josef Mengele*. Zur Typologie eines NS-Verbrechers, in: VfZ 34, 1986, pp. 245–267.

Zuelzer, Wolf, Der *Fall Nicolai*, Frankfurt/Main 1981.

Index of Persons

I

J

K

M

N

O

P

T